Operator Theory and Complex Analysis

**Workshop on Operator Theory
and Complex Analysis
Sapporo (Japan)
June 1991**

Edited by

**T. Ando
I. Gohberg**

Springer Basel AG

Editors' addresses:

Prof. T. Ando
Research Institute for Electronic Science
Hokkaido University
Sapporo
060 Japan

Prof. I. Gohberg
Raymond and Beverly Sackler Faculty of Exact Sciences
School of Mathematical Sciences
Tel Aviv University
69978 Tel Aviv, Israel

A CIP catalogue record for this book is available from the Library of Congress,
Washington D.C., USA

Deutsche Bibliothek Cataloging-in-Publication Data

Operator theory and complex analysis / Workshop on Operator
Theory and Complex Analysis, Sapporo (Japan), June 1991.
Ed. by T. Ando ; I. Gohberg. – Basel ; Boston ; Berlin : Birkhäuser, 1992
   (Operator theory ; Vol. 59)
   ISBN 978-3-0348-9699-3   ISBN 978-3-0348-8606-2 (eBook)
   DOI 10.1007/978-3-0348-8606-2
NE: Ando, Tsuyoshi [Hrsg.]; Workshop on Operator Theory and Complex
   Analysis <1991, Sapporo>; GT

Table of Contents

Editorial Introduction ix

V. Adamyan
Scattering matrices for microschemes 1
   1. General expressions for the scattering matrix 2
   2. Continuity condition 7
   References 10

D. Alpay, A. Dijksma, J. van der Ploeg, H.S.V. de Snoo
Holomorphic operators between Krein spaces and the number
of squares of associated kernels 11
   0. Introduction 11
   1. Realizations of a class of Schur functions 15
   2. Positive squares and injectivity 20
   3. Application of the Potapov-Ginzburg transform 23
   References 28

D. Alpay, H. Dym
On reproducing kernel spaces, the Schur algorithm,
and interpolation in a general class of domains 30
   1. Introduction 31
   2. Preliminaries 33
   3. $\mathcal{B}(X)$ spaces 39
   4. Recursive extractions and the Schur algorithm 47
   5. $\mathcal{H}\rho(S)$ spaces 57
   6. Linear fractional transformations 64
   7. One sided interpolation 67
   8. References 74

M. Bakonyi, H.J. Woerdeman
The central method for positive semi-definite, contractive
and strong Parrott type completion problems 78
   1. Introduction 78
   2. Positive semi-definite completions 79
   3. Contractive completions 87
   4. Linearly constrained contractive completions 89
   References 95

VI

J.A. Ball, M. Rakowski
**Interpolation by rational matrix functions and stability
of feedback systems: The 4-block case** 96
   Introduction 96
   1. Preliminaries 100
   2. A homogeneous interpolation problem 104
   3. Interpolation problem 109
   4. Parametrization of solutions 116
   5. Interpolation and internally stable feedback systems 131
   References 140

H. Bart, V.E. Tsekanovskii
Matricial coupling and equivalence after extension 143
   1. Introduction 143
   2. Coupling versus equivalence 145
   3. Examples 148
   4. Special classes of operators 153
   References 157

J.I. Fujii
Operator means and the relative operator entropy 161
   1. Introduction 161
   2. Origins of operator means 162
   3. Operator means and operator monotone functions 163
   4. Operator concave functions and Jensen's inequality 165
   5. Relative operator entropy 167
   References 171

M. Fujii, T. Furuta, E. Kamei
An application of Furuta's inequality to Ando's theorem 173
   1. Introduction 173
   2. Operator functions 175
   3. Furuta's type inequalities 176
   4. An application to Ando's theorem 177
   References 179

T. Furuta
Applications of order preserving operator inequalities 180
   0. Introduction 180
   1. Application to the relative operator entropy 181
   2. Application to some extended result of Ando's one 185
   References 190

I. Gohberg, M.A. Kaashoek
**The band extension of the real line as a limit of discrete band
extensions, I. The main limit theorem** 191
    0. Introduction 191
    I. Preliminaries and preparations 193
    II. Band extensions 201
    III. Continuous versus discrete 205
    References 219

K. Izuchi
**Interpolating sequences in the maximal ideal
space of H^∞ II** 221
    1. Introduction 221
    2. Condition (A_2) 223
    3. Condition (A_3) 227
    4. Condition (A_1) 231
    References 232

C.R. Johnson, M. Lundquist
Operator matrices with chordal inverse patterns 234
    1. Introduction 234
    2. Entry formulae 237
    3. Inertia formula 243
    References 251

P. Jonas, H. Langer, B. Textorius
**Models and unitary equivalence of cyclic selfadjoint
operators in Pontrjagin spaces** 252
    Introduction 252
    1. The class $\mathcal{F}$ of linear functionals 253
    2. The Pontrjagin space associated with $\phi \in \mathcal{F}$ 257
    3. Models for cyclic selfadjoint operators in Pontrjagin spaces 266
    4. Unitary equivalence of cyclic selfadjoint operators in Pontrjagin spaces . . 275
    References 283

T. Okayasu
The von Neumann inequality and dilation theorems for contractions . . 285
    1. The von Neumann inequality and strong unitary dilation 285
    2. Canonical representation of completely contractive maps 287
    3. An effect of generation of nuclear algebras 289
    References 290

L.A. Sakhnovich
**Interpolation problems, inverse spectral problems
and nonlinear equations** 292
    References 303

VIII

S. Takahashi
Extended interpolation problem in finitely connected domains 305
    Introduction 305
    I. Matrices and transformation formulas 306
    II. Disc Cases 309
    III. Domains of finite connectivity 318
    References 326

E.R. Tsekanovskii
**Accretive extensions and problems on the Stieltjes
operator-valued functions relations** 328
    1. Accretive and sectorial extensions of the positive operators,
    operators of the class $C(\theta)$ and their parametric representation 329
    2. Stieltjes operator-valued functions and their realization 335
    3. M.S. Livsic triangular model of the M-accretive extensions
    (with real spectrum) of the positive operators 345
    4. Canonical and generalized resolvents of QSC-extensions
    of Hermitian contractions 343
    References 344

V. Vinnikov
Commuting nonselfadjoint operators and algebraic curves 348
    1. Commuting nonselfadjoint operators and the discriminant curve 348
    2. Determinantal representations of real plane curves 350
    3. Commutative operator colligations 353
    4. Construction of triangular models: Finite-dimensional case 355
    5. Construction of triangular models: General case 359
    6. Characteristic functions and the factorization theorem 364
    References 370

P.Y. Wu
All (?) about quasinormal operators 372
    1. Introduction 372
    2. Representations 374
    3. Spectrum and multiplicity 377
    4. Special classes 379
    5. Invariant subspaces 380
    6. Commutant 382
    7. Similarity 385
    8. Quasisimilarity 387
    9. Compact perturbation 391
    10. Open problems 393
    References 394

Workshop Program 399
List of Participants 402

EDITORIAL INTRODUCTION

This volume contains the proceedings of the Workshop on Operator Theory and Complex Analysis which was held at the Hokkaido University, Sapporo, Japan, June 11 to 14, 1991. This workshop preceeded the International Symposium on the Mathematical Theory of Networks and Systems (Kobe, Japan, June 17 to 21, 1991). It was the sixth workshop of this kind, and the first to be held in Asia. Following is a list of the five preceeding workshops with references to their proceedings:

1981 Operator Theory (Santa Monica, California, USA)

1983 Applications of Linear Operator Theory to Systems and Networks

(Rehovot, Israel), OT 12

1985 Operator Theory and its Applications (Amsterdam, the Netherlands), OT 19

1987 Operator Theory and Functional Analysis (Mesa, Arizona, USA), OT 35

1989 Matrix and Operator Theory (Rotterdam, the Netherlands), OT 50

The next Workshop in this series will be on Operator Theory and Boundary Eigenvalue Problems. It will be held at the Technical University, Vienna, Austria, July 27 to 30, 1993.

The aim of the 1991 workshop was to review recent advances in operator theory and complex analysis and their interplay in applications to mathematical system theory and control theory.

The workshop had a main topic: extension and interpolation problems for matrix and operator valued functions. This topic appeared in complex analysis at the beginning of this century and now is maturing in operator theory with important applications in the theory of systems and control. Other topics discussed at the workshop were operator inequalities and operator means, matrix completion problems, operators in spaces with indefinite scalar product and nonselfadjoint operators, scattering and inverse spectral problems.

This Workshop on Operator Theory and Complex Analysis was made possible through the generous financial support of the Ministry of Education of Japan, and also of the International Information Science Foundation, the Kajima Foundation and the Japan Association of Mathematical Sciences. The organizing committee of the Mathematical Theory of Networks and Systems (MTNS) has rendered financial help for some participants of this workshop to attend MTNS also. The Research Institute for Electronic Science, Hokkaido University, provided most valuable administration assistance. All of this support is acknowledged with gratitude.

T. Ando,   I. Gohberg

August 2, 1992

Operator Theory:
Advances and Applications, Vol. 59
© 1992 Birkhäuser Verlag Basel

SCATTERING MATRICES FOR MICROSCHEMES

Vadim Adamyan

A mathematical model for a simple microscheme is constructed on the basis of the scattering theory for a pair of different self-adjoint extensions of the same symmetric ordinary differential operator on a one-dimensional manifold, which consist of a finite number of semiinfinite straight outer lines attached to a "black box" in a form of a flat connected graph. An explicit expression for the scattering is given under a continuity condition at the graph vertices.

Contemporary technologies provide for the formation of electronic microschemes composed of atomic clusters and conducting quasionedimensional tracks as "wires" on crystal surfaces. Electrons travelling along such "wires" behave like waves and not like particles. The explicit expressions for the functional characteristics of simple microschemes can be obtained using the results of the mathematical scattering theory adapted for the special case when different self-adjoint extensions of the same symmetric ordinary differential operator on a one-dimensional manifold Ω are compared. The manifold Ω simulating the visual structure of a microscheme consists of $m\ (< \infty)$ semiinfinite straight outer lines attached to a "black box" in a form of a flat connected graph. Different self-adjoint extensions of the mentioned symmetric operator distinguish only in boundary conditions at terminal points of outer lines. Having the scattering matrix for the pair of any self-adjoint operators of such kind with the special extension corresponding to the case of disjoint isolated outer lines one can immediately calculate by known formulae the high-frequency conductance of the "black box".

In the first section of this paper the general expression for the scattering matrix for two self-adjoint extensions of the second-order differential operator in $L^2(\Omega)$ is derived on the basis of more general results obtained in [1].

The detailed version of these results is given in the second part for the most important case when functions of the extension domain satisfy the condition of continuity at the points of connection of outer lines to vertices of the graph. This work was initiated by some ideas from the paper [2].

1. General expressions for the scattering matrix.

Remind the known concepts of the scattering theory. Let L be a Hilbert space and H_0, H be a pair of self-adjoint operators on L. Denote by P_0 the orthogonal projection on the absolutely continuous subspace of H_0. If the resolvent difference

$$(H - zI)^{-1} - (H_0 - zI)^{-1}$$

at some (and, consequently, at any) nonreal point z is a nuclear operator, then by the Rosenblum-Kato theorem the partial isometric wave operators

$$W_\pm(H, H_0) = \operatorname*{s-lim}_{t \to \pm\infty} e^{iHt} e^{-iH_0 t} P_0$$

exist and map the absolutely continuous subspace of H_0 onto the same subspace of H [3]. The scattering operator

$$S(H, H_0) = W_+^*(H, H_0) W_-(H, H_0)$$

is unitary on $P_0 L$ and commutes with H_0. The scattering matrix $S(\lambda)$, $-\infty < \lambda < \infty$, is the scattering operator $S(H, H_0)$ in the spectral representation of H_0.

Consider the special case when H and H_0 are different self-adjoint extensions of the same densely defined symmetric operator A in L with finite defect numbers (m, m). Let $(e_\nu)_1^m$ be any basis of the defect subspace $\ker[A^* + iI]$. Put

$$e_\nu(z) = (H_0 + iI)(H_0 - zI)^{-1} e_\nu, \quad \nu = 1, \dots, m,$$

and introduce the matrix-function $\Delta(z)$,

$$\Delta_{\mu\nu}(z) = \big((H_0 z + I)(H_0 - zI)^{-1} e_\nu, e_\mu\big), \quad \nu, \mu = 1, \dots, m.$$

According to the M. G. Krein resolvent formula for fixed H_0 and arbitrary H

$$(1) \qquad (H - zI)^{-1} = (H_0 - zI)^{-1} - \sum_{\mu,\nu=1}^{m} \big([\Delta(z) + Q]^{-1}\big)_{\mu\nu} \big(\,\cdot\,, e_\nu(\bar z)\big) e_\mu(z)$$

where a parameter Q is a Hermitian matrix [4]. The following parametric representation of the scattering matrix $S(\lambda)$ for extensions H and H_0 was derived using the formula (1) in [1].

Take any decomposition of the absolutely continuous part L_0 of H_0 into a direct integral

$$L_0 = \int_{-\infty}^{\infty} \oplus K(\lambda)\, d\lambda.$$

Without loss of generality one can assume $\dim K(\lambda) \leq m$. Let $h_\nu(\lambda)$ be a spectral image of the vector $P_0 e_\nu$. Then

$$(2) \qquad S(\lambda) = I + 2\pi i(\lambda^2 + 1) \sum_{\mu,\nu} \left([\Delta^*(\lambda + i0) + Q]^{-1}\right)_{\mu\nu} \left(\,\cdot\,, h_\nu(\lambda)\right)_{K(\lambda)} h_\mu(\lambda).$$

Now we are going to adapt (2) for the case when a given symmetric operator A is a second-order differential operator on the above mentioned one-dimensional manifold Ω of the graph Ω_{int} and m outer semiaxes.

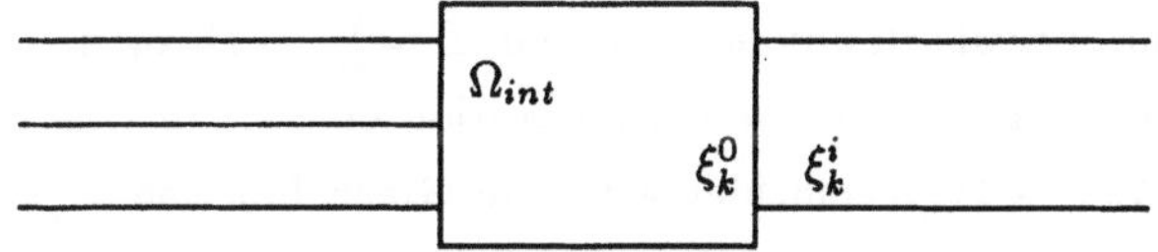

All self-adjoint extensions of A in $L^2(\Omega)$ differ only by boundary conditions at the terminals $(\xi_k^0)_1^m$ of outer lines and the corresponding connecting points $(\xi_k^i)_1^m$ from the inside of the "black box" Ω_{int}. Take as H_0 a special extension decomposing into an orthogonal sum :

$$H_0 = H_{int}^0 \oplus \sum_{k=1}^{m} \oplus H_k^0;$$

$$H_{int}^0 \;:\; L^2(\Omega_{int}) \longrightarrow L^2(\Omega_{int});$$

$$H_k^0 \;:\; L^2(E_+) \longrightarrow L^2(E_+),$$

$$(H_k^0 f)(x) = \frac{1}{\mu} f''(x), \quad f'(\xi_k^0) = 0.$$

As a consequence of the decomposition of H_0 the Green function $G_\omega^0(x, y)$, x, $y \in \Omega$, of the operator H_0, i.e. the kernel of the resolvent $[H_0 - \omega I]^{-1}$ in $L^2(\Omega)$ possesses the property: for any regular point ω $G_\omega(x, y) = 0$ if $x \in \Omega_{int}$ and y belongs to any outer line and vice versa or if x and y are points of different outer lines.

Describe the assumed properties of H_{int}^0. This operator can be considered as a self-adjoint extension of the orthogonal sum $A_{int} = \sum_\nu \oplus A_{i\nu}$ of regular symmetric differential operators of the form

$$A_{i\nu} = -\frac{d}{d\xi} p_\nu(\xi) \frac{d}{d\xi} + q_\nu(\xi)$$

on the corresponding segments (ribs) (α_ν, β_ν) of the graph with continuous real-valued functions $p_\nu(\xi)$, $q_\nu(\xi)$ and $p_\nu(\xi) > 0$. The functions $f(\xi)$ from the $A_{i\nu}$ domain in $L^2(\alpha_\nu, \beta_\nu)$ are continuously differentiable and satisfy boundary conditions

$$f(\alpha_\nu) = f(\beta_\nu) = 0; \quad f'(\alpha_\nu) = f'(\beta_\nu) = 0.$$

Note that every $A_{i\nu}$ has defect indices $(2N, 2N)$ where N is the number of the graph ribs.

Let B_ν be the "soft" self-adjoint extension of $A_{i\nu}$, i.e. the restriction of the conjugate operator $A_{i\nu}^*$ on the subset of functions $f(\xi)$ such that

$$f'(\alpha_\nu) = f'(\beta_\nu) = 0.$$

Consider the special self-adjoint extension $B = \sum_\nu \oplus B_\nu$ of A_{int}. The operator B can be taken as the part H^0_{int} in the decomposition of H_0. In this case the Green function $G^0_\omega(x, y)$ on $\Omega_{int} \times \Omega_{int}$ coincides with the Green function $E^0_\omega(x, y)$ of B and in its turn $E^0_\omega(x, y)$ is nonzero only if "x" and "y" belong the same segments (α_ν, β_ν) of Ω_{int} and on these segments $E^0_\omega(x, y)$ coincides with the Green functions of B_ν. Note that $G^0_\omega(\cdot, y) \in L^2(\Omega)$ for any regular point ω and any $y \in \Omega$. From the definition of the Green function $G^0_\omega(x, y)$ and its given properties it follows that the functions $\left(G^0_{-i}(x, \xi^0_k)\right)^m_1$ together with the functions $\left(G^0_{-i}(x, \alpha_\nu), G^0_{-i}(x, \beta_\nu)\right)^N_1$ form a basis in the defect subspace $M \equiv \ker[A^* + iI]$ of A. Put

$$\eta_{2\nu-1} = \alpha_\nu, \ \eta_{2\nu} = \beta_\nu, \ \nu = 1, \ldots, N; \quad \eta_{2N+k} = \xi^0_k, \ k = 1, \ldots, m;$$

and introduce the matrix-function $\widehat{\Gamma}(\omega)$,

$$\Gamma_{\mu\nu}(\omega) = G^0_\omega(\eta_\mu, \eta_\nu), \quad \mathrm{Im}\,\omega \neq 0, \ \mu, \nu = 1, \ldots, 2N + m.$$

According to the Hilbert identity for any regular points ω, z of H_0 and any x, $y \in \Omega$

$$(3) \qquad \int_\Omega du\, G^0_\omega(x, u) G^0_z(u, y) = \frac{1}{\omega - z}[G^0_\omega(x, y) - G^0_z(x, y)].$$

As a consequence of (3) and the relation $\overline{G^0_\omega(x, y)} = G^0_{\bar\omega}(x, y)$ we have

$$(4) \qquad \Delta_{\mu\nu}(\omega) = \int_\Omega \left([I + \omega H_0][H_0 - \omega I]^{-1} G^0_{-i}(\cdot, \eta_\nu)\right)(x) G^0_{-i}(x, \eta_\mu)\,dx$$

$$= \Gamma^0_{\mu\nu}(\omega) - \frac{1}{2}\left[\Gamma_{\mu\nu}(-i) + \overline{\Gamma_{\nu\mu}(-i)}\right].$$

Let H be an arbitrary self-adjoint extension of A in $L^2(\Omega)$. The Krein resolvent formula (1) and (4) yield the following expression for the Green function $G_\omega(x,y)$ of H through $G^0_\omega(x,y)$:

$$(5) \qquad G_\omega(x,y) = G^0_\omega(x,y) - \sum_{\mu,\nu=1}^{2N+m} \left([\Gamma^0(\omega) + Q]^{-1}\right)_{\mu\nu} G^0_\omega(x,\eta_\mu) G^0_\omega(\eta_\nu,y),$$

where Q is a Hermitian matrix.

Now to construct the scattering matrix $S(\lambda)$ for the pair H, H_0 notice that the parts B_ν of H_0 as regular self-adjoint differential operators have discrete spectra and the parts H^0_k on the outer lines form the absolutely continuous component of H_0. Consequently the first $2N$ basis vectors $G^0_{-i}(x,\eta_\nu)$ of the defect subspace M_- are orthogonal to the absolutely continuous subspace of H_0 and unlike the last m basis vectors $G^0_{-i}(x,\eta_{2N+k}) = G^0_{-i}(x,\xi^0_k)$ belong to this subspace. The natural spectral representation of the absolutely continuous part of H_0, i.e. of the orthogonal sum of the operator H^0_k, is the multiplication operator on the independent variable λ in the space of $\mathbb{C}^m$-valued functions $L^2(0,\infty;\mathbb{C}^m)$. The corresponding spectral mapping of the initial space can be defined in such way that the defect vectors $G^0_{-i}(x,\xi^0_k)$ turn into the vector-functions

$$(6) \qquad h_k(\lambda) = \frac{\sqrt[4]{\mu}}{\sqrt{\pi}\sqrt[4]{\lambda}} \frac{1}{\lambda+i} e_k, \qquad \lambda > 0,$$

where $e_k \in \mathbb{C}^m$ are the columns

$$e_1 = \begin{bmatrix} 1 \\ 0 \\ \vdots \\ 0 \end{bmatrix}, \ \ldots, \ e_m = \begin{bmatrix} 0 \\ 0 \\ \vdots \\ 1 \end{bmatrix}.$$

Using all above reasons we get immediately from (2) that $S(\lambda)$ is the $(m \times m)$-matrix-function and

$$(7) \qquad S_{kl}(\lambda) = \delta_{kl} + 2i\sqrt{\frac{\mu}{\lambda}} \left([\Gamma^0(\lambda - i0) + Q]^{-1}\right)_{2N+k,2N+l}, \qquad k,l = 1,\ldots,m.$$

Represent the parameter Q and the matrix-function $\Gamma^0(\omega)$ in the block-diagonal form

$$(8) \qquad Q = \begin{bmatrix} L & M \\ M^* & W \end{bmatrix}, \quad \Gamma^0(\omega) = \begin{bmatrix} \Gamma^0_{int}(\omega) & 0 \\ 0 & -i\sqrt{\frac{\mu}{\omega}} I_m \end{bmatrix},$$

where W is a Hermitian $(m \times m)$-matrix, I_m is the unit matrix of order m and the matrix-function $\Gamma^0_{int}(\omega)$ is determined by the Green function E^0_ω of the extension B as follows

$$(9) \qquad \left(\Gamma^0_{int}(\omega)\right)_{\mu\nu} = E^0_\omega(\eta_\mu,\eta_\nu), \qquad \mu,\nu = 1,\ldots,2N.$$

Using (7) and (8) we get the formula:

$$(10) \qquad S(\lambda) = \left\{ i\sqrt{\frac{\mu}{\lambda}} I_n + W - M^*[L + \Gamma^0_{int}(\lambda - i0)]^{-1} M \right\} \times$$

$$\times \left\{ -i\sqrt{\frac{\mu}{\lambda}} I_n + W - M^*[L + \Gamma^0_{int}(\lambda - i0)]^{-1} M \right\}^{-1}.$$

This expression describes all possible scattering matrices for microschemes comprised of given m outer lines and a given set of N ribs. The parametric matrices W, M, L contain information on the geometrical structure of the "black box" and the boundary conditions at all vertices including the connecting points to the outer lines. Without loss of generality we can consider that all connecting points are the graph vertices.

Single out now the scattering matrices for microschemes which differ only by the way the outer lines are connected to the definite inputs of the "black box", i.e. to the certain vertices of the definite graph Ω. Notice that this limiting condition generally speaking leaves the Hermitian matrix W arbitrary. The matrix can now vary only as far as the subspace $\ker MM^*$ remains unchanged. In what follows we will assume that this subspace and, respectively, its orthogonal complement in $\mathbb{C}^{2N}$ are always invariant subspaces of the L matrix. Let C_0 be the orthogonal projector on the subspace $\ker MM^*$ in $\mathbb{C}^{2N}$. Under the above condition and for various types of connection of outer lines to certain graph vertices only the block $C_0 L C_0$ of the L matrix is modified. Notice that this block for a "correct" connection is always invertible. Consider now the connections for which the L matrix in (8) remains unchanged. This matrix is a parameter in the Krein formula when resolvents of B and a definite self-adjoint extension H^0_{int} of A_{int} are compared. Let $\Gamma_{int}(\omega)$ be the matrix-function which is determined by the Green function E_ω of H^0_{int} like $\Gamma^0_{int}(\omega)$ in (9) by E^0_ω. According to the Krein formula of the form (5)

$$(11) \qquad \Gamma_{int}(\omega) = \Gamma^0_{int}(\omega) - \Gamma^0_{int}(\omega)[\Gamma^0_{int}(\omega) + L]^{-1}\Gamma^0_{int}(\omega)$$

$$= L[\Gamma^0_{int}(\omega) + L]^{-1}\Gamma^0_{int}(\omega) = L - L[\Gamma^0_{int}(\omega) + L]^{-1}L.$$

Taking into account that by the assumption that the block $C_0 L C_0$ of the Hermitian matrix L is invertible on the subspace $\ker MM^* \subset \mathbb{C}^{2N}$, denoting by Q the corresponding inverse operator in this subspace and using (11) we can write

$$(12) \qquad M^*[\Gamma^0_{int}(\omega) + L]^{-1} M = M^* Q L[\Gamma^0_{int}(\omega) + L]^{-1} L Q M$$

$$= M^* Q M - M^* Q \Gamma_{int}(\omega) Q M.$$

Inserting the last expression into (10) we find that the scattering matrix $S(\lambda)$ for the connections without changing the parameter L and the subspace $\ker MM^*$ has a form

$$(13) \qquad S(\lambda) = \left\{ i\sqrt{\frac{\mu}{\lambda}} I_n + \widetilde{W} - \widetilde{M}^* \Gamma^0_{int}(\lambda - i0)\widetilde{M} \right\} \times$$

$$\times \left\{ -i\sqrt{\frac{\mu}{\lambda}} I_n + \widetilde{W} - \widetilde{M}^* \Gamma^0_{int}(\lambda - i0)\widetilde{M} \right\}^{-1}$$

where the matrix parameters

$$\widetilde{W}\,(= W - M^*QM), \quad \widetilde{M} = QM$$

depend on the boundary conditions at those vertices of the graph Ω, which are connected with the outer lines.

If there are reasons to consider that as a result of the connection of outer lines to the graph Ω the matrices L and M are changed into M', L' so that $\ker(L - L') = \ker M'^* M' = \ker M^* M$, then it is natural to use the representation

$$(14) \qquad S(\lambda) = \left[i\sqrt{\frac{\mu}{\lambda}} I_n + H(\lambda - i0) \right] \left[-i\sqrt{\frac{\mu}{\lambda}} I_n + H(\lambda - i0) \right]^{-1},$$

$$H(\omega) = W + M'^* \left[L + \Gamma_{int}(\omega)(P_0 - QL'P_0) \right]^{-1} \times \left[P_0 \Gamma_{int}(\omega) Q_0 - I \right] M'.$$

The formula (14) can be obtained from (10) using the relation (11).

2. Continuity condition.

From the physical point of view the most natural are self-adjoint extensions of A satisfying the continuity condition at the graph vertices. This condition states that all functions from the domain of any such extension possess coinciding limiting values at any vertex along all ribs incident to this vertex. Irrespective to the present problem consider now the structure of the Krein formula (5) with parameter Q for arbitrary self-adjoint extensions satisfying the continuity condition in every vertex.

Take a vertex with s incident ribs, i.e. the vertex of degree s. It is convenient to enumerate the extreme points of the ribs at the vertex as $\eta_1, \ldots, \eta_s$. Replacing x in Eq. (5) by η_i for arbitrary y we obtain

$$(15) \qquad G_\omega(\eta_i, y) = \sum_{\mu, \nu} Q_{i\mu} \left([\Gamma^0(\omega) + Q]^{-1} \right)_{\mu\nu} G^0_\omega(\eta_\nu, y).$$

It follows from the continuity condition that

$$(16) \qquad G_\omega(\eta_1, y) = G_\omega(\eta_2, y) = \cdots = G_\omega(\eta_s, y).$$

Since y and ω in (15) and (16) are arbitrary it is obvious that the matrix Q in fact transforms any vector from $\mathbb{C}^{2N+m}$ into a vector with equal first s components. Denote by J_s the matrix of order s all components of which are unity. As Q is Hermitian, it is nothing but the following block matrix

$$Q = \begin{bmatrix} hJ_s & 0 \\ 0 & Q' \end{bmatrix},$$

where h is a real constant and Q' is a Hermitian matrix of the rank $2N + m - s$. Since the same procedure is valid for any vertex of arbitrary degree, the matrix with the suitable enumeration of the extreme points of the ribs and outer lines takes the block-diagonal form

$$(17) \qquad Q = \begin{bmatrix} h_1 J_{s_1} & 0 & \cdots & 0 \\ 0 & h_2 J_{s_2} & \cdots & 0 \\ \vdots & \vdots & \ddots & \vdots \\ 0 & 0 & \cdots & h_l J_{s_l} \end{bmatrix},$$

where l is the total number of the graph vertices and $s_1, \ldots, s_l$ are corresponding degrees of the vertices. Thus the following lemma is valid.

LEMMA. *The parameter Q in the Krein formula (5) for extensions satisfying the continuity condition is the Hermitian matrix such that nonzero elements of every its row (column) are equal and situated at the very places where the unities of the incidence matrix of the graph Ω are.*

Let the matrix Q be already reduced to the block-diagonal form (17) by a corresponding enumeration of extreme point of ribs and terminal points of outer lines. In this case the matrices W, M and $C_0 L C_0$ of the representation (8) coincide with the diagonal matrix

$$\mathbb{D} = \begin{bmatrix} h_1 & & & \mathbf{0} \\ & h_2 & & \\ & & \ddots & \\ \mathbf{0} & & & h_m \end{bmatrix},$$

where the parameters $h_1, \ldots, h_m$ are determined by the boundary conditions in vertices to which the outer lines are connected.

Using this fact and (13) we infer:

THEOREM. *Let H be a self-adjoint extension of A in $L^2(\Omega)$ satisfying the continuity condition and univalently connected with the set of parameters $h_1, \ldots, h_l$ of the corresponding matrix Q of the form (17) generating H in accordance with the Krein formula and let H_0 be the special extension of A decomposing into an orthogonal sum of*

*the self-adjoint operators on the graph and on the outer lines. The scattering matrix $S(\lambda)$
for the pair H_0, H admits the representation*

$$(18) \qquad S(\lambda) = \left[i\sqrt{\frac{\mu}{\lambda}}I_m + \Gamma(\lambda - i0)\right]\left[-i\sqrt{\frac{\mu}{\lambda}}I_m + \Gamma(\lambda - i0)\right]^{-1},$$

*where $\Gamma_{ik}(\omega) = E_\omega(\eta_i, \eta_k)$ and $E_\omega(\xi, \eta)$ is the Green function of the self-adjoint exten-
sion H_{int}^0 of A_{int} satisfying the continuity condition and determined by the same set of
parameters $h_1, \ldots, h_l$ for the same vertices like H is.*

From the representation (18) it is obvious that the analytic properties of the
scattering matrix $S(\lambda)$ are essentially determined by those of the matrix $\Gamma(\omega)$ constructed
by the Green function of the separated graph. For the regular differential operator the
matrix $\left(E_\omega(\eta_i, \eta_k)\right)_1^m$ is the meromorphic R-function. The natural problem thus arises of
the partial recovery of the graph structure and the operator on it from the matrix $S(\lambda)$ or,
equivalently, by the matrix Γ. In the case when the graph is reduced to a single segment
this problem is the well-known problem of recovery of a regular Sturm-Liouville operator
from spectra of two boundary problems. We hope to carry out the consideration of the
former problem in a more general case elsewhere.

In conclusion, as an example, consider an arbitrary graph with only two outer
lines connected to the same vertex. In this case $S(\lambda)$ is the second order matrix-function
but the determining matrix $\Gamma(\omega)$ is degenerate and takes the form

$$\Gamma(\omega) = E_\omega(\xi, \xi)\begin{bmatrix} 1 & 1 \\ 1 & 1 \end{bmatrix},$$

where ξ is the internal coordinate of the vertex of the graph tangent to the outer lines.

The scattering matrix according to (18) now can be put in the usual form

$$S(\lambda) = \begin{bmatrix} r(\lambda) & t(\lambda) \\ t(\lambda) & r(\lambda) \end{bmatrix},$$

where

$$r(\lambda) = \frac{i}{2E_\lambda(\xi, \xi)\sqrt{\lambda/\mu} - i}, \quad t(\lambda) = \frac{2E_\lambda(\xi, \xi)\sqrt{\lambda/\mu}}{2E_\lambda(\xi, \xi)\sqrt{\lambda/\mu} - i}$$

are, respectively, the reflection and transition coefficients. Notice that according to the
Landauer formula the resistance of the graph is given by

$$R(\lambda) = R_0\frac{|r|^2}{|t|^2} = R_0\frac{1}{4\lambda E_\lambda^2(\xi, \xi)},$$

where R_0 is the quantal resistance, i.e. the universal constant.

REFERENCES

1.   Adamyan, V. M.; Pavlov, B. S.: Null-range potentials and M. G. Krein's formula of generalized resolvents (in Russian), *Studies on linear operators of functions. XV. Research notes of scientific seminars of the LBMI*, 1986, v.149, pp. 7–23.

2.   Exner, P.; Šeba, P.: A new type of quantum interference transistor, *Phys. Lett. A* 129:8,9 (1988), 477–480.

3.   Reed, M.; Simon, B.: *Methods of modern mathematical physics. III: Scattering theory*, Academic Press, New York - San Francisco - London, 1979.

4.   Krein, M. G.: On the resolvents of a Hermitian operator with the defect indices (m, m) (in Russian), *Dokl. Acad. Nauk SSSR* 52:8 (1946), 657–660.

Department of Theoretical Physics
University of Odessa, 270100 Odessa
Ukraine

MSC 1991: 81U, 47A40

Operator Theory:
Advances and Applications, Vol. 59
© 1992 Birkhäuser Verlag Basel

HOLOMORPHIC OPERATORS BETWEEN KREIN SPACES AND THE
NUMBER OF SQUARES OF ASSOCIATED KERNELS

D. Alpay, A. Dijksma, J. van der Ploeg, H.S.V. de Snoo

Suppose that $\Theta(z)$ is a bounded linear mapping from the Kreĭn space $\mathfrak{F}$ to the Kreĭn space $\mathfrak{G}$, which is defined and holomorphic in a small neighborhood of $z = 0$. Then often Θ admits realizations as the characteristic function of an isometric, a coisometric and of a unitary colligation in which for each case the state space is a Kreĭn space. If the colligations satisfy minimality conditions (i.e., are controllable, observable or closely connected, respectively) then the positive and negative indices of the state space can be expressed in terms of the number of positive and negative squares of certain kernels associated with Θ, depending on the kind of colligation. In this note we study the relations between the numbers of positive and negative squares of these kernels. Using the Potapov–Ginzburg transform we give a reduction to the case where the spaces $\mathfrak{F}$ and $\mathfrak{G}$ are Hilbert spaces. For this case these relations has been considered in detail in [DLS1].

0. INTRODUCTION

Let $(\mathfrak{F}, [.,.]_{\mathfrak{F}})$ and $(\mathfrak{G}, [.,.]_{\mathfrak{G}})$, or $\mathfrak{F}$ and $\mathfrak{G}$ for short, be Kreĭn spaces and denote by $\mathbf{L}(\mathfrak{F},\mathfrak{G})$ the space of bounded linear operators from $\mathfrak{F}$ to $\mathfrak{G}$ (we write $\mathbf{L}(\mathfrak{F})$ for $\mathbf{L}(\mathfrak{F},\mathfrak{F})$). If $T \in \mathbf{L}(\mathfrak{F},\mathfrak{G})$, we write T^* ($\in \mathbf{L}(\mathfrak{G},\mathfrak{F})$) for the adjoint of T with respect to the indefinite inner products $[.,.]_{\mathfrak{F}}$ and $[.,.]_{\mathfrak{G}}$ on the spaces $\mathfrak{F}$ and $\mathfrak{G}$. We say that $T \in \mathbf{L}(\mathfrak{F},\mathfrak{G})$ is invertible (instead of boundedly invertible) if $T^{-1} \in \mathbf{L}(\mathfrak{G},\mathfrak{F},)$. By $\mathbf{S}(\mathfrak{F},\mathfrak{G})$ we denote the (generalized) Schur class of all $\mathbf{L}(\mathfrak{F},\mathfrak{G})$ valued functions Θ, defined and holomorphic on some set in the open unit disc $\mathbf{D} = \{z \in \mathbf{C} \mid |z| < 1\}$; we denote by $\mathfrak{D}(\Theta)$ the domain of holomorphy of Θ in $\mathbf{D}$. The class of Schur functions Θ for which $0 \in \mathfrak{D}(\Theta)$ will be indicated by $\mathbf{S}^\circ(\mathfrak{F},\mathfrak{G})$. If $\mathfrak{D}$ is a subset of $\mathbf{D}$, we write $\mathfrak{D}^* = \{\bar{z} \mid z \in \mathfrak{D}\}$. With each $\Theta \in \mathbf{S}(\mathfrak{F},\mathfrak{G})$ we associate the function $\tilde{\Theta}$ defined by $\tilde{\Theta}(z) = \Theta(\bar{z})^*$. Clearly, $\tilde{\Theta} \in \mathbf{S}(\mathfrak{G},\mathfrak{F})$, $\mathfrak{D}(\tilde{\Theta}) = \mathfrak{D}(\Theta)^*$ and if $\Theta \in \mathbf{S}^\circ(\mathfrak{F},\mathfrak{G})$, then $\tilde{\Theta} \in \mathbf{S}^\circ(\mathfrak{F},\mathfrak{G})$. We associate with Θ the kernels

$$\sigma_\Theta(z,w) = \frac{I - \Theta(w)^*\Theta(z)}{1 - \bar{w}z}, \quad z,w \in \mathfrak{D}(\Theta), \qquad \sigma_{\tilde{\Theta}}(z,w) = \frac{I - \Theta(\bar{w})\Theta(\bar{z})^*}{1 - \bar{w}z}, \quad z,w \in \mathfrak{D}(\Theta)^*,$$

with values in $\mathbf{L}(\mathfrak{F})$ and $\mathbf{L}(\mathfrak{G})$, respectively, and the kernel

$$S_\Theta(z,w) = \begin{pmatrix} \dfrac{I-\Theta(w)^*\Theta(z)}{1-\bar{w}z} & \dfrac{\Theta(\bar{z})^*-\Theta(w)^*}{z-\bar{w}} \\[2ex] \dfrac{\Theta(\bar{w})-\Theta(z)}{\bar{w}-z} & \dfrac{I-\Theta(\bar{w})\Theta(\bar{z})^*}{1-\bar{w}z} \end{pmatrix}, \quad z,w\in\mathfrak{D}(\Theta)\cap\mathfrak{D}(\Theta)^*, \quad z\neq\bar{w},$$

with values in $\mathbf{L}(\mathfrak{F}\oplus\mathfrak{G})$, where $\mathfrak{F}\oplus\mathfrak{G}$ stands for the orthogonal sum of the Kreĭn spaces $\mathfrak{F}$ and $\mathfrak{G}$. Here I is the generic notation for the identity operator on a Kreĭn space, so in the kernels I is the identity operator on $\mathfrak{F}$ or on $\mathfrak{G}$. If we want to be more specific we write, e.g., $I_\mathfrak{F}$ to indicate that I is the identity operator on $\mathfrak{F}$. In this paper we prove theorems about the relation between the number of positive (negative) squares of the matrix kernel S_Θ and those of the kernels σ_Θ and $\sigma_{\tilde{\Theta}}$ on the diagonal of S_Θ.

We recall the following definitions. Let $\mathfrak{K}$ be a Kreĭn space. A kernel $\mathsf{K}(z,w)$ defined for z,w in some subset $\mathfrak{D}$ of the complex plane $\mathbb{C}$ with values in $\mathbf{L}(\mathfrak{K})$, like the kernels considered above, is called nonpositive (nonnegative), if $\mathsf{K}(z,w)^*=\mathsf{K}(w,z)$, $z,w\in\mathfrak{D}$, so that all matrices of the form $([\mathsf{K}(z_i,z_j)f_i,f_j]_\mathfrak{K})^n_{i,j=1}$, where $n\in\mathbb{N}$, $z_1,\ldots,z_n\in\mathfrak{D}$ and $f_1,\ldots,f_n\in\mathfrak{K}$ are arbitrary, are hermitian, and the eigenvalues of each of these matrices are nonpositive (nonnegative, respectively). More generally, we say that $\mathsf{K}(z,w)$ has κ positive (negative) squares if $\mathsf{K}(z,w)^*=\mathsf{K}(w,z)$, $z,w\in\mathfrak{D}$, and all hermitian matrices of the form mentioned above have at most κ and at least one has exactly κ positive (negative) eigenvalues. It has infinitely many positive (negative) squares if for each κ at least one of these matrices has not less than κ positive (negative) eigenvalues. We denote the number of positive and negative squares of K by $\mathrm{sq}_+(\mathsf{K})$ and $\mathrm{sq}_-(\mathsf{K})$, respectively. If, for example, $\mathrm{sq}_-(\mathsf{K})=0$, then $\mathsf{K}(z,w)$ is nonnegative. In the sequel we denote by $\mathrm{ind}_+\mathfrak{K}$ and $\mathrm{ind}_-\mathfrak{K}$ the dimensions of the Hilbert space $\mathfrak{K}_+$ and the anti Hilbert space $\mathfrak{K}_-$ in a fundamental decomposition $\mathfrak{K}=\mathfrak{K}_+\oplus\mathfrak{K}_-$ of $\mathfrak{K}$. Then $\mathrm{ind}_\pm\mathfrak{K}=\mathrm{sq}_\pm(\mathsf{K})$ where K is the constant kernel $\mathsf{K}(z,w)=I$, and the indices are independent of the chosen fundamental decomposition of $\mathfrak{K}$. Whenever in this paper we use the term Pontryagin space we mean a Kreĭn space, $\mathfrak{K}$, say, for which $\mathrm{ind}_-\mathfrak{K}<\infty$.

The main theorems in this paper concern the relation between the values of $\mathrm{sq}_\pm(S_\Theta)$ on the one hand and the values of $\mathrm{sq}_\pm(\sigma_\Theta)$ and $\mathrm{sq}_\pm(\sigma_{\tilde{\Theta}})$ on the other hand. The most general one implies that, if $\kappa\in\mathbb{N}\cup\{0\}$, then $\mathrm{sq}_-(S_\Theta)=\kappa$ if and only if $\mathrm{sq}_-(\sigma_\Theta)=\mathrm{sq}_-(\sigma_{\tilde{\Theta}})=\kappa$, and $\mathrm{sq}_+(S_\Theta)=\kappa$ if and only if $\mathrm{sq}_+(\sigma_\Theta)=\mathrm{sq}_+(\sigma_{\tilde{\Theta}})=\kappa$. To formulate this theorem we consider two fundamental decompositions $\mathfrak{F}=\mathfrak{F}_+\oplus\mathfrak{F}_-$ and $\mathfrak{G}=\mathfrak{G}_+\oplus\mathfrak{G}_-$ of $\mathfrak{F}$ and $\mathfrak{G}$, and we denote by $P_\pm$ the orthogonal projections on $\mathfrak{F}$ onto the spaces $\mathfrak{F}_\pm$ and by $Q_\pm$ the orthogonal projections on $\mathfrak{G}$ onto the spaces $\mathfrak{G}_\pm$.

THEOREM 0.1. *Let $\mathfrak{F}$ and $\mathfrak{G}$ be Kreĭn spaces and let $\Theta\in\mathbf{S}(\mathfrak{F},\mathfrak{G})$. Then:*

(i) $\mathrm{sq}_-(S_\Theta)<\infty$ if and only if (ii) $\mathrm{sq}_-(\sigma_\Theta)<\infty$ and $\mathrm{sq}_-(\sigma_{\tilde{\Theta}})<\infty$.

If (i) and (ii) are valid, then $\mathrm{sq}_-(\sigma_\Theta)=\mathrm{sq}_-(\sigma_{\tilde{\Theta}})=\mathrm{sq}_-(S_\Theta)$ and $Q_-\Theta(z)|_{\mathfrak{F}_-}\in\mathbf{L}(\mathfrak{F}_-,\mathfrak{G}_-)$ is invertible for all except at most $2\mathrm{sq}_-(S_\Theta)$ points $z\in\mathfrak{D}(\Theta)$.

Of course Theorem 0.1 remains valid if the minus sign is everywhere replaced by the plus sign. Note that if $Q_-\Theta(z)|_{\mathfrak{F}_-}\in L(\mathfrak{F}_-,\mathfrak{G}_-)$ is invertible, then $\mathrm{ind}_-\mathfrak{F}=\mathrm{ind}_-\mathfrak{G}$. Theorem 0.1 is closely related to the theorem which states that if $T\in L(\mathfrak{F},\mathfrak{G})$ is a contraction (i.e, $T^*T\le I_{\mathfrak{F}}$), then it is a bicontraction (i.e., T and T^* are both contractions) if and only if $Q_-T|_{\mathfrak{F}_-}$ is invertible; see [DR] for a recent account of this result and related results. The existence of a contraction in a Kreĭn space which is not a bicontraction can be used to give a counterexample to the equality in Theorem 0.1 when the kernel $S_\Theta(z,w)$ has an infinite number of negative squares. Indeed, let $T\in L(\mathfrak{F})$ be a contractive but not bicontractive operator, set $\mathfrak{G}=\mathfrak{F}$ and take $\Theta(z)\equiv T$ on $\mathbb{D}$. Then $\mathrm{sq}_-(\sigma_\Theta)=0$, but we have that $\mathrm{sq}_-(S_\Theta)=\mathrm{sq}_-(\sigma_{\tilde\Theta})=\infty$. In proving these equalities we used the fact that the $n\times n$ matrix $(1/(1-z_i\bar z_j))$, where $z_1,z_2,\ldots,z_n$ are n different points in $\mathbb{D}$, is positive. This follows for instance from the formula $1/(1-z\bar w)=(2\pi)^{-1}\int_0^{2\pi}(e^{i\tau}-z)^{-1}(e^{-i\tau}-\bar w)^{-1}\mathrm{d}\tau$, $z,w\in\mathbb{D}$.

In the case where $\mathfrak{F}$ and $\mathfrak{G}$ are Kreĭn spaces such that either their positive indices $\mathrm{ind}_+\mathfrak{F}$ and $\mathrm{ind}_+\mathfrak{G}$, or their negative indices $\mathrm{ind}_-\mathfrak{F}$ and $\mathrm{ind}_-\mathfrak{G}$, are finite, Theorem 0.1 can be sharpened. We only state the version for which the last condition on the indices holds; the one for which $\mathrm{ind}_+\mathfrak{F}$ and $\mathrm{ind}_+\mathfrak{G}$ are finite is mutatis mutandis the same.

THEOREM 0.2. *Let $\mathfrak{F}$ and $\mathfrak{G}$ be Pontryagin spaces, $\Theta\in S(\mathfrak{F},\mathfrak{G})$ and let $\kappa\in\mathbb{N}\cup\{0\}$. Then the following three statements are equivalent.*

(i) $\mathrm{sq}_-(S_\Theta)=\kappa$.

(ii) $\mathrm{sq}_-(\sigma_\Theta)=\kappa$ *and* $\mathrm{ind}_-\mathfrak{F}=\mathrm{ind}_-\mathfrak{G}$.

(iii) $\mathrm{sq}_-(\sigma_{\tilde\Theta})=\kappa$ *and* $\mathrm{ind}_-\mathfrak{F}=\mathrm{ind}_-\mathfrak{G}$.

If (i)−(iii) are valid then $Q_-\Theta(z)|_{\mathfrak{F}_-}\in L(\mathfrak{F}_-,\mathfrak{G}_-)$ is invertible for all, except for at most κ, points $z\in\mathfrak{D}(\Theta)$. Moreover, if these conditions hold, then Θ can be extended to a meromorphic function on $\mathbb{D}$, and the number of negative squares of each of the three kernels associated with the extended function coincides with the number of negative squares of the corresponding kernel for the function Θ.

This theorem corresponds to the fact that if $\mathrm{ind}_-\mathfrak{F}=\mathrm{ind}_-\mathfrak{G}<\infty$, then every contraction $T\in L(\mathfrak{F},\mathfrak{G})$ is also a bicontraction; see, for example, [IKL], [DR]. Theorem 0.2 implies that if $\mathrm{ind}_-\mathfrak{F}=\mathrm{ind}_-\mathfrak{G}<\infty$, then $\mathrm{sq}_-(S_\Theta)=\mathrm{sq}_-(\sigma_\Theta)=\mathrm{sq}_-(\sigma_{\tilde\Theta})$, which for the case where $\mathfrak{F}$ and $\mathfrak{G}$ are Hilbert spaces was already proved in [DLS1]. The equality here means that either all three numbers $\mathrm{sq}_-(S_\Theta)$, $\mathrm{sq}_-(\sigma_\Theta)$, $\mathrm{sq}_-(\sigma_{\tilde\Theta})$ are infinite, or one of them is finite in which case all three are finite and equal. If $\mathrm{ind}_-\mathfrak{F}<\infty$, $\mathrm{ind}_-\mathfrak{G}<\infty$, but $\mathrm{ind}_-\mathfrak{F}\ne\mathrm{ind}_-\mathfrak{G}$, then the equality between the numbers of negative squares of the three kernels need not hold, as the following example shows. To avoid confusion with the signs we give a counterexample to Theorem 0.2 in which all signs are reversed. Take $\mathfrak{F}=\mathbb{C}^2$, $\mathfrak{G}=\mathbb{C}$ and define the 1×2 matrix function $\Theta\in S^\circ(\mathfrak{F},\mathfrak{G})$ by $\Theta(z)=(kz,1)$, $z\in\mathbb{D}$, where $k\in\mathbb{C}$ and $|k|<1$. Then σ_Θ is a 2×2 matrix kernel, $\sigma_{\tilde\Theta}$ is a scalar kernel and S_Θ is a 3×3 matrix kernel. In this case we have that $\mathrm{sq}_+(\sigma_{\tilde\Theta})=0$, but $\mathrm{sq}_+(S_\Theta)=\mathrm{sq}_+(\sigma_\Theta)=\infty$.

The proofs of Theorems 0.1 and 0.2 which we give are basically a variation of the proof of the theorem about the characterization of bicontractions in a Kreĭn space mentioned previously: we also make use of the Potapov–Ginzburg transform. We show in this paper that, if the kernels associated with Θ have a finite number of negative squares, then the Potapov–Ginzburg transformation of Θ can be defined, and, if we denote this transformation by Σ, then $\mathrm{sq}_-(\sigma_\Sigma) = \mathrm{sq}_-(\sigma_\Theta)$, $\mathrm{sq}_-(\sigma_{\tilde\Sigma}) = \mathrm{sq}_-(\sigma_{\tilde\Theta})$ and $\mathrm{sq}_-(S_\Sigma) = \mathrm{sq}_-(S_\Theta)$. Since Σ is essentially a Schur function between Hilbert spaces, the proofs of Theorems 0.1 and 0.2 are obtained through a reduction of the general case to the case where $\mathfrak{F}$ and $\mathfrak{G}$ are Hilbert spaces, which was considered in [DLS1].

Theorems 0.1 and 0.2 are related to Theorem 7.6 in the excellent lecture notes by Ando [An], which contains a detailed account of the case $\kappa = 0$. To point out the connection we state the following related result.

THEOREM 0.3. *Let $\mathfrak{F}$ and $\mathfrak{G}$ be Kreĭn spaces and let $\Theta \in \mathbf{S}(\mathfrak{F},\mathfrak{G})$. If $\mathfrak{D}(\Theta) = \mathbf{D}$, then the following assertions are equivalent.*

(a)   *For each $z \in \mathbf{D}$: $\Theta(z)$ is a bicontraction.*

(b)   *For each $z \in \mathbf{D}$: $\Theta(z)$ is a contraction and $Q_-\Theta(z)|_{\mathfrak{F}_-} \in \mathbf{L}(\mathfrak{F}_-,\mathfrak{G}_-)$ is invertible.*

(c)   *For each $z \in \mathbf{D}$: $\tilde\Theta(z)$ is a contraction and $P_-\tilde\Theta(z)|_{\mathfrak{G}_-} \in \mathbf{L}(\mathfrak{G}_-,\mathfrak{F}_-)$ is invertible.*

(d) $\mathrm{sq}_-(\sigma_\Theta) = 0$ *and* $\mathrm{sq}_-(\sigma_{\tilde\Theta}) = 0$.

(e) $\mathrm{sq}_-(S_\Theta) = 0$.

The equivalences $(a){\leftrightarrow}(b){\leftrightarrow}(c)$ are alluded to above. The implications $(e){\Rightarrow}(d)$, $(d){\Rightarrow}(a)$ are trivial: the first one follows from the fact that the kernels $\sigma_\Theta(z,w)$ and $\sigma_{\tilde\Theta}(z,w)$ are on the diagonal of $S_\Theta(z,w)$ and the second follows by taking $w = z$ in $\sigma_\Theta(z,w)$ and $\sigma_{\tilde\Theta}(z,w)$. To prove the implication $(a){\Rightarrow}(e)$, Ando in [An] uses complementation theory and the theory of operator ranges; see also [B2,3]. On the other hand on account of the invertibility of $Q_-\Theta(z)|_{\mathfrak{F}_-} \in \mathbf{L}(\mathfrak{F}_-,\mathfrak{G}_-)$ the implication can be reduced to one where the spaces involved are Hilbert spaces, and then the implication can be and has been proved in different ways; see [SF] and [BR]. In this paper we are interested in the equivalence $(d){\leftrightarrow}(e)$, which has been proved by McEnnis [Mc]. The purpose of this paper is to generalize this equivalence to the case where we allow the kernels associated with $\Theta \in \mathbf{S}(\mathfrak{F},\mathfrak{G})$ to have negative squares. These generalizations are formulated in Theorems 0.1 and 0.2. Note that we do not require that $\mathfrak{D}(\Theta) = \mathbf{D}$.

The kernels associated with $\Theta \in \mathbf{S}^\circ(\mathfrak{F},\mathfrak{G})$ have appeared in a number of papers, frequently in connection with the theory of (co–)isometric and unitary colligations; of the more recent papers, we mention for instance [A], [AD1,2,3], [AG1,2], [An], [B1,2,3], [BC], [CDLS], [DLS1,2,3], [M] and [Y]. If $\mathfrak{F}$ and $\mathfrak{G}$ are Hilbert spaces then Θ can be written as the characteristic function of (a) a closely innerconnected isometric colligation, (b) a closely outerconnected coisometric colligation, as well as of (c) a closely connected unitary colligation. If in all three cases we denote by $\Re$ the state space of the colligation, then $\Re$ is a Kreĭn space with $\mathrm{ind}_\pm(\Re) = \mathrm{sq}_\pm(\sigma_\Theta)$ in case (a), $\mathrm{ind}_\pm(\Re) = \mathrm{sq}_\pm(\sigma_{\tilde\Theta})$ in case (b) and $\mathrm{ind}_\pm(\Re) = \mathrm{sq}_\pm(S_\Theta)$ in case (c). Proofs of these results can be found in, e.g., [DLS1]; see also Remark 1.1 in Section 1 of this paper. In the general case where $\mathfrak{F}$ and $\mathfrak{G}$ are Kreĭn spaces, $\Theta \in \mathbf{S}^\circ(\mathfrak{F},\mathfrak{G})$ can also be realized on a neighborhood of 0 as the characteristic function of a minimal unitary or (co-) isometric colligation. In this case the state space of the colligation is uniquely determined by Θ up to weak isomorphisms and so its positive and negative indices are uniquely determined by Θ, cf. Section 1 below.

We briefly describe the contents of the three sections of this paper. In Section 1 (REALIZATIONS OF A CLASS OF SCHUR FUNCTIONS) we consider the case where $\mathfrak{F}$ and $\mathfrak{G}$ are Hilbert spaces, and we recall some facts from [DLS1]. For this case we show in Section 2 (POSITIVE SQUARES AND INJECTIVITY) that if the kernels have a finite number of positive squares, then the operator function $Q_-\Theta(z)|_{\mathfrak{F}_-}$ is invertible. As an aside we prove a simple result about the positivity of the Schur product of nonnegative matrices. The main theorems of this paper, Theorem 0.1 and Theorem 0.2, are proved in Section 3 (APPLICATION OF THE POTAPOV–GINZBURG TRANSFORM) by invoking the results of the previous sections. There we also recall the definition of the Potapov–Ginzburg transform.

In a sequel to this paper we give an interpretation of our results from the point of view of reproducing kernel Pontryagin spaces and operator ranges. We show how this leads to models for contractions in Pontryagin spaces in which certain maximal negative invariant subspaces can be displayed explicitly.

1. REALIZATIONS OF A CLASS OF SCHUR FUNCTIONS

A *colligation* (system or node) is a quadruple $\Delta = (\Re,\mathfrak{F},\mathfrak{G};U)$ consisting of three Kreĭn spaces $\Re$ (the *state space*), $\mathfrak{F}$ (the *incoming space*) and $\mathfrak{G}$ (the *outgoing space*) and an operator $U \in \mathbf{L}(\Re \oplus \mathfrak{F}, \Re \oplus \mathfrak{G})$ (the *connecting operator*). We often write $\Delta = (\Re,\mathfrak{F},\mathfrak{G};T,F,G,H)$ to indicate that U has the operator matrix decomposition

$$U = \begin{pmatrix} T & F \\ G & H \end{pmatrix} : \begin{pmatrix} \Re \\ \mathfrak{F} \end{pmatrix} \to \begin{pmatrix} \Re \\ \mathfrak{G} \end{pmatrix},$$

where $T \in \mathbf{L}(\Re)$ (the *basic operator*), $F \in \mathbf{L}(\mathfrak{F},\Re)$, $G \in \mathbf{L}(\Re,\mathfrak{G})$ and $H \in \mathbf{L}(\mathfrak{F},\mathfrak{G})$. The colligation is called isometric, coisometric or unitary if the connecting operator has this property. A colligation $\Delta = (\Re,\mathfrak{F},\mathfrak{G};T,F,G,H)$ is called *closely innerconnected* (controllable) if

$$\Re = \bigvee_{z \in N} \Re((I - zT)^{-1}F) = \vee\{T^n Ff \mid n \in \mathbb{N} \cup \{0\},\ f \in \mathfrak{F}\},$$

closely outerconnected (observable) if

$$\Re = \bigvee_{z \in N} \Re((I - zT^*)^{-1}G^*) = \vee\{T^{*n}G^*g \mid n \in \mathbb{N} \cup \{0\},\ g \in \mathfrak{G}\},$$

and it is called *closely connected* if

$$\mathfrak{K} = \bigvee_{z \in N} \left(\mathfrak{R}((I - zT)^{-1}F) \cup \mathfrak{R}((I - zT^*)^{-1}G^*) \right).$$

In these cases N stands for a small neighborhood around 0 and $\bigvee_{z \in N}$ signifies the closed linear span of the sets with index $z \in N$. We call an isometric (coisometric, unitary) colligation *minimal* if it is closely innerconnected (closely outerconnected, closely connected, respectively).

A *weak isomorphism* V from a Kreĭn space $\mathfrak{K}$ to a Kreĭn space $\mathfrak{K}'$ is a mapping with dense domain $\mathfrak{D}(V)$ in $\mathfrak{K}$ and dense range $\mathfrak{R}(V)$ in $\mathfrak{K}'$ such that $[Vx,Vy]_{\mathfrak{K}'} = [x,y]_{\mathfrak{K}}$, $x,y \in \mathfrak{D}(V)$. If $\mathfrak{K}$ and $\mathfrak{K}'$ are weakly isomorphic, then $\mathrm{ind}_\pm(\mathfrak{K}) = \mathrm{ind}_\pm(\mathfrak{K}')$. Two colligations $\Delta = (\mathfrak{K},\mathfrak{F},\mathfrak{G};T,F,G,H)$ and $\Delta' = (\mathfrak{K}',\mathfrak{F},\mathfrak{G};T',F',G',H')$ with the same incoming and outgoing spaces $\mathfrak{F}$ and $\mathfrak{G}$ are called *weakly isomorphic* if $H = H'$ and there exists a weak isomorphism V from $\mathfrak{K}$ to $\mathfrak{K}'$ such that

$$\begin{pmatrix} T' & F' \\ G' & H' \end{pmatrix} \begin{pmatrix} V & 0 \\ 0 & I \end{pmatrix} = \begin{pmatrix} V & 0 \\ 0 & I \end{pmatrix} \begin{pmatrix} T & F \\ G & H \end{pmatrix} \text{ on } \begin{pmatrix} \mathfrak{D}(V) \\ \mathfrak{F} \end{pmatrix}.$$

If V can be extended to an isomorphism (unitary operator) from $\mathfrak{K}'$ onto $\mathfrak{K}$, then of course the colligations Δ and Δ' are called isomorphic or unitarily equivalent. Associated with a colligation $\Delta = (\mathfrak{K},\mathfrak{F},\mathfrak{G};T,F,G,H)$ is the socalled *characteristic function*

$$\Theta(z) = \Theta_\Delta(z) = H + zG(I - zT)^{-1}F,$$

defined for all z in some neighborhood of 0. Clearly, $\Theta \in \mathbf{S}^\circ(\mathfrak{F},\mathfrak{G})$. If for $\Theta \in \mathbf{S}^\circ(\mathfrak{F},\mathfrak{G})$ there is a colligation Δ such that $\Theta(z) = \Theta_\Delta(z)$ in a neighborhood of 0, then the colligation is called a *realization* of Θ. Realizations in a general Kreĭn space case are studied in, e.g., [Az1,2], [B1,2,3], [CDLS], [DLS3], [M], [Y] and for a special class in [DLS1].

REMARK 1.1. If $\mathfrak{F}$ and $\mathfrak{G}$ are Kreĭn spaces, $\Theta \in \mathbf{S}^\circ(\mathfrak{F},\mathfrak{G})$ and $\Theta = \Theta_\Delta$ on $\mathfrak{D}(\Theta) \cap \mathfrak{D}(\Theta_\Delta)$ for some isometric (coisometric, unitary) colligation $\Delta = (\mathfrak{K},\mathfrak{F},\mathfrak{G};U)$, then $\mathrm{ind}_\pm\mathfrak{K} \geq \mathrm{sq}_\pm(\sigma_\Theta)$ ($\mathrm{ind}_\pm\mathfrak{K} \geq \mathrm{sq}_\pm(\sigma_{\tilde{\Theta}})$, $\mathrm{ind}_\pm\mathfrak{K} \geq \mathrm{sq}_\pm(S_\Theta)$, respectively). If the colligation is minimal then in the corresponding inequality the equality sign prevails. These (in-)equalities can easily be deduced from the following relations between the kernels and the inner product of the state space $\mathfrak{K}$ of Δ. For $f_1, f_2 \in \mathfrak{F}$ and $g_1, g_2 \in \mathfrak{G}$ we have

(i) $\qquad [\sigma_\Theta(z,w)f_1,f_2]_{\mathfrak{F}} = [(I - zT)^{-1}Ff_1,(I - wT)^{-1}Ff_2]_{\mathfrak{K}}, \quad$ if Δ is isometric;

(ii) $\qquad [\sigma_{\tilde{\Theta}}(z,w)g_1,g_2]_{\mathfrak{G}} = [(I - zT^*)^{-1}G^*g_1,(I - wT^*)^{-1}G^*g_2]_{\mathfrak{K}}, \quad$ if Δ is coisometric;

(iii) $\qquad [S_\Theta(z,w)\begin{pmatrix} f_1 \\ g_1 \end{pmatrix},\begin{pmatrix} f_2 \\ g_2 \end{pmatrix}]_{\mathfrak{F} \oplus \mathfrak{G}} = [(I - zT)^{-1}Ff_1 + (I - zT^*)^{-1}G^*g_1,(I - wT)^{-1}Ff_2 + (I - wT^*)^{-1}G^*g_2]_{\mathfrak{K}},$

$$\text{if } \Delta \text{ is unitary.}$$

In this section we summarize some of the results from [DLS1].

THEOREM 1.2. *Assume that $\mathfrak{F}$ and $\mathfrak{G}$ are Hilbert spaces and let $\Theta \in \mathbf{S}^\circ(\mathfrak{F}, \mathfrak{G})$. Then*

$$(1.1) \qquad \mathrm{sq}_-(\sigma_\Theta) = \mathrm{sq}_-(\sigma_{\tilde\Theta}) = \mathrm{sq}_-(\mathsf{S}_\Theta).$$

That is, either these three numbers are infinite or one of them is finite and then they all are finite and equal. If one of them is finite and equal to $\kappa \in \mathbb{N} \cup \{0\}$, say, then Θ admits the following realizations.

(a) *$\Theta = \Theta_{\Delta_1}$ on $\mathfrak{D}(\Theta)$ for some isometric colligation $\Delta_1 = (\mathfrak{K}_1, \mathfrak{F}, \mathfrak{G}; U_1)$, and then $\mathrm{ind}_+\mathfrak{K}_1 \geq \mathrm{sq}_\pm(\sigma_\Theta)$; the isometric colligation Δ_1 can be chosen closely innerconnected, in which case it is uniquely determined up to isomorphisms and $\mathrm{ind}_+\mathfrak{K}_1 = \mathrm{sq}_\pm(\sigma_\Theta)$.*

(b) *$\Theta = \Theta_{\Delta_2}$ on $\mathfrak{D}(\Theta)$ for some coisometric colligation $\Delta_2 = (\mathfrak{K}_2, \mathfrak{F}, \mathfrak{G}; U_2)$, and then $\mathrm{ind}_+\mathfrak{K}_2 \geq \mathrm{sq}_\pm(\sigma_{\tilde\Theta})$; the coisometric colligation Δ_2 can be chosen closely outerconnected, in which case it is uniquely determined up to isomorphisms and $\mathrm{ind}_+\mathfrak{K}_2 = \mathrm{sq}_\pm(\sigma_{\tilde\Theta})$.*

(c) *$\Theta = \Theta_\Delta$ on $\mathfrak{D}(\Theta)$ for some unitary colligation $\Delta = (\mathfrak{K}, \mathfrak{F}, \mathfrak{G}; U)$, and then $\mathrm{ind}_+\mathfrak{K} \geq \mathrm{sq}_\pm(\mathsf{S}_\Theta)$; the unitary colligation Δ can be chosen closely connected, in which case it is uniquely determined up to isomorphisms and $\mathrm{ind}_+\mathfrak{K} = \mathrm{sq}_\pm(\mathsf{S}_\Theta)$.*

One of the key tools in the proof of Theorem 1.2 given in [DLS1] is formulated in the following lemma. A Hilbert space version of it is given by de Branges and Shulman in [BS]. It relates isometric colligations to unitary colligations.

LEMMA 1.3. *Let $\Delta_1 = (\mathfrak{K}_1, \mathfrak{F}, \mathfrak{G}; U_1)$ be a closely innerconnected isometric colligation, in which $\mathfrak{F}$ and $\mathfrak{G}$ are Hilbert spaces and $\mathfrak{K}_1$ is a Pontryagin space. Then there exists a closely connected unitary colligation $\Delta = (\mathfrak{K}, \mathfrak{F}, \mathfrak{G}; U)$ such that $\mathrm{ind}_-\mathfrak{K} = \mathrm{ind}_-\mathfrak{K}_1$, $U|_{\mathfrak{F}} = V$ and $\Theta_\Delta = \Theta_{\Delta_1}$ on $\mathfrak{D}(\Theta_\Delta) \cap \mathfrak{D}(\Theta_{\Delta_1})$.*

Sketch of the proof of Theorem 1.2. (1) We first consider the case where $\mathrm{sq}_-(\sigma_\Theta)$ is finite, and we briefly describe the construction of the realization described in (a). Consider the linear space $\mathfrak{L}$ of finite sums $\sum_z \varepsilon_z f_z$, where $z \in \mathfrak{D}(\Theta)$, $f_z \in \mathfrak{F}$ and ε_z is a symbol associated with each $z \in \mathfrak{D}(\Theta)$ and provide $\mathfrak{L}$ with the (possibly degenerate, indefinite) inner product

$$\left[\sum_z \varepsilon_z f_z, \sum_w \varepsilon_w g_w \right] = \sum_{z,w} \left[\sigma_\Theta(z,w) f_z, g_w \right]_{\mathfrak{F}}.$$

Define the linear operators T_0, F_0, G_0 and H via the formulas

$$T_0 \varepsilon_z f = \frac{1}{z}(\varepsilon_z f - \varepsilon_0 f), \qquad\qquad F_0 f = \varepsilon_0 f,$$
$$G_0 \varepsilon_z f = \frac{1}{z}(\Theta(z)f - \Theta(0)f), \qquad\qquad H f = \Theta(0)f,$$

where $z \neq 0$ and $f \in \mathfrak{F}$. Then T_0 and G_0 are densely defined operators on $\mathfrak{L}$ with values in $\mathfrak{L}$ and $\mathfrak{G}$, respectively, $F_0 : \mathfrak{F} \to \mathfrak{L}$, $H : \mathfrak{F} \to \mathfrak{G}$,

$$U_0 = \begin{pmatrix} T_0 & F_0 \\ G_0 & H \end{pmatrix} : \begin{pmatrix} \mathfrak{L} \\ \mathfrak{F} \end{pmatrix} \to \begin{pmatrix} \mathfrak{L} \\ \mathfrak{G} \end{pmatrix}$$

is an isometric operator on a dense set and $H + zG_0(I - zT_0)^{-1}F_0 = \Theta(z)$ on $\mathfrak{F}$. Now we consider the quotient space of $\mathfrak{L}$ over its isotropic part and redefine the operators on this space in the usual way. Then completing the quotient space to a Pontryagin space and extending the operators by continuity to this completion we obtain a closely innerconnected isometric colligation with the desired properties.

(2) If it is given that $sq_-(\sigma_{\tilde{\Theta}})$ is finite we apply the above construction to $\tilde{\Theta}$ to obtain a realization in terms of a closely innerconnected isometric colligation $\Delta_1 = (\mathfrak{K}_1, \mathfrak{G}, \mathfrak{F}; U_1)$. Then $\Delta_2 = (\mathfrak{K}_2, \mathfrak{F}, \mathfrak{G}; U_2)$ with $\mathfrak{K}_2 = \mathfrak{K}_1$ and $U_2 = U_1^*$ is the closely outerconnected coisometric colligation with the properties mentioned in (b). If $sq_-(S_\Theta)$ is finite, then also $sq_-(\sigma_\Theta)$ is finite and a realization of Θ described in (c) can be obtained by constructing a realization of type (a) as done in part (1) and by invoking Lemma 1.3.

(3) We omit the details of the proof that the minimality condition implies the essential uniqueness of the colligation in the realization of Θ. This uniqueness property together with the constructions indicated in steps (1) and (2) imply that $\mathrm{ind}_-\mathfrak{K}_1 = \mathrm{ind}_-\mathfrak{K}_2 = \mathrm{ind}_-\mathfrak{K}$. The equalities in (1.1) now follow directly from the equalities in Remark 1.1. This completes the sketch of the proof.

The following factorization theorem is an application of Theorem 1.2. A proof can be found in [KL]. Below we sketch the proof given in [DLS1]. To formulate it we recall the notion of a Blaschke–Potapov product. For $\alpha \in \mathbb{D}$ and $\theta \in [0, 2\pi)$ we define the holomorphic bijection $b_{\alpha,\theta}: \mathbb{D} \to \mathbb{D}$ by $b_{\alpha,\theta}(z) = (z - \alpha)e^{i\theta}/(1 - z\bar{\alpha})$. An operator $B: \mathbb{D} \to \mathbf{L}(\mathfrak{F}, \mathfrak{F})$ is called a (finite) *Blaschke–Potapov product on $\mathfrak{F}$ of degree* $\kappa \in \mathbb{N}$ if B can be written as a (finite) product of factors of the form $(I - P) + b_{\alpha,\theta}(z)P$ for some $\alpha \in \mathbb{D}$ (α is called a zero of the product), $\theta \in [0, 2\pi)$ and some projection P on $\mathfrak{F}$ (of finite rank), and $\sum \mathrm{rank}\,P = \kappa$, where the sum is taken over the projections P appearing in the factors forming B. Note that such a product assumes unitary values on the boundary $\mathbb{T}$ of the open unit disc $\mathbb{D}$.

THEOREM 1.4. *Assume that $\mathfrak{F}$ and $\mathfrak{G}$ are Hilbert spaces and let $\Theta \in \mathbf{S}^\circ(\mathfrak{F}, \mathfrak{G})$. If one of the numbers $sq_-(\sigma_\Theta)$, $sq_-(\sigma_{\tilde{\Theta}})$ and $sq_-(S_\Theta)$ is finite and equal to $\kappa \in \mathbb{N} \cup \{0\}$, say, then Θ admits the strongly regular factorizations*

$$\Theta(z) = B_L(z)^{-1}\Theta_L(z) = \Theta_R(z)B_R(z)^{-1}, \quad z \in \mathfrak{D}(\Theta),$$

where $B_R: \mathbb{D} \to \mathbf{L}(\mathfrak{F}, \mathfrak{F})$, $B_L: \mathbb{D} \to \mathbf{L}(\mathfrak{G}, \mathfrak{G})$ are Blaschke–Potapov products of degree κ with zeros in $\mathbb{D} \backslash \{0\}$ and Θ_R, $\Theta_L: \mathbb{D} \to \mathbf{L}(\mathfrak{F}, \mathfrak{G})$ are holomorphic contraction operators. In particular, Θ can be extended to a meromorhpic operator function on $\mathbb{D}$ with κ poles (counting multiplicities) in $\mathbb{D} \backslash \{0\}$.

In the theorem strongly regular essentially means that the zero's of Θ_L (Θ_R) and the projections appearing in the Blaschke–Potapov product B_L (B_R) do not cancel any of the poles of B_L^{-1} (B_R^{-1}, respectively). For the precise definition we refer to [DLS1], Section 7, but in the sequel the above indication of what it means should be sufficient.

Sketch of the proof of Theorem 1.4. Apply Theorem 1.2 and let $\Delta = (\Re,\mathfrak{F},\mathfrak{G};T,F,G,H)$ be a closely connected unitary colligation whose characteristic function coincides with Θ. Then $\Re$ is a Pontryagin space with $\mathrm{ind}_{-}\Re = \kappa$. From the equality $T^*T + G^*G = I$ and the fact that G maps the Pontryagin space $\Re$ into the Hilbert space $\mathfrak{G}$ it follows that T is a contraction. Hence there exists a κ–dimensional nonpositive subspace Π of $\Re$ which is invariant under T and such that the spectrum of the restriction of T to Π lies outside $\mathbb{D}$; see [IKL] or [DLS1]. This leads to a decomposition of Δ as the product of two unitary colligations, one of them on the left– or right–hand side corresponding to the κ–dimensional invariant subspace Π_0, whose characteristic function is a Blaschke–Potapov product. This completes the sketch of the proof.

To remove the restriction "$0 \in \mathfrak{D}(\Theta)$" in Theorem 1.4 we use the following lemma. Its proof amounts to straightforward substitution and is therefore omitted.

LEMMA 1.5. *Let* $\Theta \in \mathbf{S}(\mathfrak{F},\mathfrak{G})$, *where* $\mathfrak{F}$ *and* $\mathfrak{G}$ *are Krein spaces. Assume that* $z_0 \in \mathfrak{D}(\Theta)$, *denote the mapping* $b_{-z_0,0} : \mathbb{D} \to \mathbb{D}$ *by* b *so that* $b(z) = (z + z_0)/(1 + z\bar{z}_0)$, $z \in \mathbb{D}$, *and define the operator function* $\Theta_0 \in \mathbf{S}(\mathfrak{F},\mathfrak{G})$ *by* $\Theta_0(z) = \Theta(b(z))$, $z \in \mathfrak{D}(\Theta_0) = b_{z_0,0}(\mathfrak{D}(\Theta))$. *Then:*

(i) $\Theta_0 \in \mathbf{S}^\circ(\mathfrak{F},\mathfrak{G})$,

(ii) $(1 - |z_0|^2)\sigma_\Theta(b(z),b(w)) = \overline{(1 + w\bar{z}_0)}\,\sigma_{\Theta_0}(z,w)(1 + z\bar{z}_0)$ *and* $\mathrm{sq}_\pm(\sigma_\Theta) = \mathrm{sq}_\pm\sigma_{\Theta_0})$,

(iii) $(1 - |z_0|^2)\sigma_{\tilde\Theta}(b(\bar{z}),b(\bar{w})) = \overline{(1 + wz_0)}\,\sigma_{\tilde\Theta_0}(z,w)(1 + zz_0)$ *and* $\mathrm{sq}_\pm(\sigma_{\tilde\Theta}) = \mathrm{sq}_\pm(\sigma_{\tilde\Theta_0})$,

(iv) $(1 - |z_0|^2)\dfrac{\Theta(b(\bar{w})) - \Theta(b(z))}{b(\bar{w}) - b(z)} = \overline{(1 + wz_0)}\,\dfrac{\Theta_0(\bar{w}) - \Theta_0(z)}{\bar{w} - z}(1 + z\bar{z}_0).$

If $z_0 \in \mathbb{R} \cap \mathfrak{D}(\Theta)$, *so that* $b(\bar{z}) = \overline{b(z)}$, *then*

(v) $(1 - |z_0|^2)S_\Theta((b(z),b(w)) = \mathrm{diag}((1 + w\bar{z}_0),(1 + wz_0))^*S_{\Theta_0}(z,w)\,\mathrm{diag}((1 + z\bar{z}_0),(1 + zz_0))$

and $\mathrm{sq}_\pm(S_\Theta) = \mathrm{sq}_\pm(S_{\Theta_0})$.

Applying Lemma 1.5 to Theorems 1.2 and 1.4, we obtain the following result.

COROLLARY 1.6. *Assume that* $\mathfrak{F}$ *and* $\mathfrak{G}$ *are Hilbert spaces and let* $\Theta \in \mathbf{S}(\mathfrak{F},\mathfrak{G})$. *Then*

$$\mathrm{sq}_-(\sigma_\Theta) = \mathrm{sq}_-(\sigma_{\tilde\Theta}) = \mathrm{sq}_-(S_\Theta)$$

(with the same interpretation as in Theorem 1.2). If these numbers are finite and equal to $\kappa \in \mathbb{N} \cup \{0\}$, *say, then* Θ *admits the strongly regular factorizations*

$$\Theta(z) = B_L(z)^{-1}\Theta_L(z) = \Theta_R(z)B_R(z)^{-1}, \quad z \in \mathfrak{D}(\Theta),$$

where $B_R : \mathbb{D} \to \mathbf{L}(\mathfrak{F},\mathfrak{F})$, $B_L : \mathbb{D} \to \mathbf{L}(\mathfrak{G},\mathfrak{G})$ *are Blaschke – Potapov products of degree* κ *and* $\Theta_R, \Theta_L : \mathbb{D} \to \mathbf{L}(\mathfrak{F},\mathfrak{G})$ *are holomorphic contraction operators. In particular,* Θ *can be extended to a meromorphic operator function on* $\mathbb{D}$ *with* κ *poles (counting multiplicities) in* $\mathbb{D}$.

2. POSITIVE SQUARES AND INJECTIVITY

For the main theorem in this section we use the following result. A more general version involving Schur products will be given at the end of this section.

LEMMA 2.1. *Let $z_1, z_2, \ldots, z_n$ be n different points in $\mathbb{D}$ and let $Q = (Q_{ij})$ be a nonnegative $n \times n$ matrix with diagonal elements $Q_{ii} > 0$, $i = 1, 2, \ldots, n$. Then the $n \times n$ matrix $(Q_{ij}/(1 - z_i \bar{z}_j))$ is positive.*

Proof. Let P be the matrix $(Q_{ij}/(1 - z_i \bar{z}_j))$. Then $P = \sum_{k=0}^{\infty} D^k Q D^{*k}$, where $D = \mathrm{diag}(z_1, z_2, \ldots, z_n)$. Hence $P \geq 0$, and to show that $P > 0$, it suffices to prove that $\mathfrak{N}(P) = \{0\}$, where $\mathfrak{N}(P) \subset \mathbb{C}^n$ is the null space of P. Let $u = (u_i) \in \mathfrak{N}(P)$. Then $0 = u^* P u = \sum_{k=0}^{\infty} (D^{*k} u)^* Q D^{*k} u$, and, since each summand in the series is nonnegative, the vector $Q D^{*k} u = 0$, $k = 0, 1, \ldots$. It follows that for every polynomial P with complex coefficients the vector $Q P(D^*) u = 0$. Let $i \in \{1, 2, \ldots, n\}$ and let P_i be a polynomial with the property that $P_i(\bar{z}_j) = \delta_{ij}$, the Kronecker delta. Then $0 = (Q P_i(D^*) u)_i = Q_{ii} u_i$, which implies that $u_i = 0$, since, by assumption, $Q_{ii} > 0$. It follows that $u = (u_i) = 0$, i.e., $\mathfrak{N}(P) = \{0\}$.

Note that in the proof of Lemma 2.1 we have used the Taylor expansion $1/(1 - z\bar{w}) = \sum_{n=1}^{\infty} (\bar{w}z)^n$. A similar proof can be given based on the integral representation of $1/(1 - z\bar{w})$ given in the Introduction just above Theorem 0.2.

THEOREM 2.2. *Let $\mathfrak{F}$ and $\mathfrak{G}$ be Hilbert spaces and let $\Theta \in S(\mathfrak{F}, \mathfrak{G})$.*
(i) If $\mathrm{sq}_+(\sigma_\Theta) < \infty$, then for each $z \in \mathcal{D}(\Theta)$, with the exception of at most $\mathrm{sq}_+(\sigma_\Theta)$ points in $\mathcal{D}(\Theta)$, there is a positive constant $\varepsilon(z)$, such that $\Theta(z)^* \Theta(z) \geq \varepsilon(z)I$.
(ii) If $\mathrm{sq}_+(\sigma_{\tilde{\Theta}}) < \infty$, then for each $z \in \mathcal{D}(\Theta)$, with the exception of at most $\mathrm{sq}_+(\sigma_{\tilde{\Theta}})$ points in $\mathcal{D}(\Theta)$, there is a positive constant $\delta(z)$, such that $\Theta(z) \Theta(z)^* \geq \delta(z)I$.

Proof. To prove (i) we let $\kappa = \mathrm{sq}_+(\sigma_\Theta)$, and we assume that there are $\kappa + 1$ different points $z_i \in \mathcal{D}(\Theta)$, $i = 1, 2, \ldots, \kappa + 1$, for which there is no positive constant ε, such that $\Theta(z_i)^* \Theta(z_i) \geq \varepsilon I$. Then, for each n we can find elements $x_i^n \in \mathfrak{F}$ such that $\|x_i^n\|_{\mathfrak{F}} = 1$ and $\|\Theta(z_i) x_i^n\|_{\mathfrak{F}} < 1/n$. Define the Gram matrix $Q_n = ((Q_n)_{ij})$ of size $(\kappa + 1) \times (\kappa + 1)$ by $(Q_n)_{ij} = [x_i^n, x_j^n]_{\mathfrak{F}}$. Then Q_n is nonnegative and $(Q_n)_{ii} = 1$. If necessary by going over to subsequences, we see that the limit $Q \equiv \lim_{n \to \infty} Q_n$ exists and is a nonnegative $(\kappa + 1) \times (\kappa + 1)$ matrix with diagonal entries $Q_{ii} = 1$. Now consider the $(\kappa + 1) \times (\kappa + 1)$ matrix $P_n = ((P_n)_{ij})$ with

$$(P_n)_{ij} = [\sigma_\Theta(z_i, z_j) x_i^n, x_j^n]_{\mathfrak{F}} = ((Q_n)_{ij}/(1 - z_i \bar{z}_j)) - ([\Theta(z_i) x_i^n, \Theta(z_j) x_j^n]_{\mathfrak{G}}/(1 - z_i \bar{z}_j)).$$

Then the limit $\lim_{n \to \infty} P_n$ exists and equals the $(\kappa + 1) \times (\kappa + 1)$ matrix $P = ((Q_{ij}/(1 - z_i \bar{z}_j)))$. On account of Lemma 2.1 the matrix P has precisely $\kappa + 1$ positive eigenvalues. However, this is impossible, since, by assumption, for each n the approximating $(\kappa + 1) \times (\kappa + 1)$ matrix P_n has at most κ positive eigenvalues. This contradiction proves (i). Part (ii) follows from (i) by replacing Θ by $\tilde{\Theta}$. This completes the proof.

LEMMA 2.3. *Let $\mathfrak{F}$ and $\mathfrak{G}$ be Kreĭn spaces and $\Theta \in S(\mathfrak{F}, \mathfrak{G})$. Assume that $\Theta(z)$ is invertible for all but finitely many $z \in \mathcal{D}(\Theta)$ and put $\Psi(z) = \Theta(z)^{-1}$. Then $\Psi \in S(\mathfrak{G}, \mathfrak{F})$ and*

(i) $\quad \sigma_{\Theta}(z,w) = -\Theta(w)^{*}\sigma_{\Psi}(z,w)\Theta(z), \quad \sigma_{\tilde{\Theta}}(z,w) = -\Theta(\bar{w})\sigma_{\tilde{\Psi}}(z,w)\Theta(\bar{z})^{*},$

(ii) $\quad S_{\Theta}(z,w) = -\operatorname{diag}(\Theta(w), \Theta(\bar{w})^{*})^{*}S_{\Psi}(z,w)\operatorname{diag}(\Theta(z), \Theta(\bar{z})^{*}),$

(iii) $\quad \operatorname{sq}_{\pm}(\sigma_{\Theta}) = \operatorname{sq}_{\mp}(\sigma_{\Psi}), \ \operatorname{sq}_{\pm}(\sigma_{\tilde{\Theta}}) = \operatorname{sq}_{\mp}(\sigma_{\tilde{\Psi}}), \ \operatorname{sq}_{\pm}(S_{\Theta}) = \operatorname{sq}_{\mp}(S_{\Psi}).$

The proof of this lemma is straightforward and therefore omitted.

COROLLARY 2.4. *Let $\mathfrak{F}$ and $\mathfrak{G}$ be Hilbert spaces and let $\Theta \in S(\mathfrak{F}, \mathfrak{G})$. Then*

$\quad (i) \ \operatorname{sq}_{+}(S_{\Theta}) < \infty$ *if and only if* $(ii) \ \operatorname{sq}_{+}(\sigma_{\Theta}) < \infty$ *and* $\operatorname{sq}_{+}(\sigma_{\tilde{\Theta}}) < \infty.$

If (i) and (ii) are valid, then $\operatorname{sq}_{+}(\sigma_{\Theta}) = \operatorname{sq}_{+}(\sigma_{\tilde{\Theta}}) = \operatorname{sq}_{+}(S_{\Theta})$ and for each $z \in \mathcal{D}(\Theta) \cap \mathcal{D}(\Theta)^{*}$, with the exception of at most $2\operatorname{sq}_{+}(S_{\Theta})$ points, $\Theta(z)$ is invertible.

Proof. Clearly, (i) implies (ii). Now assume (ii). Combining parts (i) and (ii) of Theorem 2.2, we obtain that $\Theta(z)$ is invertible for all, but at most $\operatorname{sq}_{+}(\sigma_{\Theta}) + \operatorname{sq}_{+}(\sigma_{\tilde{\Theta}})$, points z in $\mathcal{D}(\Theta)$; see, e.g., Lemma 2.5 in [An]. Hence we may apply Lemma 2.3. Using this lemma and the first statement in Corollary 1.6 with $\Psi(z) = \Theta(z)^{-1}$ instead of $\Theta(z)$, we obtain the equalities $\operatorname{sq}_{+}(\sigma_{\Theta}) = \operatorname{sq}_{+}(\sigma_{\tilde{\Theta}}) = \operatorname{sq}_{+}(S_{\Theta})$. They imply (i) and show that in the last statement of the corollary the number of exceptional points is at most $\operatorname{sq}_{+}(\sigma_{\Theta}) + \operatorname{sq}_{+}(\sigma_{\tilde{\Theta}}) = 2\operatorname{sq}_{+}(\sigma_{\Theta})$.

Using the same kind of arguments as in the proof of Corollary 2.4 we can prove the following result. Recall that an operator $T \in L(\mathfrak{F}, \mathfrak{G})$ is called expansive if $T^{*}T \geq I_{\mathfrak{F}}$; it is called biexpansive if both T and T^{*} are expansive.

COROLLARY 2.5. *Let $\mathfrak{F}$ and $\mathfrak{G}$ be Hilbert spaces and let $\Theta : D \to L(\mathfrak{F}, \mathfrak{G})$ be a holomorphic mapping. Then the following assertions are equivalent.*

(i) $\quad \Theta(z)$ *is a biexpansion for all $z \in D$.*

(ii) $\quad$ *Both kernels $\sigma_{\Theta}(z,w)$ and $\sigma_{\tilde{\Theta}}(z,w)$, $z, w \in D$, are nonpositive.*

The remaining part of this section is devoted to a generalization of Lemma 2.1 to the Schur product of two nonnegative matrices. It is not needed in the sequel. Recall that the *Schur product* of two $n \times n$ matrices $P = (P_{ij})$ and $Q = (Q_{ij})$ is the $n \times n$ matrix $P_*Q = ((P_*Q)_{ij})$ with $(P_*Q)_{ij} = P_{ij}Q_{ij}$, $i, j = 1, 2, \ldots, n$. That is, P_*Q is defined as the entry-wise product of P and Q. For example:

(1) The matrix $(Q_{ij}/(1 - z_i \bar{z}_j))$ in Lemma 2.1 is the Schur product of the $n \times n$ matrix $(1/(1 - z_i \bar{z}_j))$ and Q.

(2) If $a = (a_i)$, $b = (b_i) \in \mathbb{C}^n$ (in the sequel considered as the space of $n \times 1$ vectors) $P = aa^*$ and $Q = bb^*$, then $P_*Q = cc^*$, where $c = (c_i) \in \mathbb{C}^n$ with $c_i = a_i b_i$, $i = 1, 2, \ldots n$.

(3) If $a = \begin{pmatrix} 1 \\ 0 \\ 1 \end{pmatrix}$, $b = \begin{pmatrix} 0 \\ 1 \\ 1 \end{pmatrix}$, $c = \begin{pmatrix} 0 \\ 1 \\ 0 \end{pmatrix}$, $P = aa^* + bb^*$ and $Q = aa^* + cc^*$, then P and Q are nonnegative 3×3

matrices with $\operatorname{rank} P = \operatorname{rank} Q = 2$ and $P_*Q = \begin{pmatrix} 1 & 0 & 1 \\ 0 & 1 & 0 \\ 1 & 0 & 2 \end{pmatrix}$ is a positive matrix.

The following result concerns the decomposition of nonnegative matrices. We give a proof without recourse to the spectral decomposition theorem for Hermitian matrices. The method makes use of the nonnegativity of the matrix and is very much like the methods of Lagrange and Jacobi which apply to general Hermitian forms; see, e.g., Gantmacher [G], p. 339.

PROPOSITION 2.6. *Let* P *be a nonnegative* $n{\times}n$ *matrix with* $\operatorname{rank} P = r$. *Then there exist* r *linearly independent vectors* $p_i \in \mathbb{C}^n$ *such that* $P = \sum_{i=1}^{r} p_i p_i^*$.

Proof. Consider a block decomposition of P of the form $P = \begin{pmatrix} A & B \\ B^* & D \end{pmatrix}$, where A is a nonnegative matrix of size $k{\times}k$, say. (We actually only need the case where $k = 1$.) If $x \in \mathfrak{N}(A)$, the null space of A, and $y \in \mathbb{C}^{n-k}$ is arbitrary, then the nonnegativity of P and the Cauchy–Schwarz inequality imply that

$$|y^*B^*x|^2 = |(0\ y^*)P\begin{pmatrix} x \\ 0 \end{pmatrix}|^2 \le (x^*\ 0)P\begin{pmatrix} x \\ 0 \end{pmatrix} \cdot (0\ y^*)P\begin{pmatrix} 0 \\ y \end{pmatrix} = (x^*Ax)(y^*Dy) = 0.$$

It follows that $B^*x = 0$ and this shows that $\mathfrak{N}(A) \subset \mathfrak{N}(B^*)$. Taking orthogonal complements we see that $\mathfrak{R}(B) \subset \mathfrak{R}(A)$, where $\mathfrak{R}(A)$ denotes the range of A. Hence there exists an $k{\times}(n-k)$ matrix G such that $B = AG$. It is straightforward to verify that P can be written as

$$P = \begin{pmatrix} A & B \\ B^* & D \end{pmatrix} = \begin{pmatrix} I & 0 \\ G^* & I \end{pmatrix}\begin{pmatrix} A & 0 \\ 0 & D-G^*AG \end{pmatrix}\begin{pmatrix} I & G \\ 0 & I \end{pmatrix} = \begin{pmatrix} A^{\frac{1}{2}} \\ G^*A^{\frac{1}{2}} \end{pmatrix}\begin{pmatrix} A^{\frac{1}{2}} \\ G^*A^{\frac{1}{2}} \end{pmatrix}^* + \begin{pmatrix} 0 & 0 \\ 0 & D-G^*AG \end{pmatrix}.$$

(For more details, see for instance [Dy], Lemma A.1.) To prove the proposition we start with the block decomposition for $P = (P_{ij})$ in which $A = P_{11}$. Then we obtain

$$P = p_1 p_1^* + \begin{pmatrix} 0 & 0 \\ 0 & D-G^*AG \end{pmatrix}, \quad p_1 = \begin{pmatrix} A^{\frac{1}{2}} \\ G^*A^{\frac{1}{2}} \end{pmatrix} \in \mathbb{C}^n.$$

Note that p_1 is the zero vector if $P_{11} = 0$. We repeat the same procedure to the nonnegative $(n-1){\times}(n-1)$ matrix $D-G^*AG$. After n steps we have constructed vectors $p_i \in \mathbb{C}^n$, $i = 1,2,\ldots,n$, such that $P = \sum_{i=1}^{n} p_i p_i^*$. Since the first $i-1$ entries of each $p_i \in \mathbb{C}^n$ are zero, $1 \le i \le n$, it is clear from the construction that the nontrivial vectors among $p_1, p_2, \ldots, p_n$ are linearly independent. By deleting the trivial vectors and reordering we obtain the decomposition $P = \sum_{i=1}^{r} p_i p_i^*$ for some $r \le n$, in which $p_1, p_2, \ldots, p_r$ are linearly independent. As $P \ge 0$, we have that $x \in \mathfrak{N}(P)$ if and only if $x^*Px = 0$. Since this last relation holds if and only if $x^*p_i = 0$, $i = 1,2,\ldots,r$, it follows that $\dim \mathfrak{N}(P) = n - r$, so necessarily $r = \operatorname{rank} P$. This completes the proof.

Proposition 2.6 implies Schur's lemma; see [D], p.9. The proof of this lemma can be used to prove the desired generalization. For the sake of completeness, we give the proofs of both results.

THEOREM 2.7. *Let* P *and* Q *be two nonnegative* $n\times n$ *matrices. Then:*

(i)   (*Schur's lemma*) *The Schur product* P∗Q *is nonnegative.*

(ii)   *If* P *is positive, then* P∗Q *is positive if and only if* $Q_{ii} \neq 0$ *for all* $i = 1,\dots,n$.

Proof We first prove (i). Put $\operatorname{rank} P = r$ and $\operatorname{rank} Q = s$. Then by Proposition 2.6

$$P = \textstyle\sum_{i=1}^{r} p_i p_i^*, \quad Q = \sum_{j=1}^{s} q_j q_j^*,$$

where $p_1,\dots,p_r$ are linearly independent vectors in $\mathbb{C}^n$ and $q_1,\dots,q_s$ are also linearly independent vectors in $\mathbb{C}^n$. We denote the k–th entry of the vector p_i by $(p_i)_k$, so that $p_i = ((p_i)_k)_{k=1}^{n}$, $i = 1,2,\dots r$, and we use a similar notation for each vector q_j. It follows that

$$P{\ast}Q = \textstyle\sum_{i=1}^{r}\sum_{j=1}^{s} (p_i p_i^*){\ast}(q_j q_j^*) = \sum_{i=1}^{r}\sum_{j=1}^{s} v_{ij} v_{ij}^*,$$

where v_{ij} is vector in $\mathbb{C}^n$ and the k–th entry of v_{ij} is given by $(v_{ij})_k = (p_i)_k (q_j)_k$. This shows that P∗Q is nonnegative.

We now prove (ii). We use the same decompositions of the matrices P and Q as in the proof of part (i). Since P is positive, we have that $r = \operatorname{rank} P = n$. Note that, on account of (i), P∗Q is nonnegative.   The "only if" statement in (ii) follows easily from the fact that all diagonal elements of a positive (nonnegative) matrix are positive (nonnegative, respectively). To prove the "if" statement, we suppose that $Q_{ii} \neq 0$ and hence $Q_{ii} > 0$ for all $i = 1,\dots,n$. Since $P{\ast}Q \geq 0$, it suffices to show that P∗Q has a trivial null space $\mathfrak{N}(P{\ast}Q)$. Let $a = (a_k)_{k=1}^{n} \in \mathfrak{N}(P{\ast}Q)$. Then $a^* P{\ast}Q a = 0$, and it follows from $P{\ast}Q = \sum_{i=1}^{n}\sum_{j=1}^{s} v_{ij} v_{ij}^*$ that

$$0 = a^* v_{ij} = \textstyle\sum_{k=1}^{n} \bar{a}_k (v_{ij})_k = \sum_{k=1}^{n} (\bar{a}_k (q_j)_k)(p_i)_k, \quad i = 1,2,\dots,n, \ j = 1,2,\dots,s.$$

The linear independence of the vectors $p_1, p_2,\dots,p_n$ implies that $\bar{a}_k (q_j)_k = 0$, $j = 1,2,\dots,s$, $k = 1,2,\dots,n$, and from $\sum_{j=1}^{s} |(q_j)_k|^2 = Q_{kk} > 0$ it follows that to each k, $1 \leq k \leq n$, there corresponds an index j_k, $1 \leq j_k \leq s$, such that $(q_{j_k})_k \neq 0$. Hence for each k we have that $a_k = 0$, i.e., a is the zero vector and $\mathfrak{N}(P{\ast}Q) = \{0\}$. This completes the proof.

Lemma 2.1 is a special case of Theorem 2.7 (ii): take $P = (1/(1 - z_i \bar{z}_j))$. Example (3) given above shows that if the Schur product of two nonnegative matrices is positive then it is not true that at least one of these matrices is positive also.

3. APPLICATION OF THE POTAPOV–GINZBURG TRANSFORM

In this section we prove Theorems 0.1 and 0.2. The basic idea behind the proofs is a reduction of the Kreĭn space situation to the Hilbert space situation. This reduction is obtained by applying the Potapov–Ginzburg transform. Under various different names this transform has been used in, for example, [AG1,2], [DR], [Dy], [IKL]. We first briefly describe this transform and introduce the convenient notation used in [DR], Section 1.

Let $\mathfrak{F}$ be a Kreĭn space and let $\mathfrak{F} = \mathfrak{F}_+ \oplus \mathfrak{F}_-$ be a fixed fundamental decomposition of $\mathfrak{F}$. The operator $J_{\mathfrak{F}} = P_+ - P_-$, where $P_\pm$ is the orthogonal projection on $\mathfrak{F}$ onto $\mathfrak{F}_\pm$, is called the fundamental symmetry associated with the fundamental decomposition. Note that $J_{\mathfrak{F}}^2 = I_{\mathfrak{F}}$. The linear space $\mathfrak{F}$ provided with the inner product $[f,g]_{|\mathfrak{F}|} = [J_{\mathfrak{F}}f,g]_{\mathfrak{F}}$, $f,g \in \mathfrak{F}$, is a Hilbert space and will be denoted by $|\mathfrak{F}|$. The definition of $|\mathfrak{F}|$ depends on the fundamental decomposition, $|\mathfrak{F}| = \mathfrak{F}_+ \oplus |\mathfrak{F}_-|$ and $|\mathfrak{F}_-|$ is the anti-space of $\mathfrak{F}_-$. For the Kreĭn space $\mathfrak{G}$ we also fix a fundamental decomposition $\mathfrak{G} = \mathfrak{G}_+ \oplus \mathfrak{G}_-$ and we denote the orthogonal projection on $\mathfrak{G}$ onto $\mathfrak{G}_\pm$ by $Q_\pm$. Then $J_{\mathfrak{G}} = Q_+ - Q_-$ is the fundamental symmetry corresponding to the fundamental decomposition by means of which the Hilbert space $|\mathfrak{G}|$ is defined.

By definition $T \in \mathbf{L}(\mathfrak{F},\mathfrak{G})$ if and only if $T \in \mathbf{L}(|\mathfrak{F}|,|\mathfrak{G}|)$. If $T \in \mathbf{L}(\mathfrak{F},\mathfrak{G})$, then by T^* we denote the adjoint of T with respect to the Kreĭn space inner products on $\mathfrak{F}$ and $\mathfrak{G}$, and by $T^\times$ we denote the adjoint of T with respect to the Hilbert space inner products on $|\mathfrak{F}|$ and $|\mathfrak{G}|$. It is easy to see that T^*, $T^\times \in \mathbf{L}(\mathfrak{G},\mathfrak{F})$ and that

$$T^* = J_{\mathfrak{F}} T^\times J_{\mathfrak{G}}, \quad T^\times = J_{\mathfrak{F}} T^* J_{\mathfrak{G}}.$$

If $T \in \mathbf{L}(\mathfrak{F},\mathfrak{G})$, then with respect to the fundamental decompositions of $\mathfrak{F}$ and $\mathfrak{G}$ we often write T in following operator matrix form

$$T = \begin{pmatrix} T_{11} & T_{12} \\ T_{21} & T_{22} \end{pmatrix} : \begin{pmatrix} \mathfrak{F}_+ \\ \mathfrak{F}_- \end{pmatrix} \to \begin{pmatrix} \mathfrak{G}_+ \\ \mathfrak{G}_- \end{pmatrix},$$

where, for example, $T_{22} = Q_- T|_{\mathfrak{F}_-} \in \mathbf{L}(\mathfrak{F}_-,\mathfrak{G}_-)$. If T_{22} is invertible then the operator

$$P_+ + Q_- T = \begin{pmatrix} I & 0 \\ T_{21} & T_{22} \end{pmatrix} : \begin{pmatrix} \mathfrak{F}_+ \\ \mathfrak{F}_- \end{pmatrix} \to \begin{pmatrix} \mathfrak{F}_+ \\ \mathfrak{G}_- \end{pmatrix}$$

is also invertible and the *Potapov–Ginzburg transformation* S of T is the operator $S = (Q_+ T + P_-)(P_+ + Q_- T)^{-1}$. Clearly, $S \in \mathbf{L}(\mathfrak{F}_+ \oplus \mathfrak{G}_-, \mathfrak{G}_+ \oplus \mathfrak{F}_-)$ and S has the operator matrix form

$$S = \begin{pmatrix} T_{11} - T_{12} T_{22}^{-1} T_{21} & T_{12} T_{22}^{-1} \\ -T_{22}^{-1} T_{21} & T_{22}^{-1} \end{pmatrix} : \begin{pmatrix} \mathfrak{F}_+ \\ \mathfrak{G}_- \end{pmatrix} \to \begin{pmatrix} \mathfrak{G}_+ \\ \mathfrak{F}_- \end{pmatrix}.$$

It is straightforward to verify that if we apply the Potapov–Ginzburg transform to S we get the operator T back. For proofs of these facts we refer to [DR], where a complete survey of the Potapov–Ginzburg transform in connection with contractions is presented.

Below we consider a function $\Theta \in \mathbf{S}(\mathfrak{F},\mathfrak{G})$ and we use the same notation as above. For example, Θ_{21} stands for the operator function $\Theta_{21}(z) = Q_- \Theta(z)|_{\mathfrak{F}_+}$ defined for z in a deleted neighborhood of 0. The following theorem is an analog and a small extension of [DR], Theorem 1.3.4. We recall that $\mathfrak{F}$ and $\mathfrak{G}$ are Kreĭn spaces with fundamental decompositions $\mathfrak{F} = \mathfrak{F}_+ \oplus \mathfrak{F}_-$ and $\mathfrak{G} = \mathfrak{G}_+ \oplus \mathfrak{G}_-$ and corresponding fundamental symmetries $J_{\mathfrak{F}}$ and $J_{\mathfrak{G}}$.

THEOREM 3.1. *Let $\Theta \in S(\mathfrak{F}, \mathfrak{G})$ and assume that $\Theta_{22}(z) = Q_-\Theta(z)|_{\mathfrak{F}_-}$ is invertible for z in an open set* $\mathfrak{D} = \mathfrak{D}^* \subset \mathfrak{D}(\Theta)$. *Then Σ defined by*

(i) $\qquad \Sigma(z) = (Q_+\Theta(z) + P_-)(P_+ + Q_-\Theta(z))^{-1}$

is a well-defined and holomorphic function on $\mathfrak{D}$ with values in the space $\mathbf{L}(\mathfrak{F}_+ \oplus \mathfrak{G}_-, \mathfrak{G}_+ \oplus \mathfrak{F}_-)$ and satisfies the following identities

(ii) $\qquad \Sigma(z)^{\times} = (P_+\Theta(z)^* + Q_-)(Q_+ + P_-\Theta(z)^*)^{-1},$

(iii) $\qquad \Theta(z) = (Q_+\Sigma(z) + Q_-)(P_+ + P_-\Sigma(z))^{-1},$

(iv) $\qquad \Theta(z)^* = (P_+\Sigma(z)^{\times} + P_-)(Q_+ + Q_-\Sigma(z)^{\times})^{-1}.$

Moreover, we have

(a) $\qquad I - \Theta(w)^*\Theta(z) = J_{\mathfrak{F}}(P_+ + Q_-\Theta(w))^{\times}(I - \Sigma(w)^{\times}\Sigma(z))(P_+ + Q_-\Theta(z)),$

(b) $\qquad I - \Theta(\bar{w})\Theta(\bar{z})^* = J_{\mathfrak{G}}(Q_+ + P_-\Theta(\bar{w})^*)^{\times}(I - \Sigma(\bar{w})\Sigma(\bar{z})^{\times})(Q_+ + P_-\Theta(\bar{z})^*),$

(c) $\qquad \Theta(\bar{z})^* - \Theta(w)^* = J_{\mathfrak{F}}(P_+ + Q_-\Theta(w))^{\times}(\Sigma(\bar{z})^{\times} - \Sigma(w)^{\times})(Q_+ + P_-\Theta(\bar{z})^*),$

(d) $\qquad \Theta(\bar{w}) - \Theta(z) = J_{\mathfrak{G}}((Q_+ + P_-\Theta(\bar{w})^*)^{\times}(\Sigma(\bar{w}) - \Sigma(z))(P_+ + Q_-\Theta(z)).$

Proof. It is clear that the function Σ is well-defined and holomorphic on $\mathfrak{D}$. The relations (ii)–(iv) can be verified in the same manner as the corresponding results in [DR]. To prove (a) we consider its right-hand side and substitute for $\Sigma(z)$ and $\Sigma(w)$ the expression in (i). We obtain

$$(P_+ + Q_-\Theta(w))^{\times}(I - \Sigma(w)^{\times}\Sigma(z))(P_+ + Q_-\Theta(z)) =$$
$$= (P_+ + Q_-\Theta(w))^{\times}(P_+ + Q_-\Theta(z)) - (Q_+\Theta(w) + P_-)^{\times}(Q_+\Theta(z) + P_-)$$
$$= (P_+ + J_{\mathfrak{F}}\Theta(w)^*J_{\mathfrak{G}}Q_-)(P_+ + Q_-\Theta(z)) - (J_{\mathfrak{F}}\Theta(w)^*J_{\mathfrak{G}}Q_+ + P_-)(Q_+\Theta(z) + P_-)$$
$$= P_+ + J_{\mathfrak{F}}\Theta(w)^*J_{\mathfrak{G}}Q_-\Theta(z) - P_- - J_{\mathfrak{F}}\Theta(w)^*J_{\mathfrak{G}}Q_+\Theta(z) = J_{\mathfrak{F}}(I - \Theta(w)^*\Theta(z)),$$

which is equivalent to (a). Similarly, one can obtain (b) by substituting for $\Sigma(\bar{z})$ and $\Sigma(\bar{w})$ in its right-hand side the expression in (ii). To show (c) we consider its right-hand side and substitute for $\Sigma(w)$ the expression in (i) with z replaced by w and for $\Sigma(\bar{z})$ the expression in (ii) with z replaced by $\bar{z}$. We obtain

$$(P_+ + Q_-\Theta(w))^{\times}(\Sigma(\bar{z})^{\times} - \Sigma(w)^{\times})(Q_+ + P_-\Theta(\bar{z})^*) =$$
$$= (P_+ + Q_-\Theta(w))^{\times}(P_+\Theta(\bar{z})^* + Q_-) - (Q_+\Theta(w) + P_-)^{\times}(Q_+ + P_-\Theta(\bar{z})^*)$$
$$= (P_+ + J_{\mathfrak{F}}\Theta(w)^*J_{\mathfrak{G}}Q_-)(P_+\Theta(\bar{z})^* + Q_-) - (J_{\mathfrak{F}}\Theta(w)^*J_{\mathfrak{G}}Q_+ + P_-)(Q_+ + P_-\Theta(\bar{z})^*)$$
$$= P_+\Theta(\bar{z})^* + J_{\mathfrak{F}}\Theta(w)^*J_{\mathfrak{G}}Q_- - J_{\mathfrak{F}}\Theta(w)^*J_{\mathfrak{G}}Q_+ - P_-\Theta(\bar{z})^* = J_{\mathfrak{F}}(\Theta(\bar{z})^* - \Theta(w)^*),$$

which is equivalent to (c). Finally, (d) can be shown by substituting in its right-hand side for $\Sigma(z)$ the expression in (i) and for $\Sigma(\bar{w})$ the expression in (ii) with z replaced by $\bar{w}$. This completes the proof of the theorem.

Combining the identities (a)–(d) of Theorem 3.1 we get a useful relation between the kernels associated with the operator functions Θ and Σ.

COROLLARY 3.2. *Under the conditions of Theorem 3.1 the following kernel identities are valid for all* $z,w\in\mathfrak{D}$.

(i) $\sigma_\Theta(z,w)=J_{\mathfrak{F}}(P_++Q_-\Theta(w))^\times\sigma_\Sigma(z,w)(P_++Q_-\Theta(z))$.

(ii) $\sigma_{\tilde\Theta}(z,w)=J_{\mathfrak{G}}(Q_++P_-\Theta(\bar w)^*)^\times\sigma_{\tilde\Sigma}(z,w)(Q_++P_-\Theta(\bar z)^*)$.

(iii) $S_\Theta(z,w)=\mathrm{diag}(J_{\mathfrak{F}},J_{\mathfrak{G}})(\mathrm{diag}(P_++Q_-\Theta(w),Q_++P_-\Theta(\bar w)^*)^\times S_\Sigma(z,w)(\mathrm{diag}(P_++Q_-\Theta(z),Q_++P_-\Theta(\bar z)^*)$.

In order to use Theorem 3.1 and Corollary 3.2, we must make sure that Θ_{22} is invertible on some open set $\mathfrak{D}\subset\mathfrak{D}(\Theta)$. We first make the following observation.

LEMMA 3.3. *If A and B are Hermitian matrices and $A\leq B$, then the number of negative (positive) eigenvalues of A is larger (less) than or equal to the number of negative (positive) eigenvalues of B, where the eigenvalues are counted according to their multiplicities.*

Lemma 3.3 is an immediate consequence of the minimax characterization of the eigenvalues of a Hermitian matrix, which implies that if $A\leq B$, then $\lambda_j(A)\leq\lambda_j(B)$, $j=1,2,\ldots$, where the eigenvalues $\lambda_j(A)$ and $\lambda_j(B)$ of A and B are ordered such that if $i<j$, then $\lambda_i(A)\leq\lambda_j(A)$ and $\lambda_i(B)\leq\lambda_j(B)$; for details see [G], Anhang, and, in particular, p. 621, Satz 6. A simple and direct proof of the more global statements in the lemma here can be given by showing that if $E_-(A)$ $(E_-(B))$ stands for the linear span of the eigenspaces of A (B) corresponding to all negative eigenvalues of A $(B$, respectively) and P_A denotes the orthogonal projection onto $E_-(A)$, then the restriction $P_A|_{E_-(B)}$ to $E_-(B)$ is injective on $E_-(B)$ and hence $\dim E_-(B)\leq\dim E_-(A)$. This implies the statement about the number of negative eigenvalues of A and B; the other statement follows from the first one applied to the operators $-A$ and $-B$.

THEOREM 3.4. *Let $\Theta\in S(\mathfrak{F},\mathfrak{G})$ and set $\Theta_{22}(z)=Q_-\Theta(z)|_{\mathfrak{F}_-}$.*
(i)   *If $\mathrm{sq}_-(\sigma_\Theta)<\infty$, then for each $z\in\mathfrak{D}(\Theta)$ with the exception of at most $\mathrm{sq}_-(\sigma_\Theta)$ points in $\mathfrak{D}(\Theta)$, there is a positive constant $\varepsilon(z)$, such that $\Theta_{22}(z)^\times\Theta_{22}(z)\geq\varepsilon(z)I$.*
(ii)   *If $\mathrm{sq}_-(\sigma_{\tilde\Theta})<\infty$, then for each $z\in\mathfrak{D}(\Theta)$, with the exception of at most $\mathrm{sq}_-(\sigma_{\tilde\Theta})$ points in $\mathfrak{D}(\Theta)$, there is a positive constant $\delta(z)$, such that $\Theta_{22}(z)\Theta_{22}(z)^\times\geq\delta(z)I$.*

Proof. The following chain of (in–)equalities is valid. All kernels here are defined on the domain $\mathfrak{D}(\Theta)$ and in brackets we indicate on what inner product space they act.

$$\mathrm{sq}_-(\sigma_\Theta) \qquad\qquad \text{(on the Kreĭn space } \mathfrak{F})$$
$$\geq \mathrm{sq}_-(P_-\sigma_\Theta|_{\mathfrak{F}_-}) \qquad \text{(on the anti–Hilbert space } \mathfrak{F}_-)$$
$$= \mathrm{sq}_-(-P_-\sigma_\Theta|_{\mathfrak{F}_-}) \qquad \text{(on the Hilbert space } |\mathfrak{F}_-|)$$
$$= \mathrm{sq}_+(P_-\sigma_\Theta|_{\mathfrak{F}_-}) \qquad \text{(on the Hilbert space } |\mathfrak{F}_-|)$$
$$\geq \mathrm{sq}_+(\sigma_{\Theta_{22}}) \qquad\qquad \text{(on the Hilbert space } |\mathfrak{F}_-|).$$

The last inequality follows from the identity

$$P_-\sigma_\Theta(z,w)|_{\mathfrak{F}_-} = \frac{I - \Theta_{22}(w)^\times\Theta_{22}(z)}{1-z\bar{w}} + \frac{\Theta_{12}(w)^\times\Theta_{12}(z)}{1-z\bar{w}} = \sigma_{\Theta_{22}}(z,w) + \frac{\Theta_{12}(w)^\times\Theta_{12}(z)}{1-z\bar{w}}, \quad z,w\in\mathfrak{D}(\Theta),$$

the nonnegativity of the kernel

$$\frac{\Theta_{12}(w)^\times\Theta_{12}(z)}{1-z\bar{w}} \qquad \text{(on the Hilbert space } |\mathfrak{F}_-|\text{)}$$

and Lemma 3.3. Hence the kernel $\sigma_{\Theta_{22}}(z,w)$ has at most $\mathrm{sq}_-(\sigma_\Theta)$ positive squares on $|\mathfrak{F}_-|$. Part (i) now follows from Theorem 2.2(i). If we replace in the above argument Θ by $\tilde\Theta$, $\mathfrak{F}$ by $\mathfrak{G}$ and P by Q, we obtain that the kernel

$$\sigma_{\tilde\Theta_{22}}(z,w) = \frac{I - \Theta_{22}(\bar{w})\Theta_{22}(\bar{z})^\times}{1-z\bar{w}}, \qquad z,w\in\mathfrak{D}(\Theta),$$

has at most $\mathrm{sq}_-(\sigma_{\tilde\Theta})$ positive squares on the Hilbert space $|\mathfrak{G}_-|$. Part (ii) now follows from Theorem 2.2(ii). This completes the proof.

COROLLARY 3.5. *Let* $\Theta\in\mathbf{S}(\mathfrak{F},\mathfrak{G})$ *and assume that* $\mathrm{sq}_-(\sigma_\Theta)<\infty$ *and* $\mathrm{sq}_-(\sigma_{\tilde\Theta})<\infty$. *Then the function* $\Theta_{22}(z) = Q_-\Theta(z)|_{\mathfrak{F}_-}$ *is invertible for each* $z\in\mathfrak{D}(\Theta)\cap\mathfrak{D}(\tilde\Theta)$ *with the exception of at most* $\mathrm{sq}_-(\sigma_\Theta)+\mathrm{sq}_-(\sigma_{\tilde\Theta})$ *points.*

We now come to the proofs of the main theorems of the paper.

Proof of Theorem 0.1. Clearly, the inequality in (i) implies the inequalities in (ii). Assume (ii). Then by Corollary 3.5 the operator $\Theta_{22}(z)$ is invertible for each $z\in\mathfrak{D}(\Theta)\cap\mathfrak{D}(\tilde\Theta)$ except for at most $\mathrm{sq}_-(\sigma_\Theta)+\mathrm{sq}_-(\sigma_{\tilde\Theta})$ points. Hence the conditions of Theorem 3.1 are satisfied, and the operator function $\Sigma(z)$ in (i) of Theorem 3.1 is well defined. Corollary 3.2 implies the equalities

$$\mathrm{sq}_-(\sigma_\Sigma) = \mathrm{sq}_-(\sigma_\Theta), \quad \mathrm{sq}_-(\sigma_{\tilde\Sigma}) = \mathrm{sq}_-(\sigma_{\tilde\Theta}), \quad \mathrm{sq}_-(\mathsf{S}_\Sigma) = \mathrm{sq}_-(\mathsf{S}_\Theta).$$

Here the kernels on the left-hand side are considered as operators on Hilbert spaces. According to Corollary 1.6 their numbers of negative squares are equal. It now follows that $\mathrm{sq}_-(\sigma_\Theta) = \mathrm{sq}_-(\sigma_{\tilde\Theta}) = \mathrm{sq}_-(\mathsf{S}_\Theta)$. Hence, in particular, (i) is valid and the last statement in Theorem 0.1 holds true. This completes the proof.

Proof of Theorem 0.2. The implications $(i)\Rightarrow(ii)$ and $(i)\Rightarrow(iii)$ follow immediately from Theorem 0.1. If (ii) (or (iii)) is valid then, on account of Theorem 3.4(i) (or (ii), respectively), $\Theta_{22}(z)$ is invertible for all but at most κ points in $\mathfrak{D}(\Theta)$. Again we may apply the Potapov–Ginzburg transform to conclude, as in the proof of Theorem 0.1, that (i) is valid. If $\Sigma(z)$ is the transformation of $\Theta(z)$ then, by Corollary 1.6, $\Sigma(z)$ can be extended to a meromorphic operator function on $\mathbb{D}$ with at most κ poles. The lower right corner $\Sigma_{22}(z)$ of $\Sigma(z)$ is a square matrix function of size $\mathrm{ind}_\mathfrak{F} = \mathrm{ind}_\mathfrak{G}$ and, since $\det \Sigma_{22}(z) \not\equiv 0$ on $\mathbb{D}$, it is invertible for all z outside a discrete subset of $\mathbb{D}$. Applying the inverse Potapov–Ginzburg transform to Σ we find that Θ has the property mentioned in the last statement of the theorem. This completes the proof.

REFERENCES

[A]   D. Alpay, "Some Kreĭn spaces of analytic functions and an inverse scattering problem", Michigan Journal of Math., 34 (1987), 349–359.

[AD1]   D. Alpay, H. Dym, "Hilbert spaces of analytic functions, inverse scattering and operator models I", Integral Equations Operator Theory, 7 (1984), 589–641.

[AD2]   D. Alpay, H. Dym, "Hilbert spaces of analytic functions, inverse scattering and operator models II", Integral Equations Operator Theory, 8 (1985), 145–180.

[AD3]   D. Alpay, H. Dym, "On applications of reproducing kernel spaces to the Schur algorithm and rational J unitary factorization", Operator Theory: Adv. Appl., 18 (1986), 89–159.

[AG1]   D.Z. Arov, L.Z. Grossman, "Scattering matrices in the theory of dilations of isometric operators", Dokl. Akad. Nauk SSSR, 270 (1983), 17–20, (Russian) (English translation: Sov. Math. Dokl., 27 (1983), 518–522).

[AG2]   D.Z. Arov, L.Z. Grossman, "Scattering matrices in the theory of unitary extensions of isometric operators", manuscript.

[An]   T. Ando, *De Branges spaces and analytic operator functions*, Lecture Notes, Hokkaido University, Sapporo, 1990.

[Az1]   T. Ya. Azizov, "On the theory of extensions of isometric and symmetric operators in spaces with an indefinite metric", Preprint Voronesh University, 1982; deposited paper no. 3420–82 (Russian).

[Az2]   T.Ya. Azizov, "Extensions of J–isometric and J–symmetric operators", Funktsional. Anal. i Prilozhen, 18 (1984), 57–58 (Russian) (English translation: Functional Anal. Appl., 18 (1984), 46–48).

[B1]   L. de Branges, "Kreĭn spaces of analytic functions", J. Functional Analysis, 81 (1988), 219–259.

[B2]   L. de Branges, "Complementation in Kreĭn spaces", Trans. Amer. Math. Soc., 305 (1988), 277–291.

[B3]   L. de Branges, *Square summable power series*, manuscript.

[BC]   J.A. Ball, N. Cohen, "De Branges–Rovnyak operator models and systems theory: a survey", Operator Theory: Adv. Appl., 50 (1991), 93–136.

[BR]   L. de Branges, J. Rovnyak, *Square summable power series*, Holt, Rinehart and Winston, New York, 1966.

[BS]   L. de Branges, L.A. Shulman, "Perturbations of unitary transformations", J. Math. Anal. Appl., 23 (1968), 294–326.

[CDLS]   B. Ćurgus, A. Dijksma, H. Langer, H.S.V. de Snoo, "Characteristic functions of unitary colligations and of bounded operators in Kreĭn spaces", Operator Theory: Adv. Appl., 41 (1989), 125–152.

[D]   W.F. Donoghue, *Monotone matrix functions and analytic continuation*, Springer–Verlag, Berlin–Heidelberg–New York, 1974.

[DLS1]   A. Dijksma, H. Langer, H.S.V. de Snoo, "Characteristic functions of unitary operator colligations in Π_κ–spaces", Operator Theory: Adv. Appl., 19 (1986), 125–194.

[DLS2]   A. Dijksma, H. Langer, H.S.V. de Snoo, "Unitary colligations in Π_κ–spaces, characteristic functions and Štraus extensions", Pacific J. Math., 125 (1986), 347–362.

[DLS3]   A. Dijksma, H. Langer, H.S.V. de Snoo, "Unitary colligations in Kreĭn spaces and their role in the extension theory of isometries and symmetric linear relations in Hilbert spaces", Functional Analysis II, Proceedings Dubrovnik 1985, Lecture Notes in Mathematics, 1242 (1987), 1–42.

[DR]   M.A. Dritschel, J. Rovnyak, Extension theorems for contractions on Kreĭn spaces, Operator Theory: Adv. Appl., 47 (1990), 221–305.

[Dy]   H. Dym, *J contractive matrix functions, reproducing kernel Hilbert spaces and interpolation*, Regional conference series in mathematics, 71, Amer. Math. Soc., Providence, R.I., 1989.

[G]   F.R. Gantmacher, *Matrizentheorie*, 2nd ed., Nauka, Moscow, 1966 (Russian) (German translation: VEB Deutscher Verlag der Wissenschaften, Berlin, 1986).

[IKL] I.S. Iohvidov, M.G. Kreĭn, H. Langer, *Introduction to the spectral theory of operators in spaces with an indefinite metric*, Reihe: Mathematical Research 9, Akademie–Verlag, Berlin, 1982.

[KL]   M.G. Kreĭn, H. Langer, "Über die verallgemeinerten Resolventen und die charakteristische Funktion eines isometrischen Operators im Raume Π_κ", Hilbert Space Operators and Operator Algebras (Proc. Int. Conf., Tihany, 1970) Colloquia Math. Soc. János Bolyai, no. 5, North–Holland, Amsterdam (1972), 353–399.

[M]   S. Marcantognini, "Unitary colligations of operators in Kreĭn spaces", Integral Equations Operator Theory, 13 (1990), 701–727.

[Mc]   B.W. McEnnis, "Purely contractive analytic functions and characteristic functions of non–contractions", Acta. Sci. Math. (Szeged), 41 (1979), 161–172.

[SF]   B. Sz.–Nagy, C. Foias, *Harmonic analysis of operators on Hilbert space*, North–Holland Publishing Company, Amsterdam–London, 1970.

[Y]   A. Yang, *A construction of Kreĭn spaces of analytic functions*, Dissertation, Purdue University, 1990.

D. ALPAY
DEPARTMENT OF MATHEMATICS
BEN–GURION UNIVERSITY OF THE NEGEV
POSTBOX 653
84105 BEER–SHEVA
ISRAEL

A. DIJKSMA, J. VAN DER PLOEG, H.S.V. DE SNOO
DEPARTMENT OF MATHEMATICS
UNIVERSITY OF GRONINGEN
POSTBOX 800
9700 AV GRONINGEN
THE NETHERLANDS

MSC: Primary 47B50, Secondary 47A48

Operator Theory:
Advances and Applications, Vol. 59
© 1992 Birkhäuser Verlag Basel

ON REPRODUCING KERNEL SPACES, THE SCHUR ALGORITHM, AND INTERPOLATION IN A GENERAL CLASS OF DOMAINS

Daniel Alpay and Harry Dym*

This paper develops the theory of reproducing kernel Pontryagin spaces with reproducing kernels $\Lambda_\omega(\lambda) = X(\lambda)JX(\omega)^*/\rho_\omega(\lambda)$ based on a $k \times m$ matrix valued function $X(\lambda)$, a signature matrix J and a denominator of the general form $a(\lambda)a(\omega)^* - b(\lambda)b(\omega)^*$. This both unifies and generalizes earlier studies of such kernels wherein the denominator was taken to be either $1 - \lambda\omega^*$ or $-2\pi i(\lambda - \omega^*)$. A Schur-like algorithm is then interpreted in terms of a recursive orthogonal direct sum decomposition of such spaces. Finally, these spaces, in conjunction with a corresponding class of $\mathcal{K}(\Theta)$ spaces which were introduced earlier (in [AD4]), are used to solve a general one-sided interpolation problem in a fairly general class of domains.

CONTENTS

1. INTRODUCTION
2. PRELIMINARIES
3. $\mathcal{B}(X)$ SPACES
4. RECURSIVE EXTRACTIONS AND THE SCHUR ALGORITHM
5. $\mathcal{H}_\rho(S)$ SPACES
6. LINEAR FRACTIONAL TRANSFORMATIONS
7. ONE SIDED INTERPOLATION
8. REFERENCES

*H. Dym would like to thank Renee and Jay Weiss for endowing the chair which supports his research.

1. INTRODUCTION

In this paper we shall study reproducing kernel Pontryagin spaces of $m \times 1$ vector valued meromorphic functions with reproducing kernels of the form

$$\Lambda_\omega(\lambda) = \frac{X(\lambda)JX(\omega)^*}{\rho_\omega(\lambda)} , \tag{1.1}$$

where X is a $k \times m$ matrix valued function, J is an $m \times m$ signature matrix (i.e., $J = J^*$ and $JJ^* = I_m$), the denominator $\rho_\omega(\lambda)$ is of the form

$$\rho_\omega(\lambda) = a(\lambda)a(\omega)^* - b(\lambda)b(\omega)^* , \tag{1.2}$$

and it is further assumed that:

I. $a(\lambda)$ and $b(\lambda)$ are analytic in some open nonempty connected subset Ω of $\mathbb{C}$.

II.   The sets

$$\Omega_+ = \{\omega \in \Omega : \ \rho_\omega(\omega) > 0\} \qquad \text{and} \qquad \Omega_- = \{\omega \in \Omega : \ \rho_\omega(\omega) < 0\}$$

are both nonempty.

Because of the presumed connectedness of Ω, it follows further (see [AD4] for details) that:

III.   The set

$$\Omega_0 = \{\omega \in \Omega : \ \rho_\omega(\omega) = 0\}$$

contains at least one point μ such that $\rho_\mu(\lambda) \not\equiv 0$.

Any function $\rho_\omega(\lambda)$ of the form (1.2) which satisfies I and II (and hence III) will be said to belong to $\mathcal{D}_\Omega$.

Reproducing kernel Pontryagin spaces with reproducing kernels of the form (1.1) were extensively studied in [AD3] for the special choices $\rho_\omega(\lambda) = 1 - \lambda\omega^*$ and $\rho_\omega(\lambda) = -2\pi i(\lambda - \omega^*)$. Both of these belong to $\mathcal{D}_\Omega$ with $\Omega = \mathbb{C}$:

$1 - \lambda\omega^*$ is of the form (1.2) with $a(\lambda) = 1$, $b(\lambda) = \lambda$, $\Omega_+ = \mathbb{D}$ and $\Omega_0 = \mathbb{T}$.

$-2\pi i(\lambda - \omega^*)$ is of the form (1.2) with $a(\lambda) = \sqrt{\pi}(1 - i\lambda)$, $b(\lambda) = \sqrt{\pi}(1 + i\lambda)$, $\Omega_+ = \mathbb{C}_+$ and $\Omega_0 = \mathbb{R}$.

In this paper we shall extend some of the results reported in [AD3] to this new more general framework of kernels with denominators in $\mathcal{D}_\Omega$ and shall also solve a general one-sided interpolation problem in this setting.

Many kernels can be expressed in the general form (1.1); examples and references will be furnished throughout the text. In addition to these, which focus on $\Lambda_\omega(\lambda)$ as a reproducing kernel, the form (1.1) shows up as a bivariate generating function in the study of structured Hermitian matrices; see the recent survey by Lev-Ari [LA] and the references cited therein. In particular, Lev-Ari and Kailath [LAK] seem to have been the first to study "denominators" $\rho_\omega(\lambda)$ of the special form (1.2). They showed that Hermitian matrices with bivariate generating functions of the form (1.1) can be factored efficiently whenever $\rho_\omega(\lambda)$ is of the special form (1.2). The present analysis gives a geometric interpretation of the algorithm presented in [LAK] in terms of a direct sum orthogonal decomposition of the underlying reproducing Pontryagin spaces.

As we already noted in [AD1], the important kernel

$$K_\omega(\lambda) = \frac{J - \Theta(\lambda)J\Theta(\omega)^*}{\rho_\omega(\lambda)} \tag{1.3}$$

can also be expressed in the form (1.1), but with respect to the signature matrix

$$\hat{J} = \begin{bmatrix} J & 0 \\ 0 & -J \end{bmatrix},$$

by choosing

$$X = [I_m \quad \Theta].$$

For the most part, however, we shall take J equal to

$$J_{pq} = \begin{bmatrix} I_p & 0 \\ 0 & -I_q \end{bmatrix}$$

and shall accordingly write

$$X(\lambda) = [A(\lambda) \quad B(\lambda)]$$

with components $A \in \mathbb{C}^{k\times p}$ and $B \in \mathbb{C}^{k\times q}$, both of which are presumed to be meromorphic in Ω_+. Then (1.1) can be reexpressed as

$$\Lambda_\omega(\lambda) = \frac{A(\lambda)A(\omega)^*}{\rho_\omega(\lambda)} - \frac{B(\lambda)B(\omega)^*}{\rho_\omega(\lambda)},$$

which serves to exhibit $\Lambda_\omega(\lambda)$ as the difference of two positive kernels on Ω_+ (since $1/\rho_\omega(\lambda)$ is a positive kernel on Ω_+, as is shown in the next section). Therefore, by a result of L. Schwartz [Sch], there exists at least one (and possibly many) reproducing kernel Krein space with $\Lambda_\omega(\lambda)$ as its reproducing kernel. However, if the kernel is restricted to have only finitely many negative squares (the definition of this and a number of related notions will be provided in Section 2), then there exists a unique reproducing kernel Pontryagin space with $\Lambda_\omega(\lambda)$ as its reproducing kernel. This too was established first by Schwartz, and independently, but later, by Sorojonen [So] (and still independently, but even later, by the present authors in [AD3]).

If $\Lambda_\omega(\lambda)$ has zero negative squares, i.e., if $\Lambda_\omega(\lambda)$ is a positive kernel, then the (unique) associated reproducing kernel Pontryagin space is a Hilbert space, and the existence and uniqueness of a reproducing kernel Hilbert space with $\Lambda_\omega(\lambda)$ as its reproducing kernel also follows from the earlier work of Aronszajn [Ar].

Throughout this paper we shall let $\mathcal{B}(X)$ [resp. $\mathcal{K}(\Theta)$] denote the unique reproducing kernel Pontryagin (or Hilbert) space associated with a kernel of the form (1.1) [resp. (1.3)]. The reproducing kernel Hilbert spaces $\mathcal{B}(X)$ and $\mathcal{K}(\Theta)$, but with $\rho_\omega(\lambda)$ restricted to be equal to either $1 - \lambda\omega^*$ or $-2\pi i(\lambda - \omega^*)$, originate in the work of de Branges, partially in collaboration with Rovnyak; see [dB1], [dBR], [dB3], the references cited therein, and also Ball [Ba1]. Such reproducing kernel Hilbert spaces were applied to inverse scattering and operator models in [AD1] and [AD2], to interpolation in [D1] and

[D2], to the study of certain families of matrix orthogonal polynomials in [D3] and [D4], and to the Schur algorithm and factorization in [AD3]; the latter also extends a number of basic structural theorems from the setting of Hilbert spaces to Pontryagin spaces. In [AD4] and [AD5], the theory of $\mathcal{K}(\Theta)$ spaces was extended beyond the two special choices of ρ mentioned above, to the case of general $\rho \in \mathcal{D}_\Omega$. The parts of that extension which come into play in the present analysis (as well as some other prerequisites) are reviewed in Section 2. In this paper we shall carry out an analogous extension for the spaces $\mathcal{B}(X)$. This begins in Section 3. Recursive reductions and a Schur type algorithm are presented in Section 4. Section 5 treats the special case in which $\Lambda_\omega(\lambda)$ is positive and of the special form $\Lambda_\omega(\lambda) = \{I_p - S(\lambda)S(\omega)^*\}/\rho_\omega(\lambda)$. Section 6 deals with linear fractional transformations, and finally, in Section 7, we apply the theory developed to that point to solve a general one-sided interpolation problem in Ω_+.

The basic strategy for solving the interpolation problem in Ω_+ is much the same as for the classical choices of the disc or the halfplane except that now we seek interpolants S for which the operator M_S of multiplication by S on an appropriately defined analogue of the vector Hardy space of class 2 is contractive: $\|M_S\| \leq 1$. Although this implies that S is contractive, the converse is not generally true; see the examples in Section 5. Moreover, Ω_+ need not be connected. Interpolation problems in nonconnected domains have also been considered by Abrahamse [Ab], but both the methods and results seem to be quite different.

Finally, we wish to mention that there appear to be a number of points of contact between the interpolation problem studied in this paper and the interpolation problem described by Nudelman [N] in his lecture at the Sapporo Workshop. However, we cannot make precise comparisons because we have not yet seen a written version.

The notation is fairly standard: The symbols $\mathbb{R}$ and $\mathbb{C}$ denote the real and complex numbers, respectively; $\mathbb{D} = \{\lambda \in \mathbb{C} : |\lambda| < 1\}$, $\mathbb{T} = \{\lambda \in \mathbb{C} : |\lambda| = 1\}$, $\mathbb{E} = \{\lambda \in \mathbb{C} : |\lambda| > 1\}$ and $\mathbb{C}_+$ [resp. $\mathbb{C}_-$] stands for the open upper [resp. lower] half plane. $\mathbb{C}^{p \times q}$ denotes the set of $p \times q$ matrices with complex entries and $\mathbb{C}^p$ is short for $\mathbb{C}^{p \times 1}$. A^* will denote the adjoint of a matrix with respect to the standard inner product, and the usual complex conjugate if A is just a number.

2. PRELIMINARIES

To begin with, it is perhaps well to recall that a vector space V over the complex numbers which is endowed with an indefinite inner product $[\ ,\ ]$ is said to be a Krein space if there exist a pair of subspaces V_+ and V_- of V such that

(1) V_+ endowed with $[\ ,\ ]$ and V_- endowed with $-[\ ,\ ]$ are Hilbert spaces.

(2) $V_+ \cap V_- = \{0\}$.

(3) V_+ and V_- are orthogonal with respect to $[\ ,\ ]$ and their sum is equal to V.

V is said to be a Pontryagin space if at least one of the spaces V_+, V_- is finite dimensional. In this paper, we shall always presume that V_- is finite dimensional. In this instance, the dimension of V_- is referred to as the index of V.

A Pontryagin space $\mathcal{P}$ of $m \times 1$ vector valued meromorphic functions defined on an open nonempty subset Δ of $\mathbb{C}$, with common domain of analyticity Δ', is said to be a reproducing kernel Pontryagin space if there exists an $m \times m$ matrix valued function $L_\omega(\lambda)$ on $\Delta' \times \Delta'$ such that for every choice of $\omega \in \Delta'$, $v \in \mathbb{C}^m$ and $f \in \mathcal{P}$:

(1) $L_\omega v \in \mathcal{P}$, and

(2) $[f, L_\omega v]_{\mathcal{P}} = v^* f(\omega)$.

The matrix function $L_\omega(\lambda)$ is referred to as the reproducing kernel; there is only one such. Moreover,

$$L_\alpha(\beta) = L_\beta(\alpha)^* ,$$

for every choice of α and β in Δ'.

The kernel $L_\omega(\lambda)$ (or for that matter any Hermitian kernel) is said to have ν negative squares in Δ' if (1) for any choice of points $\omega_1, \ldots, \omega_n$ in Δ' and vectors $v_1, \ldots, v_n$ in $\mathbb{C}^m$ the $n \times n$ matrix with ij entry equal to $v_i^* L_{\omega_j}(\omega_i) v_j$ has at most ν negative eigenvalues and, (2) there is a choice of points $\omega_1, \ldots, \omega_k$ and vectors $v_1, \ldots, v_k$ for which the indicated matrix has exactly ν negative eigenvalues; it should perhaps be emphasized here that n is also allowed to vary. In a reproducing kernel Pontryagin space, the number of negative squares of the reproducing kernel is equal to the index of the space.

For additional information on Krein spaces and Pontryagin spaces, the monographs of Bognar [Bo], Iohvidov, Krein and Langer [IKL], and Azizov and Iohvidov [AI] are suggested.

Next, it is convenient to summarize some facts from [AD4] about the class $\mathcal{D}_\Omega$ which was introduced in Section 1 and on some associated reproducing kernel spaces. First, it is important to note that the definition of the class $\mathcal{D}_\Omega$ depends only upon ρ and not upon the particular choice of functions a and b in the decomposition (1.2). In particular, if $\rho_\omega(\lambda)$ can also be expressed in terms of a second pair of functions $c(\lambda)$ and $d(\lambda)$: if

$$\rho_\omega(\lambda) = c(\lambda)c(\omega)^* - d(\lambda)d(\omega)^* ,$$

then there exists a J_{11} unitary matrix M such that

$$[c(\lambda) \ \ d(\lambda)] = [a(\lambda) \ \ b(\lambda)]M$$

for every $\lambda \in \Omega$; see Lemma 5.1 of [AD4].

We have already remarked that the functions $\rho_\omega(\lambda) = 1 - \lambda\omega^*$ and $\rho_\omega(\lambda) = -2\pi i(\lambda - \omega^*)$ belong to $\mathcal{D}_\Omega$. So does the less familiar choice

$$\rho_\omega(\lambda) = -2\pi i(\lambda - \omega^*)(1 - \lambda\omega^*) . \tag{2.1}$$

The latter is of the form (1.2) with

$$a(\lambda) = \sqrt{\pi}\{\lambda + i(\lambda^2 + 1)\} \ \text{ and } \ b(\lambda) = \sqrt{\pi}\{\lambda - i(\lambda^2 + 1)\} . \tag{2.2}$$

Moreover, in this case,

$$\Omega_+ = (\mathbb{D} \cap \mathbb{C}_+) \cup (\mathbb{E} \cap \mathbb{C}_-)$$

is not connected.

Now if $\rho \in \mathcal{D}_\Omega$ with decomposition (1.2), then $a(\lambda) \neq 0$ for $\lambda \in \Omega_+$ and $s(\lambda) = b(\lambda)/a(\lambda)$ is strictly contractive in Ω_+. Therefore, the kernel

$$k_\omega(\lambda) = \frac{1}{\rho_\omega(\lambda)} = \frac{1}{a(\lambda)} \sum_{t=0}^{\infty} s(\lambda)^t s(\omega)^{*t} \frac{1}{a(\omega)^*}$$

is positive on Ω_+: for every positive integer n, and every choice of points $\omega_1, \ldots, \omega_n$ in Ω_+ and constants $c_1, \ldots, c_n$ in $\mathbb{C}$,

$$\sum_{i,j=1}^{n} c_j^* k_{\omega_i}(\omega_j) c_i \geq 0 .$$

Thus, by one of the theorems alluded to in the introduction, there exists a unique reproducing kernel Hilbert space, with reproducing kernel $k_\omega(\lambda) = 1/\rho_\omega(\lambda)$. We shall refer to this space as H_ρ and shall designate its inner product by $\langle \ , \ \rangle_{H_\rho}$. Recall that this means that, for every choice of $\omega \in \Omega_+$ and $f \in H_\rho$, (1) $1/\rho_\omega$ belongs to H_ρ, and (2) $\langle f, 1/\rho_\omega \rangle_{H_\rho} = f(\omega)$.

The space H_ρ plays the same role in the present setting as the classical Hardy spaces $H_2(\mathbb{D})$ for the disc and $H_2(\mathbb{C}_+)$ for the open upper half plane. Indeed, it is identical to the former [resp. the latter] when $\rho_\omega(\lambda) = 1 - \lambda\omega^*$ [resp. $\rho_\omega(\lambda) = -2\pi i(\lambda - \omega^*)$].

More generally, H_ρ^m will denote the space of $m \times 1$ vector valued functions

$$f = \begin{bmatrix} f_1 \\ \vdots \\ f_m \end{bmatrix} , \qquad g = \begin{bmatrix} g_1 \\ \vdots \\ g_m \end{bmatrix}$$

with coordinates f_i and g_i, $i = 1, \ldots, m$, in H_ρ and inner product

$$\langle f, g \rangle_{H_\rho^m} = \sum_{i=1}^{m} \langle f_i, g_i \rangle_{H_\rho} .$$

From now on, we shall indicate the inner product on the left by $\langle f, g \rangle_{H_\rho}$ (i.e., we drop the superscript m), in order to keep the notation simple. For any $m \times m$ signature matrix J, the symbol $H_{\rho,J}$ will denote the space H_ρ^m endowed with the indefinite inner product

$$[f, g]_{H_{\rho,J}} = \langle Jf, g \rangle_{H_\rho} .$$

The space H_ρ^m is a reproducing kernel Hilbert space with reproducing kernel $K_\omega(\lambda) = I_m/\rho_\omega(\lambda)$, whereas $H_{\rho,J}$ is a reproducing kernel Krein space with reproducing kernel $K_\omega(\lambda) = J/\rho_\omega(\lambda)$.

Because $1/\rho_\omega(\lambda)$ is jointly analytic for λ and ω^* in Ω_+, it follows (as is spelled out in more detail in [AD4]) that

$$\varphi_{\omega,k} = \frac{1}{k!}\, \frac{\partial^k}{\partial \omega^{*k}}\, \frac{1}{\rho_\omega}$$

belongs to H_ρ for every integer $k \geq 0$ and that

$$\langle f, \varphi_{\omega,k}\rangle_{H_\rho} = \frac{1}{k!} f^{(k)}(\omega)$$

for every $f \in H_\rho$ and every $\omega \in \Omega_+$.

We shall refer to the sequence

$$\varphi_{\omega,0}, \ldots, \varphi_{\omega,n-1}$$

as an elementary chain of length n based on the point $\omega \in \Omega_+$.

More generally, by a chain of length n in H_ρ^m we shall mean the columns $f_1, \ldots, f_n$ of the $m \times n$ matrix valued function

$$F(\lambda) = V\Phi_{\omega,n}(\lambda) \, ,$$

wherein V is a constant $m \times n$ matrix with nonzero first column and

$$\Phi_{\omega,n}(\lambda) \;=\; \begin{bmatrix} \varphi_{\omega,0}(\lambda) & \varphi_{\omega,1}(\lambda) & \cdots & \varphi_{\omega,n-1}(\lambda) \\ 0 & & & \\ \vdots & & & \\ 0 & & 0 & \varphi_{\omega,0}(\lambda) \end{bmatrix} \tag{2.4}$$

is the $n \times n$ upper triangular Toeplitz based on $\varphi_{\omega,0}(\lambda), \ldots, \varphi_{\omega,n-1}(\lambda)$ as indicated just above. It is important to note that

$$\Phi_{\omega,n}(\lambda) = \{a(\lambda)A_\omega - b(\lambda)B_\omega\}^{-1} \, , \tag{2.5}$$

where A_ω and B_ω are the $n \times n$ upper triangular Toeplitz operators given by the formulas

$$A_\omega = \begin{bmatrix} \alpha_0 & 0 & \cdots & 0 \\ \vdots & & & \vdots \\ & & & 0 \\ \alpha_{n-1} & \cdots & & \alpha_0 \end{bmatrix}^* \quad \text{and} \quad B_\omega = \begin{bmatrix} \beta_0 & 0 & \cdots & 0 \\ \vdots & & & \vdots \\ & & & 0 \\ \beta_{n-1} & \cdots & & \beta_0 \end{bmatrix}^* \, , \tag{2.6}$$

where

$$\alpha_j = \frac{a^{(j)}(\omega)}{j!} \quad \text{and} \quad \beta_j = \frac{b^{(j)}(\omega)}{j!} \, .$$

These chains are the proper analogue in the present setting of the chains of rational functions considered in [AD3] and [D2]. They reduce to the latter in the classical cases, i.e., when $\rho_\omega(\lambda) = 1 - \lambda\omega^*$ or $\rho_\omega(\lambda) = -2\pi i(\lambda - \omega^*)$.

In the classical cases, every finite dimensional space of vector valued meromorphic functions which is invariant under the resolvent operators

$$(R_\alpha f)(\lambda) = \frac{f(\lambda) - f(\alpha)}{\lambda - \alpha} \tag{2.7}$$

(for every α in the common domain of analyticity) is made up of such chains. An analogous fact holds for general $\rho \in \mathcal{D}_\Omega$, but now the invariance is with respect to the pair of operators

$$\{r(a, b; \alpha)f\}(\lambda) = \frac{a(\lambda)f(\lambda) - a(\alpha)f(\alpha)}{a(\alpha)b(\lambda) - b(\alpha)a(\lambda)} \tag{2.8}$$

and

$$\{r(b, a; \alpha)f\}(\lambda) = \frac{b(\lambda)f(\lambda) - b(\alpha)f(\alpha)}{b(\alpha)a(\lambda) - a(\alpha)b(\lambda)} \ , \tag{2.9}$$

see [AD5] for details.

Just as in the classical cases, a nondegenerate finite dimensional subspace of $H_{\rho,J}$ with a basis made up of chains is a reproducing kernel Pontryagin space with a reproducing kernel of the form (1.3).

More precisely, we have:

THEOREM 2.1. *Let $\rho \in \mathcal{D}_\Omega$ and let $A \in \mathbb{C}^{n \times n}$, $B \in \mathbb{C}^{n \times n}$ and $V \in \mathbb{C}^{m \times n}$ be a given set of constant matrices such that*

(1) $\det\{a(\mu)A - b(\mu)B\} \neq 0$ for some point $\mu \in \Omega_0$, and

(2)   the columns of
$$F(\lambda) = V\{a(\lambda)A - b(\lambda)B\}^{-1} \tag{2.10}$$

are linearly independent (as analytic vector valued functions of λ) in Ω'_+, the domain of analyticity of F in Ω_+.

Then, for any invertible $n \times n$ Hermitian matrix P, the space

$$\mathcal{F} = \mathrm{span}\{columns \ of \ \ F(\lambda)\} \ ,$$

endowed with the indefinite inner product

$$[F\xi \ , \ F\eta]_\mathcal{F} = \eta^* P\xi \ , \tag{2.11}$$

is an n dimensional reproducing kernel Pontryagin space with reproducing kernel

$$K_\omega(\lambda) = F(\lambda)P^{-1}F(\omega)^* \ . \tag{2.12}$$

The reproducing kernel can be expressed in the form

$$K_\omega(\lambda) = \frac{J - \Theta(\lambda)J\Theta(\omega)^*}{\rho_\omega(\lambda)}$$

for some choice of $m \times m$ signature matrix J and $m \times m$ matrix valued function $\Theta(\lambda)$ which is analytic in Ω'_+ if and only if P is a solution of the equation

$$A^* P A - B^* P B = V^* J V \ . \tag{2.13}$$

Moreover, in this instance, Θ is uniquely specified by the formula

$$\Theta(\lambda) = I_m - \rho_\mu(\lambda) F(\lambda) P^{-1} F(\mu)^* J \ , \tag{2.14}$$

with μ as in (1), up to a J unitary constant factor on the right.

PROOF. This is Theorem 4.1 of [AD4].

An infinite dimensional version of Theorem 2.1 is established in [AD5]. Therein, the matrix identity (2.13) is replaced by an operator identity in terms of the operators $r(a, b, \alpha)$ and $r(b, a, \alpha)$.

In the sequel we shall need two other versions of Theorem 2.1: Theorems 5.2 and 5.3 of [AD4], respectively. They focus on the special choice of $A = A_\omega$ and $B = B_\omega$.

THEOREM 2.2. *Let μ and F be as in Theorem 2.1, but with $A = A_\omega$ and $B = B_\omega$ for some point $\omega \in \Omega_+$. Then the columns $f_1, \ldots, f_n$ of F belong to H_ρ^m and the $n \times n$ matrix P with ij entry*

$$p_{ij} = \langle J f_j, f_i \rangle_{H_\rho}$$

is the one and only solution of the matrix equation

$$A_\omega^* P A_\omega - B_\omega^* P B_\omega = V^* J V \ . \tag{2.15}$$

THEOREM 2.3. *Let $\rho \in \mathcal{D}_\Omega$, $\mu \in \Omega_0$ and $\omega \in \Omega_+$ be such that $\rho_\mu(\omega) \neq 0$ and suppose that the $n \times n$ Hermitian matrix P is an invertible solution of (2.15) for some $m \times m$ signature matrix J and some $m \times n$ matrix V of rank m with nonzero first column. Then the columns $f_1, \ldots, f_n$ of*

$$F(\lambda) = V \{ a(\lambda) A_\omega - b(\lambda) B_\omega \}^{-1}$$

are linearly independent (as vector valued functions on Ω'_+) and the space $\mathcal{F}$ based on the span of $f_1, \ldots, f_n$ equipped with the indefinite inner product

$$[F\xi, \ F\eta]_{\mathcal{F}} = \eta^* P \xi$$

(for every choice of ξ and η in $\mathbb{C}^n$) is a $\mathcal{K}(\Theta)$ space. Moreover, Θ is analytic in Ω_+ and is uniquely specified by formula (2.14), up to a constant J unitary factor on the right.

From now on we shall say that the $m \times m$ matrix valued function Θ belongs to the class $\mathcal{P}_J^\nu(\Omega_+)$ if it is meromorphic in Ω_+ and the kernel (1.3) has ν negative squares in Ω'_+.

3. $\mathcal{B}(X)$ SPACES

Throughout this section we shall continue to assume that $\rho \in \mathcal{D}_\Omega$ and that J is an $m \times m$ signature matrix.

A $k \times m$ matrix valued function X will be termed $(\Omega_+, J, \rho)_\nu$ admissible if it is meromorphic in Ω_+ and the kernel

$$\Lambda_\omega(\lambda) = \frac{X(\lambda)JX(\omega)^*}{\rho_\omega(\lambda)} \tag{3.1}$$

has ν negative squares for λ and ω in Ω'_+, the domain of analyticity of X in Ω_+. Every such $(\Omega_+, J, \rho)_\nu$ admissible X generates a unique reproducing kernel Pontryagin space of index ν with reproducing kernel $\Lambda_\omega(\lambda)$ given by (3.1). When $\nu = 0$, X will be termed (Ω_+, J, ρ) admissible. In this case, the corresponding reproducing kernel Pontryagin space is a Hilbert space.

We shall refer to this space as $\mathcal{B}(X)$ and shall discuss some of its important properties on general grounds, in the first subsection, which is devoted to preliminaries. A second description of $\mathcal{B}(X)$ spaces in terms of operator ranges is presented in the second and final subsection. The spaces $\mathcal{B}(X)$ will play an important role in the study of the reproducing kernel space structure underlying the Schur algorithm which is carried out in the next section.

It is interesting to note that the kernel

$$K_\omega^S(\lambda) = \begin{bmatrix} \dfrac{I_p - S(\lambda)S(\omega)^*}{1 - \lambda\omega^*} & \dfrac{S(\lambda) - S(\omega^*)}{\lambda - \omega^*} \\[2ex] \dfrac{S(\lambda^*)^* - S(\omega)^*}{\lambda - \omega^*} & \dfrac{I_q - S(\lambda^*)^*S(\omega^*)}{1 - \lambda\omega^*} \end{bmatrix},$$

based on the $p \times q$ matrix valued Schur function S can be expressed in the form (3.1) by choosing

$$X(\lambda) = \sqrt{2\pi} \begin{bmatrix} \lambda I_p & I_p & \lambda S(\lambda) & S(\lambda) \\ S(\lambda^*)^* & \lambda S(\lambda^*)^* & I_q & \lambda I_q \end{bmatrix},$$

$$J = (-i) \begin{bmatrix} 0 & I_p & 0 & 0 \\ -I_p & 0 & 0 & 0 \\ 0 & 0 & 0 & -I_q \\ 0 & 0 & I_q & 0 \end{bmatrix}$$

and $\rho_\omega(\lambda)$ as in (2.1). This kernel occurs extensively in the theory of operator models; see e.g., [Ba2], [DLS] and [dBS].

3.1. Preliminaries.

THEOREM 3.1. *If the $k \times k$ matrix valued function $\Lambda_\omega(\lambda)$ defined by (3.1) has ν negative squares in Ω'_+, the domain of analyticity of X in Ω_+, then there exists a unique reproducing kernel Pontryagin space $\mathcal{P}$ with index ν of $k \times 1$ vector valued functions which are analytic in Ω'_+. Moreover, $\Lambda_\omega(\lambda)$ is the reproducing kernel of $\mathcal{P}$ and*

$$\{\Lambda_\omega v : \ \omega \in \Omega'_+ \quad and \quad v \in \mathbb{C}^k\}$$

is dense in $\mathcal{P}$.

PROOF. See e.g., Theorem 6.4 of [AD3].   ∎

Schwartz [Sch] and independently (though later) Sorojonen [So] were the first to establish a 1:1 correspondence between reproducing kernel Pontryagin spaces of index ν and kernels with ν negative squares. We shall, as we have already noted, refer to the reproducing kernel Pontryagin space $\mathcal{P}$ with reproducing kernel given by (3.1), whose existence and uniqueness is established in Theorem 3.1, as $\mathcal{B}(X)$.

THEOREM 3.2. *If X is a $k \times m$ matrix valued function in Ω_+ which is $(\Omega_+, J, \rho)_\nu$ admissible and if f belongs to the corresponding reproducing kernel Pontryagin space $\mathcal{B}(X)$ and $\omega \in \Omega'_+$, then, for $j = 0, 1, \ldots$ and every choice of $v \in \mathbb{C}^k$,*

$$\Lambda_\omega^{(j)}(\lambda)v := \left. \frac{\partial^j}{\partial \beta^{*j}} \Lambda_\beta(\lambda)v \right|_{\beta=\omega}$$

belongs to $\mathcal{B}(X)$ and

$$[f, \Lambda_\omega^{(j)} v]_{\mathcal{B}(X)} = v^* f^{(j)}(\omega) . \tag{3.2}$$

PROOF. By definition there exists a set of points $\omega_1, \ldots, \omega_n$ in Ω'_+ and vectors $v_1, \ldots, v_n$ in $\mathbb{C}^k$ such that the $n \times n$ Hermitian matrix P with ij entry

$$p_{ij} = [\Lambda_{\omega_j} v_j, \Lambda_{\omega_i} v_i]_{\mathcal{B}(X)} = v_i^* \Lambda_{\omega_j}(\omega_i) v_j$$

has ν negative eigenvalues $\lambda_1, \ldots, \lambda_\nu$. Let $u_1, \ldots, u_\nu$ be an orthonormal set of eigenvectors corresponding to these eigenvalues and let

$$f_j = [\Lambda_{\omega_1} v_1 \cdots \Lambda_{\omega_n} v_n] u_j , \qquad j = 1, \ldots, \nu .$$

Then, for any choice of constants $c_1, \ldots, c_\nu$ not all of which are zero,

$$\left[\sum_{j=1}^\nu c_j f_j, \sum_{i=1}^\nu c_i f_i \right]_{\mathcal{B}(X)} = \sum_{i,j=1}^\nu c_i^* u_i^* P u_j c_j$$

$$= \sum_{i=1}^\nu \lambda_i |c_i|^2 \; < \; 0 .$$

Thus the corresponding Gram matrix Q with ij entry

$$q_{ij} = [f_j, f_i]_{\mathcal{B}(X)} , \qquad i, j = 1, \ldots, \nu ,$$

is negative definite and the span $\mathcal{N}$ of the columns $f_1, \ldots, f_\nu$ of $F = [f_1 \cdots f_\nu]$ is a ν dimensional strictly negative subspace of $\mathcal{B}(X)$ with reproducing kernel

$$N_\omega(\lambda) = F(\lambda) Q^{-1} F(\omega)^* , \qquad \lambda, \omega \in \Omega'_+ .$$

Consequently,

$$\mathcal{H} = \mathcal{B}(X) \boxminus \mathcal{N} \,,$$

the orthogonal complement of $\mathcal{N}$ in $\mathcal{B}(X)$, is a reproducing kernel Hilbert space with reproducing kernel

$$H_\omega(\lambda) = \Lambda_\omega(\lambda) - N_\omega(\lambda) \,, \qquad \lambda, \omega \in \Omega'_+ \,.$$

Clearly $H_\omega(\lambda)$ is jointly analytic in $\Omega'_+ \times \Omega'_+$ since $\Lambda_\omega(\lambda)$ is (by its very definition) and $N_\omega(\lambda)$ is (since it involves only finite linear combinations of vector valued functions which are analytic in Ω'_+). Therefore, since $\mathcal{H}$ is a Hilbert space,

$$H_\omega^{(j)} v \in \mathcal{H} \quad \text{and} \quad [g, H_\omega^{(j)} v]_{\mathcal{B}(X)} = v^* g^{(j)}(\omega)$$

for every choice of $\omega \in \Omega'_+$, $v \in \mathbb{C}^k$ and $g \in \mathcal{H}$; see e.g., [AD4] for more information, if need be. Similar considerations apply to $\mathcal{N}$ since it is a Hilbert space with respect to $-[\ ,\ ]_{\mathcal{B}(X)}$, or, even more directly by explicit computation:

$$N_\omega^{(j)}(\lambda) = F(\lambda) Q^{-1} F^{(j)}(\omega)^*$$

and hence, since every $h \in \mathcal{N}$ can be expressed as $h = Fu$ for some $u \in \mathbb{C}^\nu$,

$$[h, N_\omega^{(j)} v]_{\mathcal{B}(X)} = [Fu, FQ^{-1}F^{(j)}(\omega)^* v]_{\mathcal{B}(X)}$$

$$= v^* F^{(j)}(\omega) Q^{-1} Q u$$

$$= v^* h^{(j)}(\omega) \,.$$

Thus

$$\Lambda_\omega^{(j)}(\lambda) v = H_\omega^{(j)}(\lambda) v + N_\omega^{(j)}(\lambda) v$$

clearly belongs to $\mathcal{B}(X)$ for every choice of $\omega \in \Omega'_+$ and $v \in \mathbb{C}^k$. Moreover, as every $f \in \mathcal{B}(X)$ admits a decomposition of the form

$$f = g + h$$

with $g \in \mathcal{H}$ and $h \in \mathcal{N}$ it follows readily that

$$[f, \Lambda_\omega^{(j)} v]_{\mathcal{B}(X)} = [g, H_\omega^{(j)} v]_{\mathcal{B}(X)} + [h, N_\omega^{(j)} v]_{\mathcal{B}(X)}$$

$$= v^* g^{(j)}(\omega) + v^* h^{(j)}(\omega)$$

$$= v^* f^{(j)}(\omega) \,,$$

as claimed. ∎

In order to minimize the introduction of extra notation, the last theorem has been formulated in the specific Pontryagin space $\mathcal{B}(X)$ which is of interest in this paper. A glance at the proof, however, reveals that it holds for arbitrary Pontryagin spaces with kernels $\Lambda_\omega(\lambda)$ which are jointly analytic in λ and ω^* for $(\lambda, \omega) \in \Delta \times \Delta$. In particular the decomposition

$$\Lambda_\omega(\lambda) = H_\omega(\lambda) - \{-N_\omega(\lambda)\} ,$$

which continues to hold in this more general setting, exhibits $\Lambda_\omega(\lambda)$ as the difference of two positive kernels both of which are jointly analytic in λ and ω^* for $(\lambda, \omega) \in \Delta \times \Delta$.

The conclusions of Theorem 3.2 remain valid for those points $\alpha \in \Omega_0$ at which $\Lambda_\omega(\lambda)$ is jointly analytic in λ and ω^* for λ and ω in a neighborhood of α. This is because, if $\omega_1, \omega_2, \ldots$ is a sequence of points in Ω'_+ which tends to α, then

$$\left[H_{\omega_n}^{\langle k \rangle} v, H_{\omega_n}^{\langle k \rangle} v \right]_{\mathcal{B}(X)} = \left[\Lambda_{\omega_n}^{\langle k \rangle} v, \Lambda_{\omega_n}^{\langle k \rangle} v \right]_{\mathcal{B}(X)} - \left[N_{\omega_n}^{\langle k \rangle} v, N_{\omega_n}^{\langle k \rangle} v \right]_{\mathcal{B}(X)}$$

stays bounded as $n \uparrow \infty$. Thus at least a subsequence of the elements $H_{\omega_n}^{\langle k \rangle} v$ tends weakly to a limit which can be identified as $H_\alpha^{\langle k \rangle} v$ since weak convergence implies pointwise convergence in a reproducing kernel Hilbert space. Thus $H_\alpha^{\langle k \rangle} v$ belongs to $\mathcal{B}(X)$, as does

$$N_\alpha^{\langle k \rangle}(\lambda) v = F(\lambda) Q^{-1} F^{(k)}(\alpha)^* v$$

and hence also

$$\Lambda_\alpha^{\langle k \rangle} v = H_\alpha^{\langle k \rangle} v + N_\alpha^{\langle k \rangle} v .$$

It remains to verify (3.2) for $\omega = \alpha$, but that is a straightforward evaluation of limits.

3.2. $\mathcal{B}(X)$ Spaces

In this subsection we give an alternative description of $\mathcal{B}(X)$ under the supplementary hypothesis that the multiplication operator

$$M_X : f \longrightarrow Xf$$

is a bounded operator from H_ρ^m into H_ρ^k. Then X is automatically analytic in Ω_+ and

$$\Gamma = M_X J M_{X^*} \tag{3.3}$$

is a bounded selfadjoint operator from H_ρ^k into itself. The construction, which is adapted from [A1], remains valid even if the kernel $\Lambda_\omega(\lambda)$ defined in (3.1) is not constrained to have a finite number of negative squares. However, in this instance the space $\mathcal{B}(X\cdot)$ will be a reproducing kernel Krein space, i.e., it will admit an orthogonal direct sum decomposition

$$\mathcal{B}(X) = \mathcal{B}_+ [\dotplus] \mathcal{B}_-$$

where $\mathcal{B}_+$ is a Hilbert space with respect to the underlying indefinite inner product $[\ , \ ]_{\mathcal{B}(X)}$ and $\mathcal{B}_-$ is a Hilbert space with respect to $-[\ , \ ]_{\mathcal{B}(X)}$ and both $\mathcal{B}_+$ and $\mathcal{B}_-$ are presumed to be infinite dimensional. Moreover, in contrast to kernels with a finite number of negative squares, there may now be many different reproducing kernel Krein spaces with the same reproducing kernel. Examples of this sort were first given by Schwartz

[Sch]. For another example see [A2], and for other constructions of reproducing kernel Krein spaces, see [A3], [dB4], [dB5] and [Y].

Let

$$\mathcal{R}_\Gamma = \{\Gamma g : \ g \in H_\rho^k\}$$

and let $\overline{\mathcal{R}}_\Gamma$ denote the closure of $\mathcal{R}_\Gamma$ with respect to the metric induced by the inner product

$$\langle \Gamma f, \Gamma g \rangle_\Gamma = \langle (\Gamma^*\Gamma)^{\frac{1}{2}} f, g \rangle_{H_\rho} \ . \tag{3.4}$$

It is readily checked that $\mathcal{R}_\Gamma$ is a pre-Hilbert space:

$$\langle \Gamma f, \Gamma f \rangle_\Gamma = 0 \quad \text{if and only if} \quad \Gamma f = 0 \ ,$$

and hence that $\overline{\mathcal{R}}_\Gamma$ is a Hilbert space. Next let $\mathcal{C}(X) = \overline{\mathcal{R}}_\Gamma$ endowed with the indefinite inner product

$$[\Gamma f, \Gamma g]_\Gamma = \langle \Gamma f, g \rangle_{H_\rho} \tag{3.5}$$

which is first defined on $\mathcal{R}_\Gamma$ and then extended to $\overline{\mathcal{R}}_\Gamma$ by limits.

LEMMA 3.1. *If $\omega \in \Omega_+$ and $v \in \mathbb{C}^k$, then*

$$M_X^* \frac{v}{\rho_\omega} = X(\omega)^* \frac{v}{\rho_\omega} \ . \tag{3.6}$$

PROOF. Let $k_\omega(\lambda) = 1/\rho_\omega(\lambda)$. Then, for every choice of $\alpha \in \Omega_+$ and $u \in \mathbb{C}^m$,

$$\langle M_X^* k_\omega v, k_\alpha u \rangle_{H_\rho} = \langle k_\omega v, M_X k_\alpha u \rangle_{H_\rho}$$

$$= \{v^* X(\omega) k_\alpha(\omega) u\}^*$$

$$= u^* k_\omega(\alpha) X(\omega)^* v \ .$$

On the other hand, by direct calculation, the left hand side of the last equality is equal to

$$u^* (M_X^* k_\omega)(\alpha) v \ .$$

This does the trick since both u and α are arbitrary. ∎

THEOREM 3.3. *$\mathcal{C}(X)$ is a reproducing kernel Krein space (of $k \times 1$ vector valued analytic functions in Ω_+) with reproducing kernel*

$$\Lambda_\omega(\lambda) = \frac{X(\lambda) J X(\omega)^*}{\rho_\omega(\lambda)} \tag{3.7}$$

for every choice of ω and λ in Ω_+.

PROOF. Since Γ is a bounded selfadjoint operator on the Hilbert space H_ρ^k it admits a spectral decomposition: $\Gamma = \int_{-\infty}^{\infty} t dE_t$ with finite upper and lower limits. Let

$$\Gamma_- = \int_{-\infty}^{0} t dE_t \quad \text{and} \quad \Gamma_+ = \int_{0}^{\infty} t dE_t \ .$$

Then Γ_- and Γ_+ are bounded selfadjoint operators on H_ρ^k,

$$\Gamma = \Gamma_- + \Gamma_+ \quad \text{and} \quad \Gamma_-\Gamma_+ = \Gamma_+\Gamma_- = 0 \ .$$

It now follows readily that

$$\overline{\mathcal{R}}_\Gamma = \overline{\mathcal{R}}_{\Gamma_+} [\dotplus] \overline{\mathcal{R}}_{\Gamma_-}$$

is a Krein space since the indicated sum decomposition is both direct and orthogonal with respect to the indefinite inner product $[\ ,\ ]_\Gamma$ given in (3.5), and $\overline{\mathcal{R}}_{\Gamma_\pm}$ is a Hilbert space with respect to $\pm[\ ,\ ]_\Gamma$. It remains to show that

(1)　$\Lambda_\omega v \in \mathcal{C}(X)$ and

(2)　$[f, \Lambda_\omega v]_\Gamma = v^* f(\omega)$

for every choice of $v \in \mathbb{C}^k$, $\omega \in \Omega_+$ and $f \in \mathcal{C}(X)$. The identification

$$\Gamma \frac{v}{\rho_\omega} = M_X J M_X^* \frac{v}{\rho_\omega}$$

$$= M_X J X(\omega)^* \frac{v}{\rho_\omega} \ ,$$

which is immediate from Lemma 3.1, serves to establish (1).

Suppose next that $f = \Gamma g$ for some $g \in H_\rho^k$. Then

$$[f, \Lambda_\omega v]_\Gamma = [\Gamma g, \Gamma \frac{v}{\rho_\omega}]_\Gamma$$

$$= \langle \Gamma g, \frac{v}{\rho_\omega} \rangle_{H_\rho}$$

$$= v^* f(\omega) \ .$$

This establishes (2) for $f \in \mathcal{R}_\Gamma$. The same conclusions may be obtained for $f \in \overline{\mathcal{R}}_\Gamma$ by a limiting argument.　∎

THEOREM 3.4.　*If X is a $k \times m$ matrix valued function which is $(\Omega_+, J, \rho)_\nu$ admissible and if also the multiplication operator M_X is bounded on H_ρ^m, then*

$$\mathcal{B}(X) = \mathcal{C}(X) \ .$$

PROOF.　Under the given assumptions, both $\mathcal{B}(X)$ and $\mathcal{C}(X)$ are reproducing kernel Pontryagin spaces with the same reproducing kernel. Therefore, by the results of Schwartz [Sch] and Sorojonen [So] cited earlier, or Theorem 3.1, $\mathcal{B}(X) = \mathcal{C}(X)$.　∎

LEMMA 3.2.　*Let X be a $k \times m$ matrix valued function which is analytic in Ω_+ such that M_X is a bounded linear operator from H_ρ^m to H_ρ^k. Then*

$$\frac{\partial^j}{\partial \omega^{*j}} M_X^* \frac{v}{\rho_\omega} = M_X^* \frac{\partial^j}{\partial \omega^{*j}} \frac{v}{\rho_\omega} \tag{3.8}$$

for $j = 0, 1, \ldots$ and every choice of $v \in \mathbb{C}^k$ and $\omega \in \Omega_+$.

PROOF. It is convenient to let

$$D^j = \frac{\partial^j}{\partial \omega^{*j}} \quad \text{and} \quad f = \frac{v}{\rho_\omega} .$$

Then, for every choice of $u \in \mathbb{C}^m$ and $\alpha \in \Omega_+$,

$$u^*(M_X^* D^j f)(\alpha) = \langle M_X^* D^j f, \frac{u}{\rho_\alpha} \rangle_{H_\rho}$$

$$= \langle D^j f, M_X \frac{u}{\rho_\alpha} \rangle_{H_\rho}$$

$$= D^j \langle f, M_X \frac{u}{\rho_\alpha} \rangle_{H_\rho}$$

$$= D^j \left\{ \frac{v^* X(\omega) u}{\rho_\alpha(\omega)} \right\}^*$$

$$= u^* D^j X(\omega)^* \frac{v}{\rho_\omega(\alpha)}$$

$$= u^* D^j (M_X^* f)(\alpha) ,$$

where the passage from line 2 to line 3 is just a special case of (3.2) applied to the Hilbert space H_ρ^k. Therefore,

$$M_X^* D^j f = D^j M_X^* f$$

as claimed. ∎

COROLLARY 1. *If X, ω and v are as in the preceding lemma and if $\Lambda_\omega(\lambda)$ and Γ are as in (3.1) and (3.3), respectively, then*

$$\Gamma \frac{\partial^j}{\partial \omega^{*j}} \frac{v}{\rho_\omega} = \frac{\partial^j}{\partial \omega^{*j}} \Gamma \frac{v}{\rho_\omega} = \frac{\partial^j}{\partial \omega^{*j}} \Lambda_\omega v . \tag{3.9}$$

COROLLARY 2. *If X, ω and v are as in the preceding lemma and if*

$$f_j = J M_X^* \varphi_{\omega,j} v , \qquad j = 0, 1, \ldots \ ,$$

then

$$f_j(\lambda) = J \sum_{s=0}^{j} \frac{X^{(j-s)}(\omega)^*}{(j-s)!} \varphi_{\omega,s}(\lambda) v \tag{3.10}$$

and

$$[f_0 \cdots f_{n-1}] = \left[J X(\omega)^* v \ \cdots \ \frac{J X^{(n-1)}(\omega)^* v}{(n-1)!} \right] \Phi_{\omega,n} . \tag{3.11}$$

PROOF. Let $D^j = \partial^j/\partial\omega^{*j}$. Then, by Lemmas 3.2 and 3.1,

$$f_j = \frac{1}{j!} D^j J M_X^* \varphi_{\omega,0} v$$

$$= \frac{1}{j!} D^j J X(\omega)^* \frac{v}{\rho_\omega}$$

which, by Leibnitz's rule, is readily seen to be equal to the right hand side of (3.10). Formula (3.11) is immediate from (3.10) and the definition (2.4) of $\Phi_{\omega,n}$.   ∎

Suitably specialized versions of formulas (3.8)–(3.10) play a useful role in the study of certain classes of matrix polynomials; see Section 11 of [D1] and Section 6 of [D4].

LEMMA 3.3. *If X is as in Lemma 3.2 and if $f = JM_X^*\varphi_{\alpha,j}u$ and $g = JM_X^*\varphi_{\beta,i}v$ for some choice of α,β in Ω_+ and u,v in $\mathbb{C}^k$, then*

$$[M_X f, M_X g]_{B(X)} = \langle Jf, g\rangle_{H_\rho} \ . \tag{3.12}$$

PROOF. This is a straightforward calculation based on the definitions:

$$[M_X f, M_X g]_{B(X)} = [\Gamma\varphi_{\alpha,j}u, \Gamma\varphi_{\beta,i}v]_{B(X)}$$

$$= \langle\Gamma\varphi_{\alpha,j}u, \varphi_{\beta,i}v\rangle_{H_\rho}$$

$$= \langle M_X J M_X^*\varphi_{\alpha,j}u, \varphi_{\beta,i}v\rangle_{H_\rho}$$

$$= \langle Jf, g\rangle_{H_\rho} \ .$$

∎

Formula (3.12) exhibits M_X as an isometry from the span $\mathcal{M}$ of $f_j = JM_X^*\varphi_{\alpha,j}u$, $j = 0,\ldots,n$ in H_ρ^k into $B(X)$. This is an important ingredient in the verification of the orthogonal direct sum decomposition

$$B(X) = B(X\Theta)[\dot{+}]X\mathcal{K}(\Theta) \tag{3.13}$$

which holds when $\mathcal{M} = \mathcal{K}(\Theta)$. In fact $\mathcal{M} = \mathcal{K}(\Theta)$ when $\mathcal{M}$ is a nondegenerate subspace of $H_{\rho,J}$, as follows from Theorems 2.2 and 2.1.

THEOREM 3.5. *Let $X = [F \ \ G]$ be a $k \times m$ matrix valued meromorphic function on Ω_+ with components $F(\lambda) \in \mathbb{C}^{k\times p}$ and $G(\lambda) \in \mathbb{C}^{k\times q}$ such that M_X is a bounded operator from H_ρ^m into H_ρ^k. Then X is analytic on Ω_+ and the following are equivalent:*

(1) X is (Ω_+, J, ρ) admissible.

(2) $\Gamma = M_F M_F^* - M_G M_G^*$ is positive semidefinite on H_ρ^k.

(3)   There is a $p \times q$ matrix valued analytic function S on Ω_+ such that $\|M_S\| \leq 1$ and $M_G = M_F M_S$.

PROOF. X is analytic on Ω_+ because

$$M_X \frac{v}{a} \in H_\rho^k$$

for every choice of $v \in \mathbb{C}^m$.

The equivalence of (1) and (2) is an easy consequence of the evaluation

$$v_i^* \Lambda_{\omega_j}(\omega_i)v_j = \left[\Gamma \frac{v_j}{\rho_{\omega_j}} \ , \ \Gamma \frac{v_i}{\rho_{\omega_i}} \right]_{\mathcal{B}(X)}$$

$$= \langle \Gamma \frac{v_j}{\rho_{\omega_j}} \ , \ \frac{v_i}{\rho_{\omega_i}} \rangle_{H_\rho}$$

and the fact that finite sums of the form $\Sigma \Lambda_{\omega_j} v_j$ are dense in $\mathcal{B}(X)$.

Suppose next that (2) holds. Then, by a (slight adaptation – to cover the case $p \neq q$ of a) theorem of Rosenblum [Ro], there exists an operator Q from H_ρ^q to H_ρ^p such that:

$$M_G = M_F Q, \quad \|Q\| \leq 1 \quad \text{and} \quad QM_s = M_s Q \ ,$$

where, in the last item M_s denotes the (isometric) operator of multiplication by $s = b/a$ on H_ρ^r, regardless of the size of the positive integer r. Thus, by Theorem 3.3 of [AD4], $Q = M_S$ for some $p \times q$ matrix valued function S which is analytic on Ω_+, as needed to complete the proof that (2) $\Longrightarrow$ (3). The converse is selfevident. ∎

A more leisurely exposition of the proof of this theorem for $\rho_\omega(\lambda) = 1 - \lambda \omega^*$ may be found e.g., in [ADD].

4. RECURSIVE EXTRACTIONS AND THE SCHUR ALGORITHM

In this section we study decompositions of the form (3.13) of the reproducing kernel Pontryagin space $\mathcal{B}(X)$ based on a $k \times m$ $(\Omega_+, J, \rho)_\nu$ admissible matrix valued function X; the existence and uniqueness of these spaces is established in Theorem 3.1. Such decompositions originate in the work of de Branges [dB2, Theorem 34], [dB3], and de Branges and Rovnyak [dBR] for the case $\nu = 0$ (which means that $\mathcal{B}(X)$ is a Hilbert space) and $\rho_\omega(\lambda) = -2\pi i(\lambda - \omega^*)$ for a number of different classes of X and Θ. Decompositions of the form (3.13) for finite ν (i.e., when $\mathcal{B}(X)$ is a reproducing kernel Pontryagin space) and the two cases $\rho_\omega(\lambda) = 1 - \lambda \omega^*$ and $\rho_\omega(\lambda) = -2\pi i(\lambda - \omega^*)$ were considered in [AD3]; such decompositions in the Krein space setting were studied in [A1] and, for Hilbert spaces of pairs, in [A4].

If $\mathcal{B}(X)$ is a nonzero Hilbert space, then it is always possible to find a one dimensional Hilbert space $\mathcal{K}(\Theta_1)$ such that $X\mathcal{K}(\Theta_1)$ is isometrically included inside $\mathcal{B}(X)$. This leads to the decomposition

$$\mathcal{B}(X) = \mathcal{B}(X\Theta_1) \oplus X\mathcal{K}(\Theta_1) \ .$$

Then, if $\mathcal{B}(X\Theta_1)$ is nonzero, there is a one-dimensional Hilbert space $\mathcal{K}(\Theta_2)$ such that $X\Theta_1\mathcal{K}(\Theta_2)$ sits isometrically inside $\mathcal{B}(X\Theta_1)$, and so forth. This leads to the supplementary sequence of decompositions

$$\mathcal{B}(X\Theta_1) = \mathcal{B}(X\Theta_1\Theta_2) \oplus X\Theta_1\mathcal{K}(\Theta_2)$$

$$\mathcal{B}(X\Theta_1\Theta_2) = \mathcal{B}(X\Theta_1\Theta_2\Theta_3) \oplus X\Theta_1\Theta_2\mathcal{K}(\Theta_3)$$

$$\vdots$$

which can be continued as long as the current space $\mathcal{B}(X\Theta_1\cdots\Theta_n)$ is nonzero. In this decomposition, the Θ_j are "elementary sections" with poles (and directions) which are allowed to vary with j.

The classical Schur algorithm corresponds such a sequence of decompositions for the special case in which $\rho_\omega(\lambda) = 1 - \lambda\omega^*$, $X = [1\ \ S]$ with S a scalar analytic contractive function in $\mathbb{D}$ and all the Θ_j have their poles at infinity. For additional discussion of the Schur algorithm from this point of view and of decompositions of the sort considered above when $\mathcal{B}(X)$ is a reproducing kernel Pontryagin space, see [AD3]. In particular, in this setting it is not always possible to choose the $\mathcal{K}(\Theta_j)$ to be one-dimensional, but, as shown in Theorem 7.2 of [AD3] for the two special choices of ρ considered there, it is possible to choose decompositions in which the $\mathcal{K}(\Theta_j)$ are Pontryagin spaces of dimension less than or equal to two. The same conclusions hold for $\rho \in \mathcal{D}_\Omega$ also as will be shown later in the section.

THEOREM 4.1. *Let X be a $k \times m$ matrix valued function which is $(\Omega_+, J, \rho)_\nu$ admissible and let $\mathcal{B}(X)$ be the (unique) associated reproducing kernel Pontryagin space with reproducing kernel given by (3.1). Let $\alpha \in \Omega'_+$, the domain of analyticity of X in Ω_+, let $\mathcal{M}$ denote the span of the functions*

$$f_j = JM_X^*\varphi_{\alpha,j-1}v = \frac{1}{(j-1)!}\ \frac{\partial^{j-1}}{\partial\omega^{*j-1}}JX(\omega)^*\frac{v}{\rho_\omega}\bigg|_{\omega=\alpha}\ , \tag{4.1}$$

$j = 1,\ldots,n$, endowed with the indefinite inner product

$$[f_j, f_i]_\mathcal{M} = \langle Jf_j, f_i\rangle_{H_\rho} = p_{ij}$$

and suppose that the $n \times n$ matrix $P = [p_{ij}]$ is invertible. Then:

(1) $\mathcal{M}$ is a $\mathcal{K}(\Theta)$ space.

(2) The operator M_X of multiplication by X is an isometry from $\mathcal{K}(\Theta)$ into $\mathcal{B}(X)$.

(3) $X\Theta$ is $(\Omega_+, J, \rho)_\mu$ admissible, where $\mu = \nu$ – the number of negative eigenvalues of P.

(4) $\mathcal{B}(X)$ admits the orthogonal direct sum decomposition

$$\mathcal{B}(X) = \mathcal{B}(X\Theta)[\dot{+}]X\mathcal{K}(\Theta)\ . \tag{4.2}$$

PROOF. Let

$$V = [v_1, \ldots, v_n]$$

denote the $m \times n$ matrix with columns

$$v_j = \frac{JX^{(j-1)}(\alpha)^*v}{(j-1)!} \, , \qquad j = 1,\ldots,n \, .$$

Then, by Leibnitz's rule, it is readily checked that

$$\begin{aligned} F &= [f_1,\ldots,f_n] \\ &= V\Phi_{\alpha,n} \, . \end{aligned}$$

By (2.4), this can be reexpressed as

$$F(\lambda) = V\{a(\lambda)A_\alpha - b(\lambda)B_\alpha\}^{-1} \, ,$$

where A_α and B_α are as in (2.6). Therefore, by Theorems 2.2 and 2.3, $\mathcal{M}$ is a $\mathcal{K}(\Theta)$ space.

Next, since

$$Xf_j = \frac{1}{(j-1)!}\Lambda_\alpha^{(j-1)}v$$

in terms of the notation introduced in Section 3, it follows from Theorem 3.2 that $Xf_j \in \mathcal{B}(X)$ and

$$[Xf_j, \, Xf_i]_{\mathcal{B}(X)} = \frac{1}{(i-1)!}\frac{1}{(j-1)!}\frac{\partial^{i-1}}{\partial\lambda^{i-1}}\frac{\partial^{j-1}}{\partial\omega^{*j-1}}\left.\frac{v^*X(\lambda)JX(\omega)^*v}{\rho_\omega(\lambda)}\right|_{\lambda=\omega=\alpha} \, . \qquad (4.3)$$

But the last evaluation is equal to

$$\langle Jf_j, f_i\rangle_{H_\rho} \, ,$$

as follows by writing

$$f_j = \sum_{t=0}^{j-1} \frac{k_\alpha^{\langle t\rangle}}{t!}v_{j-1-t}$$

with $k_\omega(\lambda) = 1/\rho_\omega(\lambda)$ and applying Theorem 3.2 to H_ρ^m. This completes the proof of (2).

The proofs of (3) and (4) are much the same as the proofs of the corresponding assertions in Theorem 6.13 of [AD3] and are therefore omitted.   ∎

If $n = 1$, then the matrix P in the last theorem is invertible if and only if $X(\alpha)^*v$ is not J neutral. In this case, the Θ which intervenes in the extraction can be expressed in the form

$$\Theta(\lambda) = I_m + \{b_\alpha(\lambda) - 1\}u(u^*Ju)^{-1}u^*J \, ,$$

where $u = JX(\alpha)^*v$ and

$$b_\alpha(\lambda) = \frac{s(\lambda) - s(\alpha)}{1 - s(\lambda)s(\alpha)^*}$$

with $s(\lambda) = b(\lambda)/a(\lambda)$; see (2.24) of [AD4], and hence

$$v^* X(\alpha)\Theta(\alpha) = 0 .$$

Thus it is possible to extract the elementary Blaschke-Potapov factor

$$G(\lambda) = I_k + \{b_\alpha(\lambda) - 1\}v(v^*v)^{-1}v^*$$

from the left of $X\Theta$ to obtain

$$X_1(\lambda) = G(\lambda)^{-1}X(\lambda)\Theta(\lambda) ,$$

which is analytic at α and is $(\Omega_+, J, \rho)_\mu$ admissible, where $\mu = \nu$ if $u^* Ju > 0$ and $\mu = \nu - 1$ if $u^* Ju < 0$.

THEOREM 4.2. *Let X be a $k \times m$ matrix valued function which is $(\Omega_+, J, \rho)_\nu$ admissible and suppose that $\mathcal{B}(X) \neq \{0\}$ and Ω_+ is connected. Then $X(\lambda)JX(\lambda)^* \not\equiv 0$ in Ω'_+, the domain of analyticity of X in Ω_+.*

PROOF. Suppose to the contrary that $X(\lambda)JX(\lambda)^* \equiv 0$ in Ω'_+. Then, for any $\omega \in \Omega'_+$, there exists a $\delta > 0$ such that the power series

$$X(\lambda) = \sum_{s=0}^{\infty} X_s(\lambda - \omega)^s$$

with $k \times m$ matrix coefficients converges, and

$$\sum_{s,t=0}^{\infty} X_s(\lambda - \omega)^s J(\lambda^* - \omega^*)^t X_t^* = 0$$

for $|\lambda - \omega| < \delta$. Therefore

$$\sum_{s,t=0}^{\infty} \varepsilon^{s+t} e^{i(s-t)\theta} X_s JX_t^* = 0 ,$$

for $0 \leq \varepsilon < \delta$ and $0 \leq \theta < 2\pi$. But this in turn implies that $X_s JX_t^* = 0$ for $s,t = 0,1,\ldots$, and hence that $X(\alpha)JX(\beta)^* = 0$ for $|\alpha - \omega| < \delta$ and $|\beta - \omega| < \delta$. Since Ω_+ is connected, this propagates to all of Ω_+ and forces $\mathcal{B}(X) = \{0\}$, which contradicts the given hypotheses. Thus $X(\lambda)JX(\lambda)^* \not\equiv 0$ in Ω'_+, as claimed.   ■

COROLLARY 1. *If, in the setting of Theorem 4.1, Ω_+ is connected, then there exists a point $\alpha \in \Omega_+$ and a vector $v \in \mathbb{C}^k$ such that $v^*X(\alpha)JX(\alpha)^*v \neq 0$.*

COROLLARY 2. *If, in the setting of Theorem 4.1, Ω_+ is connected, then the set of points in Ω'_+ at which the extraction procedure of Theorem 4.1 can be iterated arbitrarily often (up to the dimension of $\mathcal{B}(X)$) with one dimensional $\mathcal{K}(\Theta)$ spaces is dense in Ω_+.*

If Ω_+ is not connected, then it is easy to exhibit nonzero $\mathcal{B}(X)$ spaces for which $X(\lambda)JX(\lambda)^* = 0$ for every point $\lambda \in \Omega_+$. For example, if Ω_+ has two connected components: Ω_1 and Ω_2, let

$$K_1 = \frac{1}{\sqrt{2}}[I_p \ \ I_p] , \qquad K_2 = \frac{1}{\sqrt{2}}[I_p \ \ -I_p] ,$$

$$X(\lambda) = [a(\lambda)K_j \ \ b(\lambda)K_j] , \quad \text{for} \quad \lambda \in \Omega_j ,$$

and

$$\hat{J} = \begin{bmatrix} J_{pp} & 0 \\ 0 & -J_{pp} \end{bmatrix} .$$

Then, for $\lambda \in \Omega_j$ and $\omega \in \Omega_i$,

$$X(\lambda)\hat{J}X(\omega)^* = \rho_\omega(\lambda)K_j J_{pp} K_i^*$$

$$= \begin{cases} 0 & \text{if} \quad i = j \\ \rho_\omega(\lambda)I_p & \text{if} \quad i \neq j . \end{cases}$$

Thus $X(\lambda)JX(\lambda)^* = 0$ for every point $\lambda \in \Omega_+$, while $\mathcal{B}(X)$ is a $2p$ dimensional Pontryagin space of index p.

THEOREM 4.3. *Let X be a $k \times m$ matrix valued function which is $(\Omega_+, J, \rho)_\nu$ admissible such that $\mathcal{B}(X) \neq \{0\}$ and yet $X(\alpha)^*v$ is J neutral for every choice of $\alpha \in \Omega'_+$ and $v \in \mathbb{C}^k$, then there exist a pair of points α, β in Ω'_+ and vectors v, w in $\mathbb{C}^k$ such that*

(1) $w^* X(\beta)JX(\alpha)^*v \neq 0.$

(2) The two dimensional space

$$\mathcal{M} = \text{span}\left\{ \frac{JX(\alpha)^*v}{\rho_\alpha} , \ \frac{JX(\beta)^*w}{\rho_\beta} \right\}$$

endowed with the J inner product in H_ρ^m will be a $\mathcal{K}(\Theta)$ space.

(3) The space $\mathcal{B}(X)$ admits the direct orthogonal sum decomposition

$$\mathcal{B}(X) = \mathcal{B}(X\Theta)[\dot{+}]X\mathcal{K}(\Theta) .$$

(4) $X\Theta$ is $(\Omega_+, J, \rho)_{\nu-1}$ admissible.

PROOF. The proof is easily adapted from the proof of Theorem 7.2 in [AD3].

∎

There is also a nice formula for the Θ which intervenes in the statement of the last theorem: upon setting $u_1 = JX(\alpha)^*v$, $u_2 = JX(\beta)^*w$ and

$$W_{ij} = u_i(u_j^* J u_i)^{-1} u_j^* J , \quad \text{for} \quad i \neq j ,$$

we can write

$$\Theta(\lambda) = \left(I_m + \left\{\frac{s(\lambda) - s(\beta)}{1 - s(\lambda)s(\alpha)^*} - 1\right\} W_{12}\right)\left(I_m + \left\{\frac{s(\lambda) - s(\alpha)}{1 - s(\lambda)s(\beta)^*} - 1\right\} W_{21}\right) . \quad (4.4)$$

This the counterpart of formula (7.5) of [AD3] in the present more general setting; for more information on elementary factors in this setting, see [AD4].

The reader is invited to check for himself that

$$\frac{J - \Theta(\lambda)J\Theta(\omega)^*}{\rho_\omega(\lambda)} = F(\lambda)\begin{bmatrix} 0 & \gamma \\ \gamma^* & 0 \end{bmatrix}^{-1} F(\omega)^* ,$$

where

$$F(\lambda) = [u_1 \quad u_2](a(\lambda)A - b(\lambda)B)^{-1}$$

with

$$A = \begin{bmatrix} a(\alpha) & 0 \\ 0 & a(\beta) \end{bmatrix}^* , \qquad B = \begin{bmatrix} b(\alpha) & 0 \\ 0 & b(\beta) \end{bmatrix}^* ,$$

and $\gamma = u_1^* J u_2/\rho_\beta(\alpha)$. The formula

$$\frac{\rho_\omega(\lambda)\rho_\alpha(\beta)}{\rho_\alpha(\lambda)\rho_\omega(\beta)} = 1 - \left\{\frac{s(\lambda) - s(\beta)}{1 - s(\lambda)s(\alpha)^*}\right\}\left\{\frac{s(\omega) - s(\alpha)}{1 - s(\omega)s(\beta)^*}\right\}^* , \qquad (4.5)$$

which is the counterpart of (7.3) of [AD3], plays an important role in this verification.

The matrix valued function

$$\left[1 \quad c_2\lambda^2 \quad \cdots \quad c_{2\nu}\lambda^{2\nu} \quad \vdots \quad c_1\lambda \quad \cdots \quad c_{2\nu-1}\lambda^{2\nu-1}\right]$$

with $c_j = \left(\dbinom{2\nu}{j}\right)^{\frac{1}{2}}$ is $(\Omega_+, J, \rho)_\nu$ admissible with respect to $\rho_\omega(\lambda) = 1 - \lambda\omega^*$ and $J = J_{\nu+1,\nu}$. The corresponding space $\mathcal{B}(X)$ is a 2ν dimensional Pontryagin space (of polynomials) of index ν with reproducing kernel $\Lambda_\omega(\lambda) = (1 - \lambda\omega^*)^{2\nu-1}$. It does not isometrically include a one dimensional $\mathcal{K}(\Theta)$ space such that $X\Theta$ is $(\Omega_+, \rho, J)_{\nu-1}$ admissible, since $X(\lambda)JX(\lambda)^* > 0$ for every point $\lambda \in \Omega_+ = \mathbb{D}$. This serves to illustrate the need for the two dimensional $\mathcal{K}(\Theta)$ sections of Theorem 4.3.

We remark that Theorem 4.1 can be extended to include spaces $\mathcal{M}$ in which the chain f_j, $j = 1, \ldots, n$, is based on a point $\alpha \in \Omega_0$ such that $\Lambda_\omega(\lambda)$ is jointly analytic in λ and ω^* for (λ, ω) in a neighborhood of the point $(\alpha, \alpha) \in \Omega_0 \times \Omega_0$. In this neighborhood $\Lambda_\omega(\lambda)$ admits a power series expansion

$$\Lambda_\omega(\lambda) = \sum_{i,j=0}^{\infty} \Lambda_{ij}(\lambda - \alpha)^i(\omega^* - \alpha^*)^j$$

with the $k \times k$ matrix coefficients Λ_{ij}. Now if

$$P = \begin{bmatrix} v^* \Lambda_{00} v & \cdots & v^* \Lambda_{0,n-1} v \\ \vdots & & \vdots \\ v^* \Lambda_{n-1,0} v & \cdots & v^* \Lambda_{n-1,n-1} v \end{bmatrix}$$

and if A_α and B_α are defined as in (2.6), and

$$V = \left[JX(\alpha)^* v \cdots \frac{JX^{(n-1)}(\alpha)^* v}{(n-1)!} \right] ,$$

then

$$A_\alpha^* P A_\alpha - B_\alpha^* P B_\alpha = V^* J V . \tag{4.6}$$

Formula (4.6) may be verified by differentiating

$$\{a(\lambda)a(\omega)^* - b(\lambda)b(\omega)^*\}\Lambda_\omega(\lambda) = X(\lambda)JX(\omega)^*$$

i times with respect to λ, j times with respect to ω^* for $i, j = 0, \ldots, n - 1$ and then evaluating both sides with $\lambda = \omega = \alpha$. Therefore, by Theorem 2.3, the span $\mathcal{M}$ of the columns of

$$F(\lambda) = V\{a(\lambda)A_\alpha - b(\lambda)B_\alpha\}^{-1}$$

endowed with the indefinite inner product

$$[F\xi, \ F\eta]_\mathcal{M} = \eta^* P \xi$$

is a $\mathcal{K}(\Theta)$ space, whenever the hypothesis of Theorem 2.3 are satisfied. It is important to bear in mind that the indefinite inner product $\mathcal{M}$ is now defined in terms of derivatives of $\Lambda_\omega(\lambda)$ and not in terms of evaluations inside H_ρ which are no longer meaningful since the columns of F (which are just the f_j of (4.1)) do not belong to H_ρ^m when $\alpha \in \Omega_0$. Nevertheless formula (4.3) is still valid (as follows from the remarks which follow the proof of Theorem 3.2) and serves to justify the assertion that M_X maps $\mathcal{K}(\Theta)$ isometrically into $\mathcal{B}(X)$ in this case also.

Recall that a subspace $\mathcal{M}$ of an indefinite inner product space is said to be nondegenerate if zero is the only element therein which is orthogonal to all of $\mathcal{M}$. It is readily checked that if $\mathcal{M}$ is finite dimensional with basis $f_1, \ldots, f_n$, then $\mathcal{M}$ is nondegenerate if and only if the corresponding Gram matrix is invertible.

THEOREM 4.4. *Let $\mathcal{P}$ be a reproducing kernel Pontryagin space of $k \times 1$ vector valued functions defined on a set Δ with reproducing kernel $L_\omega(\lambda)$. Suppose that*

$$\mathcal{M} = \text{span}\{L_\alpha u_1, \ldots, L_\alpha u_n\}$$

is a nondegenerate subspace of $\mathcal{P}$ for some choice of $\alpha \in \Delta$ and $u_1, \ldots, u_n$ in $\mathbb{C}^k$, let $\mathcal{N} = \mathcal{M}^{[\perp]}$ and let $U = [u_1, \ldots, u_n]$ denote the $k \times n$ matrix with columns $u_1, \ldots, u_n$. Then:

(1)   The matrix $U^*L_\alpha(\alpha)U$ is invertible.

(2)   The sum decomposition

$$\mathcal{P} = \mathcal{M}[\dot{+}]\mathcal{N}$$

is direct as well as orthogonal.

(3)   Both the spaces $\mathcal{M}$ and $\mathcal{N}$ are reproducing kernel Pontryagin spaces with reproducing kernels

$$L_\omega^{\mathcal{M}}(\lambda) = L_\alpha(\lambda)U\{U^*L_\alpha(\alpha)U\}^{-1}U^*L_\omega(\alpha) \tag{4.7}$$

and

$$L_\omega^{\mathcal{N}}(\lambda) = L_\omega(\lambda) - L_\omega^{\mathcal{M}}(\lambda) \ , \tag{4.8}$$

respectively.

PROOF. The matrix $U^*L_\alpha(\alpha)U$ is the Gram matrix of the indicated basis for $\mathcal{M}$. It is invertible because $\mathcal{M}$ is nondegenerate. The fact that $\mathcal{M}$ is nondegenerate further guarantees that $\mathcal{M} \cap \mathcal{N} = \{0\}$ and hence that assertion (2) holds.

Next, it is easily checked that $L_\omega^{\mathcal{M}}(\lambda)$, as specified in formula (4.7), is a reproducing kernel for $\mathcal{M}$ and hence that $\mathcal{M}$ is a reproducing kernel Pontryagin space. Finally, since $L_\omega u - L_\omega^{\mathcal{M}} u$ belongs to $\mathcal{N}$ for every choice of $\omega \in \Delta$ and $u \in \mathbb{C}^k$ and

$$[f, L_\omega u - L_\omega^{\mathcal{M}} u]_{\mathcal{P}} = [f, L_\omega u]_{\mathcal{P}} = u^* f(\omega)$$

for every $f \in \mathcal{N}$, it follows that $\mathcal{N}$ is also a reproducing kernel Pontryagin space with reproducing kernel $L_\omega^{\mathcal{N}}(\lambda)$. ∎

THEOREM 4.5. *If, in the setting of Theorem 4.4, $\mathcal{P} = \mathcal{B}(X)$ and*

$$\Lambda_\omega(\lambda) = \frac{X(\lambda)JX(\omega)^*}{\rho_\omega(\lambda)}$$

for some choice of $\rho \in \mathcal{D}_\Omega$ and $k \times m$ matrix valued function X which is $(\Omega_+, J, \rho)_\nu$ admissible, and if $\alpha \in \Omega'_+$ (the domain of analyticity of X in Ω_+), then there exists an $m \times m$ matrix valued function $\Theta \in \mathcal{P}_J^\nu(\Omega_+)$ for some finite ν such that

(1) $\mathcal{M} = X\mathcal{K}(\Theta)$.

(2) $\mathcal{N} = \mathcal{B}(X\Theta)$.

(3) $L_\omega^{\mathcal{M}}(\lambda) = X(\lambda)\left\{\frac{J-\Theta(\lambda)J\Theta(\omega)^*}{\rho_\omega(\lambda)}\right\}X(\omega)^*.$

(4) $\Lambda_\omega^{\mathcal{N}}(\lambda) = X(\lambda)\Theta(\lambda)J\Theta(\omega)^*X(\omega)^*/\rho_\omega(\lambda).$

PROOF. To begin with, let

$$V = JX(\alpha)^*U$$

and

$$F(\lambda) = V\rho_\alpha(\lambda)^{-1} \ .$$

Then, upon setting

$$A = a(\alpha)^*I_n \quad \text{and} \quad B = b(\alpha)^*I_n \ ,$$

F can be expressed in the form

$$F(\lambda) = V\{a(\lambda)A - b(\lambda)B\}^{-1} \, ,$$

which is amenable to the analysis in Section 4 of [AD4]. The latter is partially reviewed in Section 2. In particular, hypotheses (1) and (2) of Theorem 2.1 (above) are clearly met: (1) holds for any $\mu \in \Omega_0$ for which $|a(\mu)| = |b(\mu)| \neq 0$ since $\alpha \notin \Omega_0$, whereas (2) holds because $\mathcal{M}$ (which equals the span of the columns of XF) is nondegenerate by assumption. Thus since

$$P := \frac{V^* J V}{\rho_\alpha(\alpha)}$$

is a Hermitian invertible solution of the matrix equation

$$A^* P A - B^* P B = V^* J V \, ,$$

it follows from Theorem 2.1 that the span $\mathcal{F}$ of the columns of F endowed with the indefinite inner product

$$[Fu, \; Fv]_{\mathcal{F}} = v^* P u$$

is a $\mathcal{K}(\Theta)$ space. This proves (1) and further implies that

$$F(\lambda) P^{-1} F(\omega) = \frac{J - \Theta(\lambda) J \Theta(\omega)^*}{\rho_\omega(\lambda)} \; .$$

Thus

$$\Lambda_\omega^{\mathcal{M}}(\lambda) = \frac{X(\lambda)V}{\rho_\alpha(\lambda)} \left\{ \frac{V^* J V}{\rho_\alpha(\alpha)} \right\}^{-1} \frac{V^* X(\omega)^*}{\rho_\alpha(\omega)}$$

$$= X(\lambda) F(\lambda) P^{-1} F(\omega)^* X(\omega)^* \, ,$$

which proves (3).

The remaining two assertions follow easily from the first two. ∎

We remark that if the matrix U which appears in the statement of Theorem 4.4 is invertible, then

$$\Lambda_\omega^{\mathcal{M}}(\lambda) = \Lambda_\alpha(\lambda) \Lambda_\alpha(\alpha)^{-1} \Lambda_\omega(\alpha) \tag{4.9}$$

and

$$\Lambda_\omega^{\mathcal{N}}(\lambda) = \Lambda_\omega(\lambda) - \Lambda_\alpha(\lambda) \Lambda_\alpha(\alpha)^{-1} \Lambda_\omega(\alpha) \, . \tag{4.10}$$

Conclusion (4) of the last theorem exhibits the fact that (whether U is invertible or not) $\Lambda_\omega^{\mathcal{N}}(\lambda)$ has the same form as $\Lambda_\omega(\lambda)$. Fast algorithms for matrix inversion are based upon this important property. Lev-Ari and Kailath [LAK] showed that if a kernel $\Lambda_\omega(\lambda)$ is of the form

$$\Lambda_\omega(\lambda) = \frac{X(\lambda) J X(\omega)^*}{\rho_\omega(\lambda)}$$

for some $\rho_\omega(\lambda)$ with $\rho_\omega(\lambda)^* = \rho_\lambda(\omega)$, then the right hand side of (4.10) will be of the same form if and only if $\rho_\omega(\lambda)$ admits a representation of the form (1.2). The present analysis gives the geometric picture in terms of the reproducing kernel spaces which underlie the purely algebraic methods used in [LAK].

We complete this section with a generalization of Theorem 3.5.

THEOREM 4.6. *Let* $X = [C \ \ D]$ *be a* $k \times m$ *matrix valued function which is* $(\Omega_+, J, \rho)_\nu$ *admissible and for which* M_X *is a bounded operator from* H_ρ^m *into* H_ρ^k. *Then there exists a* $p \times q$ *matrix valued meromorphic function* S *on* Ω_+ *such that*

(1) $[I_p \ \ -S]$ *is* $(\Omega_+, J, \rho)_\nu$ *admissible,*

and

(2) $D = -CS$.

PROOF. If $\nu = 0$, then the assertion is immediate from Theorem 3.5.

If $\nu > 0$, then, by repeated applications of either Theorem 4.1 or 4.3, whichever is applicable, there exists an $m \times m$ matrix valued function $\Theta \in \mathcal{P}_J^\nu(\Omega_+)$ which is analytic in Ω_+ such that $X\Theta$ is (Ω_+, J, ρ) admissible and the multiplication operator M_Θ is bounded on H_ρ^m. The last assertion follows from Theorem 6.1 and the formulas for Θ which are provided in and just after the proofs of Theorems 4.1 and 4.3, respectively. Thus, the multiplication operator $M_{X\Theta}$ is also bounded on H_ρ^m and so, by Theorem 3.5, there exists a $p \times q$ matrix valued analytic function S_o on Ω_+ with $\|M_{S_o}\| \leq 1$ such that

$$C\Theta_{12} + D\Theta_{22} = -(C\Theta_{11} + D\Theta_{21})S_o \ .$$

Therefore,

$$D(\Theta_{21}S_o + \Theta_{22}) = -C(\Theta_{11}S_o + \Theta_{12}) \ ,$$

which in turn implies that

$$D = -CS$$

with

$$S = (\Theta_{11}S_o + \Theta_{12})(\Theta_{21}S_o + \Theta_{22})^{-1} \ .$$

The indicated inverse exists in Ω_+ except for at most a countable set of isolated points because $\det(\Theta_{21}S_o + \Theta_{22})$ is analytic and not identically equal to zero in Ω_+. Indeed, since Θ is both analytic and J unitary at any point $\mu \in \Omega_0$ at which $|a(\mu)| = |b(\mu)| \neq 0$, it follows by standard arguments that $\Theta_{22}^{-1}\Theta_{21}$ is strictly contractive at μ and so too in a little disc centered at μ. This does the trick, since every such disc has a nonempty intersection with Ω_+ (otherwise $|a(\lambda)/b(\lambda)| \leq 1$ in some such disc with equality at the center; this forces $b(\lambda) = ca(\lambda)$, first throughout the disc by the maximum modulus principle, and then throughout all of Ω since it is connected) and S_o is contractive in Ω_+.

Now, let $F = [f_1 \cdots f_n]$ be an $m \times n$ matrix valued function whose columns form a basis for $\mathcal{K}(\Theta)$, let Q denote the invertible $n \times n$ Hermitian matrix with ij entry

$$q_{ij} = \langle f_j, f_i \rangle_{\mathcal{K}(\Theta)} \ ,$$

and finally, let

$$Y = [I_p \ \ -S] \quad \text{and} \quad G = \Theta_{11} - S\Theta_{21} \ .$$

Then it follows readily from the decomposition

$$\frac{Y(\lambda)JY(\omega)^*}{\rho_\omega(\lambda)} = Y(\lambda)\frac{J - \Theta(\lambda)J\Theta(\omega)^*}{\rho_\omega(\lambda)}Y(\omega)^* + Y(\lambda)\frac{\Theta(\lambda)J\Theta(\omega)^*}{\rho_\omega(\lambda)}X(\omega)^*$$

$$= Y(\lambda)F(\lambda)Q^{-1}F(\omega)^*Y(\omega)^* + G(\lambda)\frac{I_p - S_o(\lambda)S_o(\omega)^*}{\rho_\omega(\lambda)}G(\omega)^*$$

that the difference between the kernel on the left and the first kernel on the right is a
positive kernel. Therefore, for any set of points $\alpha_1,\ldots,\alpha_t$ in the domain of analyticity of
S in Ω_+ and any set of vectors $\xi_1,\ldots,\xi_t$ in $\mathbb{C}^k$, the $t \times t$ matrices

$$P_1 = \left[\frac{\xi_i^*Y(\alpha_i)JY(\alpha_j)^*\xi_j}{\rho_{\alpha_j}(\alpha_i)}\right] \quad\text{and}\quad P_2 = \left[\frac{\xi_i^*Y(\alpha_i)F(\alpha_i)Q^{-1}F(\alpha_j)^*Y(\alpha_j)^*\xi_j}{\rho_{\alpha_j}(\alpha_i)}\right] ,$$

$i,j = 1,\ldots,t$, are ordered: $P_1 \geq P_2$. Thus, by the minimax characterization of the
eigenvalues of a Hermitian matrix,

$$\lambda_j(P_1) \geq \lambda_j(P_2) , \qquad j = 1,\ldots,t ,$$

in which λ_j denotes the j'th eigenvalue of the indicated matrix, indexed in increasing
size. In particular, $\lambda_{\nu+1}(P_2) \geq 0$ and hence the kernel based on S has at most ν negative
squares.

On the other hand, since X is $(\Omega_+, J, \rho)_\nu$ admissible, there exists a set of
points $\beta_1,\ldots,\beta_r$ in Ω_+ and vectors $\eta_1,\ldots,\eta_r$ in $\mathbb{C}^k$ such that the $r \times r$ matrix with ij
entry equal to

$$\frac{\eta_i^*X(\beta_i)JX(\beta_j)^*\eta_j}{\rho_{\beta_j}(\beta_i)} = \eta_i^*C(\beta_i)\frac{I_p - S(\beta_i)S(\beta_j)^*}{\rho_{\beta_j}(\beta_i)}C(\beta_j)^*\eta_j$$

has exactly ν negative eigenvalues. This shows that the kernel based on S has at least ν
negative eigenvalues, providing that the exhibited equality is meaningful, i.e., providing
that the points $\beta_1,\ldots,\beta_r$ lie in the domain of analyticity of S. But if this is not already
the case, it can be achieved by arbitrarily small perturbations of the points $\beta_1,\ldots,\beta_r$
because S has at most countably many isolated poles in Ω_+. This can be accomplished
without decreasing the number of negative eigenvalues of the matrix on the left of the
last equality because the matrix will only change a little since X is analytic in Ω_+, and
therefore its eigenvalues will also only change a little. In particular, negative eigenvalues
will stay negative, positive eigenvalues will stay positive, but zero eigenvalues could go
either way. This can be verified by Rouché's theorem, or by easy estimates; see e.g.,
Corollary 12.2 of Bhatia [Bh] for the latter. ∎

5. $\mathcal{H}_\rho(S)$ SPACES

In this section we shall first obtain another characterization of the space $\bar{\mathcal{R}}_\Gamma$
endowed with the indefinite inner product (3.5) in the special case that $\Gamma = M_XJM_X^*$ is
positive semidefinite. We shall then specialize these results to

$$X = [I_p \quad -S] \tag{5.1}$$

and

$$J = J_{pq} = \begin{bmatrix} I_p & 0 \\ 0 & -I_q \end{bmatrix} , \qquad (5.2)$$

where S is a $p \times q$ matrix valued function which is analytic in Ω_+ such that the multiplication operator M_S from H_ρ^q to H_ρ^p is contractive. The resulting space $\mathcal{C}([I_p \quad - S])$ will then be designated by the symbol $\mathcal{H}_\rho(S)$.

THEOREM 5.1. *If X is a $k \times m$ matrix valued function which is analytic in Ω_+ such that the multiplication operator M_X from H_ρ^m to H_ρ^k is bounded and if*

$$\Gamma = M_X J M_X^* \geq 0 ,$$

then

(1) $\mathcal{C}(X) = \operatorname{ran} \Gamma^{\frac{1}{2}}$ with norm

$$\|\Gamma^{\frac{1}{2}} g\|_\Gamma = \|(I - P)g\|_{H_\rho} ,$$

where P denotes the orthogonal projection of H_ρ^k onto the kernel of Γ.

(2) $\operatorname{ran} \Gamma$ is dense in $\operatorname{ran} \Gamma^{\frac{1}{2}}$ and

$$\langle \Gamma g, \Gamma h \rangle_\Gamma = \langle \Gamma g, h \rangle_{H_\rho}$$

for every choice of g and h in H_ρ^k.

(3) $\mathcal{C}(X)$ is the reproducing kernel Hilbert space with reproducing kernel given by (3.1).

(4) X is (Ω_+, ρ, J) admissible.

(5) $\mathcal{C}(X) = \mathcal{B}(X)$.

PROOF. Since $\ker \Gamma = \ker \Gamma^{\frac{1}{2}}$, it is readily checked that $\| \; \|_\Gamma$, as defined in (1), is indeed a norm on $\operatorname{ran} \Gamma^{\frac{1}{2}}$. Moreover, if $\Gamma^{\frac{1}{2}} f_n$, $n = 1, 2, \ldots$, is a Cauchy sequence in $\operatorname{ran} \Gamma^{\frac{1}{2}}$, then $(I - P)f_n$ is a Cauchy sequence in the Hilbert space H_ρ^k, and hence tends to a limit g in H_ρ^k as $n \uparrow \infty$. Therefore, since $I - P$ is an orthogonal projector, it follows by standard arguments that

$$g = \lim_{n \uparrow \infty} (I - P)f_n = \lim_{n \uparrow \infty} (I - P)^2 f_n = (I - P)g$$

and hence that

$$\|\Gamma^{\frac{1}{2}} f_n - \Gamma^{\frac{1}{2}} g\|_\Gamma = \|(I - P)(f_n - g)\|_{H_\rho}$$

$$= \|(I - P)f_n - g\|_{H_\rho} .$$

Thus $\operatorname{ran} \Gamma^{\frac{1}{2}}$ is closed with respect to the indicated norm; it is in fact a Hilbert space with respect to the inner product

$$\langle \Gamma^{\frac{1}{2}} f, \Gamma^{\frac{1}{2}} g \rangle_\Gamma = \langle (I - P)f, g \rangle_{H_\rho} .$$

For the particular choice $g = v/\rho_\omega$ with $v \in \mathbb{C}^k$ and $\omega \in \Omega_+$, the identity

$$\langle \Gamma^{\frac{1}{2}} f, \Gamma^{\frac{1}{2}} g \rangle_\Gamma = \langle (I - P)f, \Gamma^{\frac{1}{2}} g \rangle_{H_\rho}$$

$$= \langle \Gamma^{\frac{1}{2}} f, g \rangle_{H_\rho}$$

$$= v^* (\Gamma^{\frac{1}{2}} f)(\omega) \, ,$$

serves to exhibit

$$\Gamma g = \frac{X J X(\omega)^*}{\rho_\omega} v = \Lambda_\omega v$$

as the reproducing kernel for ran $\Gamma^{\frac{1}{2}}$. This completes the proof of (1), since there is only one such space. (2) is immediate from (1) and the fact that ker $\Gamma^{\frac{1}{2}}$ = ker Γ; (3), (4) and (5) are covered by Theorems 3.3, 3.5 and 3.4, respectively. $\blacksquare$

For ease of future reference we summarize the main implications of the preceding theorem directly in the language of the $p \times q$ matrix valued function S introduced at the beginning of this section.

THEOREM 5.2. *If S is a $p \times q$ matrix valued function which is analytic in Ω_+ such that the multiplication operator M_S from H_ρ^q to H_ρ^p is contractive and if X and J are given by (5.1) and (5.2), respectively, then:*

(1) $\Gamma = M_X J M_X^* = I - M_S M_S^*$

(2) $\mathcal{H}_\rho(S) = \mathrm{ran}\ \Gamma^{\frac{1}{2}}$ *with*

$$\|\Gamma^{\frac{1}{2}} f\|_{\mathcal{H}_\rho(S)} = \|(I - P)f\|_{H_\rho} \, ,$$

where P designates the orthogonal projection of H_ρ^p onto ker $\Gamma^{\frac{1}{2}}$.

(3) ran Γ *is dense in* ran $\Gamma^{\frac{1}{2}}$ *and*

$$\langle \Gamma g, \Gamma h \rangle_{\mathcal{H}_\rho(S)} = \langle \Gamma g, h \rangle_{H_\rho}$$

for every choice of g and h in H_ρ^p.

(4) $\mathcal{H}_\rho(S)$ *is a reproducing kernel Hilbert space with reproducing kernel*

$$\Lambda_\omega(\lambda) = \frac{I_p - S(\lambda)S(\omega)^*}{\rho_\omega(\lambda)} \, . \tag{5.3}$$

The next theorem is the analogue in the present setting of general $\rho \in \mathcal{D}_\Omega$ of a theorem which originates with de Branges and Rovnyak [dBR1] for $\rho_\omega(\lambda) = 1 - \lambda\omega^*$.

THEOREM 5.3. *Let S be a $p \times q$ matrix valued function which is analytic in Ω_+ such that the multiplication operator M_S from H_ρ^q to H_ρ^p is contractive and for $f \in H_\rho^p$ let*

$$\kappa(f) = \sup\{\|f + M_S g\|_{H_\rho}^2 - \|g\|_{H_\rho}^2 : g \in H_\rho^q\} \, .$$

Then

$$\mathcal{H}_\rho(S) = \{f \in H_\rho^p : \ \kappa(f) < \infty\}$$

and

$$\|f\|^2_{\mathcal{H}_\rho(S)} = \kappa(f) \ . \tag{5.4}$$

PROOF. Let X and J be given by (5.1) and (5.2), respectively. Then clearly Theorem 5.1 is applicable since

$$\Gamma = M_X J M_X^* = I - M_S M_S^* \geq 0 \ .$$

Moreover, since $\Gamma \leq I$, it follows that $\Gamma^{\frac{1}{2}}$ is a contraction and hence, by Theorem 4.1 of Fillmore and Williams [FW], that $f \in \text{ran } \Gamma^{\frac{1}{2}}$ if and only if

$$\sup\{\|f + (I - \Gamma^{\frac{1}{2}}\Gamma^{\frac{1}{2}})^{\frac{1}{2}}g\|^2_{H_\rho} - \|g\|^2_{H_\rho} : \ g \in H_\rho^p\} < \infty \ ,$$

or equivalently, if and only if

$$\sup\{\|f + (M_S M_S^*)^{\frac{1}{2}}g\|^2_{H_\rho} - \|g\|^2_{H_\rho} : \ g \in H_\rho^p\} < \infty \ . \tag{5.5}$$

Therefore, since

$$H_\rho^p = \overline{\text{ran}(M_S M_S^*)^{\frac{1}{2}}} + \ker(M_S M_S^*)^{\frac{1}{2}} \ ,$$

any $g \in H_\rho^p$ can be approximated arbitrarily well by the elements of the form

$$(M_S M_S^*)^{\frac{1}{2}}u + v$$

with $u \in H_\rho^p$ and $v \in \ker(M_S M_S^*)^{\frac{1}{2}}$. Thus the sup in (5.5) can be reexpressed as

$$\sup\{\|f + (M_S M_S^*)^{\frac{1}{2}}(M_S M_S^*)^{\frac{1}{2}}u\|^2_{H_\rho} - \|(M_S M_S^*)^{\frac{1}{2}}u + v\|^2_{H_\rho} :$$

$$u \in H_\rho^p \text{ and } v \in \ker(M_S M_S^*)^{\frac{1}{2}}\}$$

$$= \sup\{\|f + M_S M_S^* u\|^2_{H_\rho} - \|(M_S M_S^*)^{\frac{1}{2}}u\|^2_{H_\rho} : \ u \in H_\rho^p\}$$

$$= \sup\{\|f + M_S M_S^* u\|^2_{H_\rho} - \|M_S^* u\|^2_{H_\rho} : \ u \in H_\rho^p\}$$

$$= \sup\{\|f + M_S(M_S^* u + w)\|^2_{H_\rho} - \|M_S^* u + w\|^2_{H_\rho} : \ u \in H_\rho^p \ \text{ and } \ w \in \ker M_S\}$$

$$= \sup\{\|f + M_S h\|^2_{H_\rho} - \|h\|^2_{H_\rho} : \ h \in H_\rho^q\} \ .$$

This proves that $f \in \text{ran } \Gamma^{\frac{1}{2}} = \mathcal{H}_\rho(S)$ if and only if $\kappa(f) < \infty$.

Next, if $f = \Gamma^{\frac{1}{2}} h$ for some $h \in H_\rho^p$, then (as is also shown in [FW])

$$\kappa(f) = \inf\{\|h\|_{H_\rho}^2 : (I - M_S M_S^*)^{\frac{1}{2}} h = f\}$$

$$= \|(I - P)h\|_{H_\rho}^2 ,$$

where P denotes the orthogonal projection of H_ρ^p onto the kernel of Γ. This serves to establish (5.4), thanks to item 2 of Theorem 5.2. ∎

THEOREM 5.4. *Let S be a $p \times q$ matrix valued function which is analytic in Ω_+ such that $\|M_S\| \leq 1$ and let J be given by (5.2). Then*

(1) $[I \quad - M_S]f \in \mathcal{H}_\rho(S)$, *and*

(2) $\|[I \quad - M_S]f\|_{\mathcal{H}_\rho(S)} \leq \langle Jf, f\rangle_{H_\rho}$

for every choice of

$$f = \begin{bmatrix} g \\ h \end{bmatrix} \in H_\rho^m ,$$

with components $g \in H_\rho^p$ and $h \in H_\rho^q$, if and only if

$$h = M_S^* g . \tag{5.6}$$

PROOF. Suppose first that (1) and (2) hold. Then, by Theorem 5.3,

$$\|[I \quad - M_S]f\|_{\mathcal{H}_\rho(S)}^2 = \|g - M_S h\|_{\mathcal{H}_\rho(S)}^2$$

$$= \sup\{\|g - M_S h + M_S u\|_{H_\rho}^2 - \|u\|_{H_\rho}^2 : u \in H_\rho^q\}$$

$$= \sup\{\|g + M_S v\|_{H_\rho}^2 - \|h + v\|_{H_\rho}^2 : v \in H_\rho^q\} .$$

Therefore, by the prevailing assumptions,

$$\|g\|_{H_\rho}^2 + 2Re\langle g, M_S v\rangle_{H_\rho} + \|M_S v\|_{H_\rho}^2 - \|h\|_{H_\rho}^2 - 2Re\langle h, v\rangle_{H_\rho} - \|v\|_{H_\rho}^2$$

$$=\leq \|g\|_{H_\rho}^2 - \|h\|_{H_\rho}^2$$

for every $v \in H_\rho^q$. But this in turn implies that

$$\|M_S v\|_{H_\rho}^2 + 2Re\langle M_S^* g - h, v\rangle_{H_\rho} - \|v\|_{H_\rho}^2 \leq 0$$

for every $v \in H_\rho^q$ and hence in particular, upon choosing

$$v = \varepsilon(M_S^* g - h) \quad \text{with} \quad \varepsilon > 0 ,$$

that

$$\varepsilon^2 \|M_S(M_S^* g - h)\|_{H_\rho}^2 + 2\varepsilon\|M_S^* g - h\|_{H_\rho}^2 - \varepsilon^2\|M_S^* g - h\|_{H_\rho}^2 \leq 0$$

for every $\varepsilon > 0$. The desired conclusion (5.6) now follows easily upon first dividing through by ε and then letting $\varepsilon \downarrow 0$.

Next, to obtain the converse, suppose that (5.6) holds. Then, by Theorem 5.2,

$$[I - M_S]f = (I - M_S M_S^*)g = \Gamma g$$

belongs to $\mathcal{H}_\rho(S)$ and

$$\|[I - M_S]f\|^2_{\mathcal{H}_\rho(S)} = \langle \Gamma g, \Gamma g \rangle_{\mathcal{H}_\rho(S)}$$
$$= \langle \Gamma g, g \rangle_{H_\rho}$$
$$= \|g\|^2_{H_\rho} - \|M_S^* g\|^2_{H_\rho}$$
$$= \langle Jf, f \rangle_{H_\rho} .$$

Thus (1) and (2) hold and the proof is complete. ∎

We remark that the inequality in (2) of the last theorem can be replaced by equality.

Finally, to complete this section we observe that if S is a $p \times q$ matrix valued function which is analytic in Ω_+, then the multiplication operator M_S is contractive if and only if the kernel (5.3) is positive, i.e., if and only if it has zero negative squares.

THEOREM 5.5. *Let S be a $p \times q$ matrix valued function which is analytic on Ω_+. Then the kernel*

$$\Lambda_\omega(\lambda) = \frac{I_p - S(\lambda)S(\omega)^*}{\rho_\omega(\lambda)}$$

is positive on Ω_+ if and only if $\|M_S\| \leq 1$.

PROOF. Suppose first that $\|M_S\| \leq 1$ and let $f = \Sigma_{j=1}^n \xi_j / \rho_{\omega_j}$ for any choice of points $\omega_1, \ldots, \omega_n \in \Omega_+$ and vectors $\xi_1, \ldots, \xi_n \in \mathbb{C}^p$. Then it is readily checked that

$$\sum_{i,j=1}^n \xi_j^* \Lambda_{\omega_i}(\omega_j) \xi_i = \|f\|^2_{H_\rho} - \|M_S^* f\|^2_{H_\rho} \geq 0$$

which establishes the positivity of the kernel.

Next, to go the other way, we define a linear operator T on finite sums of the form f given above by the rule

$$T \frac{\xi}{\rho_\omega} = S(\omega)^* \frac{\xi}{\rho_\omega} .$$

By the presumed positivity of the kernel $\Lambda_\omega(\lambda)$, T is well defined and contractive on finite sums of this form and hence, since such sums are dense in H_ρ, T can be extended by limits in the usual way to a contractive operator (which we continue to call T) on all of H_ρ. Finally the evaluation

$$\xi^*(T^*g)(\omega) = \langle T^*g, \frac{\xi}{\rho_\omega} \rangle_{H_\rho}$$

$$= \langle g, T\frac{\xi}{\rho_\omega} \rangle_{H_\rho}$$

$$= \langle g, S(\omega)^* \frac{\xi}{\rho_\omega} \rangle_{H_\rho}$$

$$= \xi^* S(\omega)g(\omega) \, ,$$

which is valid for every choice of $\xi \in \mathbb{C}^p$, $\omega \in \Omega_+$ and $g \in H_\rho^q$ serves to identify T^* with the multiplication operator M_S. Therefore

$$\|M_S\| = \|T^*\| \le 1 \, ,$$

as claimed.   ∎

COROLLARY. *If S is a $p \times q$ matrix valued function which is analytic on Ω_+ such that $\|M_S\| \le 1$, then*

$$I_p - S(\omega)S(\omega)^* \ge 0 \tag{5.7}$$

for every choice of $\omega \in \Omega_+$.

It is important to bear in mind that even through (5.7) implies that $\|M_S\| \le 1$ for $\rho_\omega(\lambda) = 1 - \lambda\omega^*$ and $\rho_\omega(\lambda) = -2\pi i(\lambda - \omega^*)$, the last corollary does not have a valid converse for every choice of $\rho \in \mathcal{D}_\Omega$, as we now illustrate by a pair of examples.

EXAMPLE 1. *Let $a(\lambda) = 1$ and $b(\lambda) = \lambda^2$ so that $\Omega_+ = \mathbb{D}$ and let S be any scalar contractive analytic function from $\mathbb{D}$ into $\mathbb{D}$ such that $S(\frac{1}{2}) = -S(-\frac{1}{2}) \ne 0$. Then M_S is not a contraction.*

DISCUSSION. It follows from the standard Nevanlinna-Pick theory (see e.g. [D2]) that there exists an S of the desired type with $S(\frac{1}{2}) = c$ if and only if the 2×2 matrix

$$\left[\frac{1 - S(\omega_i)S(\omega_j)^*}{\rho_{\omega_j}(\omega_i)} \right] \, , \qquad i,j = 1,2 \, , \tag{5.8}$$

with $\omega_1 = -\omega_2 = \frac{1}{2}$, $S(\omega_1) = -S(\omega_2) = c$, and $\rho_\omega(\lambda) = 1 - \lambda\omega^*$ is positive semidefinite. The matrix of interest:

$$\begin{bmatrix} \frac{1-|c|^2}{1-1/4} & \frac{1+|c|^2}{1+1/4} \\[2mm] \frac{1+|c|^2}{1+1/4} & \frac{1-|c|^2}{1-1/4} \end{bmatrix} \, ,$$

is readily seen to be positive semidefinite if and only if $|c| \le \frac{1}{2}$.

On the other hand, if $\|M_S\| \leq 1$, then the matrix (5.8) must be positive semidefinite for the same choice of points and assigned values as before but with $\rho_\omega(\lambda) = 1 - \lambda^2 \omega^{*2}$. But this matrix:

$$\begin{bmatrix} \dfrac{1-|c|^2}{1-1/16} & \dfrac{1+|c|^2}{1-1/16} \\[2ex] \dfrac{1+|c|^2}{1-1/16} & \dfrac{1-|c|^2}{1-1/16} \end{bmatrix} \, ,$$

is not positive semidefinite for any $c \neq 0$ as is readily seen by computing its determinant.

EXAMPLE 2. *Let* $\rho_\omega(\lambda) = -2\pi i(\lambda - \omega^*)(1 - \lambda\omega^*)$ *with* $a(\lambda)$ *and* $b(\lambda)$ *as in (2.2), and let* $S(\omega) = c$ *for* $\omega \in \mathbb{D} \cap \mathbb{C}_+$ *and* $S(\omega) = -c$ *for* $\omega \in \mathbb{E} \cap \mathbb{C}_-$, *where* $|c| \leq 1$. *Then* S *is a contractive analytic function in* Ω_+ *but* M_S *is not a contractive mapping of* H_ρ *into itself for* $c \neq 0$.

DISCUSSION. If $\|M_S\| \leq 1$, then the matrix (5.8) will be positive semidefinite for any pair of points ω_1, ω_2 in Ω_+. For the particular choice $\omega_1 = i/\alpha$, $\omega_2 = -i\alpha$ with $\alpha > 1$, the matrix of interest is equal to

$$\begin{bmatrix} \dfrac{1-|c|^2}{4\pi(\alpha^2-1)/\alpha^3} & \dfrac{1+|c|^2}{4\pi(1-\alpha^2)/\alpha} \\[2ex] \dfrac{1+|c|^2}{4\pi(1-\alpha^2)/\alpha} & \dfrac{1-|c|^2}{4\pi(\alpha^2-1)\alpha} \end{bmatrix}$$

which is not positive semidefinite if $c \neq 0$.

6. LINEAR FRACTIONAL TRANSFORMATIONS

In this section we shall recall a number of well-known properties of the linear fractional transformation

$$T_\Theta[S_o] = (\Theta_{11}S_o + \Theta_{12})(\Theta_{21}S_o + \Theta_{22})^{-1} \tag{6.1}$$

for

$$\Theta = \begin{bmatrix} \Theta_{11} & \Theta_{12} \\ \Theta_{21} & \Theta_{22} \end{bmatrix}$$

which is analytic and $J = J_{pq}$ contractive in Ω_+. In particular it is well known that the indicated inverse exists if (the $p \times q$ matrix valued function) S_o is analytic and contractive in Ω_+ and that in this case $T_\Theta[S_o]$ is also analytic and contractive in Ω_+.

THEOREM 6.1. *If* Θ *is given by (2.14) and the matrices* A *and* B *in (2.10) are such that* A *is invertible and the spectral radius of* BA^{-1} *is less than one, then* Θ *is analytic in* Ω_+ *and the multiplication operator* M_Θ *is bounded on* H_ρ^m.

PROOF. By (2.14) and (2.10),

$$\Theta(\lambda) = I_m - a(\mu)^*\{1 - s(\lambda)s(\mu)^*\}VA^{-1}\{I_m - s(\lambda)BA^{-1}\}^{-1}P^{-1}F(\mu)^*J \, ,$$

where, as before, $s(\lambda) = b(\lambda)/a(\lambda)$. In particular

$$\{I_m - s(\lambda)BA^{-1}\}^{-1} = \sum_{n=0}^{\infty} s(\lambda)^n (BA^{-1})^n$$

and since by assumption

$$\sum_{n=0}^{\infty} \|(BA^{-1})^n\| < \infty$$

it is easily seen that $\Theta(\lambda)$ admits a representation of the form

$$\Theta(\lambda) = \sum_{n=0}^{\infty} s(\lambda)^n \Theta_n$$

with $m \times m$ matrix coefficients Θ_n which are summable:

$$\sum_{n=0}^{\infty} \|\Theta_n\| < \infty .$$

It now follows readily by standard estimates that if

$$f(\lambda) = \frac{1}{a(\lambda)} \sum_{j=0}^{\infty} s(\lambda)^j u_j$$

belongs to H_ρ^m, i.e., if the $m \times 1$ vector coefficients u_j are norm square summable, then

$$\|M_\Theta f\|_{H_\rho}^2 = \sum_{n=0}^{\infty} \left\| \sum_{j=0}^{n} \Theta_{n-j} u_j \right\|^2$$

$$\leq \sum_{n=0}^{\infty} \left\{ \sum_{j=0}^{n} \|\Theta_{n-j}\| \right\} \left\{ \sum_{k=0}^{n} \|\Theta_{n-k}\| \, \|u_k\|^2 \right\}$$

$$\leq \sum_{k=0}^{\infty} \|u_k\|^2 \left\{ \sum_{n=0}^{\infty} \|\Theta_n\| \right\}^2 .$$

But this proves that

$$\|M_\Theta\| \leq \sum_{n=0}^{\infty} \|\Theta_n\| < \infty ;$$

as claimed. ∎

THEOREM 6.2. *Let S_o be a $p \times q$ matrix valued function which is analytic in Ω_+ such that the multiplication operator M_{S_o} is contractive, let $\Theta \in \mathcal{P}_J^0(\Omega_+)$ be analytic in Ω_+, and let $S = T_\Theta[S_o]$. Then $\|M_S\| \leq 1$.*

PROOF. If $S = T_\Theta[S_o]$, then it is readily checked that

$$[I_p \quad - S]\Theta = E[I_p \quad - S_o] ,$$

where

$$E = \Theta_{11} - S\Theta_{21} .$$

Now let

$$X = [I_p \quad - S] .$$

Then

$$\frac{I_p - S(\alpha)S(\beta)^*}{\rho_\beta(\alpha)} = \frac{X(\alpha)JX(\beta)^*}{\rho_\beta(\alpha)}$$

$$= X(\alpha)\frac{J - \Theta(\alpha)J\Theta(\beta)^**}{\rho_\alpha(\beta)}X(\beta)^* + \frac{X(\alpha)\Theta(\alpha)J\Theta(\beta)^*X(\beta)^*}{\rho_\beta(\alpha)}$$

$$= X(\alpha)\frac{J - \Theta(\alpha)J\Theta(\beta)^**}{\rho_\alpha(\beta)}X(\beta)^* + E(\alpha)\left\{\frac{I_p - S_o(\alpha)S_o(\beta)^*}{\rho_\beta(\alpha)}\right\}E(\beta)^* .$$

$$(6.2)$$

By Theorem 5.5, $\|M_S\| \leq 1$ if and only if

$$\sum_{i,j=1}^{n} \xi_i^* \left\{\frac{I_p - S(\alpha_i)S(\alpha_j)^*}{\rho_{\alpha_j}(\alpha_i)}\right\} \xi_j \geq 0 \tag{6.3}$$

for every choice of points $\alpha_1, \ldots, \alpha_n$ in Ω_+ and vectors $\xi_1, \ldots, \xi_n$ in $\mathbb{C}^p$. Since $\|M_{S_o}\| \leq 1$ by assumption, the inequality (6.3) holds with S_o in place of S. Therefore, by (6.2) it holds for S also, since $\Theta \in \mathcal{P}_J^0(\Omega_+)$. Thus $\|M_S\| \leq 1$.   ∎

Now as

$$\Theta = \begin{bmatrix} \Theta_{11} & \Theta_{12} \\ \Theta_{21} & \Theta_{22} \end{bmatrix}$$

is $J = J_{pq}$ contractive in Ω_+, it can be expressed in the form

$$\Theta = \begin{bmatrix} I_p & \sigma_1 \\ 0 & I_q \end{bmatrix} \begin{bmatrix} \varepsilon_1 & 0 \\ 0 & \varepsilon_2 \end{bmatrix} \begin{bmatrix} I_p & 0 \\ \sigma_2 & I_q \end{bmatrix} , \tag{6.4}$$

where

$\varepsilon_2 = \Theta_{22}$ is invertible in Ω_+,

$\sigma_1 = \Theta_{12}\Theta_{22}^{-1}, \quad \sigma_2 = \Theta_{22}^{-1}\Theta_{21} ,$

$\varepsilon_1 = \Theta_{11} - \Theta_{12}\Theta_{22}^{-1}\Theta_{21}$

and ε_1, σ_1, σ_2 and ε_2^{-1} are all contractive in Ω_+; see e.g., p.15 of [D2].

THEOREM 6.3. *If Θ is as in Theorem 6.2, then the multiplication operators* M_{σ_2}, M_{σ_1} *and* $M_{\varepsilon_2^{-1}}$ *are all contractive. Moreover, M_{σ_2} is a strict contraction.*

PROOF. Since

$$K_\omega(\lambda) = \frac{J - \Theta(\lambda)J\Theta(\omega)^*}{\rho_\omega(\lambda)}$$

has zero negative squares in Ω_+, the same holds true for its 22 block:

$$\sum_{i,j=1}^{k} \eta_i^* \left\{ \frac{-I_q - \Theta_{21}(\alpha_i)\Theta_{21}(\alpha_j)^* + \Theta_{22}(\alpha_i)\Theta_{22}(\alpha_j)^*}{\rho_{\alpha_j}(\alpha_i)} \right\} \eta_j \geq 0 \qquad (6.5)$$

for every choice of $\alpha_1,\ldots,\alpha_k$ in Ω_+ and $\eta_1,\ldots,\eta_k$ in $\mathbb{C}^q$. Thus clearly

$$\sum_{i,j=1}^{k} \eta_i^* \Theta_{22}(\alpha_i) \left\{ \frac{I_q - \sigma_2(\alpha_i)\sigma_2(\alpha_j)^*}{\rho_{\alpha_j}(\alpha_i)} \right\} \Theta_{22}(\alpha_j)^* \eta_j \geq 0$$

and

$$\sum_{i,j=1}^{k} \eta_i^* \Theta_{22}(\alpha_i) \left\{ \frac{I_q - \varepsilon_2(\alpha_i)^{-1}\varepsilon_2(\alpha_j)^{-*}}{\rho_{\alpha_j}(\alpha_i)} \right\} \Theta_{22}(\alpha_j)^* \eta_j \geq 0$$

for every such choice of $\alpha_1,\ldots,\alpha_k$ and $\eta_1,\ldots,\eta_k$. Thus, by Theorem 5.5, the multiplication operators $M_{\varepsilon_2^{-1}}$ and M_{σ_2} are both contractive.

The proof that M_{σ_1} is contractive is immediate from Theorem 6.2 and the observation that

$$\sigma_1 = T_\Theta[0] \ .$$

To check that M_{σ_2} is strictly contractive, it suffices to note that (6.5) implies that

$$I - M_{\sigma_2} M_{\sigma_2}^* \geq M_{\varepsilon_2}^{-1} M_{\varepsilon_2}^{-*} \ . \qquad \blacksquare$$

7. ONE-SIDED INTERPOLATION

In this section we study the following general one-sided interpolation problem for a fixed $\rho \in \mathcal{D}_\Omega$. The data for this problem consists of a set of (not necessarily distinct) points

$$\omega_1,\ldots,\omega_n \quad \text{in} \quad \Omega_+$$

and two sets of associated vectors

$$\xi_1,\ldots,\xi_n \quad \text{in} \quad \mathbb{C}^p$$

$$\eta_{11},\ldots,\eta_{1r_1}; \eta_{21},\ldots,\eta_{2r_2}; \ldots; \eta_{n1},\ldots,\eta_{nr_n} \quad \text{in} \quad \mathbb{C}^q$$

such that

$$\xi_1 \varphi_{\omega_1,0},\ldots,\xi_1 \varphi_{\omega_1,r_1-1}; \ldots; \xi_n \varphi_{\omega_n,0},\ldots,\xi_n \varphi_{\omega_n,r_n-1}$$

are linearly independent. The problem is to find necessary and sufficient conditions on the data for the existence of a $p \times q$ matrix valued function S such that:

(1) S is analytic in Ω_+ (which is actually automatic from (3)),

(2) $\qquad \xi_j^* \dfrac{S^{(t-1)}(\omega_j)}{(t-1)!} = \eta_{jt}^* \ , \quad t = 1,\ldots,r_j \ ; \quad j = 1,\ldots,n \ ,$

(3)   The operator M_S of multiplication by S is a contractive map of H_ρ^q into H_ρ^p.

It is also desirable to give a description of the set of all solutions to this problem when these conditions are met.

We shall solve this interpolation problem by adapting the strategy of Section 7 of [D3] to the present setting. We begin by introducing the auxiliary notation

$$v_{j1} = \begin{bmatrix} \xi_j \\ \eta_{j1} \end{bmatrix} \quad \text{and} \quad v_{jt} = \begin{bmatrix} 0 \\ \eta_{jt} \end{bmatrix} , \qquad t = 2, \ldots, r_j \; ; \quad j = 1, \ldots, n$$

$$V_j = [v_{j1} \cdots v_{jr_j}] , \qquad j = 1, \ldots, n ,$$

$$F_j = V_j \Phi_{\omega_j, r_j} , \qquad j = 1, \ldots, n .$$

$$[f_1 \cdots f_\nu] = [F_1 \cdots F_n] , \qquad \nu = r_1 + \cdots + r_n ,$$

and

$$f_t = \begin{bmatrix} g_t \\ h_t \end{bmatrix} , \qquad t = 1, \ldots, \nu ,$$

where $g_t \in H_\rho^p$ and $h_t \in H_\rho^q$.

LEMMA 7.1. *Let S be a $p \times q$ matrix valued function which is analytic in Ω_+ such that $\|M_S\| \leq 1$. Then S is a solution of the stated interpolation problem if and only if*

$$M_S^* g_t = h_t \tag{7.1}$$

for $t = 1, \ldots, \nu$.

PROOF.   In order to keep the notation simple we shall deal with the first block, i.e., the columns in F_1, only. The rest goes through in just the same way.

By definition,

$$f_t = \sum_{j=1}^{t} \varphi_{\omega_1, t-j} v_{1j} , \qquad t = 1, \ldots, r_1 .$$

with components

$$g_t = \varphi_{\omega_1, t-1} \xi_1$$

and

$$h_t = \sum_{j=1}^{t} \varphi_{\omega_1, t-j} \eta_{1j} .$$

Thus, by the evaluation in Corollary 2 to Lemma 3.2,

$$M_S^* g_t - h_t = \sum_{j=1}^{t} \left\{ \frac{S^{(j-1)}(\omega_1)^* \xi_1}{(j-1)!} - \eta_{1j} \right\} \varphi_{\omega_1, t-j} .$$

The rest is selfevident since the functions $\varphi_{\omega_1, 0}, \ldots, \varphi_{\omega_1, t-1}$ are assumed to be linearly independent in the formulation of the problem.   ∎

THEOREM 7.1. *The one-sided interpolation problem which was formulated at the beginning of this section is solvable if and only if the $\nu \times \nu$ matrix P with ij entry*

$$p_{ij} = \langle J f_j, f_i \rangle_{H_\rho} \tag{7.2}$$

is positive semidefinite.

PROOF. Suppose first that the interpolation problem is solvable, and let S be a solution. Then, by the preceding lemma, the components g_t and h_t of f_t are related by formula (7.1). Thus if

$$f = \sum_{j=1}^{\nu} c_j f_j \quad \text{and} \quad g = \sum_{j=1}^{\nu} c_j g_j$$

for some choice of constants $c_1, \ldots, c_\nu$, then

$$f = \begin{bmatrix} g \\ M_S^* g \end{bmatrix} .$$

Therefore

$$\sum_{i,j=1}^{\nu} c_i^* p_{ij} c_j = \langle J f, f \rangle_{H_\rho}$$

$$= \|g\|_{H_\rho}^2 - \|M_S^* g\|_{H_\rho}^2$$

$$\geq 0 ,$$

since $\|M_S^*\| = \|M_S\| \leq 1$.

Now suppose conversely that $P > 0$ and let

$$\mathcal{M} = \text{span}\{f_1, \ldots, f_\nu\}$$

endowed with the J inner product

$$\langle f_j, f_i \rangle_{\mathcal{M}} = \langle J f_j, f_i \rangle_{H_\rho} = p_{ij} .$$

Then, with the help of Theorem 2.3, it is readily checked that conditions (1) and (2) of Theorem 2.1 are met. Now let

$$A = \text{diag}\{A_{\omega_1}, \ldots, A_{\omega_n}\} \quad \text{and} \quad B = \text{diag}\{B_{\omega_1}, \ldots, B_{\omega_n}\}$$

be the block diagonal matrices with entries A_{ω_j} and B_{ω_j} of size $r_j \times r_j$ given by (2.6). Then

$$F(\lambda) = V\{a(\lambda)A - b(\lambda)B\}^{-1} \tag{7.3}$$

and, by Theorem 2.2,

$$A^* P A - B^* P B = V^* J V . \tag{7.4}$$

Therefore, by Theorem 2.1, $\mathcal{M}$ is a finite dimensional reproducing kernel Hilbert space with reproducing kernel

$$K_\omega(\lambda) = \frac{J - \Theta(\lambda)J\Theta(\omega)^*}{\rho_\omega(\lambda)}$$

where Θ is uniquely specified up to a right J unitary factor by (2.14). Moreover, since

$$\det\{a(\lambda)A - b(\lambda)B\} = \rho_{\omega_1}(\lambda)^{r_1} \cdots \rho_{\omega_n}(\lambda)^{r_n} ,$$

it follows easily from (2.14) that Θ is analytic in all of Ω except at the union of the zeros of the functions $\rho_{\omega_1}, \ldots, \rho_{\omega_n}$. In particular, Θ is analytic in Ω_+ and is both analytic and invertible at the special point $\mu \in \Omega_0$ which is used in the definition of Θ. Thus Θ is invertible in all of Ω except for an at most countable set of isolated points. Moreover, it is readily checked that Theorem 6.1 is applicable for the current choices of A and B, and hence that the multiplication operator M_Θ is a bounded mapping of H_ρ^m into itself.

Next, since

$$k_\alpha(\lambda) = \frac{I_m}{\rho_\alpha(\lambda)} , \qquad \alpha, \lambda \in \Omega_+ ,$$

is a reproducing kernel for H_ρ^m, it is easily seen that

$$v^* f_t(\alpha) = \langle Jf_t, \frac{J - \Theta J\Theta(\alpha)^* v}{\rho_\alpha} \rangle_{H_\rho}$$

$$= \langle f_t, \frac{v}{\rho_\alpha} \rangle_{H_\rho} - \langle Jf_t, \frac{\Theta J\Theta(\alpha)^* v}{\rho_\alpha} \rangle_{H_\rho}$$

$$= v^* f_t(\alpha) - \langle Jf_t, \frac{\Theta J\Theta(\alpha)^* v}{\rho_\alpha} \rangle_{H_\rho} ,$$

and hence that

$$v^* J\Theta(\alpha)J(M_\Theta^* Jf_t)(\alpha) = \langle M_\Theta^* Jf_t, \frac{J\Theta(\alpha)^* Jv}{\rho_\alpha} \rangle_{H_\rho}$$

$$= \langle Jf_t, \frac{\Theta J\Theta(\alpha)^* v}{\rho_\alpha} \rangle_{H_\rho}$$

$$= 0$$

for every choice of $\alpha \in \Omega_+$ and $v \in \mathbb{C}^m$. Therefore, since Θ is invertible except for at most a set of isolated points in Ω_+, it follows that (the analytic vector valued function)

$$M_\Theta^* Jf_t = 0 . \tag{7.5}$$

Now, by (6.4),

$$\Theta = \Psi_1 \Psi_2 ,$$

where

$$\Psi_1 = \begin{bmatrix} \varepsilon_1 & \sigma_1\varepsilon_2 \\ 0 & \varepsilon_2 \end{bmatrix} \quad \text{and} \quad \Psi_2 = \begin{bmatrix} I_p & 0 \\ \sigma_2 & I_q \end{bmatrix} .$$

By Theorems 6.1 and 6.3, M_{Ψ_1} and M_{Ψ_2} are both bounded multiplication operators on H_ρ^m. Thus it is meaningful to write

$$M_\Theta^* = M_{\Psi_2}^* M_{\Psi_1}^*$$

and therefore, since $M_{\Psi_2}^*$ is clearly invertible, it follows from (7.5) that

$$M_{\Psi_1}^* J f_t = 0 .$$

But this in turn reduces to the pair of constraints

$$M_{\varepsilon_1}^* g_t = 0 \tag{7.6}$$

and

$$M_{\varepsilon_2}^* \{ M_{\sigma_1}^* g_t - h_t \} = 0 .$$

But now as $M_{\varepsilon_2}^*$ is invertible (with inverse $(M_{\varepsilon_2^{-1}})^*$) the latter constraint implies that

$$M_{\sigma_1}^* g_t = h_t .$$

This exhibits σ_1 as a solution to the given interpolation problem, thanks to Lemma 7.1 and Theorem 6.3. This completes the proof of the existence of a solution to the stated interpolation problem when $P > 0$.

It remains to show that the interpolation problem is also solvable when $P \geq 0$, by passage to limit. To this end, let us first write

$$P = G - H$$

with entries corresponding to the decomposition

$$p_{ij} = \langle g_j, g_i \rangle_{H_\rho} - \langle h_j, h_i \rangle_{H_\rho} .$$

By assumption $G > 0$. Now, just as in [D2], let us replace the η_{ij} in the formulation of the problem by $\eta_{ij}(1 + \varepsilon)^{-\frac{1}{2}}$ with $\varepsilon > 0$. Then the new Pick matrix, corresponding to the perturbed data,

$$P_\varepsilon = G - \frac{1}{1 + \varepsilon} H$$

$$= \frac{\varepsilon}{1 + \varepsilon} G + \frac{1}{1 + \varepsilon} P$$

$$\geq \frac{\varepsilon}{1 + \varepsilon} G > 0$$

for every choice of $\varepsilon > 0$. Therefore, by the preceding analysis, there exists a $p \times q$ matrix valued analytic function S_ε with $\|M_{S_\varepsilon}\| \leq 1$ such that

$$M_{S_\varepsilon}^* g_j = (1 + \varepsilon)^{-\frac{1}{2}} h_j$$

for $j = 1, \ldots, \nu$. The proof is completed by letting $\varepsilon \downarrow 0$ and invoking the fact that the unit ball in the set of bounded operators from H_ρ^p to H_ρ^q is compact in the weak operator topology (see e.g., Lemma 2.3 on p.102 of [Be]). This means that there exists a contractive operator Q from H_ρ^p to H_ρ^q such that

$$\lim_{\varepsilon \downarrow 0} \langle M_{S_\varepsilon}^* f, g \rangle_{H_\rho} = \langle Qf, g \rangle_{H_\rho}$$

for every choice of $f \in H_\rho^p$ and $g \in H_\rho^q$ as $\varepsilon \downarrow 0$ through an appropriately chosen subsequence (which we shall not indicate in the notation). In particular,

$$\langle Qg_t, u \rangle_{H_\rho} = \lim_{\varepsilon \downarrow 0} \langle M_{S_\varepsilon}^* g_t, u \rangle_{H_\rho}$$

$$= \lim_{\varepsilon \downarrow 0} (1 + \varepsilon)^{-\frac{1}{2}} \langle h_t, u \rangle_{H_\rho}$$

$$= \langle h_t, u \rangle_{H_\rho} .$$

Now let $T = M_s$ denote the operator of multiplication by $s = b/a$ on H_ρ^r, regardless of the size of r. Then since $M_{S_\varepsilon} T = T M_{S_\varepsilon}$ for every $\varepsilon > 0$, it is readily checked that $Q^* T = T Q^*$ and hence, by Theorem 3.3 of [AD4], that there exists a $p \times q$ matrix valued function S which is analytic in Ω_+ such that $Q^* = M_S$. This completes the proof, since S is a solution to the one-sided interpolation problem.   ■

THEOREM 7.2. *If the $\nu \times \nu$ matrix P with entries given by (7.2) is positive definite and if Θ is specified by (2.14) for the space $\mathcal{M}$ considered in the proof of the last theorem, then*

$$\{T_\Theta[S_o] : \ S_o \ \text{is analytic in} \ \ \Omega_+ \ \ \text{and} \ \ \|M_{S_o}\| \leq 1\}$$

is a complete list of the set of solutions to the given interpolation problem.

PROOF. The proof is divided into steps.

STEP 1. *If S is a solution of the interpolation problem, then*

$$Y = [I_p \ \ - S]\Theta$$

is (Ω_+, J, ρ) admissible.

PROOF OF STEP 1. The argument is adapted from the proof of Theorem 6.2 in [D2]. Let $X = [I_p \ \ - S]$ and let

$$\Lambda_\omega(\lambda) = \frac{X(\lambda) J X(\omega)^*}{\rho_\omega(\lambda)}$$

denote the reproducing kernel for $\mathcal{H}_\rho(S)$. Then it is readily seen that

$$\frac{Y(\alpha) J Y(\beta)^*}{\rho_\beta(\alpha)} = \frac{X(\alpha) J X(\beta)^*}{\rho_\beta(\alpha)} - \frac{X(\alpha)\{J - \Theta(\alpha) J \Theta(\beta)^*\} X(\beta)^*}{\rho_\beta(\alpha)}$$

$$= \Lambda_\beta(\alpha) - X(\alpha) F(\alpha) P^{-1} F(\beta)^* X(\beta)^*$$

for α and β in Ω_+. Thus if

$$f = \sum_{t=1}^{n} \Lambda_{\alpha_t} \xi_t$$

for any choice of $\alpha_1, \ldots, \alpha_n$ in Ω_+ and $\xi_1, \ldots, \xi_n$ in $\mathbb{C}^m$ and if γ_{ij} denotes the ij entry of P^{-1}, then

$$\sum_{s,t=1}^{n} \frac{\xi_s^* Y(\alpha_s) J Y(\alpha_t)^* \xi_t}{\rho_{\alpha_t}(\alpha_s)} = \sum_{s,t=1}^{n} \xi_s^* \Lambda_{\alpha_t}(\alpha_s) \xi_t$$

$$- \sum_{s,t=1}^{n} \sum_{i,j=1}^{\nu} \xi_s^* X(\alpha_s) f_i(\alpha_s) \gamma_{ij} f_j(\alpha_t)^* X(\alpha_t)^* \xi_t \ .$$

Moreover, since S is an interpolant, it follows from Lemma 7.1 and Theorem 5.4 that $X f_i \in \mathcal{H}_\rho(S)$ and hence that the right hand side of the last equality is equal to

$$\|f\|_{\mathcal{H}_\rho(S)}^2 - \sum_{s,t=1}^{n} \sum_{i,j=1}^{\nu} \langle X f_i, \Lambda_{\alpha_s} \xi_s \rangle_{\mathcal{H}_\rho(S)} \gamma_{ij} \langle \Lambda_{\alpha_t} \xi_t, X f_j \rangle_{\mathcal{H}_\rho(S)}$$

$$= \|f\|_{\mathcal{H}_\rho(S)}^2 - \sum_{i,j=1}^{\nu} \langle X f_i, f \rangle_{\mathcal{H}_\rho(S)} \gamma_{ij} \langle f, X f_j \rangle_{\mathcal{H}_\rho(S)} \ .$$

But this in turn is equal to

$$\|f\|_{\mathcal{H}_\rho(S)}^2 - \|\Pi f\|_{\mathcal{H}_\rho(S)}^2$$

in which Π denotes the orthogonal projection of f onto the span of the $X f_i$ in $\mathcal{B}(X)$. This last evaluation uses the fact that M_X is an isometry from $\mathcal{K}(\Theta)$ into $\mathcal{B}(X)$, as noted in the remark following Theorem 5.4. The rest is selfevident since

$$\|f\|_{\mathcal{H}_\rho(S)}^2 - \|\Pi f\|_{\mathcal{H}_\rho(S)}^2 \geq 0 \ .$$

STEP 2. *If S is an interpolant, then*

$$[I_p \quad - S]\Theta = C[I_p \quad - S_o]$$

where C and S_o are analytic in Ω_+, and M_C is bounded and $\|M_{S_o}\| \leq 1$.

PROOF OF STEP 2. This is immediate from Theorem 3.5 since M_Y is bounded (because both M_S and M_Θ are; the former by assumption and the latter by Theorem 6.1), and Y is (Ω_+, J, ρ) admissible, by Step 1.

STEP 3. *If S is a solution of the given interpolation problem, then $S = T_\Theta[S_o]$ for some $p \times q$ matrix valued function S_o with $\|M_{S_o}\| \leq 1$.*

PROOF OF STEP 3. By Step 2,

$$\Theta_{11} - S\Theta_{21} = C$$

and

$$\Theta_{12} - S\Theta_{22} = -CS_o \ .$$

Therefore

$$S(\Theta_{21}S_o + \Theta_{22}) = \Theta_{11}S_o + \Theta_{12} \ ,$$

which is the same as the asserted formula. The term $\Theta_{21}S_o + \Theta_{22}$ is invertible in Ω_+ by standard arguments which utilize the fact that Θ is J contractive and S_o is contractive, both in Ω_+.

STEP 4. *If $S = T_\Theta[S_o]$ for some $p\times q$ matrix valued function S_o with $\|M_{S_o}\| \le 1$, then S is a solution of the given interpolation problem.*

PROOF OF STEP 4. By Theorem 6.2, $\|M_S\| \le 1$. In the proof of Theorem 7.1 it is shown that

$$\sigma_1 = T_\Theta[0]$$

is a solution of the interpolation problem. This means that

$$M_{\sigma_1}^* g_t = h_t$$

for $t = 1,\ldots,\nu$, and hence that S is a solution of the given interpolation problem if and only if

$$(M_S^* - M_{\sigma_1}^*)g_t = 0$$

for $t = 1,\ldots,\nu$. But now as

$$S - \sigma_1 = \varepsilon_1 S_o(I_q + \sigma_2 S_o)^{-1}\varepsilon_2^{-1} \ ,$$

$$(M_S^* - M_{\sigma_1}^*)g_t = M_{\varepsilon_2^{-1}}^* M_{(I_q+\sigma_2 S_o)^{-1}}^* M_{S_o}^* M_{\varepsilon_1}^* g_t$$

$$= 0 \ ,$$

by (7.6). The calculation is meaningful because all of the indicated multiplication operators are bounded thanks to Theorem 6.3. This completes the proof of the step and the theorem.   ∎

8. REFERENCES

[Ab]   M.B. Abrahamse, *The Pick interpolation theorem for finitely connected domains*, Michigan Math. J. **26** (1979), 195–203.

[A1]   D. Alpay, *Krein spaces of analytic functions and an inverse scattering problem*, Michigan Math. J. **34** (1987), 349–359.

[A2]   ——, *Some remarks on reproducing kernel Krein spaces*, Rocky Mountain Journal of Mathematics, in press.

[A3]   ——, *Some reproducing kernel spaces of continuous functions*, J. Math. Anal. Appl. **160** (1991), 424–433.

[A4]   ——, *On linear combinations of positive functions, associated reproducing kernel spaces and a non-Hermitian Schur algorithm*, Archiv der Mathematik (Basel), in press.

[AD1]   D. Alpay and H. Dym, *Hilbert spaces of analytic functions, inverse scattering, and operator models I*, Integral Equations and Operator Theory **7** (1984), 589–641.

[AD2]   ——, *Hilbert spaces of analytic functions, inverse scattering, and operator models II*, Integral Equations and Operator Theory **8** (1985), 145–180.

[AD3]   ——, *On applications of reproducing kernel spaces to the Schur algorithm and rational J unitary factorization*, in: I. Schur Methods in Operator Theory and Signal Processing (I. Gohberg, ed.), Operator Theory: Advances and Applications **OT18**, Birkhäuser Verlag, Basel, 1986, pp. 89-159.

[AD4]   ——, *On a new class of reproducing kernel spaces and a new generalization of the Iohvidov laws*, Linear Algebra Appl., in press.

[AD5]   ——, *On a new class of structured reproducing kernel spaces*, J. Functional Anal., in press.

[ADD]   D. Alpay, P. Dewilde and H. Dym, *On the existence and convergence of solutions to the partial lossless inverse scattering problem with applications to estimation theory*, IEEE Trans. Inf. Theory, **35** (1981), 1184–1205.

[Ar]   N. Aronszajn, *Theory of reproducing kernels*, Trans. Amer. Math. Soc. **68** (1950), 337–404.

[AI]   T.Ya. Azizov and I.S. Iohvidov, *Linear Operators in Spaces with an Indefinite Metric*, Wiley, New York, 1989.

[Ba1]   J.A. Ball, *Models for non contractions*, J. Math. Anal. Appl. **52** (1975), 235–254.

[Ba2]   ——, *Factorization and model theory for contraction operators with unitary part*, Memoirs Amer. Math. Soc. **198**, 1978.

[Be]   B. Beauzamy, *Introduction to Operator Theory and Invariant Subspaces*, North Holland, 1988.

[Bh]   R. Bhatia, *Perturbation bounds for matrix eigenvalues*, Pitman Research Notes in Math., **162**, Longman, Harlow, UK, 1987.

[Bo]   J. Bognar, *Indefinite Inner Product Spaces*, Springer-Verlag, Berlin, 1974.

[dB1]   L. de Branges, *Hilbert spaces of analytic functions, I*, Trans. Amer. Math. Soc. **106** (1963), 445–468.

[dB2]   ——, *Hilbert Spaces of Entire Functions*, Prentice Hall, Englewood Cliffs, N.J. 1968.

[dB3]   ——, *The expansion theorem for Hilbert spaces of entire functions*, in: Entire Functions and Related Topics of Analysis, Proc. Symp. Pure Math., Vol. 11, Amer. Math. Soc., Providence, R.I., 1968.

[dB4]   —, *Complementation in Krein spaces*, Trans. Amer. Math. Soc. **305** (1988), 277–291.

[dB5]   —, *Krein spaces of analytic functions*, J. Functional Anal. **81** (1988), 219–259.

[dBR]   L. de Branges and J. Rovnyak, *Canonical models in quantum scattering theory*, in: Perturbation Theory and its Applications in Quantum Mechanics (C. Wilcox, ed.), Wiley, New York, 1966, pp. 295–392.

[dBS]   L. de Branges and L. Shulman, *Perturbations of unitary transformations*, J. Math. Anal. Appl. **23** (1968), 294–326.

[DLS]   A. Dijksma, H. Langer and H. de Snoo, *Characteristic functions of unitary operator colligations in π_k spaces*, in: Operator Theory and Systems (H. Bart, I. Gohberg and M.A. Kaashoek, eds.), Operator Theory: Advances and Applications **OT19**, Birkhäuser Verlag, Basel, 1986, pp. 125–194.

[D1]   H. Dym, *Hermitian block Toeplitz matrices, orthogonal polynomials, reproducing kernel Pontryagin spaces, interpolation and extension*, in: Orthogonal Matrix-Valued Polynomials and Applications (I. Gohberg, ed.), Operator Theory: Advances and Applications **OT34**, Birkhäuser Verlag, Basel, 1988, pp. 79-135.

[D2]   —, *J Contractive Matrix Functions, Reproducing Kernel Hilbert Spaces and Interpolation*, CBMS Regional Conference Series in Mathematics, No. 71, Amer. Math. Soc., Providence, 1989.

[D3]   —, *On reproducing kernel spaces, J unitary matrix functions, interpolation and displacement rank*, in: The Gohberg Anniversary Collection (H. Dym, S. Goldberg, M.A. Kaashoek and P. Lancaster, eds.), Operator Theory: Advances and Applications **OT41**, Birkhäuser Verlag, Basel, 1989, pp. 173-239.

[D4]   —, *On Hermitian block Hankel matrices, matrix polynomials, the Hamburger moment problem, interpolation and maximum entropy*, Integral Equations and Operator Theory **12** (1989), 757-812.

[FW]   P.A. Fillmore and J.P. Williams, *On operator ranges*, Adv. Math. **7** (1971), 254–281.

[IKL]   I.S. Iohvidov, M.G. Krein and H. Langer, *Introduction to the Spectral Theory of Operators in Spaces with an Indefinite Metric*, Mathematische Forschung, Vol. 9, Akademie-Verlag, Berlin, 1982.

[LAK]   H. Lev-Ari and T. Kailath, *Triangular factorization of structured Hermitian matrices*, in: I. Schur Methods in Operator Theory and Signal Processing (I. Gohberg, ed.), Operator Theory: Advances and Applications **OT18** (1986), 301–324.

[Nu]   A.A. Nudelman, *Lecture at Workshop on Operator Theory and Complex Analysis*, Sapporo, Japan, June 1991.

[Ro]   M. Rosenblum, *A corona theorem for countably many functions*, Integral Equations and Operator Theory **3** (1980), 125–137.

[Sch]   L. Schwartz, *Sous espaces hilbertiens d'espaces vectoriels topologiques et noyaux associés (noyaux reproduisants)*, J. Analyse Math. **13** (1964), 115–256.

[So]   P. Sorjonen, *Pontrjaginräume mit einem reproduzierenden Kern*, Ann. Acad.
       Sci. Fenn. Ser. A Math. **594** (1973), 1–30.

[Y]   A. Yang, *A construction of Krein spaces of analytic functions*, Ph.D. Thesis,
       Purdue University, May, 1990.

D. Alpay   H. Dym
Department of Mathematics   Department of Theoretical Mathematics
Ben Gurion University of the Negev   The Weizmann Institute of Science
Beer Sheva 84105, Israel   Rehovot 76100, Israel

MSC: Primary 47A57, Secondary 47A56

Operator Theory:
Advances and Applications, Vol. 59
© 1992 Birkhäuser Verlag Basel

THE CENTRAL METHOD FOR POSITIVE SEMI-DEFINITE, CONTRACTIVE AND STRONG PARROTT TYPE COMPLETION PROBLEMS

Mihály Bakonyi and Hugo J. Woerdeman

In this paper we obtain a new linear fractional parametrization for the set of all positive semi-definite completions of a generalized banded partial operator matrix. As applications we obtain a cascade transform parametrization for the set of all contractive completions of a triangular partial operator matrix satisfying possibly an extra linear constraint (thus extending the results on the Strong Parrott problem). In each of the problems also a maximum entropy principle appears.

1. Introduction. In this paper we establish a new parametrization for the set of all positive semi-definite completions of a given "generalized banded" operator matrix. Before we elaborate on this problem we start with an important application, namely a generalization of the Strong Parrott problem introduced by C. Foiaş and A. Tannenbaum. It concerns the following.

For $1 \leq i \leq j \leq n$ let $B_{ij} : \mathcal{H}_j \to \mathcal{K}_i$ be given bounded linear operators acting between Hilbert spaces. Further, let also be given the operators

$$(1.1) \qquad S = \begin{pmatrix} S_1 \\ \vdots \\ S_n \end{pmatrix} : \mathcal{H} \to \oplus_{i=1}^n \mathcal{K}_i, \quad T = \begin{pmatrix} T_1 \\ \vdots \\ T_n \end{pmatrix} : \mathcal{H} \to \oplus_{i=1}^n \mathcal{H}_i.$$

We want to find contractive completions of the following problem:

$$(1.2) \qquad \begin{pmatrix} B_{11} & B_{12} & \cdots & B_{1n} \\ ? & B_{22} & \cdots & B_{2n} \\ \vdots & \vdots & & \vdots \\ ? & ? & \cdots & B_{nn} \end{pmatrix} \begin{pmatrix} S_1 \\ \vdots \\ S_n \end{pmatrix} = \begin{pmatrix} T_1 \\ \vdots \\ T_n \end{pmatrix},$$

i.e., we want to find B_{ij}, $1 \leq j < i \leq n$, so that $B = (B_{ij})_{i,j=1}^n$ is a contraction satisfying the linear constraint $BS = T$. The introduction of the Strong Parrott problem was a consequence of questions arising in the theory of contractive intertwining dilations (see, e.g., the recent book by C. Foiaş and A.E. Frazho [10]).

For the problem (1.2) we derive neccessary and sufficient conditions for the existence of a contractive solution. In case these conditions are met we build a so-called "central completion", a solution with several distingueshed properties. From the central completion we construct a cascade transform parametrization for the set of all solutions. As we mentioned before the above results appear as application of our results on positive semi-definite completions.

The (strictly) positive definite completion problem is a well studied subject. The first results in this domain were by H. Dym and I. Gohberg in [8]. These results were in many ways generalized by H. Dym and I. Gohberg (see the references in [12]) and I. Gohberg, M.A. Kaashoek and H.J. Woerdeman in [12], [13], [14]. A complete Schur analysis of positive semi-definite operator matrices was given by T. Constantinescu in [7], and these results were later used by Gr. Arsene, Z. Ceauşescu and T. Constantinescu in [1] in positive semi-definite completion problems. An analysis of positive semi-definite completions in the classes of so-called U^*DU- and L^*DL-factorable positive semi-definite operator matrices was recently given by M. Bakonyi and H.J. Woerdeman in [3]. The methods in [3] cover the finite dimensional case, but do not extend to the general case described in this paper.

In the study of positive semi-definite "generalized banded" completions, we first develop some distingueshed properties which uniquely characterize a so called central completion, a notion that appeared in different settings and with different names in [8], [1], [12] and [3]. Next we present a result on which the rest of the paper is based, namely a linear fractional transform parametrization for the set of all solutions. The coefficients of the transformation are obtained from the Cholesky factorizations of the central completion. This is a generalization of results in [12], where the positive definite case was considered, and of results in [3].

Our paper is organized as follows. In Section 2 we treat positive semi-definite completions and in Section 3 contractive completions. Section 4 is dedicated to the study of a generalized Strong Parrott problem.

2. Positive Semi-Definite Completions. Consider the following 3×3 problem:

$$(2.1) \qquad \begin{pmatrix} A_{11} & A_{12} & ? \\ A_{21} & A_{22} & A_{23} \\ ? & A_{32} & A_{33} \end{pmatrix} \geq 0,$$

where

$$(2.2) \qquad \begin{pmatrix} A_{11} & A_{12} \\ A_{21} & A_{22} \end{pmatrix} \geq 0, \begin{pmatrix} A_{22} & A_{23} \\ A_{32} & A_{33} \end{pmatrix} \geq 0.$$

By this we mean that we want to find the (1,3) entry A_{13} of the operator matrix in (2.1) such that with this choice (and with $A_{31} = A_{13}^*$) we obtain a positive semi-definite 3×3 operator matrix. Note that the positivity of the 2×2 operator matrices in (2.2) implies that

$$A_{12} = A_{11}^{1/2} G_1 A_{22}^{1/2}, A_{23} = A_{22}^{1/2} G_2 A_{33}^{1/2}$$

where $G_1 : \bar{\mathcal{R}}(A_{22}) \to \bar{\mathcal{R}}(A_{11})$ and $G_2 : \bar{\mathcal{R}}(A_{33}) \to \bar{\mathcal{R}}(A_{22})$ are contractions. For a linear operator A we denote by $\mathcal{R}(A)$ its range and by $\bar{\mathcal{R}}(A)$ the closure of its range, and if $A \geq 0$ then $A^{1/2}$ is its square root with $A^{1/2} \geq 0$. Let also for a contraction $G : \mathcal{L} \to \mathcal{K}$ denote $D_G = (I_{\mathcal{L}} - G^*G)^{1/2} : \mathcal{L} \to \mathcal{L}$ and $\mathcal{D}_G = \bar{\mathcal{R}}(D_G)$. It was proved in [1] that there exists a one-to-one correspondence between the set of all positive semi-definite completions of (2.1) and the set of all contractions $G : \mathcal{D}_{G_2} \to \mathcal{D}_{G_1^*}$ via

$$(2.3) \qquad A_{13} = A_{11}^{1/2}(G_1 G_2 + D_{G_1^*} G D_{G_2})A_{33}^{1/2}.$$

With the choice $G = 0$ we obtain the particular completion

$$(2.4) \qquad A_{13} = A_{11}^{1/2} G_1 G_2 A_{33}^{1/2}.$$

We shall call this the *central completion* of (2.1), referring to the fact that in the operator ball in which A_{13} lies (namely the one described by (2.3)) we choose the centre.

If F is a positive semi-definite operator matrix it is known that there exist an upper triangular operator matrix V and a lower triangular matrix W such that

$$(2.5) \qquad F = V^*V = W^*W.$$

The factorizations (2.5) are called *lower-upper and upper-lower Cholesky factorizations*, respectively. Moreover if $\hat{V}$ and $\hat{W}$ are upper (lower) triangulars with $F = \hat{V}^*\hat{V} = \hat{W}^*\hat{W}$, then there exists block diagonal unitaries $U : \bar{\mathcal{R}}(V) \to \bar{\mathcal{R}}(\hat{V})$ and $\hat{U} : \bar{\mathcal{R}}(W) \to \bar{\mathcal{R}}(\hat{W})$ with $UV = \hat{V}$ and $\hat{U}W = \hat{W}$. This implies that if F is a positive semi-definite $n \times n$ operator matrix, then the operators

$$(2.6) \qquad \Delta_U(F) := diag(V_{ii}^*V_{ii})_{i=1}^n$$

and

$$(2.7) \qquad \Delta_L(F) := diag(W_{ii}^*W_{ii})_{i=1}^n$$

do not depend upon the particular choice of V and W in (2.5).

Returning to our problem (2.1), if F is an arbitrary completion corresponding to the parameter G in (2.3) then F admits the factorization (2.5) with

$$(2.8) \qquad V = \begin{pmatrix} A_{11}^{1/2} & G_1 A_{22}^{1/2} & (G_1 G_2 + D_{G_1^*} G D_{G_2})A_{33}^{1/2} \\ 0 & D_{G_1} A_{22}^{1/2} & (D_{G_1} G_2 - G_1^* G D_{G_2})A_{33}^{1/2} \\ 0 & 0 & D_G D_{G_2} A_{33}^{1/2} \end{pmatrix}$$

and

$$(2.9) \qquad W = \begin{pmatrix} D_{G^*} D_{G_1^*} A_{11}^{1/2} & 0 & 0 \\ (D_{G_2^*} G_1^* - G_2 G^* D_{G_1^*})A_{11}^{1/2} & D_{G_2^*} A_{22}^{1/2} & 0 \\ (G_2^* G_1^* + D_{G_2} G^* D_{G_1^*})A_{11}^{1/2} & G_2^* A_{22}^{1/2} & A_{33}^{1/2} \end{pmatrix}.$$

Further, using relations like $G_1^*(\mathcal{D}_{G_1^*}) \subseteq \mathcal{D}_{G_1}$, one easily obtains that $\bar{\mathcal{R}}(V_{ij}) \subseteq \bar{\mathcal{R}}(V_{ii})$ and $\bar{\mathcal{R}}(W_{ij}) \subseteq \bar{\mathcal{R}}(W_{ii})$, for all i and j. The triangularity of V and W now yields

$$(2.10) \qquad \bar{\mathcal{R}}(V) = \bar{\mathcal{R}}(A_{11}^{1/2}) \oplus \mathcal{D}_{G_1} \oplus \mathcal{D}_G, \bar{\mathcal{R}}(W) = \mathcal{D}_{G^*} \oplus \mathcal{D}_{G_2^*} \oplus \bar{\mathcal{R}}(A_{33}^{1/2}).$$

One immediately sees from these equalities that when $G = 0$ the closures of the ranges of the Cholesky factors of the completion are as large as possible.

Relation (2.5) implies the existence of a unitary $U : \bar{\mathcal{R}}(W) \to \bar{\mathcal{R}}(V)$ with $UW = V$. A straightforward computation gives us the explicit expression of U, namely

$$(2.11) \qquad U = \begin{pmatrix} D_{G_1^*} D_{G^*} & G_1 D_{G_2^*} - D_{G_1^*} G G_2^* & G_1 G_2 + D_{G_1^*} G D_{G_2} \\ -G_1^* D_{G^*} & D_{G_1} D_{G_2^*} - G_1^* G G_2^* & D_{G_1} G_2 - G_1^* G D_{G_2} \\ -G^* & -D_G G_2^* & D_G D_{G_2^*} \end{pmatrix}.$$

Note that the $(3,1)$ entry in U is zero if and only if $G = 0$. As it will turn out, this will be a characterization for the central completion, thus providing a generalization of the banded inverse characterization in the invertible case, discovered in [8]. We will state the result precisely in the $n \times n$ case. Before we can do this we have to recall the following.

We remind the reader of the Schur type structure of positive semi-definite matrices obatined in [7]: *There exists an one-to-one correspondence between the set of positive semi-definite matrices $(A_{ij})_{i,j=1}^n$ with fixed block diagonal entries and the set of all upper triangular families of contractions $\mathcal{G} = \{\Gamma_{ij}\}_{1 \le i \le j \le n}$, where $\Gamma_{ii} = I_{\mathcal{R}(A_{ii})}$, $i = 1, ..., n$, and $\Gamma_{ij} : \mathcal{D}_{\Gamma_{i+1,j}} \to \mathcal{D}_{\Gamma_{i,j-1}}$ for $1 \le i < j \le n$.* The family of contractions $\mathcal{G}$ is referred to as the *choice triangle* corresponding to $(A_{ij})_{i,j=1}^n$. In [7] it is also proven that if $A \ge 0$ and $\mathcal{G} = \{\Gamma_{ij}\}_{1 \le i \le j \le n}$ is its choice triangle, then A admits Cholesky factorizations $A = V^*V = W^*W$ with

$$(2.12) \qquad V : \oplus_{i=1}^n \bar{\mathcal{R}}(A_{ii}) \to \bar{\mathcal{R}}(A_{11}) \oplus (\oplus_{k=2}^n \mathcal{D}_{\Gamma_{1k}})$$

upper triangluar and

$$(2.13) \qquad W : \oplus_{i=1}^n \bar{\mathcal{R}}(A_{ii}) \to \oplus_{k=1}^{n-1} \mathcal{D}_{\Gamma_{kn}^*} \oplus \bar{\mathcal{R}}(A_{nn})$$

lower triangular. The operators V and W have dense range, and their block diagonal entries are given by

$$(2.14) \qquad V_{ii} = D_{\Gamma_{1i}} \cdots D_{\Gamma_{i-1,i}} A_{ii}^{1/2}$$

and

$$(2.15) \qquad W_{ii} = D_{\Gamma_{in}^*} \cdots D_{\Gamma_{i-1,i}^*} A_{ii}^{1/2}.$$

Let us now consider the $n \times n$ generalized banded positive semi-definite completion problem. Recall that $S \subseteq \underline{n} \times \underline{n}$ $(\underline{n} = \{1, ..., n\})$ is called a *generalized banded pattern* if (1) $(i,i) \in S$, $i = 1, ..., n$; (2) if $(i,j) \in S$ then $(j,i) \in S$; and (3) $(i,j) \in S$ and

$i \leq p, q \leq j$ imply $(p, q) \in S$. The problem is the following. Given are $A_{ij} : \mathcal{H}_j \to \mathcal{H}_i$ for (i, j) in a prescribed generalized banded pattern S. We want to find $A_{ij}, (i, j) \in (\underline{n} \times \underline{n}) \backslash S$ such that $A = (A_{ij})_{i,j=1}^n \geq 0$. Such an operator matrix A will be called a *positive semi-definite completion* of the band $\{A_{ij}, (i, j) \in S\}$. It is known (see [8]) that a positive definite completion of $\{A_{ij}, (i, j) \in S\}$ exists if and only if

$$(2.16) \qquad\qquad\qquad (A_{ij})_{i,j \in J} \geq 0$$

for all $J \subseteq \underline{n}$ with $J \times J \subseteq S$. When $\{A_{ij}, (i, j) \in S\}$ verifies condition (2.16) we shall call this band *positive semi-definite*.

In [1] a parametrization was given for the set of all positive semi-definite completions of $\{A_{ij}, (i, j) \in S\}$ as follows. This parametrization is based on the result in [7] quoted above and the fact that making a completion of $\{A_{ij}, (i, j) \in S\}$ precisely corresponds to choosing the parameters $\{\Gamma_{ij}, 1 \leq i \leq j \leq n, (i, j) \notin S\}$. Thus there exists an one-to-one correspondence between the set of all positive semi-definite completions of $\{A_{ij}, (i, j) \in S\}$ and the completions of $\{\Gamma_{ij}, 1 \leq i \leq j \leq n, (i, j) \in S\}$ to a $(A_{ii})_{i=1}^n$ choice triangle. This parametrization is recursive in nature, because of the way the choice triangles are constructed.

The completion corresponding to the choice $\Gamma_{ij} = 0$ whenever $1 \leq i \leq j \leq n$ with $(i, j) \notin S$ is called the *central completion* of $\{A_{ij}, (i, j) \in S\}$. It shall be denoted by F_c, where the subscript "c" stands for central.

An alternative way to obtain the central completion is described below. For a given $n \times n$ positive generalized band $\{A_{ij}, (i, j) \in S\}$ one can proceed as follows: choose a position $(i_0, j_0) \notin S$, $i_0 \leq j_0$, such that $S \cup \{(i_0, j_0), (j_0, i_0)\}$ is also generalized banded. Choose A_{i_0,j_0} such that $(A_{ij})_{i,j=i_0}^{j_0}$ is the central completion of $\{A_{ij}, (i, j) \in S$ and $i_0 \leq i, j \leq j_0\}$. This is a 3×3 problem and A_{i_0,j_0} can be found via a formula as in (2.4). Proceed in the same way with the thus obtained partial matrix until all positions are filled. It turns out (see [1]) that the resulting positive semi-definite completion is the central completion F_c. Note that for $(i_0, j_0) \notin S$, $i_0 \leq j_0$, the entry A_{i_0,j_0} only depends upon $\{A_{ij}, (i, j) \in S$ and $i_0 \leq i, j \leq j_0\}$. This implies that the submatrix of F_c located in the rows and columns $\{k, k+1, ..., l\}$ is precisely the central completion of $\{A_{ij}, (i, j) \in S \cap \{k, k+1, ..., l\} \times \{k, k+1, ..., l\}\}$. This principle is referred to as the "inheritance principle".

Our first result gives four equivalent conditions which characterize the central completion. This is a positive semi-definite operator analogue of Theorem 6.2 in [8].

THEOREM 2.1. *Let S be generalized banded pattern and F a positive semi-definite completion of $\{A_{ij}, (i, j) \in S\}$. Let $F = V^*V = W^*W$ be the lower-upper and upper-lower Cholesky factorizations of F. Then the following are equivalent:*

(i)   *F is the central completion of $\{A_{ij}, (i, j) \in S\}$.*

(ii)   *$\Delta_U(F) \geq \Delta_U(\tilde{F})$ for all positive semi-definite completions $\tilde{F}$ of $\{A_{ij}, (i, j) \in S\}$;*

(iii) *$\Delta_L(F) \geq \Delta_L(\tilde{F})$ for all positive semi-definite completions $\tilde{F}$ of $\{A_{ij}, (i, j) \in S\}$;*

(iv) *The unitary $U : \bar{\mathcal{R}}(W) \to \bar{\mathcal{R}}(V)$ with $UW = V$ verifies $U_{ij} = 0$ for $i \geq j, (i, j) \notin S$.*

Note that the uniqueness of the central completion implies that $\Delta_U(F) = \Delta_U(\tilde{F})$ (or $\Delta_L(F) = \Delta_L(\tilde{F})$) yields $F = \tilde{F}$. The maximality of $\Delta_U(F)$ ($\Delta_L(F)$) can be viewed as a *maximum entropy principle* (see, e.g., [6]).

Proof. The equivalence of (i) and (ii) can be read off immediately from (2.14), and similarly the equivalence of (i) and (iii) can be read off immediately from (2.15).

We prove the equivalence if (i) and (iv) by induction on the number of missing entries in the pattern S. For the 3×3 problem (2.1), discussed at the beginning of this section, formula (2.11) proves immediately the equivalence.

Let $S \subseteq \underline{n} \times \underline{n}$ be a generalized banded pattern and $\{A_{ij}, (i,j) \in S\}$ positive semi-definite. Let F_c denote the central completion of $\{A_{ij}, (i,j) \in S\}$, and let V_c and W_c be upper and lower triangular operator matrices such that

$$(2.17) \qquad\qquad F_c = V_c^* V_c = W_c^* W_c.$$

Consider the unitary operator matrix $U : \bar{\mathcal{R}}(W_c) \to \bar{\mathcal{R}}(V_c)$ so that $U W_c = V_c$. Let $\hat{S}$ denote the pattern $\hat{S} = S \cap (\underline{n-1} \times \underline{n-1})$, and $\hat{F} = \left(\hat{F}_{ij}\right)_{ij=1}^{n-1}$ obtained from F_c by compressing its last two rows and columns. So, $\hat{F}_{ij} = (F_c)_{ij}$ for $i, j \leq n - 1$,

$$\hat{F}_{i,n-1} = \hat{F}_{n-1,i}^* = \left((F_c)_{i,n-1} \ (F_c)_{in}\right), i < n - 1,$$

and

$$\hat{F}_{n-1,n-1} = \begin{pmatrix} (F_c)_{n-1,n-1} & (F_c)_{n-1,n} \\ (F_c)_{n,n-1} & (F_c)_{nn} \end{pmatrix}.$$

Consider the data $\{\hat{F}_{ij}, (i,j) \in \hat{S}\}$. From the way the central completion is defined one sees that $\hat{F}(= F_c)$ is the central completion of $\{\hat{F}_{ij}, (i,j) \in \hat{S}\}$. Now, in the same way, consider the operator matrices $\hat{U} = (\hat{U}_{ij})_{i,j=1}^n$, $\hat{V} = (\hat{V}_{ij})_{i,j=1}^n$ and $\hat{W} = (\hat{W}_{ij})_{i,j=1}^n$ obtained by the compression of the last two rows and columns of U, V_c and W_c, respectively. We obtain by the induction hypothesis that $\hat{U}_{ij} = 0$ for $(i,j) \notin \hat{S}$ with $i > j$. Thus it remains to show that $U_{nj} = 0$ for j with $(n,j) \notin S$ and $(n-1,j) \in S$. For this purpose let $\gamma = \min\{j, (n,j) \in S\}$ and consider the decomposition

$$U = \hat{U} = \begin{pmatrix} \Sigma_{11} & \Sigma_{12} & \Sigma_{13} \\ \Sigma_{21} & \Sigma_{22} & \Sigma_{23} \\ \Sigma_{31} & \Sigma_{32} & \Sigma_{33} \end{pmatrix}$$

with $\Sigma_{11} = (U_{ij})_{i,j=1}^{\gamma-1}$, $\Sigma_{22} = (U_{ij})_{i,j=\gamma}^{n-1}$ and $\Sigma_{33} = U_{nn}$. Consider also the corresponding decomposition of $F_c = (\phi_{ij})_{i,j=1}^3$. Again we have that F_c is also the central completion of

$$\begin{pmatrix} \phi_{11} & \phi_{12} & ? \\ \phi_{21} & \phi_{22} & \phi_{23} \\ ? & \phi_{32} & \phi_{33} \end{pmatrix}$$

But then from the 3×3 case we obtain that $\Sigma_{13} = 0$ and, consequently, $U_{nj} = 0$ for $j \leq \gamma - 1$, proving (iii).

Implication (iii) $\rightarrow$ (i) can be proved by the same type of induction process. One needs to use the observation that if S_1 and S_2 are two generalized banded patterns and F is the central completion of both $\{A_{ij}, (i,j) \in S_1\}$ and $\{A_{ij}, (i,j) \in S_2\}$, then F is the central completion of $\{A_{ij}, (i,j) \in S_1 \cap S_2\}$. We omit the details. $\square$

THEOREM 2.2. *Let $S \subseteq \underline{n} \times \underline{n}$ be a generalized banded pattern and $\{A_{ij}, (i,j) \in S\}$ be positive semi-definite. Let F_c denote the central completion of $\{A_{ij}, (i,j) \in S\}$, and V_c and W_c be upper and lower triangular operator matrices such that*

$$(2.18) \qquad\qquad F_c = V_c^* V_c = W_c^* W_c.$$

Further, let $U : \bar{\mathcal{R}}(W_c) \rightarrow \bar{\mathcal{R}}(V_c)$ be the unitary operator matrix so that

$$(2.19) \qquad\qquad U W_c = V_c.$$

Then each positive semi-definite completion of $\{A_{ij}, (i,j) \in S\}$ is of the form

$$(2.20) \qquad T(G) = V_c^* (I + UG)^{*-1}(I - G^*G)(I + UG)^{-1} V_c$$

$$= W_c^* (I + GU)^{-1}(I - GG^*)(I + GU)^{*-1} W_c,$$

where $G = (G_{ij})_{i,j=1}^{n} : \bar{\mathcal{R}}(V_c) \rightarrow \bar{\mathcal{R}}(W_c)$ is a contraction with $G_{ij} = 0$ whenever $i \geq j$ or $(i,j) \in S$. Moreover, the correspondence between the set of all positive semi-definite completions and all such contractions G is one-to-one.

The decompositions of $\bar{\mathcal{R}}(V)$ and $\bar{\mathcal{R}}(W)$ are given by

$$(2.21) \qquad\qquad \bar{\mathcal{R}}(V) = \bar{\mathcal{R}}(A_{11}) \oplus (\oplus_{k=2}^{n} \mathcal{D}_{\Gamma_{1k}})$$

and

$$(2.22) \qquad\qquad \bar{\mathcal{R}}(W) = \oplus_{k=1}^{n-1} D_{\Gamma_{kn}^*} \oplus \bar{\mathcal{R}}(A_{nn}).$$

Before starting the proof, we need additional results.

PROPOSITION 2.3. *Let $S \subseteq \underline{n} \times \underline{n}$ be a generalized banded pattern and $\{A_{ij}, (i,j) \in S\}$ positive semi-definite. Let F_c denote the central completion and F an arbitrary positive semi-definite completion of $\{A_{ij}, (i,j) \in S\}$. Then*

$$(2.23) \qquad\qquad \mathcal{R}(F^{1/2}) \subseteq \mathcal{R}(F_c^{1/2}).$$

We remark first that if ϕ is an operator on $\mathcal{H}$ and $A = \phi^*\phi$ then there exists a unitary U on $\mathcal{H}$ such that $A^{1/2} = U\phi$, and thus $\mathcal{R}(\phi^*) = \mathcal{R}(A^{1/2})$. Thus (2.23) is equivalent with the fact that if $F_c = \Sigma_c^* \Sigma_c$ and $F = \Sigma^* \Sigma$ with Σ_c and Σ upper (lower) triangular then $\mathcal{R}(\Sigma^*) \subseteq \mathcal{R}(\Sigma_c^*)$.

Proof. We start the proof with the 3×3 problem (2.1). Let F be the positive semi-definite completion of (2.1) corresponding to the parameter G in (2.3). Then, as we have already seen, $F = V^*V$ where V is given by (2.8). Thus

$$(2.24) \qquad V = \begin{pmatrix} I & 0 & D_{G_1^*}G \\ 0 & I & -G_1^*G \\ 0 & 0 & D_G \end{pmatrix} V_c,$$

which yields $\mathcal{R}(V^*) \subseteq \mathcal{R}(V_c^*)$. The result now follows from the remark preceeding the proof.

Consider now a given generalized banded pattern $S \subseteq \underline{n} \times \underline{n}$. We prove our result by induction assuming the statement is correct for all generalized banded patterns $\hat{S}$ which have S as a proper subset. The case $S = \underline{n} \times \underline{n} \backslash \{(1,n),(n,1)\}$ reduces to the 3×3 problem. Let $(i_0, j_0) \notin S$, $i_0 < j_0$, be such that $\hat{S} = S \cup \{(i_0, j_0),(j_0, i_0)\}$ is also generalized banded. Let $\{A_{ij},(i,j) \in S\}$, F_c and F be as in the statement of the proposition. Consider the partial matrix $\{B_{ij},(i,j) \in \hat{S}\}$, where $B_{ij} = F_{ij}$ for $(i,j) \in \hat{S}$. Let $\hat{F}_c$ denote the central completion of this latter partial matrix. By the induction hypothesis, since clearly F is a completion of $\{B_{ij},(i,j) \in \hat{S}\}$ we have that

$$(2.25) \qquad \mathcal{R}(F^{1/2}) \subseteq \mathcal{R}(\hat{F}_c^{1/2}).$$

Observe that the matrices $\hat{F}_c$ and F_c differ only on the positions (i,j) and (j,i), where $1 \leq i \leq i_0$ and $j_0 \leq j \leq n$. Moreover, defining $\tilde{S} = (\underline{n} \times \underline{n}) \backslash \{(i,j),(j,i), 1 \leq i \leq i_0, j_0 \leq j \leq n\}$ and the partial matrix $\{C_{ij},(i,j) \in \tilde{S}\}$, where $C_{ij} = (\hat{F}_c)_{ij}$ for $(i,j) \in \tilde{S}$, we have that F_c is also the central completion of this latter partial matrix. We can now use the 3×3 case to conclude that

$$(2.26) \qquad \mathcal{R}(\hat{F}_c^{1/2}) \subseteq \mathcal{R}(F_c^{1/2}),$$

since $\hat{F}_c$ can be viewed as a completion of $\{C_{ij},(i,j) \in \tilde{S}\}$.

Now (2.23) is a consequence of (2.25) and (2.26), and the remark preceeding the proof. $\square$

PROPOSITION 2.4. *Let $S \subseteq \underline{n} \times \underline{n}$ be a generalized banded pattern and $\{A_{ij},(i,j) \in S\}$ positive semi-definite. Let F_c denote the central completion and F an arbitrary positive semi-definite completion of $\{A_{ij},(i,j) \in S\}$, and let*

$$(2.27) \qquad F_c = V_c^*V_c = W_c^*W_c$$

be lower-upper and upper-lower Cholesky factorizations of F_c. Then, if $F - F_c = \Omega^* + \Omega$, where $\Omega = (\Omega_{ij})_{ij=1}^n$ with $\Omega_{ij} = 0$ whenever $i \geq j$ or $(i,j) \in S$, there exists an operator matrix $Q = (Q_{ij})_{i,j=1}^n : \bar{\mathcal{R}}(V_c) \to \bar{\mathcal{R}}(W_c)$, with

$$(2.28) \qquad \Omega = W_c^*QV_c$$

and $Q_{ij} = 0$ whenever $i \geq j$ or $(i,j) \in S$.

Proof. We prove the proposition by induction in a similar way as Proposition 2.3. The 3×3 case is straightforward to check. (Using (2.3), (2.8) and (2.9), we obtain that only the $(1,3)$ entry of Q is nonzero, and equals G.)

Consider an arbitrary generalized banded pattern $S \subseteq \underline{n} \times \underline{n}$ and assume that the proposition is true for all generalized banded patterns $\hat{S}$ which have S as a proper subset. The case $S = (\underline{n} \times \underline{n})\backslash\{(1,n),(n,1)\}$ reduces to the 3×3 problem. Let $(i_0, j_0) \notin S$, $i_0 < j_0$, be such that $\hat{S} = S \cup \{(i_0, j_0),(j_0, i_0)\}$ is also generalized banded. Let $\{A_{ij},(i,j) \in S\}$, F_c, F, W_c and V_c be as in the statement of the proposition. Consider the partial matrix $\{B_{ij},(i,j) \in \hat{S}\}$, where $B_{ij} = F_{ij}$ for $(i,j) \in \hat{S}$. Let $\hat{F}_c$ denote the central completion of this latter partial matrix. By the induction hypothesis,

$$(2.29) \qquad \hat{\Omega} = \hat{W}_c^* \hat{Q} \hat{V}_c$$

where $\hat{\Omega}$ and $\hat{Q}$ are upper triangular with support outside the band $\hat{S}$, $\hat{\Omega}^* + \hat{\Omega} = F - \hat{F}_c$, $\hat{Q} : \bar{\mathcal{R}}(\hat{V}_c) \to \bar{\mathcal{R}}(\hat{W}_c)$, and $\hat{F}_c = \hat{V}_c^* \hat{V}_c = \hat{W}_c^* \hat{W}_c$ are lower-upper and upper-lower Cholesky factorizations of $\hat{F}_c$. By Proposition 2.3 and the remark preceeding proof of Proposition 2.3 we have that $\mathcal{R}(\hat{V}_c^*) \subseteq \mathcal{R}(V_c^*)$ and $\mathcal{R}(\hat{W}_c^*) \subseteq \mathcal{R}(W_c^*)$. But this yields that there exists an upper triangular α and a lower triangular β such that $\hat{V}_c = \alpha V_c$ and $\hat{W}_c = \beta W_c$. Now, taking $Q_1 = \beta^* \hat{Q} \alpha$ we obtain from (2.29) that

$$(2.30) \qquad \hat{\Omega} = W_c^* Q_1 V_c$$

and clearly Q_1 is upper triangular with support outside $\hat{S}$.

As in the proof of Proposition 2.3, F_c is also the central completion of the partial matrix $\{C_{ij},(i,j) \in \tilde{S}\}$, where $\tilde{S} = (n \times n)\backslash\{(i,j),(j,i), 1 \le i \le i_0, j_0 \le j \le n\}$ and $C_{ij} = (\hat{F}_c)_{ij}$ for $(i,j) \in \tilde{S}$. By the 3×3 case we may conclude that

$$(2.31) \qquad \tilde{\Omega} = W_c^* Q_2 V_c$$

where $\tilde{\Omega}$ and Q_2 are upper triangular with support outside the band $\tilde{S}$, $\tilde{\Omega}^* + \tilde{\Omega} = F - \hat{F}_c$, $Q_2 : \bar{\mathcal{R}}(V_c) \to \bar{\mathcal{R}}(W_c)$.

Since $F - F_c = (F - \hat{F}_c) + (\hat{F}_c - F_c)$, we have that $\Omega = \hat{\Omega} + \tilde{\Omega}$, and thus (2.30) and (2.31) imply the desired relation (2.28) with $Q = Q_1 + Q_2$, which clearly is of the desired form. $\square$

We are now ready to prove the parametrization result.

Proof of Theorem 2.2. Write $F_c = C + C^*$, where C is upper triangular with $C_{ii} = 1/2F_{ii}$, $i = 1,...,n$, and define for a contraction $G = (G_{ij})_{i,j=1}^n : \bar{\mathcal{R}}(W_c) \to \bar{\mathcal{R}}(V_c)$ with $G_{ij} = 0$ whenever $i > j$ or $(i,j) \in S$,

$$(2.32) \qquad \mathcal{L}(G) = C - W_c^*(I + GU)^{-1}GV_c.$$

Since $U_{ij} = 0$ for $(i,j) \notin S$ with $i > j$, one easily sees that GU is strictly upper triangular and so $(I + GU)^{-1}$ exists and is upper triangular. Since W_c^* and V_c are both also upper triangular one readily obtains that

$$(2.33) \qquad (\mathcal{L}(G))_{ij} = C_{ij},(i,j) \in S.$$

Further, using (2.32) and the unitarity of U it is straightforward to check that $\mathcal{L}(G) + \mathcal{L}(G)^* = \mathcal{T}(G)$. This together with (2.33) yields that $\mathcal{T}(G)$ is a completion of $\{A_{ij}, (i,j) \in S\}$ and since $\|G\| \le 1$ the operator matrix $\mathcal{T}(G)$ is positive semi-definite.

Assume that for two contractions G_1 and G_2 (of the required form) we have that $\mathcal{T}(G_1) = \mathcal{T}(G_2)$. Then also $\mathcal{L}(G_1) = \mathcal{L}(G_2)$ and since W_c^* and V_c^* are injective on $\bar{\mathcal{R}}(W_c)$ and $\bar{\mathcal{R}}(V_c)$, respectively, equation (2.32) implies that $(I + G_1 U)^{-1} G_1 = (I + G_2 U)^{-1} G_2$. Thus $G_1(I + UG_2) = (I + G_1 U)G_2$ which yields $G_1 = G_2$.

Conversely, let F be an arbitrary positive semi-definite completion of $\{A_{ij}, (i,j) \in S\}$. Consider $\Omega = (\Omega_{ij})_{i,j=1}^{n}$ such that $\Omega_{ij} = 0$ whenever $i \le j$ or $(i,j) \in S$, and $F_c - F = \Omega + \Omega^*$. Then by Proposition 2.4 there exists an operator $Q = (Q_{ij})_{ij}^{n} : \bar{\mathcal{R}}(W_c) \to \bar{\mathcal{R}}(V_c)$ with $Q_{ij} = 0$ whenever $i > j$ or $(i,j) \notin S$ and $\Omega = W_c^* Q V_c$. Since UQ is strictly upper triangular, we can define

$$G = Q(I - UQ)^{-1},$$

which will give that $\Omega = W_c^*(I + GU)^{-1}GV_c$. Since $F = F_c - \Omega - \Omega^*$, and taking into account (2.32) we obtain that $F = \mathcal{T}(G)$. Since $F = \mathcal{T}(G)$ is positive semi-definite, the relation (2.20) implies that G is a contraction. This finishes our proof. ⊓

3. Contractive Completions. Consider the following 2×2 problem:

$$(3.1) \qquad \left\| \begin{pmatrix} B_{11} & ? \\ B_{21} & B_{22} \end{pmatrix} \right\| \le 1$$

where

$$\left\| \begin{pmatrix} B_{11} \\ B_{21} \end{pmatrix} \right\| \le 1, \left\| \begin{pmatrix} B_{21} & B_{22} \end{pmatrix} \right\| \le 1.$$

Note that the contractivity of the latter operator matrices implies that

$$B_{11} = G_1 D_{B_{21}}, B_{22} = D_{B_{21}^*} G_2$$

where G_1 and G_2 are contractions. It was proved in [2] and [9] that there exists a one-to-one correspondence between the set of all contractive completions of (3.1) and the set of all contractions $G : \mathcal{D}_{G_2} \to \mathcal{D}_{G_1^*}$ given by

$$(3.2) \qquad B_{12} = -G_1 B_{21}^* G_2 + D_{G_1^*} G D_{G_2}.$$

With the choice $G = 0$ we obtain the particular completion $B_{12} = -G_1 B_{21}^* G_2$. We shall call this the *central completion* of (3.1).

Let $\{B_{ij}, 1 \le j \le i \le n\}$ be a $n \times n$ *contractive triangle*, i.e., let $B_{ij} : \mathcal{K}_j \to \mathcal{H}_i$, $1 \le j \le i \le n$, be operators acting between Hilbert spaces with the property that

$$\|(B_{ij})_{i=p,j=1}^{n,p}\| \le 1, p = 1, \dots n.$$

In order to make a contractive completion one can proceed as follows: choose a position (i_0, j_0) with $i_0 = j_0 - 1$, and choose B_{i_0, j_0} such that $(B_{ij})_{i=i_0, j=1}^{n, j_0}$ is the central completion of $\{B_{ij}, i \geq i_0, j \leq j_0\}$ as in the 2×2 case. Proceed in the same way with the thus obtained partial matrix (some compressing of columns and rows is needed) until all positions are filled. We shall refer to F_c as the *central completion* of $\{B_{ij}, (i,j) \in T\}$.

THEOREM 3.1. *Let $\{B_{ij}, 1 \leq j \leq i \leq n\}$ be a contractive triangle. Let F_c denote the central completion of $\{B_{ij}, 1 \leq j \leq i \leq n\}$ and let Φ_c and Ψ_c be upper and lower triangular operator matrices such that*

$$(3.3) \qquad \Phi_c^* \Phi_c = I - F_c^* F_c, \Psi_c^* \Psi_c = I - F_c F_c^*.$$

Further, let $\omega_1 : \mathcal{D}_{F_c} \to \bar{\mathcal{R}}(\Phi_c)$ and $\omega_2 : \mathcal{D}_{F_c^*} \to \bar{\mathcal{R}}(\Psi_c)$ be unitary operator matrices so that

$$(3.4) \qquad \Phi_c = \omega_1 D_{F_c}, \Psi_c = \omega_2 D_{F_c^*},$$

and put

$$(3.5) \qquad \tau_c = -\omega_1 F_c^* \omega_2^*.$$

Then each contractive completion of $\{B_{ij}, 1 \leq j \leq i \leq n\}$ is of the form

$$(3.6) \qquad S(G) = F_c - \Psi_c^* G(I + \tau_c G)^{-1} \Phi_c.$$

where $G = (G_{ij})_{i,j=1}^n : \bar{\mathcal{R}}(\Phi_c) \to \bar{\mathcal{R}}(\Psi_c)$ is a contraction with $G_{ij} = 0$ whenever $(i,j) \notin T$. Moreover, the correspondence between the set of all positive semi-definite completions and all such contractions G is one-to-one.

Furthermore, $S(G)$ is isometric (co-isometric, unitary) if and only if G is.

The decompositions of $\bar{\mathcal{R}}(\Phi_c)$ and $\bar{\mathcal{R}}(\Psi_c)$ are simply given by

$$\bar{\mathcal{R}}(\Phi_c) = \oplus_{i=1}^n \bar{\mathcal{R}}((\Phi_c)_{ii}), \bar{\mathcal{R}}(\Psi_c) = \oplus_{i=1}^n \bar{\mathcal{R}}((\Psi_c)_{ii}).$$

Proof. We apply Theorem 2.2 using the correspondence

$$(3.7) \qquad \begin{pmatrix} I & B \\ B^* & I \end{pmatrix} \geq 0 \text{ if and only if } \|B\| \leq 1.$$

Consider the $(n+n) \times (n+n)$ positive semi-definite band which one obtains by embedding the contractive triangle $\{B_{ij}, 1 \leq j \leq i \leq n\}$ in a large matrix via (3.7). It is easy to check that when applying Theorem 2.2 on this $(n+n) \times (n+n)$ positive semi-definite band one obtains

$$V_c = \begin{pmatrix} I & F_c \\ 0 & \Phi_c \end{pmatrix}, W_c = \begin{pmatrix} \Psi_c & 0 \\ F_c^* & I \end{pmatrix}, U = \begin{pmatrix} \Psi_c^* & F_c \\ \tau_c & \Phi_c \end{pmatrix}$$

(use $F_c^* D_{F_c^*} = D_{F_c} F_c$). It follows now from Theorem 2.1 that $(\tau_c)_{ij} = 0$ for $i > j$. Further, it is easy to compute that

$$(3.8) \qquad \mathcal{T}\left(\begin{pmatrix} 0 & G \\ 0 & 0 \end{pmatrix} \right) = \begin{pmatrix} I & S(G) \\ S(G)^* & Q(G) \end{pmatrix} = \begin{pmatrix} \hat{Q}(G) & S(G) \\ S(G)^* & I \end{pmatrix},$$

where we have

$$(3.9) \qquad I = Q(G) = \mathcal{S}(G)^*\mathcal{S}(G) + \Phi_c(I + \tau_c G)^{*-1}(I - G^*G)(I + \tau_c G)^{-1}\Phi_c,$$

and

$$(3.10) \qquad I = \hat{Q}(G) = \mathcal{S}(G)\mathcal{S}(G)^* + \Psi_c(I + G\tau_c)^{-1}(I - GG^*)(I + G\tau_c)^{*-1}\Psi_c.$$

We obtain the first part of the theorem from (3.8) and Theorem 2.2. From relation (3.9) one immediately sees that G is an isometry if and only if $\mathcal{S}(G)$ is. Similarly, one obtains from (3.10) that G is a co-isometry if and only if $\mathcal{S}(G)$ is. This proves the last statement in the theorem. $\square$

The existence of an isometric (co-isometric, unitary) completion is reduced to the existence of a strictly upper triangular isometry (co-isometry, unitary) acting between the closures of the ranges of Φ_c and Ψ_c. Taking into account the specific structures of Φ_c and Ψ_c one recovers the characterizations of existence of such completions given in [5] and [1] (see also [3]).

REMARK 3.2. We can apply Theorem 2.1 to characterize the central completion. We first mention that for an arbitrary completion F of $\{B_{ij}, 1 \le j \le i \le n\}$ one can define Φ, Ψ and τ analogously as in (3.3), (3.4), and (3.5). The equivalence of (i), (ii) and (iii) in Theorem 2.1 implies that the central completion is characterized by the maximality of $diag(\Phi_{ii}^*\Phi_{ii})_{i=1}^n$ or $diag(\Psi_{ii}^*\Psi_{ii})_{i=1}^n$. This is a so-called "maximum entropy principle". From the equivalence of (i) and (iv) in Theorem 2.1 one also easily obtains that the uppertriangularity of τ characterizes the central completion.

4. Linearly Constrained Contractive Completions. We return to the problem (1.2). The next lemma will reduce this linearly constrained contractive completion problem to a positive semi-definite completion problem. The lemma is a slight variation of an observation by D. Timotin [15].

LEMMA 4.1. *Let* $B : \mathcal{H} \to \mathcal{K}$, $S : \mathcal{G} \to \mathcal{H}$ *and* $T : \mathcal{G} \to \mathcal{K}$ *be linear operators acting between Hilbert spaces. Then* $\|B\| \le 1$ *and* $BS = T$ *if and only if*

$$(4.1) \qquad \begin{pmatrix} I & S & B^* \\ S^* & S^*S & T^* \\ B & T & I \end{pmatrix} \ge 0.$$

Proof. The operator matrix (4.1) is positive semi-definite if and only if

$$(4.2) \qquad \begin{pmatrix} S^*S & T^* \\ T & I \end{pmatrix} - \begin{pmatrix} S^* \\ B \end{pmatrix} \begin{pmatrix} S & B^* \end{pmatrix} = \begin{pmatrix} 0 & T^* - S^*B^* \\ T - BS & I - BB^* \end{pmatrix} \ge 0,$$

and this latter inequality is satisfied if and only if $\|B\| \le 1$ and $BS = T$. $\square$

THEOREM 4.2. *Let* $B_{ij} : \mathcal{H}_j \to \mathcal{K}_i$, $1 \le i \le j \le n$, $S_i : \mathcal{H} \to \mathcal{H}_i$, $i = 1, ..., n$ *and* $T_j : \mathcal{H} \to \mathcal{K}_j$ *be given linear operators acting between Hilbert spaces and S and T be as* *in (1.1). Then there exist contractive completions B of $\{B_{ij}, 1 \leq i \leq j \leq n\}$ satisfying the linear constraint $BS = T$ if and only if*

$$(4.3) \qquad \begin{pmatrix} S^*S - S^{(i)*}S^{(i)} & T^{(i)*} - S^{(i)*}B^{(i)*} \\ T^{(i)} - B^{(i)}S^{(i)} & I - B^{(i)}B^{(i)*} \end{pmatrix} \geq 0$$

for $i = 1, .., n,$ where

$$(4.4) \qquad B^{(i)} = \begin{pmatrix} B_{1i} & \cdots & B_{1n} \\ \cdot & \cdots & \cdot \\ \cdot & \cdots & \cdot \\ B_{ii} & \cdots & B_{in} \end{pmatrix}, \ S^{(i)} = \begin{pmatrix} S_i \\ \cdot \\ \cdot \\ S_n \end{pmatrix}, \ T^{(i)} = \begin{pmatrix} T_1 \\ \cdot \\ \cdot \\ T_i \end{pmatrix}$$

for $i = 1, ..., n.$

    Proof. By Lemma 4.1 there exists a contractive completion B of $\{B_{ij}, 1 \leq i \leq j \leq n\}$ satisfying the linear constrained $BS = T$ if and only if there exists a positive semi-definite completion of the partial matrix

$$(4.5) \qquad \begin{pmatrix} I & 0 & \cdots & 0 & S_1 & B_{11}^* & ? & \cdots & ? \\ 0 & I & \cdots & 0 & S_2 & B_{12}^* & B_{22}^* & \cdots & ? \\ \vdots & \vdots & & \vdots & \vdots & \vdots & \vdots & & \vdots \\ 0 & 0 & \cdots & I & S_n & B_{1n}^* & B_{2n}^* & \cdots & B_{nn}^* \\ S_1^* & S_2^* & \cdots & S_n^* & S^*S & T_1^* & T_2^* & \cdots & T_n^* \\ B_{11} & B_{12} & \cdots & B_{1n} & T_1 & I & 0 & \cdots & 0 \\ ? & B_{21} & \cdots & B_{2n} & T_2 & 0 & I & \cdots & 0 \\ \vdots & \vdots & & \vdots & \vdots & \vdots & \vdots & & \vdots \\ ? & ? & \cdots & B_{nn} & T_n & 0 & 0 & \cdots & I \end{pmatrix}$$

As it is known, the existence of a positive semi-definite completion of (4.5) is equivalent to the positive semi-definiteness of the principal submatrices of (4.5) formed with known entries. This latter condition is equivalent with (4.3). $\square$

    Let us examine the 2×2 case a little further, i.e.,

$$(4.6) \qquad \begin{pmatrix} B_{11} & B_{12} \\ ? & B_{22} \end{pmatrix} \begin{pmatrix} S_1 \\ S_2 \end{pmatrix} = \begin{pmatrix} T_1 \\ T_2 \end{pmatrix}$$

The necessary and sufficient conditions (4.3) for this case reduce to

$$(4.7) \qquad B_{11}S_1 + B_{12}S_2 = T_1, \left\| \begin{pmatrix} B_{11} & B_{12} \end{pmatrix} \right\| \leq 1$$

and

$$(4.8) \qquad \begin{pmatrix} I - B_{12}^*B_{12} - B_{22}^*B_{22} & S_2 - B_{12}^*T_1 - B_{22}^*T_2 \\ S_2^* - T_1^*B_{12} - T_2^*B_{22} & S_1^*S_1 + S_2^*S_2 - T_1^*T_1 - T_2^*T_2 \end{pmatrix} \geq 0.$$

Assume that (4.7) and (4.8) are satisfied. Similar to Section 3, let $G_1 : \mathcal{H}_1 \to \mathcal{D}_{B_{12}^*}$ and $G_2 : \mathcal{D}_{B_{12}} \to \mathcal{K}_2$ be contractions such that

$$(4.9) \qquad B_{11} = D_{B_{12}^*} G_1, \, B_{22} = G_2 D_{B_{12}}.$$

Any solution of the constrained problem (4.6) is in particular a solution of the unconstrained problem (the lower triangular analogue (3.1)), and therefore we must have that (use the analogue of (3.2))

$$(4.10) \qquad B_{21} = -G_2 B_{12}^* G_1 + D_{G_2^*} \Gamma D_{G_1},$$

where $\Gamma : \mathcal{D}_{G_1} \to \mathcal{D}_{G_2^*}$ is some contraction. The equation $B_{21} S_1 + B_{22} S_2 = T_2$ implies that Γ is uniquely defined on $\bar{\mathcal{R}}(D_{G_1} S_1)$ by

$$(4.11) \qquad D_{G_2^*} \Gamma D_{G_1} S_1 := T_2 - B_{22} S_2 + G_2 B_{12}^* G_1 S_1.$$

We define $\Gamma_0 : \mathcal{D}_{G_1} \to \mathcal{D}_{G_2^*}$ to be the contraction defined on $\bar{\mathcal{R}}(D_{G_1} S_1)$ as above, and 0 on the orthogonal complement, i.e.,

$$(4.12) \qquad \Gamma_0 \mid \mathcal{D}_{G_1} \ominus \bar{\mathcal{R}}(D_{G_1} S_1) = 0$$

We let $B_{21}^{(0)}$ denote the corresponding choice for B_{21}, that is,

$$(4.13) \qquad B_{21}^{(0)} = -G_2 B_{12}^* G_1 + D_{G_2^*} \Gamma_0 D_{G_1}.$$

We shall refer to

$$(4.14) \qquad \begin{pmatrix} B_{11} & B_{12} \\ B_{21}^{(0)} & B_{22} \end{pmatrix}$$

as the *central completion* of problem (4.6).

In the $n \times n$ problem (1.2) (assuming conditions (4.3) are met) we construct step by step the central completion of (1.2) as follows. Start by making the central completion of the 2×2 problem

$$(4.15) \qquad \begin{pmatrix} B_{11} & B_{12} & \cdots & B_{1n} \\ ? & B_{22} & \cdots & B_{2n} \end{pmatrix} \begin{pmatrix} S_1 \\ S_2 \\ \vdots \\ S_n \end{pmatrix} = \begin{pmatrix} T_1 \\ T_2 \end{pmatrix}$$

and obtain in this way $B_{21}^{(0)}$. Continue by induction and obtain at step p, $1 \le p \le n-1$, $B_{p1}^{(0)}, \cdots, B_{p,p-1}^{(0)}$ by taking the central completion of the 2×2 problem

$$(4.16) \qquad \begin{pmatrix} B_{11} & \cdots & B_{1,p-1} & B_{1p} & \cdots & B_{1n} \\ \vdots & & \vdots & \vdots & & \vdots \\ B_{p-1,p}^{(0)} & \cdots & B_{p-1,p-1}^{(0)} & B_{p-1,p} & \cdots & B_{p-1,n} \\ ? & \cdots & ? & B_{pp} & \cdots & B_{pn} \end{pmatrix} \begin{pmatrix} S_1 \\ \vdots \\ S_{p-1} \\ S_p \\ \vdots \\ S_n \end{pmatrix} = \begin{pmatrix} T_1 \\ \vdots \\ T_{p-1} \\ T_p \end{pmatrix}.$$

The final result B_0 of this process is the *central completion* of the problem (1.2).

LEMMA 4.3. *Let B_0 be a contractive completion of (1.2). Then B_0 is the central completion of (1.2) if and only if*

$$(4.17) \qquad \begin{pmatrix} I & S & B_0^* \\ S^* & S^*S & T^* \\ B_0 & T & I \end{pmatrix}$$

is the central completion of the positive semi-definite completion problem (4.5).

Proof. By the inheritance principle and the way the central completion is defined it suffices to prove the lemma in the 2×2 case. Take an arbitrary contractive completion B of (4.6), corresponding to the parameter Γ in (4.10), say. The lower-upper Cholesky factorization of the corresponding positive semi-definite completion problem is given by

$$(4.18) \qquad V^*V = \begin{pmatrix} I & S & B^* \\ S^* & S^*S & T^* \\ B & T & I \end{pmatrix},$$

where

$$(4.19) \qquad V = \begin{pmatrix} I & S & B^* \\ 0 & 0 & 0 \\ 0 & 0 & \Phi^* \end{pmatrix}$$

and Φ is lower triangular such that $I - BB^* = \Phi\Phi^*$. It is straightforward to check that

$$(4.20) \qquad \Phi = \begin{pmatrix} D_{B_{12}^*}D_{G_1^*} & 0 \\ -G_2 B_{12} D_{G_1^*} - D_{G_2^*}\Gamma G_1^* & D_{G_2^*}D_{\Gamma^*} \end{pmatrix}.$$

Since for $\Gamma = \Gamma_0$ the operator $D_{\Gamma^*}^2$ is maximal among all Γ satisfying (4.11), the lemma follows from the equivalence of (i) and (ii) in Theorem 2.1. $\square$

THEOREM 4.4. *Let B_0 be the central completion of the linearly constrained contractive completion problem (1.2) (for which the conditions (4.3) are satisfied). Let $\rho : \mathcal{H}_1 \oplus \mathcal{H}_2 \to \bar{\mathcal{R}}((S^*S - T^*T)^{1/2})$ be such that*

$$(4.21) \qquad (S^*S - T^*T)^{1/2}\rho = S^* D_{B_0}^2,$$

and Ψ and Φ lower triangulars such that

$$(4.22) \qquad \Psi^*\Psi = I - \rho^*\rho - B_0^*B_0$$

and

$$(4.23) \qquad \Phi\Phi^* = I - B_0 B_0^*.$$

Consider the contraction $\omega_1 : \mathcal{D}_{B_0} \to \bar{\mathcal{R}}(\Psi)$ and the unitary $\omega_2 : \bar{\mathcal{R}}(\Phi^*) \to \mathcal{D}_{B_0^*}$ with the properties

$$(4.24) \qquad \Psi = \omega_1 D_{B_0}$$

and

(4.25)
$$\Phi = D_{B_0^*}\omega_2$$

Finally, define

(4.26)
$$\tau = -\omega_1 B_0^* \omega_2.$$

Then there exists an one-to-one correspondence between the set of all contractive solutions of the problem (1.2) and the set of all strictly lower triangular contractions $G : \bar{\mathcal{R}}(\Psi) \to \bar{\mathcal{R}}(\Phi^*)$ given by

(4.27)
$$V(G) = B_0 - \Phi(I + G\tau)^{-1}G\Psi$$

Moreover, $V(G)$ is a co-isometry if and only if G is a co-isometry and $V(G)$ is an isometry if and only if $S^*S = T^*T$ and G is an isometry.

The decompositions of $\bar{\mathcal{R}}(\Phi^*)$ and $\bar{\mathcal{R}}(\Psi)$ are simply given by

$$\bar{\mathcal{R}}(\Phi^*) = \oplus_{i=1}^n \bar{\mathcal{R}}(\Phi_{ii}^*), \bar{\mathcal{R}}(\Psi) = \oplus_{i=1}^n \bar{\mathcal{R}}(\Psi_{ii}).$$

Proof. We shall obtain our results by applying Theorem 2.2 for the positive semi-definite completion problem (4.5). Straightforward computation yield that

(4.28)
$$V_c = \begin{pmatrix} I & S & B_0^* \\ 0 & 0 & 0 \\ 0 & 0 & \Phi^* \end{pmatrix}$$

and

(4.29)
$$W_c = \begin{pmatrix} \Psi & 0 & 0 \\ \rho & (S^*S - T^*T)^{1/2} & 0 \\ B_0 & T & I \end{pmatrix}$$

We remark here that the relation

(4.30)
$$S^*S - T^*T = S^* D_{B_0}^2 S \geq S^* D_{B_0}^4 S$$

gives the existence of the contraction ρ with (4.21).

Now we have to determine the unitary $U = (U_{ij})_{i,j=1}^3$ so that $UW_c = V_c$. Note that the existence of ω_1 and ω_2 is assured by the relations (4.22) and (4.23). An immediate computation shows that

$$U = \begin{pmatrix} \Psi^* & \rho^* & B_0^* \\ 0 & 0 & 0 \\ -\omega_2^* B_0 \omega_1^* & -\omega_2^* B_0 \hat{\omega}_1^* & \Phi^* \end{pmatrix}$$

where $\begin{pmatrix} \omega_1 \\ \hat{\omega}_1 \end{pmatrix}$ is unitary with

$$\begin{pmatrix} \Psi \\ \rho \end{pmatrix} = D_{B_0} \begin{pmatrix} \omega_1 \\ \hat{\omega}_1 \end{pmatrix}.$$

Substituting these data in the first equality of (2.20) gives

$$(4.31) \qquad T\left(\begin{pmatrix} 0 & 0 & G^* \\ 0 & 0 & 0 \\ 0 & 0 & 0 \end{pmatrix}\right) = \begin{pmatrix} I & S & V(G)^* \\ S^* & S^*S & T^* \\ V(G) & T & Q(G) \end{pmatrix},$$

where $V(G)$ is given by (4.27) and

$$(4.32) \qquad I = Q(G) = V(G)V(G)^* + \Phi(I + G\tau)^{-1}(I - GG^*)(I + G\tau)^{*-1}\Phi^*$$

The first part of the theorem now follows from (4.31) and Lemma 4.1. Further, (4.32) implies that $V(G)$ is a co-isometry if and only if G is.

If the contractive solution $V(G)$ to the constrained problem (1.2) is isometric, then clearly we must have that $S^*S = T^*T$ and thus $\rho = 0$. In this case,

$$(4.33) \qquad W_c = \begin{pmatrix} \Psi & 0 & 0 \\ 0 & 0 & 0 \\ B_0 & T & I \end{pmatrix}.$$

Using the second inequality in (2.20) in this special case, we obtain that

$$(4.34) \qquad T\left(\begin{pmatrix} 0 & 0 & G^* \\ 0 & 0 & 0 \\ 0 & 0 & 0 \end{pmatrix}\right) = \begin{pmatrix} \hat{Q}(G) & S & V(G)^* \\ S^* & S^*S & T^* \\ V(G) & T & I \end{pmatrix}$$

where

$$(4.35) \qquad I = \hat{Q}(G) = V(G)^*V(G) + \Psi^*(I + G\tau)^{-1}(I - G^*G)(I + G\tau)^{*-1}\Psi.$$

Relation (4.35) implies that when $S^*S = T^*T$, the spaces $\mathcal{D}_{V(G)}$ and $\mathcal{D}_G$ have the same dimensions and thus $V(G)$ is isometric if and only if G is. This finishes the proof. $\square$

In the 2×2 case another parametrization was derived in [4].

REMARK 4.5. By Theorem 4.4 we can reduce the existence of a co-isometric completion of the problem (1.2) to the existence of a strictly lower triangular co-isometry acting between $\bar{\mathcal{R}}(\Psi)$ and $\bar{\mathcal{R}}(\Phi^*)$. Also, when $S^*S = T^*T$, the existence of a isometric completion of the problem (1.2) reduces to the existence of a strictly lower triangular isometry acting between $\bar{\mathcal{R}}(\Psi)$ and $\bar{\mathcal{R}}(\Phi^*)$.

REMARK 4.6. There exists a unique solution to (1.2) if and only if 0 is the only strictly lower triangular contraction acting $\bar{\mathcal{R}}(\Psi) \to \bar{\mathcal{R}}(\Phi^*)$. This can be translated in the following. If i_0 denotes the minimal index for which $\Psi_{i_0 i_0} \neq 0$, then there exists a unique solution if and only if $\Phi_{kk} = 0$ for $k = i_0 + 1, ..., n$.

REMARK 4.7. As in Remark 3.2 the upper triangularity of τ characterizes the central completion. For this one can simply use Theorem 2.1 and Lemma 4.3. Also the maximality of $diag(\Phi_{ii}\Phi_{ii}^*)_{i=1}^n$ or $diag(\Psi_{ii}^*\Psi_{ii})_{i=1}^n$ characterizes the central completion (a maximum entropy principle).

For a different analysis in the 2×2 case we refer to [4].

REFERENCES

[1] Gr. Arsene, Z. Ceauşescu, and T. Constantinescu. Schur Analysis of Some Completion Problems. *Linear Algebra and its Applications.* **109**: 1–36, 1988.

[2] Gr. Arsene and A. Gheondea. Completing Matrix Contractions. *J. Operator Theory.* **7**: 179–189, 1982.

[3] M. Bakonyi and H.J. Woerdeman. Positive Semi-Definite and Contractive Completions of Operator Matrices, submitted.

[4] M. Bakonyi and H.J. Woerdeman. On the Strong Parrott Completion Problem, to appear in *Proceedings of the AMS.*

[5] J.A. Ball and I. Gohberg. Classification of Shift Invariant Subspaces of Matrices With Hermitian Form and Completion of Matrices. *Operator Theory: Adv. Appl.* **19**: 23–85, 1986.

[6] J.P. Burg, Maximum Entropy Spectral Analysis, Doctoral Dissertation, Department of Geophysics, Stanford University, 1975.

[7] T. Constantinescu, A Schur Analysis of Positive Block Matrices. in: *I. Schur Methods in Operator Theory and Signal Processing* (Ed. I. Gohberg). Operator Theory: Advances and Applications **18**, Birkhauser Verlag, 1986, 191-206.

[8] H. Dym and I. Gohberg. Extensions of Band Matrices with Band Inverses. *Linear Algebra Appl.* **36**: 1-24, 1981.

[9] C. Davis, W.M. Kahan, and H.F. Weinberger. Norm Preserving Dilations and Their Applications to Optimal Error Bounds. *SIAM J. Numer. Anal.* **19**: 444-469, 1982.

[10] C. Foias and A. E. Frazho. *The Commutant Lifting Approach to Interpolation Problems.* Operator Theory: Advances and Applications, Vol. 44. Birkhäuser, 1990.

[11] C. Foias and A. Tannenbaum. A Strong Parrott Theorem. *Proceedings of the AMS* **106**: 777-784, 1989.

[12] I. Gohberg, M. A. Kaashoek and H. J. Woerdeman. The Band Method For Positive and Contractive Extension Problems. *J. Operator Theory* **22**: 109-155, 1989.

[13] I. Gohberg, M. A. Kaashoek and H. J. Woerdeman. The Band Method For Positive and Contractive Extension Problems: an Alternative Version and New Applications. *Integral Equations Operator Theory* **12**: 343-382, 1989.

[14] I. Gohberg, M. A. Kaashoek and H. J. Woerdeman. A Maximum Entropy Priciple in the General Framework of the Band Method. *J. Funct. Anal.* **95**: 231-254, 1991.

[15] D. Timotin, A Note on Parrott's Strong Theorem, preprint.

Department of Mathematics
The College of William and Mary
Williamsburg, Virginia 23187-8795

MSC: Primary 47A20, Secondary 47A65

Operator Theory:
Advances and Applications, Vol. 59
© 1992 Birkhäuser Verlag Basel

INTERPOLATION BY RATIONAL MATRIX FUNCTIONS AND STABILITY OF FEEDBACK SYSTEMS: THE 4-BLOCK CASE

Joseph A. Ball and Marek Rakowski

Abstract. We consider the problem of constructing rational matrix functions which satisfy a set of finite order directional interpolation conditions on the left and right, as well as a collection of infinite order directional interpolation conditions on both sides. We set down consistency requirements for solutions to exist as well as a normalization procedure to make the conditions independent, and show how the general standard problem of H^∞ control fits into this framework. We also solve an inverse problem: given an admissible set of interpolation conditions, we characterize the collection of plants for which the associated H^∞-control problem is equivalent to the prescribed interpolation problem.

Key words: Lumped and generic interpolation, homogeneous interpolation problem, stabilizing compensators, 4-block problem, H^∞ control.

Introduction

The tangential (also called directional) interpolation problem for rational matrix functions (with or without an additional norm constraint) has attracted a lot of interest in the past few years (see [ABDS, BGR1-6, BH, BRan, D, FF, Ki]); much of this work was spurred on by the connections with the original frequency domain approach to H^∞-control theory (see [BGR4, BGR5, DGKF, Fr, Ki, V]). The simplest case of the interpolation problem is of the following sort. We are given points $z_1, \cdots, z_M$ in some subset σ of the complex plane $\mathbb{C}$, nonzero $1 \times m$ row vectors $x_1, \cdots, x_M$ and $1 \times n$ row vectors $y_1, \cdots, y_M$ and seek a rational $m \times n$ matrix function $W(z)$ analytic on σ which satisfies

$$(0.1) \qquad x_i W(z_i) = y_i, \quad i = 1, \cdots, M.$$

In the two-sided version of the problem, we are given additional points $w_1, \cdots, w_N$ in σ, nonzero $n \times 1$ column vectors $u_1, \cdots, u_N$ and $m \times 1$ column vectors $v_1, \cdots, v_N$ and demand in addition that W satisfy

$$(0.2) \qquad W(w_j)u_j = v_j, \quad j = 1, \cdots, N.$$

If for some pair of indices (i, j) it happens that $z_i = w_j$, in applications a third type of interpolation condition arises

$$(0.3) \qquad x_i W'(\xi_{ij})u_j = \rho_{ij} \ whenever \ z_i = w_j =: \xi_{ij}$$

and where ρ_{ij} is a given number. By introducing matrices

$$A_\zeta = \begin{bmatrix} z_1 & & \\ & \ddots & \\ & & z_M \end{bmatrix}, B_+ = \begin{bmatrix} x_1 \\ \vdots \\ x_M \end{bmatrix}, B_- = - \begin{bmatrix} y_1 \\ \vdots \\ y_M \end{bmatrix}$$

$$C_- = [u_1 \cdots u_N], \ C_+ = [v_1 \cdots v_N], \ A_\pi = \begin{bmatrix} w_1 & & \\ & \ddots & \\ & & w_N \end{bmatrix}$$

and $\Gamma = [\gamma_{ij}]_{1 \leq i \leq M, 1 \leq j \leq N}$ where

$$\gamma_{ij} = \begin{cases} \rho_{ij} & if \ z_i = w_j \\ (w_i - z_j)^{-1} x_i u_j & if \ z_i \neq w_j \end{cases}$$

the conditions (0.1)–(0.3) can be written in the streamlined compact form

$$(0.1') \qquad \sum_{z_0 \in \sigma} Res_{z=z_0}(z - A_\zeta)^{-1} B_+ W(z) = -B_-$$

$$(0.2') \qquad \sum_{z_0 \in \sigma} Res_{z=z_0} W(z) C_-(z - A_\pi)^{-1} = C_+$$

$$(0.3') \qquad \sum_{z_0 \in \sigma} Res_{z=z_0}(z - A_\zeta)^{-1} B_+ W(z) C_-(z - A_\pi)^{-1} = \Gamma$$

where $Res_{z=z_0} X(z)$ is the residue of the meromorphic matrix function $X(z)$ at z_0. Interpolation conditions of higher multiplicity can be encoded by allowing the matrices A_ζ and A_π to have a more general Jordan form. This is the formalism developed in [BGR4]. If σ is either the unit disk or the right half plane, one can also consider the problem with an additional norm constraint

$$(0.4) \qquad \sup_{z \in \sigma} \|W(z)\| < 1$$

to arrive at a matrix version of the classical Nevanlinna-Pick or Hermite-Fejer interpolation problem. In [BGR4] a systematic analysis of the problem (0.1')–(0.3'), with or without the norm constraint (0.4), is presented, including realization formulas (in the sense of systems theory) for the set of all solutions.

In [BR5], in addition to the **lumped** interpolation conditions (0.1)–(0.3) or (0.1')–(0.3') there was imposed a generic (or infinite order) interpolation condition

$$(P_+ W)^{(j)}(z_0) = P_-^{(j)}(z_0) \ for \ j = 0, 1, 2, \cdots$$

for some $z_0 \in \sigma$, or equivalently

$$(0.5) \qquad P_+(z)W(z) = P_-(z) \ for \ all \ z,$$

where P_+ and P_- are given polynomial matrices of respective sizes $K \times m$ and $K \times n$. In [BR5] the theory developed in [BGR4] for the problem (0.1')–(0.3') was extended to handle the problem (0.1')–(0.3') together with (0.5) (with or without the additional norm constraint (0.4)), with the exception of the explicit realization formulas for the linear fractional parametrization of the set of all solutions. The features obtained include: an admissibility criterion for data sets $(C_+, C_-, A_\pi, A_\zeta, B_+, B_-, \Gamma, P_+(z), P_-(z))$ which guarantees consistency and minimizes redundancy in the set of interpolation conditions (0.1')–(0.3') and (0.5), reduction of the construction of the linear fractional parametrizer of the set of all solutions to the solution of a related homogeneous interpolation problem, and a constructive procedure for solving this latter homogeneous interpolation problem.

The associated homogeneous interpolation problem is of the following form. In general, we let $\mathcal{R}$ denote the field of rational functions and $\mathcal{R}^{m \times n}$ denotes the set of $m \times n$ matrices over $\mathcal{R}$ (i.e. rational $m \times n$ matrix functions). For $\sigma \subset \mathbb{C}$, $\mathcal{R}(\sigma)$ is the subring of $\mathcal{R}$ consisting of rational functions analytic on σ, and $\mathcal{R}^{m \times n}(\sigma)$ is the set of $m \times n$ matrices over $\mathcal{R}(\sigma)$. From the data set $\omega = (C_+, C_-, A_\pi, A_\zeta, B_+, B_-, \Gamma, P_+(z), P_-(z))$ we construct an $\mathcal{R}(\sigma)$-module $S \subset \mathcal{R}^{(m+n) \times 1}$ given by

$$
S = \left\{ \begin{bmatrix} C_+ \\ C_- \end{bmatrix} (z - A_\pi)^{-1} x + \begin{bmatrix} h_+(z) \\ h_-(z) \end{bmatrix} : x \in \mathbb{C}^{n_\pi}, \right.
$$
$$
h_+ \in \mathcal{R}^{m \times 1}(\sigma), h_- \in \mathcal{R}^{n \times 1}(\sigma) \text{ such that}
$$
$$
\sum_{z_0 \in \sigma} Res_{z=z_0} (z - A_\zeta)^{-1} [B_+ \ B_-] \begin{bmatrix} h_+(z) \\ h_-(z) \end{bmatrix} = \Gamma x \}
$$
$$
(0.6) \qquad\qquad \left. \cap \{ r \in \mathcal{R}^{(m+n) \times 1} : [P_+(z) \ P_-(z)] r(z) = 0 \right\}.
$$

Then the module form of the homogeneous interpolation problem is to find a rational $(m + n) \times (m - K + n)$ matrix function $\Theta(z)$ such that

$$
(0.7) \qquad\qquad S = \Theta \mathcal{R}^{(m-K+n) \times 1}(\sigma).
$$

More concretely, the condition (0.7) can be viewed as prescribing the zeros and poles of Θ on σ (including partial multiplicities) from the local Smith form together with some additional directional information, as well as prescribing a left kernel polynomial for Θ. If the norm constraint (0.4) is also part of the original (nonhomogeneous) interpolation problem, then Θ is required to satisfy additional conditions

$$
(0.8a) \qquad\qquad \Theta(z)^* (I_m \oplus -I_n) \Theta(z) = I_{m-K} \oplus -I_n \ for \ z \in \partial\sigma
$$

and

$$(0.8b) \qquad \Theta(z)^*(I_m \oplus -I_n)\Theta(z) \le I_{m-K} \oplus -I_n \ \text{for } z \in \sigma.$$

The linear fractional parametrization of the set of all solutions takes the form

$$(0.9) \qquad W = (\Theta_{11}Q_1 + \Theta_{12}Q_2)(\Theta_{21}Q_1 + \Theta_{22}Q_2)^{-1}$$

where $Q_1 \in \mathcal{R}^{(m-K)\times n}(\sigma)$ and $Q_2 \in \mathcal{R}^{n \times n}(\sigma)$ are appropriate parameters, and where

$$\Theta = \begin{bmatrix} \Theta_{11} & \Theta_{12} \\ \Theta_{21} & \Theta_{22} \end{bmatrix} \text{ with } \Theta_{11} \in \mathcal{R}^{m \times (m-K)}.$$

For more complete details we refer to [BR5].

The purpose of this paper is to consider the interpolation problem (0.1')–(0.3') and (0.5) with an additional right generic interpolation condition

$$(0.10) \qquad W(z)Q_-(z) = Q_+(z)$$

where Q_- and Q_+ are given matrix polynomials of respective sizes $n \times L$ and $m \times L$. Here we obtain a canonical extension of all the results in [BR5] to handle the additional interpolation condition (0.10). In this more general setting the relevant analogue of S in (0.6) is the $\mathcal{R}(\sigma)$-module $\tilde{S} \subset \mathcal{R}^{(m+n)\times 1}$ given by

$$(0.11) \qquad \tilde{S} = S + \begin{bmatrix} Q_+ \\ Q_- \end{bmatrix} \mathcal{R}^{L \times 1}$$

where S is given by (0.6). The associated homogeneous interpolation problem is to find a rational $(m + n) \times (m - K + n)$ matrix function Θ such that

$$(0.12) \qquad \tilde{S} = \Theta \begin{bmatrix} \mathcal{R}^{(m-K+n-L)\times 1}(\sigma) \\ \mathcal{R}^{L \times 1} \end{bmatrix}.$$

If the norm constraint (0.4) is included in the original interpolation problem, then Θ is also required to satisfy

$$(0.13a) \qquad \Theta(z)^*(I_m \oplus -I_n)\Theta(z) = I_{m-K} \oplus -I_n \ \text{for } z \in \partial\sigma$$

$$(0.13b) \qquad \Theta(z)^*(I_m \oplus -I_n)\Theta(z) \le I_{m-K} \oplus -I_{n-L} \oplus 0 \ \text{for } z \in \sigma.$$

and the linear fractional parametrization has the degenerate form

$$W = [\Theta_{11}Q_1 + \Theta_{12}Q_2 \; \Theta_{13}][\Theta_{21}Q_1 + \Theta_{22}Q_2 \; \Theta_{23}]^{-1}$$

where $Q_1 \in \mathcal{R}^{(m-K)\times(n-L)}(\sigma)$ and $Q_2 \in \mathcal{R}^{(n-L)\times(n-L)}(\sigma)$ are appropriate parameters and where $\Theta = \begin{bmatrix} \Theta_{11} & \Theta_{12} & \Theta_{13} \\ \Theta_{21} & \Theta_{22} & \Theta_{23} \end{bmatrix}$ with $\Theta_{11} \in \mathcal{R}^{m\times(m-K)}$ and $\Theta_{12} \in \mathcal{R}^{m\times(n-L)}$.

In addition we show here how the standard problem of H^∞ control theory (see [Fr]) fits into this framework; this also extends the work in [BR5] to incorporate the extra condition (0.10). Specifically, given a rational matrix function $\mathcal{P} = \begin{bmatrix} \mathcal{P}_{11} & \mathcal{P}_{12} \\ \mathcal{P}_{21} & \mathcal{P}_{22} \end{bmatrix}$ representing the plant for an H^∞ problem, we show that the set of interpolation conditions (0.1')–(0.3'), (0.5), (0.10) are satisfied by the closed loop transfer function $\mathcal{P}_{11} + \mathcal{P}_{12}K(I - \mathcal{P}_{22}K)^{-1}\mathcal{P}_{21}$ if and only if the closed loop system with compensator K is internally stable. The characterization is in terms of null-pole data of a related matrix function $\tilde{\mathcal{P}}$ (a partial inversion or chain formalism transform of $\mathcal{P}$). For the case considered here the more general chain formalism transform introduced in [BHV] is required. We also solve an inverse problem of describing which plants are associated with a prescribed set of interpolation conditions, and thereby establish an equivalence between interpolation and feedback stabilization. It turns out that the set of interpolation conditions (0.1')–(0.3') corresponds to the 1-block case, the set (0.1')–(0.3'), (0.5) corresponds to the 2-block case, and the general interpolation problem (0.1')–(0.3'), (0.5), (0.10) corresponds to the general 4-block case in the H^∞ theory. A significant limitation of the interpolation approach in the early development of the H^∞ theory was the lack of an interpolation theory incorporating generic interpolation conditions (0.5) and (0.10) in addition of (0.1')–(0.3') (but see [Hu]). Part of the motivation of [BR5] and this paper is to address this situation.

The paper is organized as follows. Section 1 recalls preliminaries concerning null-pole structure from [BGR4] and [BR1]–[BR3] which will be needed in the sequel. Section 2 formulates and solves the homogeneous interpolation problem of the type (0.12). Section 3 sorts out admissibility conditions on an interpolation data set $(C_+, C_-, A_\pi, A_\zeta, B_+, B_-, \Gamma, P_+(z), P_-(z), Q_+(z), Q_-(z))$ to guarantee consistency and minimize redundancy. In Section 4 we obtain the linear fractional parametrization for the set of all solutions and Section 5 delineates the connection with the H^∞-control theory.

1. Preliminaries.

In this paper, we will use the concepts of the null-pole structure of a rational matrix function W over a fixed subset σ of the complex plane $\mathbb{C}$. They have been developed for a regular rational matrix function (that is, a rational matrix function which is square and whose determinant does not vanish identically) in [BGR1]-[BGR3]; for a comprehensive treatment see [BGR4]. These concepts have been generalized to an arbitrary rational matrix function in [BR1]-[BR3] (see also [BR4]). We recall now the basic definitions and facts which will be used later.

Let $\mathcal{R}$ denote the field of scalar rational functions, and let $\mathcal{R}(\sigma)$ be the subring of $\mathcal{R}$ formed by functions analytic on σ. Let $\mathcal{R}^{m \times n}$ denote the space of $m \times n$ rational matrix functions, and let $\mathcal{R}^{m \times n}(\sigma)$ be the subspace of $\mathcal{R}^{m \times n}$ formed by functions analytic on σ.

An observable pair of matrices (C_π, A_π) of sizes $m \times n_\pi$ and $n_\pi \times n_\pi$, respectively, is said to be a <u>right pole pair</u> for W over σ if

(i) $\sigma(A_\pi) \subset \sigma$,

(ii) for every $x \in \mathbb{C}^{n_\pi \times 1}$ there exists $\varphi \in \mathcal{R}^{n \times 1}(\sigma)$ such that

$$(1.1) \qquad C_\pi(z - A_\pi)^{-1}x - W(z)\varphi(z) \in \mathcal{R}^{m \times 1}(\sigma),$$

(iii) for every $\varphi \in \mathcal{R}^{n \times 1}(\sigma)$ there exists $x \in \mathbb{C}^{n_\pi \times 1}$ such that (1.1) holds.
A right pole pair (C_π, A_π) can also be related to a canonical set of right pole chains or pole functions for $W(z)$ over σ, and to a piece of a realization $W(z) = D + C(z - A)^{-1}B$ for $W(z)$; this is discussed in [BR5]. For a more complete discussion, see [BGK] or [BGR4].

To discuss null or zero structure in the nonregular case, we need to introduce a definition. We define a real non-archimedean valuation $|\cdot|_{z=\lambda}$ of $\mathcal{R}$ by putting

$$|r|_{z=\lambda} = \begin{cases} 0, & \textit{if } r = 0 \\ e,^{-\eta} & \textit{if } r \neq 0 \end{cases}$$

where η is such that $r(z) = (z-\lambda)^\eta \tilde{r}(z)$ with $\tilde{r}(z)$ analytic and nonzero at λ. If $x = (x_1, x_2, \cdots, x_n) \in \mathcal{R}^n$, where $\mathcal{R}^n$ can be identified with $\mathcal{R}^{n \times 1}$ or $\mathcal{R}^{1 \times n}$, let

$$\|x\|_{z=\lambda} = \max\{|x_1|_{z=\lambda}, |x_2|_{z=\lambda}, \cdots, |x_n|_{z=\lambda}\}.$$

$(\mathcal{R}^n, \|\cdot\|_{z=\lambda})$ is a non-Archimedean normed vector space over the real valued field $(\mathcal{R}, |\cdot|_{z=\lambda})$. Subspaces Λ and Ω of $(\mathcal{R}^n, \|\cdot\|_{z=\lambda})$ are said to be orthogonal (see [M]) if

$$(1.2) \qquad \|x + y\|_{z=\lambda} = \max\{\|x\|_{z=\lambda}, \|y\|_{z=\lambda}\}$$

for all $x \in \Lambda$ and $y = \Omega$. We say that subspaces Λ and Ω of $\mathcal{R}^n$ are orthogonal on $\sigma \subset \mathbb{C}$ if (2.2) holds for all $x \in \Lambda, y \in \Omega, \lambda \in \sigma$. We say that an element x of $\mathcal{R}^n$ is orthogonal to a subspace of Ω of $\mathcal{R}^n$ (resp. an element $y \in \mathcal{R}^n$) on σ if $\mathcal{R}x$ is orthogonal to Ω (resp. to $\mathcal{R}y$) on σ.

The orthogonality on σ of subspaces Ω and Λ of $\mathcal{R}^n$ has a straightforward characterization in terms of linear algebra. Let $\Omega(\lambda)$ (resp. $\Lambda(\lambda)$) denote the subspace of $\mathbb{C}^n$ formed by the values at λ of those functions in Ω which are analytic at λ. By Proposition 2.3 in [BR2], Ω and Λ are orthogonal on σ if and only if

$$\Omega(\lambda) \cap \Lambda(\lambda) = (0)$$

for each $\lambda \in \sigma$.

Let $W \in \mathcal{R}^{m \times n}$, and let W^{ol} denote the left kernel of W, that is

$$W^{ol} = \{\varphi \in \mathcal{R}^{1 \times m} : \varphi W = 0\}.$$

A controllable pair of matrices (A_ζ, B_ζ) of sizes $n_\zeta \times n_\zeta$ and $n_\zeta \times m$, respectively, is said to be a left null pair for W over σ (see [BR2]) if

  (i) $\sigma(A_\zeta) \subset \sigma$,

  (ii) for each $x \in \mathbb{C}^{1 \times n_\zeta}$ the function $x(z - A_\zeta)^{-1} B_\zeta$ is orthogonal to W^{ol} on $\sigma(A_\zeta)$,

  (iii) the function $(z - A_\zeta)^{-1} B_\zeta W(z) h(z)$ is analytic on σ whenever $h \in \mathcal{R}^{n \times 1}(\sigma)$ and $W(z)h(z)$ is analytic on σ,

  (iv) the size of A_ζ is maximal subject to the above conditions.

A left null pair can be related to a canonical set of left null chains or left null functions for $W(z)$ over σ (see [BR2], [BR3] for the nonregular case, [BGK], [BGR4] for the regular case) and to a piece of a realization $W^\times(z) = D^\ddagger - D^\ddagger C(z - A + BD^\ddagger C)^{-1} BD^\ddagger$ for a generalized inverse $W^\times(z)$ for $W(z)$ (see [BCRR], [R] for the nonregular case and [BGK], [BGR4] for the regular case).

Complete information regarding left null-pole structure of $W \in \mathcal{R}^{m \times n}$ over σ is given by the null-pole subspace

$$S_\sigma(W) = W\mathcal{R}^{n \times 1}(\sigma).$$

An indication of this statement is that the null-pole subspace is a complete invariant for right equivalence over W; more precisely, two rational $m \times n$ matrix functions (assumed for simplicity to have trivial right annhilators) W_1 and W_2 are related by $W_1 = W_2 Q$ where Q is analytic with nonsingular values on σ if and only if $S_\sigma(W_1) = S_\sigma(W_2)$.

Let (C_π, A_π) be a right pole pair and let (A_ζ, B_ζ) be a left null pair for a function $W \in \mathcal{R}^{m \times n}$. Then there exists a unique matrix Γ (see [BR3]) such that

$$(1.3) \quad S_\sigma(W) = (W\mathcal{R}^{n \times 1}) \cap \{C_\pi(z - A_\pi)^{-1} x + h(z) : x \in \mathbb{C}^{n_\pi \times 1}, h \in \mathcal{R}^{m \times 1}(\sigma)$$
$$\text{and} \sum_{z_0 \in \sigma} Res_{z=z_0}(z - A_\zeta)^{-1} B_\zeta h(z) = \Gamma x\}$$

The matrix Γ is called the <u>coupling matrix</u> associated with the right pole pair (C_π, A_π) and a left null pair (A_ζ, B_ζ) for W over σ. The triple

$$(1.4) \qquad\qquad \omega = ((C_\pi, A_\pi), (A_\zeta, B_\zeta), \Gamma)$$

is called a <u>left null-pole triple</u> for W over σ or a <u>left σ-spectral triple</u> of W. More detailed motivation for these concepts and connections with more established notions in systems theory can be found in [BGR1] and [BGR4] for the regular case and [BR5] for the nonregular case.

Let $W \in \mathcal{R}^{m \times n}$. Any matrix polynomial P_κ whose rows form a minimal polynomial basis for W^{ol} (see [F]) is said to be a <u>left kernel polynomial</u> for W. The following proposition characterizes matrix polynomials P_κ and data sets ω arising as a left kernel polynomial and a left σ-spectral triple for some rational matrix function.

PROPOSITION 1.1. *(cf. Proposition 4.1 in [BR2] and Theorem 3.1 in [BR3]) Let $W \in \mathcal{R}^{m \times n}$ be a function with a left kernel polynomial P_κ and left σ-spectral triple ω as in (1.4). Then ω and P_κ satisfy the following conditions:*

(NPDSi) *the pair (C_π, A_π) is observable and $\sigma(A_\pi) \subset \sigma$,*

(NPDSii) *the pair (A_ζ, C_ζ) is controllable and $\sigma(A_\zeta) \subset \sigma$,*

(NPDSiii) $\Gamma A_\pi - A_\zeta \Gamma = B_\zeta C_\pi$,

(NPDSiv) *P_κ has no zeros in $\mathbb{C}$ and the leading coefficients of the rows of P_κ are linearly independent,*

(NPDSv) *the function $P_\kappa(z)C_\pi(z - A_\pi)^{-1}$ is analytic on $\mathbb{C}$,*

(NPDSvi) *if $\sigma(A_\zeta) = \{\lambda_1, \lambda_2, \cdots, \lambda_r\}$, where the points $\lambda_1, \lambda_2, \cdots, \lambda_r$ are distinct, then the pair*

$$\left(\begin{bmatrix} A_\zeta & & & & \\ & \lambda_1 I & & & \\ & & \lambda_2 I & & \\ & & & \ddots & \\ & & & & \lambda_r I \end{bmatrix}, \begin{bmatrix} B_\zeta \\ P_\kappa(\lambda_1) \\ P_\kappa(\lambda_2) \\ \vdots \\ P_\kappa(\lambda_r) \end{bmatrix} \right)$$

is controllable.

Conversely, if $\{C_\pi, A_\pi, A_\zeta, B_\zeta, \Gamma\}$ is a collection of matrices and P_κ is a matrix polynomial such that (NPDSi)-(NPDSvi) hold, then there exists a rational matrix function W with a left σ-spectral triple ω as in (1.4) and with a left kernel polynomial equal to P_κ.

The construction of such a function W is indicated in [BR2] (see also Section 6 in [BR3]).

Given a matrix polynomial P_κ and a triple $\omega = ((C_\pi, A_\pi), (A_\zeta, B_\zeta), \Gamma)$, where $C_\pi, A_\pi, A_\zeta, B_\zeta, \Gamma$ are matrices of appropriate sizes, we may associate with ω and P_κ a linear space

$$(1.5) \quad S_\sigma(\omega, P_\kappa) = \{h \in \mathcal{R}^{m \times 1} : P_\kappa h = 0\} \cap \{C_\pi(z - A_\pi)^{-1}x + h(z) : x \in \mathbb{C}^{n_\pi \times 1}, h \in \mathcal{R}^{m \times 1}(\sigma)$$

$$\text{and } \sum_{z_0 \in \sigma} Res_{z=z_0}(z - A_\zeta)^{-1}B_\zeta h(z) = \Gamma x\}.$$

It follows from formulas (1.3) and (1.5) that if ω is a left σ-spectral triple and P_κ is a left kernel polynomial for a function $W \in \mathcal{R}^{m \times n}$, then $S_\sigma(\omega, P_\kappa) = S_\sigma(W)$. We will use the notation $S_\sigma(\omega, P_\kappa)$ and $S_\sigma(W)$ interchangeably. We shall refer to the collection $(\omega, P_\kappa) = (C_\pi, A_\pi, A_\zeta, B_\zeta, \Gamma, P_\kappa(z))$ as a <u>complete (left) null-pole data set</u> for W over σ.

2. A homogeneous interpolation problem.

In the previous section we saw that a complete null-pole data set (ω, P_κ) over σ for a rational matrix function W is a useful set of invariants for describing the null-pole subspace $S_\sigma(W) = W\mathcal{R}^{n\times 1}(\sigma)$ associated with W. However, for the 4-block interpolation problem, a more general submodule of $\mathcal{R}^m$ arises in a natural way. Specifically, assume that W in $\mathcal{R}^{m\times n}$ has a block row decomposition $W = [W_1 \, W_2]$ with $W_1 \in \mathcal{R}^{m\times(n-k)}$ and $W_2 \in \mathcal{R}^{m\times k}$, and consider the $\mathcal{R}(\sigma)$-submodule of $\mathcal{R}^{m\times 1}$

$$(2.1) \qquad S_\sigma(W_1, W_2) = W_1\mathcal{R}^{(n-k)\times 1}(\sigma) + W_2\mathcal{R}^{k\times 1}$$

We refer to $S_\sigma(W_1, W_2)$ as the <u>extended null-pole subspace</u> associated with the block row matrix function $[W_1 \, W_2]$ over σ. For our applications $[W_1 \, W_2]$ has trivial right annhilator in $\mathcal{R}^{n\times 1}$, i.e., the columns of $[W_1 \, W_2]$ are linearly independent over $\mathcal{R}$. Hence we make the assumption on $W = [W_1 \, W_2]$ that the decomposition in (2.1) is direct. Note that we can recover $W_2\mathcal{R}^{k\times 1}$ from $S_\sigma(W_1, W_2)$ as

$$W_2\mathcal{R}^{k\times 1} = \bigcap_{r\in\mathcal{R}} rS_\sigma(W_1, W_2)$$

The submodule $W_1\mathcal{R}^{(n-k)\times 1}(\sigma)$ is not as uniquely determined by $S_\sigma(W_1, W_2)$; all one can say is that it is a direct sum complement to $\bigcap_{r\in\mathcal{R}} rS_\sigma(W_1, W_2)$ inside the $\mathcal{R}(\sigma)$- submodule $S_\sigma(W_1, W_2)$. In any case the direct sum decomposition of $S_\sigma(W_1, W_2)$ can be viewed as an affine version of the Wold decomposition for an isometric operator on a Hilbert space (see [NF]). From another point of view, note that $S_\sigma(W_1', W_2') = S_\sigma(W_1, W_2)$ if and only if

$$[W_1' \, W_2'] = [W_1 \, W_2]\begin{bmatrix} F & 0 \\ G & H \end{bmatrix}$$
$$= [W_1F + W_2G \quad W_2H]$$

where F and F^{-1} are in $\mathcal{R}^{(n-k)\times(n-k)}(\sigma)$, $G \in \mathcal{R}^{k\times(n-k)}$ and H and H^{-1} are in $\mathcal{R}^{k\times k}$. Indeed, to see this, note that both

$$\begin{bmatrix} F & 0 \\ G & H \end{bmatrix} \text{ and } \begin{bmatrix} F^{-1} & 0 \\ -H^{-1}GF^{-1} & H^{-1} \end{bmatrix} = \begin{bmatrix} F & 0 \\ G & H \end{bmatrix}^{-1}$$

map $\mathcal{R}^{(n-k)\times 1}(\sigma) \oplus \mathcal{R}^{k\times 1}$ onto itself, and conversely, any multiplier in $\mathcal{R}^{n\times n}$ with this property necessarily must be of the form $\begin{bmatrix} F & 0 \\ G & H \end{bmatrix}$ with $F^{\pm 1} \in \mathcal{R}^{(n-k)\times(n-k)}(\sigma), G \in \mathcal{R}^{k\times(n-k)}, H^{\pm 1} \in \mathcal{R}^{k\times k}$. In particular we may always choose H so that the columns of $W_2' = W_2H$ form a minimal polynomial basis for their span, namely, $\bigcap_{r\in\mathcal{R}} rS_\sigma(W_1, W_2)$. Note that the zeros and poles of $W_1' = W_1F + W_2G$ can be quite different from those of W_1; the invariant is the left zero-pole structure in a subspace complementary to the $\mathcal{R}$-span of the columns of W_2. We may always choose $G \in \mathcal{R}^{k \times (n-k)}$ so that the span of the columns of $W_1' = W_1 + W_2 G$ is orthogonal over the set of zeros and poles of W_1' to the span of the columns of W_2 (or, equivalently, of W_2'). Let us call any σ-spectral triple $\omega = ((C_\pi, A_\pi), (A_\zeta, B_\zeta), \Gamma)$ for W_1' obtained from the pair (W_1, W_2) in this way a W_2-normalized left σ-spectral triple for W_1. The $\mathcal{R}$-span of $S_\sigma(W_1, W_2)$, $\bigcup_{r \in \mathcal{R}} r S_\sigma(W_1, W_2) = [W_1 \ W_2]\mathcal{R}^{n \times 1}$, can be specified via a left kernel polynomial P_κ for the function $[W_1 \ W_2]$. Given a block row matrix function $W = [W_1 \ W_2]$, we have now introduced (1) a W_2-normalized left σ-spectral triple ω for W_1, (2) a left kernel polynomial P_κ for $[W_1 \ W_2]$ and (3) a polynomial matrix $Q \in \mathcal{R}^{m \times L}$ whose columns form a minimal polynomial basis for the $\mathcal{R}$-span of the columns of W_2. Let us call the whole collection (ω, P_κ, Q) a <u>complete extended null-pole data set</u> for $W = [\ W_1 \ W_2]$. The analysis above shows that two block row matrix functions $[W_1 \ W_2]$ and $[W_1' \ W_2']$ for which the associated extended null-pole subspaces are the same $(S_\sigma(W_1', \ W_2') = S_\sigma(W_1, W_2))$ share the same complete extended null-pole data sets. Moreover, we can recover the extended null-pole subspace $S_\sigma(W_1, W_2)$ from the data set via the formula

$$
\begin{aligned}
&(\{f \in \mathcal{R}^{m \times 1} : P_\kappa f = 0\} \cap \\
&\{C_\pi(z - A_\pi)^{-1}x + h(z) : x \in \mathbb{C}^{n_\pi \times 1} \text{ and } h \in \mathcal{R}^{m \times 1}(\sigma) \text{ such that} \\
&\sum_{z_0 \in \sigma} Res_{z=z_0}(z - A_\zeta)^{-1}B_\zeta h(z) = \Gamma x\}) + Q\mathcal{R}^{k \times 1}
\end{aligned}
$$

(2.2)

For our application to interpolation problems in Section 4, we need to understand the inverse problem, namely: which data sets (ω, P_κ, Q) arise as the extended complete null-pole data set over σ of a block row $W = [W_1 \ W_2] \in \mathcal{R}^{m \times n}$. We view this problem as a homogeneous interpolation problem. The answer is given by the following result.

THEOREM 2.1. *Suppose* (ω, P_κ, Q) *(where* $\omega = ((C_\pi, A_\pi), (A_\zeta, B_\zeta), \Gamma))$ *is the extended complete null-pole data set for a block row function* $W = [W_1 \ W_2] \in \mathcal{R}^{m \times n}$ *(where* $W_1 \in \mathcal{R}^{m \times (n-k)}$ *and* $W_2 \in \mathcal{R}^{m \times k}$*). Then the data set* (ω, P_κ, Q) *satisfies the following conditions:*

(NPDSi) the pair (C_π, A_π) *is observable and* $\sigma(A_\pi) \subset \sigma$,

(NPDSii) the pair (A_ζ, B_ζ) *is controllable and* $\sigma(A_\zeta) \subset \sigma$,

(NPDSiii) $\Gamma A_\pi - A_\zeta \Gamma = B_\zeta C_\pi$,

(NPDSiv) P_κ *has no zeros in* $\mathbb{C}$ *and the leading coefficients of the rows of* P_κ *are linearly independent,*

(NPDSv) the function $P_\kappa(z)C_\pi(z - A_\pi)^{-1}$ *is analytic on* $\mathbb{C}$,

(NPDSvi) if $\sigma(A_\zeta) = \{\lambda_1, \lambda_2, \cdots, \lambda_r\}$, *where the points* $\lambda_1, \lambda_2, \cdots \lambda_r$ *are distinct, then the pair*

$$\left(\begin{bmatrix} A_\zeta & & & & \\ & \lambda_1 I & & & \\ & & \lambda_2 I & & \\ & & & \ddots & \\ & & & & \lambda_r I \end{bmatrix}, \begin{bmatrix} B_\zeta \\ P_\kappa(\lambda_1) \\ P_\kappa(\lambda_2) \\ \vdots \\ P_\kappa(\lambda_r) \end{bmatrix} \right)$$

is controllable.

(NPDSvii) Q *has no zeros in* $\mathbb{C}$ *and the leading coefficients of the columns of* Q *are linearly independent*

(NPDSviii) the function $(z - A_\zeta)^{-1} B_\zeta Q(z)$ *is analytic on* $\mathbb{C}$

(NPDSix) if $\sigma(A_\pi) = \{w_1, w_2, \cdots, w_s\}$, *then the pair*

$$\left([C_\pi \; Q(w_1) \; Q(w_2) \; \cdots \; Q(w_s)], \begin{bmatrix} A_\pi & & & & \\ & w_1 I & & & \\ & & w_2 I & & \\ & & & \ddots & \\ & & & & w_s I \end{bmatrix} \right)$$

is observable;

(NPDSx) $P_\kappa Q = 0$.

Conversely, if $\{C_\pi, A_\pi, A_\zeta, B_\zeta, \Gamma\}$ *is a collection of matrices and* P_κ *and* Q *are matrix polynomials such that conditions (NPDSi) - (NPDSx) hold, then there is a block row matrix function* $W = [W_1 \; W_2] \in \mathcal{R}^{m \times n}$ *such that* (ω, P_κ, Q) *is an extended complete null-pole data set for* W *over* σ.

REMARK. Note that there is some redundancy in the conditions (NPDSi) - (NPDSx). In particular (NPDSvi) implies the first part of (NPDSii) and (NPDSix) implies the first part of (NPDSi). Nevertheless, we list them in this form to make the connections with the 1-block case (NPDSi) - (NPDSiii) and the 2-block case (NPDSi) - (NPDSvi) clear.

PROOF: To prove the first statement, we note that condition (NPDSvii) follows from Proposition 3.18 in [BR2], and condition (NPDSix) follows from the fact that Q is analytic on $\mathbb{C}$ and the column spans of W_1 and Q are orthogonal on σ. We give now the constructive proof of the second statement. The construction is a modification of the construction in [BR2] (see also Section 6 in [BR3]).

Let $\{C_\pi, A_\pi, A_\zeta, B_\zeta, \Gamma\}$ be a collection of matrices and let P_κ and Q be matrix polynomials such that conditions (NPDSi) - (NPDSx) hold.

Step 1 Using the results from [GK], find a regular rational matrix function H with the σ-spectral triple $\omega = ((C_\pi, A_\pi), (A_\zeta, B_\zeta), \Gamma)$. Find a Smith-McMillan factorization EDF of H and set $W_1 = ED$.

Step 2 Let ν be the largest geometric multiplicity of a pole of H in σ, let μ be the largest geometric multiplicity of a zero of H in σ, and let η be the largest sum of the geometric multiplicity of a pole and the geometric multiplicty of a zero of H at any single point of σ. Let d_i denote the ith diagonal entry of D. For $i = \eta - \mu + 1, \eta - \mu + 2, \cdots, \nu$, let p_i be the minimal degree monic polynomial such that if $d_{m-\eta+i}$ has a zero point $\lambda \in \sigma$ of order k then $p_i d_i$ has a zero at λ of order k. Define an $m \times \eta$ matrix polynomial $Q = [q_{ij}]$ by

$$q_{ij} = \begin{cases} 1, & \text{if } i = j \leq \eta - \mu \\ p_i, & \text{if } \eta - \mu < i = j \leq \nu \\ 1, & \text{if } i = j - \eta + m > m - \mu \\ 0, & \text{otherwise,} \end{cases}$$

and put

$$W_2 = W_1 Q.$$

Step 3 For $i = 1, 2, \cdots, \mu$, let ϕ_i be the $(m - i - 1)^{th}$ row of E^{-1}, so that if H has a zero at a point $\lambda \in \sigma$ of the geometric multiplicity κ then $\{\phi_1, \phi_2, \cdots, \phi_\kappa\}$ is a canonical set of left null functions for H at λ. Note that the left null pair constructed from the functions $\phi_1, \cdots, \phi_\mu$ is left-similar to the pair (A_ζ, B_ζ). We may, in fact, assume that both pairs are equal. Modify the functions $\phi_1, \phi_2, \cdots, \phi_\mu$ as follows. Suppose that $\sigma(A_\pi) \cup \sigma(A_\zeta) = \{\lambda_1, \lambda_2, \cdots, \lambda_s\}$ and the largest multiplicity of a zero of H in σ is μ, and consider the point λ_i $(i = 1, 2, \cdots, s)$. Let the geometric multiplicity of a zero of H at λ_i be κ. By condition (NPDSvi), the matrix

$$A_i = \begin{bmatrix} \phi_1(\lambda_i) \\ \phi_2(\lambda_i) \\ \vdots \\ \phi_\mu(\lambda_i) \\ P_\kappa(\lambda_i) \end{bmatrix}$$

has full row rank. By conditions (NPDSviii) and (NPDSx), $A_i Q(\lambda_i) = 0$. Since μ is the largest geometric multiplicity of a zero of H in σ, there is a point λ such that the matrix

$$A = \begin{bmatrix} \phi_1(\lambda) \\ \phi_2(\lambda) \\ \vdots \\ \phi_\mu(\lambda) \\ P_\kappa(\lambda) \end{bmatrix}$$

has full row rank and $AQ(\lambda) = 0$. Hence we can add to ϕ_j $(j = \kappa + 1, \kappa + 2, \cdots, \mu)$ a function

$$p(z)c,$$

where c is a vector and $p(z)$ is a scalar polynomial vanishing at $\lambda_1, \lambda_2, \cdots, \lambda_s$ to the order μ, so that the modified $\phi_{\kappa+1}, \phi_{\kappa+2}, \cdots, \phi_\mu$ (which are again called $\phi_{\kappa+1}, \cdots, \phi_\mu$) are such that the matrix

$$
B_i = \begin{bmatrix} \phi_1(\lambda_i) \\ \phi_2(\lambda_i) \\ \vdots \\ \phi_\mu(\lambda_i) \\ P_\kappa(\lambda_i) \end{bmatrix}
$$

has full row rank and $B_i Q(\lambda_i) = 0$. In this manner we obtain functions $\phi_1, \phi_2, \cdots, \phi_\mu$ whose span is orthogonal to the row span of P_κ on $\sigma(A_\pi) \cup \sigma(A_\zeta)$ and such that the function

$$
\Phi(z) = \begin{bmatrix} \phi_1(z) \\ \phi_2(z) \\ \vdots \\ \phi_\mu(z) \end{bmatrix} Q(z)
$$

vanishes on $\sigma(A_\pi) \cup \sigma(A_\zeta)$. Now it follows from the construction and condition (NPDSviii) that if ϕ_i is a left null function for H at $\lambda_1, \lambda_2, \cdots, \lambda_s$ of orders $l_1, l_2, \cdots, l_s$, then $\phi_i(z)Q(z)$ vanishes at λ_j to the order at least k_j where $k_j \geq max\{1, l_j\}$. Suppose $k_j < \infty$. By condition (NPDSvii),

$$
\begin{aligned}
\phi_i(z)Q(z) &= (\prod_{j=1}^{s} (z - \lambda_i)^{k_j}) v_i(z) Q(z) \\
&= \tilde{\phi}_i(z) Q(z).
\end{aligned}
$$

where v_i is analytic, and does not vanish, at $\lambda_1, \lambda_2, \cdots, \lambda_s$.

Let $\psi_i = \phi_i - \tilde{\phi}_i$ $(i = 1, 2, \cdots, \mu)$. Then the set $\{\psi_1, \psi_2, \cdots, \psi_\mu\}$ contains a canonical set of left null functions for H at each zero of H in σ, the span of $\{\psi_1, \psi_2, \cdots, \psi_\mu\}$ is orthogonal to the row span of P_κ on $\sigma(A_\pi) \cup \sigma(A_\zeta)$, and $\psi_i Q = 0$ $(i = 1, 2, \cdots, \mu)$.

Extend the span of $\{\psi_1, \psi_2, \cdots, \psi_\mu\}$ to an orthogonal complement Ξ of the row span of P_κ in $(Q^{ol}, \sigma(A_\pi) \cup \sigma(A_\zeta))$, and project each column of W_2 onto an orthogonal complement of the row span of Q in $(P_\kappa^{or}, \sigma(A_\pi) \cup \sigma(A_\zeta))$ along

$$
\Xi^{or} + \{column\ span\ of\ Q\}
$$

to get W_3.

Step 4 Multiply W_3 on the right by a regular rational matrix function without poles or zeros in $\sigma(A_\pi) \cup \sigma(A_\zeta)$, so that the resulting function W_4 has no zeros nor poles in $\sigma \setminus (\sigma(A_\pi) \cup \sigma(A_\zeta))$.

Step 5 Find a minimal polynomial basis $\{u_1, u_2, \cdots, u_r\}$ for an orthogonal complement of the column span of $[W_4\ Q]$ in $(P_\kappa^{or}, \sigma(A_\pi) \cup \sigma(A_\zeta))$. Set

$$
W = [W_4\ u_1\ u_2\ \cdots\ u_r\ Q].
$$

The extended null-pole subspace $S_\sigma(\omega, P_\kappa, Q)$ given in (2.2) has a special relationship with the null-pole subspace $S(\omega, P_\kappa)$ studied in [BR3] as the following result shows.

THEOREM 2.2. *Suppose* $(\omega, P_\kappa(z), Q(z)) = (C_\pi, A_\pi, A_\zeta, B_\zeta, \Gamma, P_\kappa(z), Q(z))$ *is a σ-admissible extended null-pole data set and let $S_\sigma(\omega, P_\kappa)$ given by (1.5) and $S_\sigma(\omega, P_\kappa, Q)$ given by (2.2) be the associated $\mathcal{R}(\sigma)$-modules. Then*

$$S_\sigma(\omega, P_\kappa) \cap \mathcal{R}^{m\times 1}(\sigma) = S_\sigma(\omega, P_\kappa, Q) \cap \mathcal{R}^{m\times 1}(\sigma).$$

PROOF: The containment $\subset$ is trivial from the definition. Conversely, suppose that $f \in S_\sigma(\omega, P_\kappa, Q)$ is analytic on σ. By (2,2)

$$(2.3) \qquad f(z) = C_\pi(z - A_\pi)^{-1}x + h(z) + Q(z)r(z)$$

where $x \in \mathbb{C}^{n_\pi}, h \in \mathcal{R}^{m\times 1}(\sigma), r \in \mathcal{R}^{k\times 1}$ are such that $P_\kappa f = 0$ and $\sum_{z_0 \in \sigma} Res_{z=z_0}(z - A_\zeta)^{-1}B_\zeta h(z) = \Gamma x$. In general, if $g \in \mathcal{R}^{m\times 1}$ has partial fraction decomposition $g = g_- + g_+$ where $g_- \in \mathcal{R}_0^{m\times 1}(\sigma^c)$ (elements of $\mathcal{R}^{m\times 1}$ with all poles inside σ and vanishing at infinity) and $g_+ \in \mathcal{R}^{m\times 1}(\sigma)$, we define $P_{\sigma^c}^0(g) = g_-$. Since $C_\pi(z - A_\pi)^{-1}x \in \mathcal{R}_0^{m\times 1}(\sigma^c)$ and $h \in \mathcal{R}^{m\times 1}(\sigma)$ in (2.2), we have

$$(2.4) \qquad 0 = (P_{\sigma^c}^0(f))(z) = C_\pi(z - A_\pi)^{-1}x + (P_{\sigma^c}^0 Qr)(z)$$

But a consequence of (NPDSix) is that (2.4) can happen only if

$$C_\pi(z - A_\pi)^{-1}x = 0, \ Qr \in \mathcal{R}^{m\times 1}(\sigma).$$

From (NPDSi) and Lemma 12.2.2 in [BGR4], we conclude that $x = 0$. Hence (2.3) becomes

$$f(z) = h(z) + Q(z)r(z)$$

where $P_\kappa f = 0$ and $\sum_{z_0 \in r} Res_{z=z_0}(z - A_\pi)^{-1}B_\zeta h(z) = 0$. Since Qr is analytic on σ and by (NPDSvii) Q has no zeros on σ, we must have that r is analytic on σ. Furthermore, from (NPDSviii) we see that $\sum_{z_0 \in \sigma} Res_{z=z_0}(z - A_\zeta)^{-1}B_\zeta Q(z)r(z) = 0$. Then $f = h + Qr$ in fact is an element of $S_\sigma(\omega, P_\kappa) \cap \mathcal{R}^{m\times 1}(\sigma)$ as asserted.

3. Interpolation Problem.

Let σ be a subset of the complex plane $\mathbb{C}$. We look for a rational matrix function W which is analytic on σ and satisfies the following five interpolation conditions.

Let $z_1, z_2, \cdots, z_M$ be given (not necessarily distinct) points in σ. Let x_j and y_j be prescribed $1 \times m$ and $1 \times n$ vector functions analytic at z_j $(j = 1, 2, \cdots, M)$, and let $k_1, k_2, \cdots, k_M$ be positive integers. We require that

$$(3.1) \qquad \frac{d^{i-1}}{dz^{i-1}}(x_j(z)W(z))\Big|_{z=z_j} = \frac{d^{i-1}}{dz^{i-1}}y_j(z)\Big|_{z=z_j}$$

for $i = 1, 2, \cdots, k_j$ and $j = 1, 2, \cdots, M$.

Let $w_1, w_2, \cdots, w_N$ be given (not necessarily distinct) points on σ. Let u_j and v_j be prescribed $n \times 1$ and $m \times 1$ vector functions analytic at w_j $(j = 1, 2, \cdots, N)$ and let $l_1, l_2, \cdots, l_N$ be positive integers. The second condition is that

$$(3.2) \qquad \frac{d^{i-1}}{dz^{i-1}}(W(z)u_j(z))\Big|_{z=w_j} = \frac{d^{i-1}}{dz^{i-1}}f_j(z)\Big|_{z=w_j}$$

for $i = 1, 2, \cdots, l_j$ and $j = 1, 2, \cdots, N$.

For each pair of points z_i and w_j such that $z_i = w_j$, we are given numbers $\gamma_{fg}(f = 1, 2, \cdots, k_i$ and $g = 1, 2, \cdots, l_j)$ so that

$$(3.3) \qquad \frac{1}{(f + g - 1)!}\frac{d^{f+g-1}}{dz^{f+g-1}}(x_i^{(f)}(z)\, W(z)u_j^{(g)}(z))\Big|_{z=z_i} = \gamma_{fg},$$

where $h^{(\eta)}(z)$ is a polynomial obtained by discarding all but the first η coefficients in the Taylor expansion of a function h at z_i; that is, if $h(z) = \sum_{j=1}^{\infty} h^{\{j\}}(z - z_i)^{j-1}$ in a neighborhood of z_i, $h^{(\eta)}(z) = \sum_{j=1}^{\eta} h^{\{j\}}(z - z_i)^{j-1}$.

The fourth interpolation condition is as follows. We are given $1 \times m$ and $1 \times n$ rational vector functions p_{j+} and p_{j-} $(j = 1, 2, \cdots, K)$ analytic on σ and require that

$$(3.4) \qquad \frac{d^{i-1}}{dz^{i-1}}(p_{j+}(z)W(z))\Big|_{z=\lambda} = \frac{d^{i-1}}{dz^{i-1}}p_{j-}(z)\Big|_{z=\lambda}$$

for all positive integers i and for at least one (and hence for all) $\lambda \in \sigma$.

Finally, we are given $n \times 1$ and $m \times 1$ rational vector functions q_{j-} and q_{j+} $(j = 1, 2, \cdots, L)$ analytic on σ and demand that

$$(3.5) \qquad \frac{d^{i-1}}{dz^{i-1}}(W(z)q_{j-}(z))\Big|_{z=\lambda} = \frac{d^{i-1}}{dz^{i-1}}q_{j+}(z)\Big|_{z=\lambda}$$

for all positive integers i and for at least one (and hence all) points $\lambda \in \sigma$.

Below, we reformulate the conditions (3.1) - (3.5) and show when they are consistent. The combination (3.1) - (3.4) was considered in detail in [BR5]; we summarize this case (the so-called 2-block case) first. We note that the problem of finding a rational matrix function which is analytic on σ and satisfies conditions (3.1) - (3.3) is called the two-sided Lagrange-Sylvester interpolation problem. Its complete solution is presented in Chapter 16 of [BGR4]. The problem of finding a rational matrix function which is analytic on σ and satisfies conditions (3.1) - (3.4) has been solved in [BR5].

3.1 2-block Interpolation Problem.

In [BR5] the following general interpolation problem was considered. We are given a subset σ of the complex plane $\mathbb{C}$ and an interpolation data set

$$(3.6) \qquad \omega = (C_+, C_-, A_\pi, A_\zeta, B_+, B_-, \Gamma, P_+(z), P_-(z))$$

consisting of matrices $C_+ \in \mathbb{C}^{m \times n_\pi}, C_- \in \mathbb{C}^{n \times n_\pi}, A_\pi \in \mathbb{C}^{n_\pi \times n_\pi}, A_\zeta \in \mathbb{C}^{n_\zeta \times n_\zeta}, B_+ \in \mathbb{C}^{n_\zeta \times m}$, $B_- \in \mathbb{C}^{n_\zeta \times n}, \Gamma \in \mathbb{C}^{n_\zeta \times n_\pi}$ and matrix polynomials $P_+(z) \in \mathcal{R}^{K \times m}$ and $P_-(z) \in \mathcal{R}^{K \times n}$ which satisfy the admissibility requirements

(IDSi) the pair (C_-, A_π) is observable and $\sigma(A_\pi) \subset \sigma$;

(IDSii) the pair (A_ζ, B_+) is controllable and $\sigma(A_\zeta) \subset \sigma$;

(IDSiii) $\Gamma A_\pi - A_\zeta \Gamma = B_+ C_+ + B_- C_-$

(IDSiv) $P_-(\mathcal{R}^{n \times 1}) \subset P_+(\mathcal{R}^{m \times 1})$, $P_+(z)$ has no zeros in σ and the rows of $[P_+(z)\ P_-(z)]$ form a minimal polynomial basis;

(IDSv) the function

$$[P_+(z)\ P_-(z)] \begin{bmatrix} C_+ \\ C_- \end{bmatrix} (z - A_\pi)^{-1}$$

is analytic on $\mathbb{C}$.

(IDSvi) if $\sigma(A_\pi) = \{\lambda_1, \lambda_2, \cdots, \lambda_r\}$, the pair

$$\left(\begin{bmatrix} A_\zeta & & & & \\ & \lambda_1 I & & & \\ & & \lambda_2 I & & \\ & & & \ddots & \\ & & & & \lambda_r I \end{bmatrix}, \begin{bmatrix} B_+ \\ P_+(\lambda_1) \\ P_+(\lambda_2) \\ \vdots \\ P_+(\lambda_r) \end{bmatrix} \right)$$

is controllable. A collection of data ω in (3.1.1) satisfying (IDSi) - (IDSvi) is said to be a σ-admissible interpolation data set. The problem then is to describe all rational matrix functions with $W(z)$ analytic on σ which satisfy the interpolation conditions

$$(3.1') \qquad \sum_{z_0 \in \sigma} Res_{z=z_0}(z - A_\zeta)^{-1} B_+ W(z) = -B_-$$

$$(3.2') \qquad \sum_{z_0 \in \sigma} Res_{z=z_0} W(z) C_-(z - A_\pi)^{-1} = C_+$$

$$(3.3') \qquad \sum_{z_0 \in \sigma} Res_{z=z_0}(z - A_\zeta)^{-1} B_+ W(z) C_-(z - A_\pi)^{-1} = \Gamma$$

and

$$(3.4') \qquad P_+ W = -P_-$$

We mention that the problem (3.1') - (3.4') is simply a more compact way of writing conditions (3.1) - (3.4). Indeed (3.1') is equivalent to (3.1) if we let, for each $j = 1, 2, \cdots, M$,

$$
(3.7) \qquad A_{\zeta_j} = \begin{bmatrix} z_j & & & \\ 1 & z_j & & \\ & \ddots & \ddots & \\ & & 1 & z_j \end{bmatrix}, \, B_{j+} = \begin{bmatrix} x_j^{\{1\}} \\ x_j^{\{2\}} \\ \vdots \\ x_j^{\{k_j\}} \end{bmatrix}, \, B_{j-} = - \begin{bmatrix} y_j^{\{1\}} \\ y_j^{\{2\}} \\ \vdots \\ y_j^{\{k_j\}} \end{bmatrix},
$$

where $f^{\{i\}}$ denotes the ith coefficient in the Taylor expansion of a function f at z_j, and set

$$
(3.8) \qquad A_\zeta = \begin{bmatrix} A_{\zeta_1} & & & \\ & A_{\zeta_2} & & \\ & & \ddots & \\ & & & A_{\zeta_M} \end{bmatrix}, \, B_+ = \begin{bmatrix} B_{1+} \\ B_{2+} \\ \vdots \\ B_{M+} \end{bmatrix}, \, B_- = \begin{bmatrix} B_{1-} \\ B_{2-} \\ \vdots \\ B_{M-} \end{bmatrix}.
$$

Conversely, if matrices A_ζ, B_+, B_- of appropriate sizes are given, there exists a nonsingular matrix S such that $SA_\zeta S^{-1}$ is a block diagonal matrix with the diagonal blocks in lower Jordan form. After replacing A_ζ by $SA_\zeta S^{-1}$, B_+ by SB_+ and B_- by SB_-, we can convert the interpolation condition (3.1) to the equivalent condition (3.1'). Similarly, if for each $j = 1, 2, \cdots, N$ we let

$$
(3.9) \qquad C_{j+} = \begin{bmatrix} v_j^{\{1\}} v_j^{\{2\}} \cdots v_j^{\{l_j\}} \end{bmatrix}, C_{j-} = [u_j^{\{1\}} u_j^{\{2\}} \cdots u_j^{\{l_j\}}]
$$

$$
A_{\pi_j} = \begin{bmatrix} w_j & 1 & & \\ & w_j & \ddots & \\ & & \ddots & 1 \\ & & & w_j \end{bmatrix}
$$

and then set

$$
C_+ = [C_{1+} \, C_{2+} \cdots C_{N+}], \; C_- = [C_{1-} \, C_{2-} \cdots C_{N-}],
$$

$$
A_\pi = \begin{bmatrix} A_{\pi_1} & & & \\ & A_{\pi_2} & & \\ & & \ddots & \\ & & & A_{\pi_N} \end{bmatrix}
$$

then (3.2') collapses to (3.2). Conversely, if the matrices C_+, C_-, A_π are given, we can find a nonsingular matrix S such that $S^{-1}A_\pi S$ is in Jordan form. After replacing A_π by $S^{-1}A_\pi S$, C_- by $C_- S$, and C_+ by $C_+ S$, we can reformulate the interpolation condition (3.2') in the more detailed form (3.2).

A similar analysis gives an equivalence between (3.4) and (3.4'). Suppose first that $z_i = w_j$ and let $\Gamma_{ij} = [\gamma_{fg}]$ where $\gamma_{fg}(f = 1, 2, \cdots k_i$ and $g = 1, 2, \cdots, l_j)$ are numbers associated with z_i and w_j as in (3.3). Then the interpolation conditions (3.3) hold if and only if

$$Res_{z=z_i}(z - A_{\zeta_i})^{-1}B_{i+}W(z)C_{j-}(z - A_{\pi_j})^{-1} = \Gamma_{ij}$$

If $z_i \neq w_j$ let Γ_{ij} be the unique solution of the Sylvester equation in X

$$XA_{\pi_j} - A_{\zeta_i}X = B_{i-}C_{j-} + B_{i+}C_{j+}$$

where A_{π_j} and A_{ζ_i} are as in (3.7) and (3.8) (see [BGR4], Appendix A.1) and set $\Gamma = [\Gamma_{ij}]$ with $1 \leq i \leq M$ and $1 \leq j \leq N$. Then the set of interpolation conditions (3.3) is equivalent to the single block interpolation condition (3.3'). We also mention that validity of the Sylvester equation

$$\Gamma A_{\pi} - A_{\zeta}\Gamma = B_-C_- + B_+C_+$$

is a necessary and sufficient condition for the consistency of (3.1') - (3.3') (see [BGR4] or [BR5]). Conditions (3.3) can be recovered from (3.3') by multiplying both sides of the equality by non-singular matrices S and T such that $SA_{\zeta}S^{-1}$ and $T^{-1}A_{\zeta}T$ are in lower and upper Jordan forms, respectively, and reading the numbers γ_{fg} from the appropriate blocks of the matrix $S\Gamma T$.

Finally, if P_+ and P_- are rational matrix functions whose j^{th} rows $(j = 1, \cdots, K)$ are equal to p_{j+} and $-p_{j-}$ respectively, then it is easy to see that conditions (3.4) are equivalent to (3.4'). Demanding that P_+ and P_- are polynomials is with no loss of generality. The admissibility conditions (IDSi) - (IDSvi) guarantee consistency and minimize redundancies in the interpolation conditions. For complete details we refer to [BR5].

3.2 The general 4-block case.

Let Q_- and Q_+ be rational matrix functions whose j^{th} columns $(j = 1, 2, \cdots, L)$ are equal to q_{j-} and q_{j+}, respectively. Conditions (3.5) are equivalent to

$$(3.5') \qquad\qquad\qquad W(z)Q_-(z) \equiv Q_+(z).$$

Similarly as with P_+ and P_-, we will assume first Q_- and Q_+ are matrix polynomials and the columns of

$$(3.10) \qquad\qquad\qquad \begin{bmatrix} Q_+ \\ Q_- \end{bmatrix}$$

form a minimal basis.

We consider now the consistency of condition (3.5'). It follows immediately from (3.5') that

$$\mathcal{R}^{1\times m}Q_+(z) \subset \mathcal{R}^{1\times n}Q_-(z).$$

Also, since W is analytic on σ, Q_- has no zeros in σ. Indeed, if Q_- had a zero in σ, then $Q_+ = WQ_-$ would have a zero in σ, and so the rank of the matrix polynomial (3.10) would drop at some point of σ, contradicting the fact that the columns of (3.10) form a minimal polynomial basis.

Condition (3.1) says that the first k_j Taylor coefficients at z_j of $x_j(z)W(z)$ coincide with the first k_j Taylor coefficients at z_j of $y_j(z)$. Therefore, by (3.5'), the first k_j Taylor coefficients at z_j of $x_j(z)Q_+(z)$ coincide with the first k_j Taylor coefficients at z_j of $y_j(z)Q_-(z)$. Consequently, the function

$$(z - A_{\zeta_j})^{-1}[B_{j+} \ B_{j-}]\begin{bmatrix} Q_+(z_j) \\ Q_-(z_j) \end{bmatrix}$$

is analytic on $\mathbb{C}$ where A_{ζ_j}, B_{j+} and B_{j-} are as in (3.7). Hence the function

$$(z - A_\zeta)^{-1}[B_+ \ B_-]\begin{bmatrix} Q_+(z) \\ Q_-(z) \end{bmatrix}$$

is analytic on $\mathbb{C}$.

Suppose that a vector $u_j(w_j)$ is in the column span of $Q_-(w_j)$, where u_j is one of the functions in (3.2). Then $u_j(w_j) = Q_-(w_j)u_0$, and it follows from (3.2) and (3.5') that

$$Q_+(w_j)u_0 = v_j(w_j).$$

So in this case conditions (3.2) and (3.5) are either redundant or contradictory. To exclude such situations we require that the matrix

$$\left[u_{j_1}(w_{j_1}) \ u_{j_2}(w_{j_2}) \cdots u_{j_k}(w_{j_k}) \ Q_-(w_{j_1}) \right]$$

have full column rank whenever $w_{j_1} = w_{j_2} = \cdots = w_{j_k}$. In terms of the data A_π, C_+, C_-, this assumption is equivalent to the controllability of the pair

$$(3.11) \qquad \left(\left[C_- \ Q_-(w_{j_1}) \cdots Q_-(w_{j_k}) \right], \begin{bmatrix} A_\pi & & & \\ & w_{j_1}I & & \\ & & \ddots & \\ & & & w_{j_k}I \end{bmatrix} \right)$$

where $\{w_{j_1}, w_{j_2}, \cdots, w_{j_k}\}$ is the set of distinct points among $\{w_1, w_2, \cdots, w_N\}$. We will call this assumption the consistency of right generic and right lumped interpolation conditions.

Finally, in view of (3.4') and (3.5'),

$$P_+ Q_+ = P_+ W Q_-$$
$$= -P_- Q_-.$$

So conditions (3.4') and (3.5') are consistent if

$$[P_+ \ P_-]\begin{bmatrix} Q_+ \\ Q_- \end{bmatrix} = 0.$$

We summarize various consistency and minimality properties our interpolation data must have in the following definition. We will call the data

$$(3.12) \qquad \omega = \left(C_+, C_-, A_\pi, A_\zeta, B_+, B_-, \Gamma, P_+(z), P_-(z), Q_-(z), Q_+(z) \right)$$

a <u>σ-admissible interpolation data set</u> if $C_+ \in \mathbb{C}^{m \times n_\pi}, C_- \in \mathbb{C}^{n \times n_\pi}, A_\pi \in \mathbb{C}^{n_\pi \times n_\pi}, A_\zeta \in \mathbb{C}^{n_\zeta \times n_\zeta}, B_+ \in \mathbb{C}^{n_\zeta \times m}, B_- \in \mathbb{C}^{n_\zeta \times n}, \Gamma \in \mathbb{C}^{n_\zeta \times n_\pi}, P_+(z) \in \mathcal{R}^{K \times m}, P_-(z) \in \mathcal{R}^{K \times n}, Q_- \in \mathcal{R}^{m \times L},$ and $Q_+ \in \mathcal{R}^{n \times L}$ are such that

(IDSi) the pair (C_-, A_π) is observable and $\sigma(A_\pi) \subset \sigma$;

(IDSii) the pair (A_ζ, B_+) is controllable and $\sigma(A_\pi) \subset \sigma$;

(IDSiii) $\Gamma A_\pi - A_\zeta \Gamma = B_+ C_+ + B_- C_-$;

(IDSiv) $P_- \mathcal{R}^{n \times 1} \subset P_+ \mathcal{R}^{m \times 1}, P_+(z)$ has no zeros in σ and the rows of $[P_+(z) \ P_-(z)]$ form a minimal polynomial basis;

(IDSv) the function

$$[P_+(z) \ P_-(z)]\begin{bmatrix} C_+ \\ C_- \end{bmatrix}(z - A_\pi)^{-1}$$

is analytic on $\mathbb{C}$.

(IDSvi) if $\sigma(A_\pi) = \{\lambda_1, \lambda_2, \cdots, \lambda_r\}$, the pair

$$\left(\begin{bmatrix} A_\pi & & & & \\ & \lambda_1 I & & & \\ & & \lambda_2 I & & \\ & & & \ddots & \\ & & & & \lambda_r I \end{bmatrix}, \begin{bmatrix} B_+ \\ P_+(\lambda_1) \\ P_+(\lambda_2) \\ \vdots \\ P_+(\lambda_r) \end{bmatrix} \right)$$

is controllable;

(IDSvii) $\mathcal{R}^{1 \times m} Q_+(z) \subset \mathcal{R}^{1 \times n} Q_-(z), Q_-(z)$ has no zeros in σ, and the columns of

$$\begin{bmatrix} Q_+(z) \\ Q_-(z) \end{bmatrix}$$

form a minimal polynomial basis;

(IDSviii) the function

$$(z - A_\zeta)^{-1}[B_+ \ B_-]\begin{bmatrix} Q_+(z) \\ Q_-(z) \end{bmatrix}$$

is analytic on $\mathbb{C}$.

(IDSix) if $\sigma(A_\pi) = \{w_1, w_2, \cdots, w_s\}$, then the pair

$$\left([C_- \ Q_-(w_1) \ Q_-(w_2) \cdots Q_-(w_s)], \begin{bmatrix} A_\pi & & & & \\ & w_1 I & & & \\ & & w_2 I & & \\ & & & \ddots & \\ & & & & w_s I \end{bmatrix} \right)$$

is observable.

(IDSx) $[P_+ \ P_-]\begin{bmatrix} Q_+ \\ Q_- \end{bmatrix} = 0.$

Given a σ-admissible interpolation data set, the interpolation problem is to find which $m \times n$ rational matrix functions W analytic on σ satisfy conditions (3.1') - (3.5').

4. Parametrization of solutions.

Theorems 4.1 and 4.2 below present our results on parameterization of solutions of the interpolation problem (3.1') - (3.5') (respectively with or without an additional norm constraint). We first need the following observation. If $\omega = (C_+, C_-, A_\pi, A_\zeta, B_+, B_-, \Gamma, P_+(z), P_-(z), Q_-(z), Q_+(z))$ is a σ-admissible interpolation data set, then the collection of matrices

$$(4.1) \qquad \tilde{\omega} = \left(\left(\begin{bmatrix} C_+ \\ C_- \end{bmatrix}, A_\pi \right), \left(A_\zeta, [B_+ \ B_-] \right), \Gamma \right)$$

together with the matrix polynomials $P_\kappa = [P_+ \ P_-]$ and $Q = \begin{bmatrix} Q_+ \\ Q_- \end{bmatrix}$ satisfy conditions (NPDSi) - (NPDSx) in Theorem 2.1. Therefore there exists a rational block row matrix function $\Theta = [\Theta_1 \ \Theta_3] = \begin{bmatrix} \Theta_{11} & \Theta_{12} & \vdots & \Theta_{13} \\ \Theta_{21} & \Theta_{22} & \vdots & \Theta_{23} \end{bmatrix}$ in $\mathcal{R}^{(m+n)\times(m-K+n)}$ (where $\Theta_1 \in \mathcal{R}^{(m+n)\times(m-K+n-L)}$, $\Theta_2 \in \mathcal{R}^{(m+n)\times L}$ and $\Theta_1 = \begin{bmatrix} \Theta_{11} & \Theta_{12} \\ \Theta_{21} & \Theta_{22} \end{bmatrix}$ with $\Theta_{11} \in \mathcal{R}^{m\times(m-K)}$, $\Theta_3 = \begin{bmatrix} \Theta_{13} \\ \Theta_{23} \end{bmatrix}$ with $\Theta_{13} \in \mathcal{R}^{m\times L}$) which has $\left(\omega, [P_+(z) \ P_-(z)], \begin{bmatrix} Q_+(z) \\ Q_-(z) \end{bmatrix} \right)$ as an extended complete null-pole data set over σ. The 2-block version of the following appears as Theorem 4.1 in [BR5].

THEOREM 4.1. *Let $\sigma \in \mathbb{C}$ and let ω as in (3.12) be a σ-admissible interpolation data set. Then there exist functions $W \in \mathcal{R}^{m\times n}(\sigma)$ which satisfy the interpolation conditions (3.1') - (3.5'). Moreover, if*

$$\Theta = \begin{bmatrix} \Theta_{11} & \Theta_{12} & \vdots & \Theta_{13} \\ \Theta_{21} & \Theta_{22} & \vdots & \Theta_{23} \end{bmatrix}$$

is a block row rational matrix function with extended complete null-pole data set over σ equal to $\left(\tilde{\omega}, [P_+(z)\ P_-(z)], \begin{bmatrix} Q_+(z) \\ Q_-(z) \end{bmatrix} \right)$, then a function $W \in \mathcal{R}^{m \times n}$ is analytic on σ and satisfies interpolation conditions (3.1') - (3.5') if and only if

$$W = [\Theta_{11}Q_1 + \Theta_{12}Q_2\ \Theta_{13}][\Theta_{21}Q_1 + \Theta_{22}Q_2\ \Theta_{23}]^{-1}$$

where $Q_1 \in \mathcal{R}^{(m-K) \times (n-L)}(\sigma)$ and $Q_2 \in \mathcal{R}^{(n-L) \times (n-L)}$ are such that

$$[\Theta_{21}Q_1 + \Theta_{22}Q_2\ \Theta_{23}] \left(\mathcal{R}^{(n-L) \times 1}(\sigma) \oplus \mathcal{R}^{L \times 1} \right)$$

(4.2)
$$= [\Theta_{21}\ \Theta_{22}\ \Theta_{23}] \left(\mathcal{R}^{(m-K+n-L) \times 1}(\sigma) \oplus \mathcal{R}^{L \times 1} \right).$$

We note that condition (4.2) holds for a generic pair (Q_1, Q_2) of rational matrix functions which are coprime over σ. One simply must arrange that the n-dimensional $\mathcal{R}$-vector subspace $\begin{bmatrix} Q_1 & 0 \\ Q_2 & 0 \\ 0 & I \end{bmatrix} \mathcal{R}^{n \times 1}$ is σ-orthogonal to the right annihilator of the $n \times (m - K + n)$ rational matrix function $[\Theta_{21}\ \Theta_{22}\ \Theta_{23}]$, an $\mathcal{R}$-subspace inside $\mathcal{R}^{(m-K+n) \times 1}$ of dimension $m - K$.

For the H^∞-control problem, one takes σ to be either the right half plane $\Pi^+ = \{z : \text{Re } z > 0\}$ or the unit disk $\mathcal{D} = \{z : |z| < 1\}$ and asks that W, in addition to meeting a set of interpolation conditions on σ, satisfy $\|W\|_\infty := rmsup_{z \in \sigma}\|W(z)\| < \gamma$ for some prescribed tolerance level γ. Without loss of generality we assume that γ has been normalized to $\gamma = 1$. The following refinement of Theorem 4.1 gives the solution of this generalized Nevanlinna Pick problem (i.e. interpolation problem with norm constraint). The 2-block version of the following result is Theorem 4.2 in [BR5].

THEOREM 4.2. *Let σ be either the right half plane Π^+ or the unit disk $\mathcal{D}$ and let ω as in (3.12) be a σ-admissible interpolation data set. Suppose that there exists a rational $(m + n) \times (m - K + n)$ block row matrix function*

$$\Theta(z) = \begin{bmatrix} \Theta_{11}(z) & \Theta_{12}(z) & \vdots & \Theta_{13}(z) \\ \Theta_{21}(z) & \Theta_{22}(z) & \vdots & \Theta_{23}(z) \end{bmatrix}$$

(with $\Theta_{11} \in \mathcal{R}^{m \times (m-K)}$, $\Theta_{12} \in \mathcal{R}^{m \times (n-L)}$) such that

(a) $\left(\begin{bmatrix} C_+ \\ C_- \end{bmatrix}, A_\pi, A_\zeta, [B_+\ B_-], \Gamma, [P_+(z)\ P_-(z)], \begin{bmatrix} Q_+(z) \\ Q_-(z) \end{bmatrix} \right)$ *is an extended complete null-pole data set for Θ over σ*

and

(b) Θ *is analytic on $\partial\sigma$ (including at infinity if $\sigma = \Pi^+$) and satisfies $\Theta(z)^* J \Theta(z) = j \oplus -I_L, z \in \partial\sigma$ where $J = I_m \oplus -I_n, j = I_{m-K} \oplus -I_{n-L}$.*

Then the following are equivalent.

    (i) *There exist functions $W \in \mathcal{R}^{m \times n}(\sigma)$ which satisfy the interpolation conditions (3.1') - (3.5') and in addition satisfy $\|W\|_\infty < 1$.*

    (ii) *Θ satisfies $\Theta(z)^* J \Theta(z) \leq j \oplus 0_{(n-L) \times (n-L)}$ at all points z of analyticity in σ.*

    Moreover, if (i) and (ii) hold, function $W \in \mathcal{R}^{m \times n}$ is analytic on σ, satisfies the interpolation conditions (3.1') - (3.5') and has $\|W\|_\infty < 1$ if and only if

$$(4.3) \qquad W = \left[\Theta_{11} H + \Theta_{12} \; \Theta_{13} \right] \left[\Theta_{21} H + \Theta_{22} \; \Theta_{23} \right]^{-1} =: \mathcal{G}_\Theta[H]$$

for an $H \in \mathcal{R}^{(m-K) \times (n-L)}(\sigma)$ with $\|H\|_\infty < 1$.

REMARK. In fact the existence of a matrix function Θ satisfying conditions (a), (b) in Theorem 4.2 is also necessary for interpolants W with $\|W\|_\infty < 1$ to exist, but we do not prove this here.

    To make Theorem 4.2 useful we need a systematic way of computing the rational matrix function Θ appearing in Theorem 4.2. For practical purposes we need only compute a rational matrix function Θ of the form

$$\Theta = \tilde{\Theta} \begin{bmatrix} I & 0 & 0 \\ 0 & I & 0 \\ \gamma_1 & \gamma_2 & \gamma_3 \end{bmatrix}$$

where γ_1, γ_2, and γ_3 are any rational matrix functions (of the appropriate sizes) with γ_3 invertible and $\tilde{\Theta}$ is as in Theorem 4.2, since in such a case Θ and $\tilde{\Theta}$ induce the same linear fractional map $\mathcal{G}_\Theta = \mathcal{G}_{\tilde{\Theta}}$ (see [BHV]). One can obtain such a matrix function Θ from a preliminary $\tilde{\tilde{\Theta}}$ constructed to meet the specifications in Theorem 4.1 as follows. If $\tilde{\tilde{\Theta}} = \begin{bmatrix} \tilde{\tilde{\Theta}}_{11} & \tilde{\tilde{\Theta}}_{12} & \tilde{\tilde{\Theta}}_{13} \\ \tilde{\tilde{\Theta}}_{21} & \tilde{\tilde{\Theta}}_{22} & \tilde{\tilde{\Theta}}_{23} \end{bmatrix}$,

define $W(z) = \left[W_{ij}(z) \right]_{1 \leq i,j \leq 3}$ by $W_{ij}(z) = \tilde{\tilde{\Theta}}_{1i}(z)^* \tilde{\tilde{\Theta}}_{1i}(z) - \tilde{\tilde{\Theta}}_{2i}(z)^* \tilde{\tilde{\Theta}}_{2j}(z)$.
Partition W as a block 2×2 matrix function by

$$W = \begin{bmatrix} W_0 & W_{30}^* \\ W_{30} & W_{33} \end{bmatrix}$$

where $W_0 = [W_{ij}]_{1 \leq i,j \leq 2}$ and set V equal to the Schur complement

$$V = W_0 - W_{30}^* W_{33}^{-1} W_{30}.$$

Finally find a square outer matrix function $R_0(z)$ such that

$$V(z) = R_0(z)^* j R_0(z).$$

Then

$$\Theta = \tilde{\tilde{\Theta}} \begin{bmatrix} R_0^{-1} & 0 \\ 0 & I \end{bmatrix}$$

is the desired matrix function such that the associated linear fractional map $\mathcal{G}_\Theta$ parametrizes interpolants with norm less than 1 as in Theorem 4.2. Hence, once one has constructed a $\tilde{\tilde{\Theta}}$ as in Theorem 4.1, one arrives at an appropriate Θ via one j-spectral factorization $V = R^*jR$. For more complete details we refer to Theorem 4.3 in [BHV].

Alternatively, once one has a matrix function $\tilde{\tilde{\Theta}}$ meeting the specifications in Theorem 4.1, one could transform to a plant $\mathcal{P}$ such that the transform $\tilde{\mathcal{P}}$ of $\mathcal{P}$ to the generalized chain formalism is equal to $\tilde{\tilde{\Theta}}$ (see Section 5). Then the interpolation problem with norm constraint $\|W\|_\infty < 1$ is equivalent to the H^∞-problem for the plant $\mathcal{P}$ (with tolerance level 1); one can then use any of the known solutions of the H^∞-problem (e.g. [DGKF], [G], [PAJ]) to produce the parameterizer Θ of contractive (or "bounded real" in engineering parlance) interpolants.

Of course, it would be desirable to have the criterion for existence of interpolants W with $\|W\|_\infty < 1$ and the construction of the linear fractional map $\mathcal{G}_\Theta$ expressed more directly in terms of the interpolation data ω, as has been done for the 1-block case in [BGR4]; this together with a more explicit realization formula for Θ in terms of the data set σ, is the remaining open problem in this area.

As was remarked above, $\left(\tilde{\omega}, [P_+(z)\ P_-(z)], \begin{bmatrix} Q_+(z) \\ Q_-(z) \end{bmatrix}\right)$ is a σ-admissible extended null-pole data set whenever $\omega = \left(C_+, C_-, A_\pi, A_\zeta, B_+, B_-, \Gamma, P_+(z), P_-(z), Q_-(z), Q_+(z)\right)$ is a σ-admissible interpolation data set. The converse statement is not true in general; the following theorem, which will be used in our characterization of stabilizable linear feedback systems in the next section, characterizes which σ-admissible extended null-pole data sets $\left(\tilde{\omega}, [P_+(z)\ P_-(z)], \begin{bmatrix} Q_+(z) \\ Q_-(z) \end{bmatrix}\right)$ arise from σ-admissible interpolation data sets in terms of the associated extended null-pole subspace $S\left(\tilde{\omega}, [P_+\ P_-], \begin{bmatrix} Q_+ \\ Q_- \end{bmatrix}\right)$; see Theorem 4.3 [BR5] for the 2-block case and Theorem 13.2.3 in [BGR4] for the 1-block case.

THEOREM 4.3. *Let $(\tilde{\omega}, P, Q)$ be a σ-admissible extended null-pole data set of the form*

$$(\tilde{\omega}, P, Q) = \left(\begin{bmatrix} C_+ \\ C_- \end{bmatrix}, A_\pi, A_\zeta, [B_+\ B_-], \Gamma, [P_+(z)\ P_-(z)], \begin{bmatrix} Q_+(z) \\ Q_-(z) \end{bmatrix}\right)$$

and set

$$\omega = \left(C_+, C_-, A_\pi, A_\zeta, B_+, B_-, \Gamma, P_+(z), P_-(z), Q_-(z), Q_+(z)\right).$$

Then the following are equivalent.

(i) *ω is a σ-admissible interpolation data set.*

(ii) *The extended null-pole subspace $S = S_\sigma(\tilde{\omega}, P, Q) \subset \mathcal{R}^{(m-K+n)\times 1}$ (see (2.2)) has the following properties:*

(a) *$S \cap \left[\mathcal{R}^{m\times 1} \oplus 0\right] \subset \mathcal{R}^{m\times 1}(\sigma) \oplus 0,$*

$$(b) \quad P_{0 \oplus \mathcal{R}^{n \times 1}}\left(S \cap \mathcal{R}^{(m+n) \times 1}(\sigma) \right) = 0 \oplus \mathcal{R}^{n \times 1}(\sigma) \text{ where } P_{0 \oplus \mathcal{R}^{n \times 1}}\left(\begin{bmatrix} a \\ b \end{bmatrix} \right) = \begin{bmatrix} 0 \\ b \end{bmatrix}.$$

PROOF: Suppose first that ω is a σ-admissible interpolation data set. By definition, any function $f \in S$ has the form

$$f(z) = \begin{bmatrix} f_+(z) \\ f_-(z) \end{bmatrix} = \begin{bmatrix} C_+ \\ C_- \end{bmatrix} (z - A_\pi)^{-1} x + \begin{bmatrix} h_+(z) \\ h_-(z) \end{bmatrix} + \begin{bmatrix} Q_+(z) \\ Q_-(z) \end{bmatrix} r(z)$$

where $x \in \mathbb{C}^{n_\pi}$, $h_+ \in \mathcal{R}^{m \times 1}(\sigma), h_- \in \mathcal{R}^{n \times 1}(\sigma), r \in \mathcal{R}^{L \times 1}$. Suppose that $f_-(z) = C_-(z - A_\pi)^{-1} x + h_-(z) + Q_-(z) r(z) = 0$. Let $P_{\sigma^c}^0$ denote the projection map $P_{\sigma^c}^0 : k \to k_-$ if $k \in \mathcal{R}^{n \times 1}$ has partial fraction decomposition $k = k_- + k_+$ where k_- has all poles in σ and vanishes at infinity (i.e. $k_- \in \mathcal{R}_0^{n \times 1}(\sigma^c)$) and $k_+ \in \mathcal{R}^{n \times 1}(\sigma)$. Then from $f_- = 0$ we get that $P_{\sigma^c}^0(f_-) = 0$; since $C_-(z - A_\pi)^{-1} x \in \mathcal{R}_0^{n \times 1}(\sigma^c)$ and $h_- \in \mathcal{R}^{n \times 1}(\sigma)$, we see that

$$(4.4) \qquad\qquad 0 = P_{\sigma^c}^0(f_-) = C_-(z - A_\pi)^{-1} x + P_{\sigma^c}^0(Q_- r).$$

But now by condition (IDSix), (4.4) forces

$$(4.5) \qquad\qquad C_-(z - A_\pi)^{-1} x = 0, \quad P_{\sigma^c}^0(Q_- r) = 0.$$

By Lemma 12.2.2 in [BGR4], the first condition in (4.5) forces $x = 0$. Also, since by (IDSvii) Q_- has no zeros in σ, the second condition in (4.5) forces r to be analytic on σ. But then $f_+(z) = h_+(z) + Q_+(z) r(z)$ is analytic on σ. Thus S satisfies conditions (ii-a) in Theorem 4.3.

To verify (ii-b) we must show that for any given $f_- \in \mathcal{R}^{m \times 1}(\sigma)$ we can find $f_+ \in \mathcal{R}^{m \times 1}(\sigma)$ so that $\begin{bmatrix} f_+ \\ f_- \end{bmatrix} \in S$. But by Theorem 2.2, $S_\sigma(\tilde{\omega}, P) \cap \mathcal{R}^{(m+n) \times 1}(\sigma) = S_\sigma(\tilde{\omega}, P, Q) \cap \mathcal{R}^{(m+n) \times 1}(\sigma)$. Hence this result follows in the same way as for the 2-block case (see Theorem 4.3 in [BR5]).

Conversely, suppose $\left(\tilde{\omega}, [P_+ \ P_-], \begin{bmatrix} Q_+ \\ Q_- \end{bmatrix} \right)$ is a σ-admissible extended null-pole data set for which $S = S_\sigma\left(\tilde{\omega}, [P_+ \ P_-], \begin{bmatrix} Q_+ \\ Q_- \end{bmatrix} \right)$ satisfies (ii-a) and (ii-b). Thus $\left(\tilde{\omega}, [P_+ \ P_-], \begin{bmatrix} Q_+ \\ Q_- \end{bmatrix} \right)$ satisfies (NPDSi)-(NPDSx) and we must verify that ω satisfies (IDSi) - (IDSx). First note that (IDSiii), (IDSv), (IDSviii) and (IDSx) follow from their counterparts (NPDSiii), (NPDSv), (NPDSviii) and (NPDSx) respectively without use of any assumptions on $S_\sigma\left(\tilde{\omega}, [P_+ \ P_-], \begin{bmatrix} Q_+ \\ Q_- \end{bmatrix} \right)$. Next, one can use Theorem 2.3, assumption (ii-b) and the same arguments as those for the 2-block case (see Theorem 4.3 in [BR5]) to deduce (IDSii), (IDSiv) and (IDSvi) from their counterparts (NPDSii), (NPDSiv) and (NPDSvi). It remains now only to verify (IDSi), (IDSvii) and (IDSix).

Suppose now that (C_-, A_π) is not observable. Then by Lemma 12.2.2. in [BGR4] there is an $x \in \mathbb{C}^{n_\pi}$ not zero with $C_-(z - A_\pi)^{-1} x = 0$ identically in z. From Theorem 4.5 (i) in [BR3], we can solve for $h_+ \in \mathcal{R}^{m \times 1}(\sigma)$ so that

$$(4.6) \qquad f(z) = \begin{bmatrix} C_+ \\ C_- \end{bmatrix} (z - A_\pi)^{-1} x + \begin{bmatrix} h_+(z) \\ 0 \end{bmatrix} = \begin{bmatrix} C_+(z - A_\pi)^{-1} x + h_+(z) \\ 0 \end{bmatrix}$$

is in $S_\sigma(\tilde{\omega}, P) \subset S_\sigma(\tilde{\omega}, P, Q)$ (here we use that (IDSvi) has already been verified). Furthermore, since $\left(\begin{bmatrix} C_+ \\ C_- \end{bmatrix}, A_\pi \right)$ is observable by (NPDSi), and since $C_-(z - A_\pi)^{-1} x = 0$, by Lemma 12.2.2. in [BGR4] again necessarily $C_+(z - A_\pi)^{-1} x$ is not identically zero. Then f in (4.6) is an element of $S_\sigma(\tilde{\omega}, P, Q)$ in violation of (ii-a). Hence we must have (C_-, A_π) observable, i.e. (IDSi) holds.

Next suppose (IDSvii) is not true. Then we can find $r \in \mathcal{R}^{L \times 1}$ such that $Q_+ r \notin \mathcal{R}^{m \times 1}(\sigma)$ but $Q_- r \in \mathcal{R}^{n \times 1}(\sigma)$. Again by an argument based on Theorem 4.5 (i) in [BR3], we can then produce an element $h_+ \in \mathcal{R}^{m \times 1}(\sigma)$ so that $\begin{bmatrix} h_+ \\ Q_- r \end{bmatrix} \in S_\sigma(\tilde{\omega}, P) \subset S_\sigma(\tilde{\omega}, P, Q)$. The n

$$f = \begin{bmatrix} Q_+ \\ Q_- \end{bmatrix} r - \begin{bmatrix} h_+ \\ Q_- r \end{bmatrix} = \begin{bmatrix} Q_+ r - h_+ \\ 0 \end{bmatrix}$$

is an element of $S_\sigma(\tilde{\omega}, P, Q)$ in violation of (ii-a). Hence (IDSvii) must hold.

Finally, suppose (IDSix) is violated. Then we can find $x \neq 0$ in $\mathbb{C}^{n_\pi}$ and $r \neq 0$ in $\mathcal{R}^{L \times 1}$ so that $h_-(z) = C_-(z - A_\pi)^{-1} x + Q_-(z) r(z) \in \mathcal{R}^{n \times 1}(\sigma)$. From Theorem 4.5 (iii) in [BR3] we can produce $h_+ \in \mathcal{R}^{m \times 1}(\sigma)$ so that $\begin{bmatrix} h_+ \\ h_- \end{bmatrix} \in S_\sigma(\tilde{\omega}, P) \subset S_\sigma(\tilde{\omega}, P, Q)$ where $h_- = (I - P_{\sigma\bullet}^0)(Q_- r)$. Also by Theorem 4.5 (i) from [BR3] there is $g \in \mathcal{R}^{n \times 1}(\sigma)$ so that $\begin{bmatrix} C_+ \\ C_- \end{bmatrix} (z - A_\pi)^{-1} x + \begin{bmatrix} g(z) \\ 0 \end{bmatrix} \in S_\sigma(\tilde{\omega}, P) \subset S_\sigma(\tilde{\omega}, P, Q)$ (here we use that (IDSvi) has already been verified). But then

$$\begin{aligned} f &= \begin{bmatrix} C_+ \\ C_- \end{bmatrix} (z - A_\pi)^{-1} x + \begin{bmatrix} g(z) \\ 0 \end{bmatrix} - \begin{bmatrix} h_+(z) \\ h_-(z) \end{bmatrix} + \begin{bmatrix} Q_+(z) \\ Q_-(z) \end{bmatrix} r(z) \\ &= \begin{bmatrix} C_+(z - A_\pi)^{-1} x + g(z) + Q_+(z) r(z) - h_+(z) \\ 0 \end{bmatrix} \in S_\sigma(\tilde{\omega}, P, Q) \end{aligned}$$

However, since $x \neq 0$, $C_-(z - A_\pi)^{-1} x + Q_-(z) r(z)$ is analytic on σ and (NPDSix) is in force, we are guaranteed that $C_+(z - A_\pi)^{-1} x + Q_+(z) r(z)$ is not analytic on σ. Hence f is an element of $S_\sigma(\tilde{\omega}, P, Q)$ in violation of (ii - a). Thus, (IDSix) must hold and Theorem 4.3 follows.

The proof of Theorem 4.1 requires the following lemma, an adaptation of one of the main ideas in [BC].

LEMMA 4.4. Let $\sigma \subset \mathbb{C}$, let ω as in (3.12) be a σ-admissible interpolation data set, and let

$$\Theta = \begin{bmatrix} \Theta_{11} & \Theta_{12} & \vdots & \Theta_{13} \\ \Theta_{21} & \Theta_{22} & \vdots & \Theta_{23} \end{bmatrix}$$

be a block row rational matrix function with extended complete null-pole data set over σ equal to $\left(\tilde{\omega}, [P_+ \ P_-], \begin{bmatrix} Q_+ \\ Q_- \end{bmatrix}\right)$ where $\tilde{\omega}$ is as in (4.1). Then a function $W \in \mathcal{R}^{m\times n}$ is analytic on σ and satisfies the interpolation conditions (3.1')-(3.5') if and only if

$$\begin{bmatrix} W(z) \\ I \end{bmatrix} [\Theta_{21} \ \Theta_{22} \ \Theta_{23}] \left(\mathcal{R}^{(m-K+n-L)\times 1}(\sigma) \oplus \mathcal{R}^{L\times 1} \right)$$

(4.7)
$$\subset \Theta\left(\mathcal{R}^{(m-K+n-L)\times 1}(\sigma) \oplus \mathcal{R}^{L\times 1} \right).$$

PROOF: As in the proof of Theorem 4.3, one can show that

$$[\Theta_{21} \ \Theta_{22} \ \Theta_{23}]\left(\mathcal{R}^{(m-K+n-L)\times 1}(\sigma) \oplus \mathcal{R}^{L\times 1} \right)$$
$$= \{C_-(z - A_\pi)^{-1}x : x \in C^{n_\pi}\} + \mathcal{R}^{m\times 1}(\sigma) + Q_- \mathcal{R}^{L\times 1}$$
$$= S_\sigma\left(\left((C_-, A_\pi), (\emptyset, \emptyset), \emptyset\right), \emptyset, Q_- \right).$$

Suppose W is analytic on σ and satisfies the interpolation conditons (3.1')-(3.5'). By Lemma 4.4 in [BR5],

$$\begin{bmatrix} W \\ I \end{bmatrix} \left(\{C_-(z - A_\pi)^{-1}x : x \in \mathbb{C}^{n_\pi}\} + \mathcal{R}^{m\times 1}(\sigma) \right) \subset$$
$$S_\sigma(\tilde{\omega}, P) \subset S_\sigma(\tilde{\omega}, P, Q)$$
$$= \Theta\left(\mathcal{R}^{(m-K+n-L)\times 1}(\sigma) \oplus \mathcal{R}^{L\times 1} \right).$$

But by condition (3.5'),

$$\begin{bmatrix} W \\ I \end{bmatrix} Q_- = \begin{bmatrix} Q_+ \\ Q_- \end{bmatrix}$$

and hence the inclusion (4.7) follows.

Conversely, suppose that the inclusion (4.7) holds. Then

$$\begin{bmatrix} W \\ I \end{bmatrix} Q_- \mathcal{R}^{L\times 1} \subset \bigcap_{r\in\mathcal{R}} r\Theta\left(\mathcal{R}^{(m-K+n-L)\times 1}(\sigma) \oplus \mathcal{R}^{L\times 1} \right)$$
$$= \Theta\left(0 \oplus \mathcal{R}^{L\times 1} \right)$$
$$= \begin{bmatrix} Q_+ \\ Q_- \end{bmatrix} \mathcal{R}^{L\times 1}$$

and hence W satisfies (3.5'). Moreover, by an argument in the proof of Theorem 4.3 we know that

$$\{C_-(z - A_\pi)^{-1}x : x \in \mathbb{C}^{n_\pi}\} \cap Q_- \mathcal{R}^{L \times 1} = \{0\}$$

Using also that $Q_- r = 0$ implies that $Q_+ r = 0$ (a consequence of (IDSvii)), we see that the inclusion (4.7) implies that

$$\begin{bmatrix} W \\ I \end{bmatrix} \left(\{C_-(z - A_\pi)^{-1}x : x \in \mathbb{C}^{n_\pi}\} + \mathcal{R}^{n \times 1}(\sigma) \right)$$
$$\subset S_\sigma \left(\tilde{\omega}, [P_+ \ P_-] \right) = \Theta \mathcal{R}^{(m-K+n) \times 1}(\sigma).$$

Now Lemma 4.4 from [BR5] implies that W is analytic on σ and satisfies the remaining interpolation conditions (3.1') - (3.4').

PROOF OF THEOREM 4.1. By Lemma 4.4 it suffices to show that $W \in \mathcal{R}^{m \times n}$ satisfies

$$\begin{bmatrix} W \\ I \end{bmatrix} [\Theta_{21} \ \Theta_{22} \ \Theta_{23}] \left(\mathcal{R}^{(m-K+n-L) \times 1}(\sigma) \oplus \mathcal{R}^{L \times 1} \right)$$
$$\text{(4.8)} \qquad \subset \Theta \left(\mathcal{R}^{(m-K+n-L) \times 1}(\sigma) \oplus \mathcal{R}^{L \times 1} \right)$$

if and only if W has a representation of the form

$$\text{(4.9)} \qquad W = [\Theta_{11}Q_1 + \Theta_{12}Q_2 \ \Theta_{13}][\Theta_{21}Q_1 + \Theta_{22}Q_2 \ \Theta_{23}]^{-1}$$

where $Q_1 \in \mathcal{R}^{(m-K) \times (n-L)}(\sigma), Q_2 \in \mathcal{R}^{(n-L) \times (n-L)}(\sigma)$ are such that

$$[\Theta_{21}Q_1 + \Theta_{22}Q_2 \ \Theta_{23}] \left(\mathcal{R}^{(n-L) \times 1}(\sigma) \oplus \mathcal{R}^{L \times 1} \right)$$
$$\text{(4.10)} \qquad = [\Theta_{21} \ \Theta_{22} \ \Theta_{23}] \left(\mathcal{R}^{(m-K+n-L) \times 1}(\sigma) \oplus \mathcal{R}^{L \times 1} \right).$$

Suppose first the (4.9) and (4.10) hold. Rewrite (4.9) in the form

$$\begin{bmatrix} W \\ I \end{bmatrix} [\Theta_{21}Q_1 + \Theta_{22}Q_2 \ \Theta_{23}] \Theta \begin{bmatrix} Q_1 & 0 \\ Q_2 & 0 \\ 0 & I \end{bmatrix}.$$

Apply both sides to $\mathcal{R}^{(n-L) \times 1}(\sigma) \oplus \mathcal{R}^{L \times 1}$ and use (4.10) to get

$$\begin{bmatrix} W \\ I \end{bmatrix} [\Theta_{21} \ \Theta_{22} \ \Theta_{23}] \left(\mathcal{R}^{(m-K+n-L) \times 1}(\sigma) \oplus \mathcal{R}^{L \times 1} \right)$$
$$= \Theta \begin{bmatrix} Q_1 & 0 \\ Q_2 & 0 \\ 0 & I \end{bmatrix} \left(\mathcal{R}^{(n-L) \times 1}(\sigma) \oplus \mathcal{R}^{L \times 1} \right).$$

Since Q_1 and Q_2 are analytic on σ, we have

$$\begin{bmatrix} Q_1 & 0 \\ Q_2 & 0 \\ 0 & I \end{bmatrix} \left(\mathcal{R}^{(n-L)\times 1}(\sigma) \oplus \mathcal{R}^{L\times 1} \right) \subset \mathcal{R}^{(m-K+n-L)\times 1}(\sigma) \oplus \mathcal{R}^{L\times 1}$$

and (4.8) follows.

Conversely, suppose (4.8) holds. Note that $\bigcap\limits_{r\in\mathcal{R}} r \begin{bmatrix} W \\ I \end{bmatrix} [\Theta_{21}\ \Theta_{22}\ \Theta_{23}] \left(\mathcal{R}^{(m-K+n-L)\times 1}(\sigma) \oplus \mathcal{R}^{L\times 1} \right) = \begin{bmatrix} W \\ I \end{bmatrix} \Theta_{23}\mathcal{R}^{L\times 1}$ is a $\mathcal{R}$-subspace of $\mathcal{R}^{n\times 1}$ of dimension L which (by (4.8)) is contained in $\bigcap\limits_{r\in\mathcal{R}} r\Theta\left(\mathcal{R}^{(m-K+n-L)\times 1}(\sigma) \oplus \mathcal{R}^{L\times 1} \right) = \begin{bmatrix} \Theta_{13} \\ \Theta_{23} \end{bmatrix} \mathcal{R}^{L\times 1}$, also an $\mathcal{R}$-subspace of dimension L. By dimension count,

$$\begin{bmatrix} W \\ I \end{bmatrix} \Theta_{23}\mathcal{R}^{L\times 1} = \begin{bmatrix} \Theta_{13} \\ \Theta_{23} \end{bmatrix} \mathcal{R}^{L\times 1}.$$

Next note that the quotient module

$$\begin{bmatrix} W \\ I \end{bmatrix} [\Theta_{21}\ \Theta_{22}\ \Theta_{23}] \left(\mathcal{R}^{(m-K+n-L)\times 1}(\sigma) \oplus \mathcal{R}^{L\times 1} \right) \Big/ \begin{bmatrix} W \\ I \end{bmatrix} \Theta_{23}\mathcal{R}^{L\times 1}$$

is a free $\mathcal{R}(\sigma)$-submodule of

$$\Theta\left(\mathcal{R}^{(m-K+n-L)\times 1}(\sigma) \oplus \mathcal{R}^{L\times 1} \right) \Big/ \begin{bmatrix} \Theta_{13} \\ \Theta_{23} \end{bmatrix} \mathcal{R}^{L\times 1}$$

with $n - L$ generators; we may choose $n - L$ generators of the form $\Theta \begin{bmatrix} q_{1,+} \\ q_{1,-} \\ 0 \end{bmatrix}, \cdots \Theta \begin{bmatrix} q_{n-L,+} \\ q_{n-L,-} \\ 0 \end{bmatrix}$ and then set

$$Q_+ = \begin{bmatrix} q_{1,+} \cdots q_{n-L,+} \end{bmatrix} \in \mathcal{R}^{(m-K)\times(n-L)}(\sigma),$$

$$Q_- = \begin{bmatrix} q_{1,-} \cdots q_{n-L,-} \end{bmatrix} \in \mathcal{R}^{(n-L)\times(n-L)}(\sigma).$$

Then the left side of (4.8) has the representation

$$\begin{bmatrix} W \\ I \end{bmatrix} [\Theta_{21}\ \Theta_{22}\ \Theta_{23}] \left(\mathcal{R}^{(m-K+n-L)\times 1}(\sigma) \oplus \mathcal{R}^{L\times 1} \right)$$

$$= \Theta \begin{bmatrix} Q_1 & 0 \\ Q_2 & 0 \\ 0 & I \end{bmatrix} \left(\mathcal{R}^{(n-L)\times 1}(\sigma) \oplus \mathcal{R}^{L\times 1} \right)$$

$$= \begin{bmatrix} \Theta_{11}Q_1 + \Theta_{12}Q_2 & \Theta_{13} \\ \Theta_{21}Q_1 + \Theta_{22}Q_2 & \Theta_{23} \end{bmatrix} \left(\mathcal{R}^{(n-L)\times 1}(\sigma) \oplus \mathcal{R}^{L\times 1} \right)$$

From equality of the bottom components in (4.11) we deduce that the pair (Q_1, Q_2) satisfies (4.10) and then (4.11) can be rewritten as

$$\begin{bmatrix} W \\ I \end{bmatrix} [\Theta_{21}Q_1 + \Theta_{22}Q_2 \ \Theta_{23}]\left(\mathcal{R}^{(n-L)\times 1}(\sigma) \oplus \mathcal{R}^{L\times 1} \right)$$

$$(4.12) \qquad = \begin{bmatrix} \Theta_{11}Q_1 + \Theta_{12}Q_2 & \Theta_{13} \\ \Theta_{21}Q_1 + \Theta_{22}Q_2 & \Theta_{23} \end{bmatrix} \left(\mathcal{R}^{(n-L)\times 1}(\sigma) \oplus \mathcal{R}^{L\times 1} \right)$$

Also from (4.10) we see that the square matrix function $[\Theta_{21}Q_1 + \Theta_{22}Q_2 \ \Theta_{23}]$ cannot have determinant vanishing indentically, and hence $[\Theta_{21}Q_1 + \Theta_{22}Q_2 \ \Theta_{23}]^{-1}$ exists. Now (4.12) leads immediately to the representation (4.9) for W.

The proof of Theorem 4.2 requires the following lemma.

LEMMA 4.5. *Suppose* σ, ω *and* $\Theta = \begin{bmatrix} \Theta_{11} & \Theta_{12} & \Theta_{13} \\ \Theta_{21} & \Theta_{22} & \Theta_{23} \end{bmatrix}$ *are as in the hypothesis of Theorem 4.2. Then* Θ *satisfies*

$$\Theta(z)^* J \Theta(z) \le j \oplus 0, z \in \sigma$$

(where $J = I_m \oplus -I_n$ *and* $j = I_{m-K} \oplus I_{n-L}$*) if and only if* $[I_{n-L} \ 0][\Theta_{22} \ \Theta_{23}]^{-1}\Theta_{21}$ *has analytic continuation to all of* σ.

PROOF: By assumption $\Theta(z)^* J \Theta(z) = j \oplus -I_L$ for $z \in \partial\sigma$. In particular

$$\begin{bmatrix} \Theta_{12}^* \\ \Theta_{13}^* \end{bmatrix} [\Theta_{12} \ \Theta_{13}] - \begin{bmatrix} \Theta_{22}^* \\ \Theta_{23}^* \end{bmatrix} [\Theta_{22} \ \Theta_{23}] = -I_n, z \in \partial\sigma.$$

Hence $\begin{bmatrix} \Theta_{22}^* \\ \Theta_{23}^* \end{bmatrix} [\Theta_{22} \ \Theta_{23}] = I_{2n} + \begin{bmatrix} \Theta_{12}^* \\ \Theta_{13}^* \end{bmatrix} [\Theta_{12} \ \Theta_{13}]$ for $z \in \partial\sigma$. As $[\Theta_{22} \ \Theta_{23}]$ is square, we conclude that $[\Theta_{22}(z) \ \Theta_{23}(z)]$ is invertible for $z \in \partial\sigma$. Then for each fixed z we can rewrite the system of equations

$$(4.13) \qquad \begin{bmatrix} \Theta_{11} & \Theta_{12} & \Theta_{13} \\ \Theta_{21} & \Theta_{22} & \Theta_{23} \end{bmatrix} (z) \begin{bmatrix} u \\ y \\ y^0 \end{bmatrix} = \begin{bmatrix} v \\ w \end{bmatrix}$$

(for $z \in \partial\sigma$) in the equivalent form

$$(4.14) \qquad \begin{bmatrix} U_{11} & U_{12} \\ U_{21} & U_{22} \\ U_{31} & U_{32} \end{bmatrix} (z) \begin{bmatrix} w \\ u \end{bmatrix} = \begin{bmatrix} v \\ y \\ y^0 \end{bmatrix}$$

where

$$U_{11} = [\Theta_{12} \ \Theta_{13}][\Theta_{22} \ \Theta_{23}]^{-1}$$

$$U_{12} = \Theta_{11} - [\Theta_{12} \ \Theta_{13}][\Theta_{22} \ \Theta_{23}]^{-1}\Theta_{21}$$

$$\begin{bmatrix} U_{21} \\ U_{31} \end{bmatrix} = [\Theta_{22} \ \Theta_{23}]^{-1}$$

$$\begin{bmatrix} U_{22} \\ U_{23} \end{bmatrix} = -[\Theta_{22} \ \Theta_{23}]^{-1}\Theta_{21}.$$

Note that $\Theta(z)^* J \Theta(z) = j \oplus -I_L$ on $\partial\sigma$ is equivalent to

$$\|v\|^2 - \|w\|^2 = \|u\|^2 - \|y\|^2 - \|y^0\|^2$$

or equivalently, to

$$\|v\|^2 + \|y\|^2 + \|y^0\|^2 = \|w\|^2 + \|y\|^2$$

whenever $z \in \partial\sigma$ and (u, y, y^0, v, w) satisfies (4.13). Since (4.13) and (4.14) are equivalent, we see that $\Theta(z)^* J \Theta(z) = j \oplus -I_L$ on $\partial\sigma$ is equivalent to $U(z) = \begin{bmatrix} U_{11} & U_{12} \\ U_{21} & U_{22} \\ I_{31} & U_{32} \end{bmatrix}(z)$ being isometric on $\partial\sigma$. Similarly, the validity of $\Theta(z)^* J \Theta(z) \leq j \oplus 0$ on σ is equivalent to

$$\|v\|^2 - \|w\|^2 \leq \|u\|^2 - \|y\|^2,$$

or otherwise stated, to

$$\|v\|^2 + \|y\|^2 \leq \|w\|^2 + \|u\|^2$$

whenever $z \in \sigma$ and (u, y, y^0, v, w) satisfy (4.13). Since (4.13) and (4.14) are equivalent, we read off that this in turn is equivalent to $[I_{m+n-L}\ 0]U(z)$ being contractive on σ. In particular

$$-U_{22} = [I_{n-L}\ 0][\Theta_{22}\ \Theta_{23}]^{-1}\Theta_{21}$$

must be contractive at all points of analyticity of σ, and hence has analytic continuation to all of the set σ.

Conversely, to show that $\Theta(z)^* J \Theta(z) \leq j \oplus 0$ on σ is equivalent to showing that $[I_{m+n-L}\ 0]$ $U(z)$ has contractive values at all points of analyticity in σ, or, equivalently by the maximum modulus theorem, that

$$[I_{m+n-L}\ 0]U(z) = \begin{bmatrix} U_{11}(z) & U_{12}(z) \\ U_{21}(z) & U_{22}(z) \end{bmatrix}$$

have analytic continuation to all of σ. By assumption we know only that

$$U_{22} = -[I_{n-L}\ 0][\Theta_{22}\ \Theta_{23}]^{-1}\Theta_{21}$$

has such an analytic continuation. The fact that this forces U_{11}, U_{12}, and U_{21} to have such an analytic continuation as well relies on the special structure of the extended null-pole subspace $\Theta\left(\mathcal{R}^{(m-K+n-L)\times 1}(\sigma) \oplus \mathcal{R}^{L\times 1}\right) = S_\sigma\left(\tilde{\omega}, [P_+, P_-], \begin{bmatrix} Q_+ \\ Q_- \end{bmatrix}\right)$ given by Theorem 4.3. Details for the 1-block case (the case where $K = 0$ and $L = 0$) are given in Theorem 13.2.3 in [BGR4]; we leave it to the reader to check that the general case here can be verified by analogous arguments.

PROOF OF THEOREM 4.2. Let σ, ω and Θ be as in the statement of Theorem 4.2. Suppose first that $\Theta(z)^* J \Theta(z) \leq j \oplus 0_{(n-L)\times(n-L)}$ at all points $z \in \sigma$ where $\Theta(z)$ is analytic and that $H \in \mathcal{R}^{(m-K)\times(n-L)}(\sigma)$ with $\|H\|_\infty < 1$. Then

$$(4.15) \qquad [\Theta_{21} H + \Theta_{22}\ \Theta_{23}] = [\Theta_{22}\ \Theta_{23}]\left([\Theta_{22}\ \Theta_{23}]^{-1}\Theta_{21}[H\ 0] + I_n\right)$$

By Lemma 4.5 we know that $[I_{n-L}\ 0][\Theta_{22}\ \Theta_{23}]^{-1}\Theta_{21}$ is analytic on σ and, from the proof of Lemma 4.5, that $\|[\Theta_{21}\ \Theta_{23}]^{-1}\Theta_{21}(z)\| < 1$ for $z \in \partial\sigma$. Hence $t[I_{n-L}\ 0][\Theta_{22}\ \Theta_{23}]^{-1}\Theta_{21} H + I_{n-L}$ is invertible for all $z \in \partial\sigma$ and for all t with $0 \leq t \leq 1$. Hence $wno\ det\left([I_{n-L}\ 0][\Theta_{22}\ \Theta_{23}]^{-1}\Theta_{21} H + I_{n-L}\right) = wno\ det\ I_{n-L} = 0$, where $wno\ X$ is the change of the argument of $X(z)$ along $\partial\sigma$. As we observed above that $[I_{n-L}\ 0][\Theta_{22}\ \Theta_{23}]^{-1}\Theta_{21} H + I_{n-L}$ is analytic on σ, we conclude that $det\left([I_{n-L}\ 0][\Theta_{22}\ \Theta_{23}]^{-1}\Theta_{21} H + I_{n-L}\right)$ has no zero in σ. From this we see that

$$\left([\Theta_{22}\ \Theta_{23}]^{-1}\Theta_{21}[H\ 0] + I_n\right)\left(\mathcal{R}^{(n-L)\times 1}(\sigma) \oplus \mathcal{R}^{L\times 1}\right) = \mathcal{R}^{(n-L)\times 1}(\sigma) \oplus \mathcal{R}^{L\times 1}.$$

From (4.15) we conclude that

$$[\Theta_{21} H + \Theta_{22}\ \Theta_{23}]\left(\mathcal{R}^{(n-L)\times 1}(\sigma) \oplus \mathcal{R}^L\right)$$

$$(4.16) \qquad = [\Theta_{22}\ \Theta_{23}]\left(\mathcal{R}^{(n-L)\times 1}(\sigma) \oplus \mathcal{R}^L\right).$$

On the other hand,

$$[\Theta_{21}\ \Theta_{22}\ \Theta_{23}]\left(\mathcal{R}^{(m-K+n-L)\times 1}(\sigma) \oplus \mathcal{R}^L\right)$$

$$(4.17) \qquad = [\Theta_{22}\ \Theta_{23}]\left[[\Theta_{22}\ \Theta_{23}]^{-1}\Theta_{21}\ I_n\right]\left(\mathcal{R}^{(m-K+n-L)\times 1}(\sigma) \oplus \mathcal{R}^L\right).$$

But by Lemma 4.5 $[I_{n-L}\ 0][\Theta_{22}\ \Theta_{23}]^{-1}\Theta_{21}$ is analytic on σ, so

$$\left[[\Theta_{22}\Theta_{23}]^{-1}\Theta_{21}\ I_n\right]\left(\mathcal{R}^{(m-K+n-L)\times 1}(\sigma) \oplus \mathcal{R}^{L\times 1}\right)$$

$$(4.18) \qquad = \mathcal{R}^{(n-L)\times 1}(\sigma) \oplus \mathcal{R}^{L\times 1}.$$

Combining (4.17) and (4.18), we obtain

$$[\Theta_{21}\ \Theta_{22}\ \Theta_{23}]\left(\mathcal{R}^{(m-K+n-L)\times 1}(\sigma)\oplus\mathcal{R}^{L\times 1}\right)$$

$$(4.19)\qquad = [\Theta_{22}\ \Theta_{23}]\left(\mathcal{R}^{(n-L)\times 1}(\sigma)\oplus\mathcal{R}^{L\times 1}\right).$$

Combine (4.19) and (4.16) to see that the pair (H, I) is an admissible parameter pair in the sense of Lemma 4.4. Hence, by Lemma 4.4 we know that $W = [\Theta_{11}H + \Theta_{12}\ \Theta_{13}][\Theta_{21}H + \Theta_{22}\ \Theta_{23}]^{-1}$ is analytic on σ and satisfies the interpolation conditions (3.1') - (3.5'). Moreover, for $z \in \partial\sigma$,

$$W(z)^*W(z) - I = [W(z)^*\ I]J\begin{bmatrix}W(z)\\I\end{bmatrix}$$

$$= \psi^*\begin{bmatrix}H & 0\\I & 0\\0 & I\end{bmatrix}^*\Theta^*J\Theta\begin{bmatrix}H & 0\\I & 0\\0 & I\end{bmatrix}\psi$$

$$= \psi^*\begin{bmatrix}H^*H - I & 0\\0 & -I\end{bmatrix}\psi < 0$$

where $\psi = [\Theta_{21}H + \Theta_{22}\ \Theta_{23}]^{-1}$. Hence $\|W(z)\| < 1$ for $z \in \partial\sigma$. By the maximum modulus principle, $\|W(z)\| < 1$ for $z \in \sigma$.

Conversely, suppose that there exists a function $W \in \mathcal{R}^{m\times n}$ which is analytic on σ, satisfies the interpolation conditions (3.1') - (3.5') and has $\|W\|_\infty < 1$. Then by Theorem 4.1 we know that there are $Q_1 \in \mathcal{R}^{(m-K)\times(n-L)}(\sigma)$ and $Q_2 \in \mathcal{R}^{(n-L)\times(n-L)}(\sigma)$ such that

$$[\Theta_{21}Q_1 + \Theta_{22}Q_2\ \Theta_{23}]\left(\mathcal{R}^{(n-L)\times 1}(\sigma)\oplus\mathcal{R}^{L\times 1}\right)$$

$$(4.20)\qquad = [\Theta_{21}\ \Theta_{22}\ \Theta_{23}]\left(\mathcal{R}^{(m-K+n-L)\times 1}(\sigma)\oplus\mathcal{R}^{L\times 1}\right),$$

from which we recover W as

$$(4.21)\qquad W = [\Theta_{11}Q_1 + \Theta_{12}Q_2\ \Theta_{13}][\Theta_{21}Q_1 + \Theta_{22}Q_2\ \Theta_{23}]^{-1}.$$

An alternative way to write (4.21) is

$$\begin{bmatrix}W\\I\end{bmatrix}\psi = \Theta\begin{bmatrix}Q_1 & 0\\Q_2 & 0\\0 & I\end{bmatrix}$$

where $\psi = [\Theta_{21}Q_1 + \Theta_{22}Q_2\ \Theta_{23}]$. Since $\|W\|_\infty < 1$ and Θ satisfies condition (b) in Theorem 4.2 on $\partial\sigma$ by assumption, we have, for $z \in \sigma$,

$$0 < \psi^*(W^*W - I)\psi$$

$$= \begin{bmatrix} Q_1 & 0 \\ Q_2 & 0 \\ 0 & I \end{bmatrix}^* \Theta^* J \Theta \begin{bmatrix} Q_1 & 0 \\ Q_2 & 0 \\ 0 & I \end{bmatrix}$$

$$= \begin{bmatrix} Q_1^* & Q_2^* & 0 \\ 0 & 0 & I \end{bmatrix} \left(j \oplus -I_L \right) \begin{bmatrix} Q_1 & 0 \\ Q_2 & 0 \\ 0 & I \end{bmatrix}$$

$$(4.22) \qquad = \begin{bmatrix} Q_1^*Q_1 - Q_2^*Q_2 & 0 \\ 0 & -I_L \end{bmatrix}.$$

Thus $Q_2(z)$ cannot have any null space for $z \in \sigma$, and hence, as it is square, $Q_2(z)^{-1}$ exists for $z \in \partial\sigma$. Also (4.22) then implies that $H(z) = Q_1(z)Q_2(z)^{-1}$ is well defined and satisfies $\|H(z)\| < 1$ for $z \in \partial\sigma$.

Next note from (4.20) the inclusion

$$[\Theta_{22} \ \Theta_{23}]\left(\mathcal{R}^{(n-L)\times 1}(\sigma) \oplus \mathcal{R}^{L\times 1} \right)$$

$$(4.23) \qquad \subset [\Theta_{21}Q_1 + \Theta_{22}Q_2 \ \Theta_{23}]\left(\mathcal{R}^{(n-L)\times 1}(\sigma) \oplus \mathcal{R}^{L\times 1} \right).$$

We also know from the proof of Lemma 4.5 that $[\Theta_{22} \ \Theta_{23}]$ is invertible on $\partial\sigma$. Then (4.23) can be written in the form

$$\mathcal{R}^{(n-L)\times 1}(\sigma) \oplus \mathcal{R}^{L\times 1}$$

$$(4.24) \qquad \subset \left([\Theta_{22} \ \Theta_{23}]^{-1}\Theta_{21}[Q_1 \ 0] + \begin{bmatrix} Q_2 & 0 \\ 0 & I_L \end{bmatrix} \right)\left(\mathcal{R}^{(n-L)\times 1}(\sigma) \oplus \mathcal{R}^{L\times 1} \right).$$

Note that the $n \times n$ matrix function $[\Theta_{22} \ \Theta_{23}]^{-1}\Theta_{21}[Q_1 \ 0] + \begin{bmatrix} Q_2 & 0 \\ 0 & I_L \end{bmatrix}$ has the form $\begin{bmatrix} A & 0 \\ B & I_L \end{bmatrix}$ where $A = [I_{n-L} \ 0][\Theta_{22} \ \Theta_{23}]^{-1}\Theta_{21}Q_1 + Q_2$ has size $(n - L) \times (n - L)$ and $B = [0 \ I_L][\Theta_{22} \ \Theta_{23}]^{-1}\Theta_{21}Q_1$ has size $L \times (n - L)$. In this notation (4.24) assumes the form

$$\mathcal{R}^{(n-L)\times 1}(\sigma) \oplus \mathcal{R}^{L\times 1} \subset \begin{bmatrix} A & 0 \\ B & I_L \end{bmatrix}\left(\mathcal{R}^{(n-L)\times 1}(\sigma) \oplus \mathcal{R}^{L\times 1} \right).$$

This in turn requires

$$(4.25) \qquad \mathcal{R}^{(n-L)\times 1}(\sigma) \subset A\mathcal{R}^{(n-L)\times 1}(\sigma).$$

Note that

$$A = \left([I_{n-L}\ 0][\Theta_{22}\ \Theta_{23}]^{-1}\Theta_{21}H + I\right)Q_2$$ where $\|[\Theta_{22}\ \Theta_{23}]^{-1}\Theta_{21}\| < 1$ on $\partial\sigma$ (from the proof of Lemma 4.5) and where $\|H(z)\| < 1$ and $Q_2(z)$ is invertible on $\partial\sigma$ from observations made above. Hence $A(z)$ is invertible on ∂z and $wno\ det\ A(z)$ is well defined. The inclusion (4.25) gives us that

$$(4.26) \qquad wno\ det\ A(z) := wno\ det\left([I_{n-L}\ 0][\Theta_{22}\ \Theta_{23}]^{-1}\Theta_{21}Q_1 + Q_2\right) \leq 0.$$

On the other hand,

$$(4.27) \qquad A(z) = \left([I_{n-L}\ 0][\Theta_{22}\ \Theta_{23}]^{-1}\Theta_{21}H + I\right)Q_2.$$

As was observed above, $\|[I_{n-L}\ 0][\Theta_{22}\ \Theta_{23}]^{-1}\Theta_{21}H\| < 1$ on $\partial\sigma$. Then by a standard homotopy argument $wno\ det\left([I_{n-L}\ 0][\Theta_{22}\ \Theta_{23}]^{-1}\Theta_{21}H + I\right) = 0$, so

$$(4.28) \qquad wno\ det\ A = wno\ det\ Q_2.$$

Since Q_2 is analytic on σ, by the argument principle we deduce

$$(4.29) \qquad wno\ det\ Q_2 \geq 0.$$

Combining (4.26), (4.28) and (4.29) leads to

$$(4.30) \qquad wno\ det\ A = 0 = wno\ det\ Q_2.$$

Condition (4.30) has several consequences. First of all, $det\ Q_2$ has no zeros on σ so Q_2^{-1} and $H := Q_1 Q_2^{-1}$ are analytic on σ. Secondly, we deduce that equality must hold in (4.24) and hence in (4.23). This has a consequence that

$$[\Theta_{22}\ \Theta_{23}]\left(\mathcal{R}^{(n-L)\times 1}(\sigma) \oplus \mathcal{R}^{L\times 1}\right) = [\Theta_{21}\ \Theta_{22}\ \Theta_{23}]\left(\mathcal{R}^{(m-K+n-L)\times 1}(\sigma) \oplus \mathcal{R}^{L\times 1}\right),$$

or equivalently

$$\mathcal{R}^{(n-L)\times 1}(\sigma) \oplus \mathcal{R}^{L\times 1} = [[\Theta_{22}\ \Theta_{23}]^{-1}\Theta_{21}\ I_n]\left(\mathcal{R}^{(m-K+n-L)\times 1}(\sigma) \oplus \mathcal{R}^{L\times 1}\right).$$

This condition in turn is equivalent to

$$[I_{n-L}\ 0][\Theta_{22}\ \Theta_{23}]^{-1}\Theta_{21} \text{ is analytic on } \sigma.$$

By Lemma 4.5 this last condition is equivalent to $\Theta(z)^*J\Theta(z) \le j \oplus 0$ for $z \in \sigma$. Thus we have verified that condition (ii) in Theorem 4.2 is necessary for interpolants W with $\|W\|_\infty < 1$ to exist.

It remains only to verify that W has the form (4.3) for an $H \in \mathcal{R}^{(m-K)\times(n-L)}(\sigma)$ with $\|W\|_\infty < 1$. We have already verified that $H := Q_1 Q_2^{-1}$ is analytic on σ with norm < 1 on $\partial\sigma$. By the maximum modulus theorem it follows that $\sup_{z\in\sigma}\|H(z)\| < 1$. Moreover, from (4.21) we deduce that

$$W = ([\Theta_{11}H + \Theta_{12}\ \Theta_{13}]\begin{bmatrix} Q_2 & 0 \\ 0 & I_L \end{bmatrix})([\Theta_{21}H + \Theta_{22}\ \Theta_{13}]\begin{bmatrix} Q_2 & 0 \\ 0 & I_L \end{bmatrix})^{-1}$$
$$= [\Theta_{11}H + \Theta_{12}\ \Theta_{13}][\Theta_{21}H + \Theta_{22}\ \Theta_{23}]^{-1}$$

as needed.

5. Interpolation and internally stable feedback systems.

In this section we establish the connections between the interpolation theory presented in the previous sections and the problem of designing a compensator to stabilize a given plant in a standard feedback configuration which has been studied in the control literature (see [Fr], [DGKF]).

We emphasize that the connection between internal stability and interpolation has been a recurring theme in the systems theory literature (see e. g. [YBL], [V], [Ki]). The matrix version of the result is usually derived via a coprime factorization of the plant and the Youla parametrization of stabilizing compensators (see [YJB]). Our contribution here is to relate the interpolation conditions directly to the original plant $\mathcal{P}$; the connection is given in terms of the extended complete null-pole data set of a related block row rational matrix function $\tilde{\mathcal{P}}$. We also solve the inverse problem of describing which plants $\mathcal{P}$ go with a prescribed set of interpolation conditions

5.1 Preliminaries on feedback systems.

Suppose we are given a rational block matrix function $\mathcal{P} = \begin{bmatrix} \mathcal{P}_{11} & \mathcal{P}_{12} \\ \mathcal{P}_{21} & \mathcal{P}_{22} \end{bmatrix}$, where $\mathcal{P}_{11}, \mathcal{P}_{12},$ $\mathcal{P}_{21}, \mathcal{P}_{22}$ have respective sizes $n_z \times n_w, n_z \times n_u, n_y \times n_w, n_y \times n_u$, and σ is a subset of the extended complex plane $\mathbb{C}^\infty$. (For discrete time systems σ is usually taken to be closed unit disk while for continuous time systems σ is usually taken to be the closed right half plane including infinity.) The problem is to design a rational $n_u \times n_y$ matrix function K (the compensator) so that the closed loop system depicted in Figure 5.1 is <u>internally stable</u>, a notion which we shall make precise in a moment. (Here we assume that all input-output maps are causal, linear, time invariant and finite dimensional, and that the Laplace transform has already been implemented so all input-output maps are represented as multiplication by rational matrix functions.)

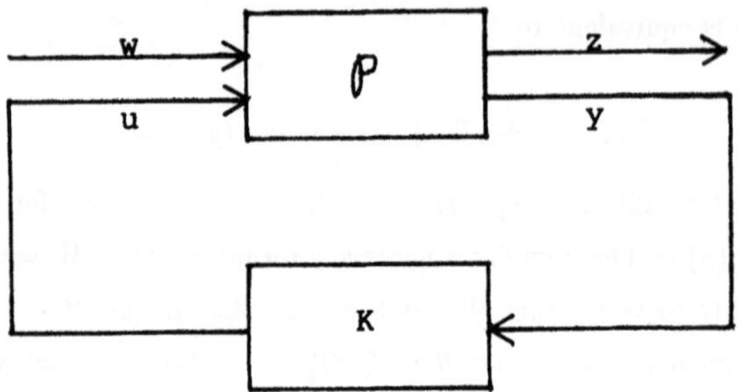

Figure 5.1

In Figure 5.1, w, z, u, y are functions with values in $\mathbb{C}^{n_w \times 1}, \mathbb{C}^{n_z \times 1}, \mathbb{C}^{n_u \times 1}$ and $\mathbb{C}^{n_y \times 1}$, respectively, which are analytic in some right half plane (for the continuous time case) or in some disk centered at the origin (for the discrete time case); in our discussion here we also assume that all the functions are rational, although this assumption is not necessary. The configuration depicted in Figure 5.1 is equivalent to the system of algebraic equations

$$\mathcal{P}_{11}w + \mathcal{P}_{12}u = z$$
$$\mathcal{P}_{21}w + \mathcal{P}_{22}u = y$$

(5.1) $$Ky = u.$$

In the control theory context, the function w is the <u>disturbance</u> or <u>reference signal</u>, u is the <u>control signal</u>, z is the <u>error signal</u> and y is the <u>measurement signal</u>. The idea is to design a compensator K which computes the control signal u based on the measurement y so as to make the overall system $\sum(\mathcal{P}, K) : w \to z$ perform better. The standard H^∞-control problem is to design K which minimizes the largest error z (in the sense of L^2-norm) over all disturbances w of L^2-norm at most 1, subject to the additional constraint that K stabilizes the system:

(5.2) $$\min_{K\ stabilizing} \max_{\|w\|_2 \leq 1} \|z\|_2.$$

Here the L^2-norm is over the imaginary line for continuous time systems and over the unit circle for discrete time systems.

In this section we put aside the norm constraint (5.2) and analyze only the connection between the stability of the system in Fig. 5.1 and interpolation theory. To define precisely internal stability and the related notion of well-posedness for the system in Figure 5.1, following [Fr] we introduce auxiliary signals $v_1 \in \mathcal{R}^{n_u \times 1}$ and $v_2 \in \mathcal{R}^{n_y \times 1}$ as in Figure 5.2.

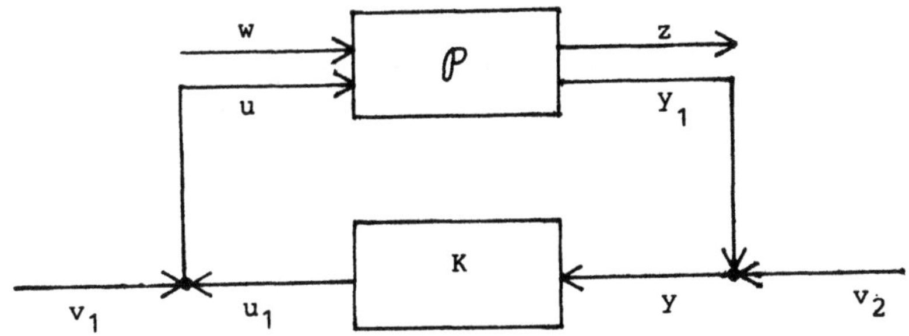

Figure 5.2

The diagram in Figure 5.2 is equivalent to the system of equations

$$P_{11}w + P_{12}u = z$$

$$P_{21}w + P_{22}u = y - v_2$$

(5.3)
$$Ky = u - v_1.$$

<u>Well-posedness</u> means that the system (5.3) can be solved for (z, u, y) in terms of (w, v_1, v_2) and that the resulting map $\begin{bmatrix} w \\ v_1 \\ v_2 \end{bmatrix} \rightarrow \begin{bmatrix} z \\ u \\ y \end{bmatrix}$ is given by multiplication by a proper rational matrix function $\mathcal{H}$ (for the continuous time case) or by a rational matrix function $\mathcal{H}$ analytic at zero (for the discrete time case). <u>Internal stability</u> means that in addition this function $\mathcal{H}$ is analytic on all of σ. Thus internal stability amounts to the assertion that the output signal z and the internal signals u and y are uniquely determined and stable for any choice of stable disturbance signals w, v_1, v_2; for a more complete discussion we refer to [Fr] and [V]. By elementary algebra one can show that H is given by the explicit formula

(5.4)
$$\mathcal{H} = \begin{bmatrix} P_{11} + P_{12}K\Delta^{-1}P_{22} & P_{12} + P_{12}K\Delta^{-1}P_{22} & P_{12}K\Delta^{-1} \\ K\Delta^{-1}P_{21} & I + K\Delta^{-1}P_{22} & K\Delta^{-1} \\ \Delta^{-1}P_{21} & \Delta^{-1}P_{22} & \Delta^{-1} \end{bmatrix}$$

where $\Delta = I - P_{22}K$. In particular the closed-loop transfer function T_{zw} from disturbance (and/or reference) signal w to error z is given by

$$T_{zw} = P_{11} + P_{12}K(I - P_{22}K)^{-1}P_{21}$$

(5.5)
$$=: \mathcal{F}_p[K].$$

A standard assumption in the literature is that $\mathcal{P}_{12}$ is injective (as a multiplication operator from $\mathcal{R}^{n_u \times 1}$ into $\mathcal{R}^{n_z \times 1}$) and that $\mathcal{P}_{21}$ is surjective (as a multiplication operator from $\mathcal{R}^{n_w \times 1}$ into $\mathcal{R}^{n_y \times 1}$); in particular, $n_u \leq n_z$ and $n_y \leq n_w$. An equivalent assumption is that the linear fractional map $K \to \mathcal{F}_p[K]$ defined by (5.5) is injective.

If $\mathcal{P}_{21}$ in fact is square and invertible, one can solve the system of equations

$$\mathcal{P}_{11} w + \mathcal{P}_{12} u = z$$
$$\mathcal{P}_{21} w + \mathcal{P}_{22} u = y$$

for (z, w) in terms of (u, y); the result is

$$\tilde{\mathcal{P}}_{11} u + \tilde{\mathcal{P}}_{12} y = z$$
$$\tilde{\mathcal{P}}_{21} u + \tilde{\mathcal{P}}_{22} y = w$$

where

$$\tilde{\mathcal{P}} = \begin{bmatrix} \mathcal{P}_{12} - \mathcal{P}_{11} \mathcal{P}_{21}^{-1} \mathcal{P}_{22} & \mathcal{P}_{11} \mathcal{P}_{21}^{-1} \\ -\mathcal{P}_{21}^{-1} \mathcal{P}_{22} & \mathcal{P}_{21}^{-1} \end{bmatrix}$$

In the language of circuit theory the transform $\mathcal{P} \to \tilde{\mathcal{P}}$ amounts to the transform from the scattering to the chain formalism (see [B]), and was the basis for the analysis in [BR5]. To remove the assumption that $\mathcal{P}_{21}$ be invertible, we use here the generalized transform to the chain formalism introduced in [BHV].

Since $\mathcal{P}_{21}$ by assumption is surjective, we may add extra rows $\mathcal{P}_{21}^0$ to $\mathcal{P}_{21}$ so that the augmented matrix function $\begin{bmatrix} \mathcal{P}_{21} \\ \mathcal{P}_{21}^0 \end{bmatrix}$ is square and invertible. Define $\mathcal{P}_{22}^0$ of a compatible size arbitrarily and set

$$\mathcal{P} = \begin{bmatrix} \mathcal{P}_{11} & \vdots & \mathcal{P}_{12} \\ \cdots\cdots\cdots\cdots \\ \mathcal{P}_{21} & \vdots & \mathcal{P}_{22} \\ \mathcal{P}_{21}^0 & \vdots & \mathcal{P}_{22}^0 \end{bmatrix} : \begin{bmatrix} w \\ u \end{bmatrix} \to \begin{bmatrix} z \\ y \\ y^0 \end{bmatrix}$$

Here y^0 is to be considered as a physically meaningless fictitious signal created for mathematical convenience. Since $\begin{bmatrix} \mathcal{P}_{21} \\ \mathcal{P}_{21}^0 \end{bmatrix}$ is invertible, we may solve the system of equations

$$\mathcal{P}_{11} w + \mathcal{P}_{12} u = z$$
$$\mathcal{P}_{21} w + \mathcal{P}_{22} u = y$$
$$\mathcal{P}_{21}^0 w + \mathcal{P}_{22}^0 u = y^0.$$

for (z, w) in terms of (u, y, y^0). The result is

$$(5.6) \qquad \tilde{\mathcal{P}} \begin{bmatrix} u \\ y \\ y^0 \end{bmatrix} = \begin{bmatrix} z \\ w \end{bmatrix}$$

where $\tilde{\mathcal{P}} = \begin{bmatrix} \tilde{\mathcal{P}}_{11} & \tilde{\mathcal{P}}_{12} & \tilde{\mathcal{P}}_{13} \\ \tilde{\mathcal{P}}_{21} & \tilde{\mathcal{P}}_{22} & \tilde{\mathcal{P}}_{23} \end{bmatrix}$ is given by

$$(5.6a) \qquad \tilde{\mathcal{P}}_{11} = \mathcal{P}_{12} - \mathcal{P}_{11} \begin{bmatrix} \mathcal{P}_{21} \\ \mathcal{P}_{21}^0 \end{bmatrix}^{-1} \begin{bmatrix} \mathcal{P}_{22} \\ \mathcal{P}_{22}^0 \end{bmatrix}$$

$$(5.6b) \qquad [\tilde{\mathcal{P}}_{12} \ \tilde{\mathcal{P}}_{13}] = \mathcal{P}_{11} \begin{bmatrix} \mathcal{P}_{21} \\ \mathcal{P}_{21}^0 \end{bmatrix}^{-1}$$

$$(5.6c) \qquad \tilde{\mathcal{P}}_{21} = - \begin{bmatrix} \mathcal{P}_{21} \\ \mathcal{P}_{21}^0 \end{bmatrix}^{-1} \begin{bmatrix} \mathcal{P}_{22} \\ \mathcal{P}_{22}^0 \end{bmatrix}$$

and

$$(5.6d) \qquad [\tilde{\mathcal{P}}_{22} \ \tilde{\mathcal{P}}_{23}] = \begin{bmatrix} \mathcal{P}_{21} \\ \mathcal{P}_{21}^0 \end{bmatrix}^{-1} .$$

The system of equations (5.1) associated with the feedback configuration in Figure 5.1 can be expressed in terms of $\tilde{\mathcal{P}}$ as

$$(5.7) \qquad \begin{bmatrix} 0 & 0 & 0 \\ -I & 0 & 0 \\ 0 & I & -K \end{bmatrix} \begin{bmatrix} w \\ v_1 \\ -v_2 \end{bmatrix} + \begin{bmatrix} -I & \tilde{\mathcal{P}}_{11} & \tilde{\mathcal{P}}_{12} & \tilde{\mathcal{P}}_{13} \\ 0 & \tilde{\mathcal{P}}_{21} & \tilde{\mathcal{P}}_{22} & \tilde{\mathcal{P}}_{23} \\ 0 & -I & K & 0 \end{bmatrix} \begin{bmatrix} z \\ u \\ y_1 \\ y^0 \end{bmatrix} .$$

In particular the closed-loop transfer function T_{zw} from w to z works out to be

$$(5.8) \qquad T_{zw} = \begin{bmatrix} \tilde{\mathcal{P}}_{11} K + \tilde{\mathcal{P}}_{12}, \tilde{\mathcal{P}}_{13} \end{bmatrix} \begin{bmatrix} \tilde{\mathcal{P}}_{21} K + \tilde{\mathcal{P}}_{22}, \tilde{\mathcal{P}}_{23} \end{bmatrix}^{-1} .$$

We also remark that the assumptions on $\mathcal{P}$ imply that $\begin{bmatrix} \tilde{\mathcal{P}}_{22} & \tilde{\mathcal{P}}_{23} \end{bmatrix}$ is square and invertible and that $\tilde{\mathcal{P}}$ is injective as a multiplication operator from $\mathcal{R}^{(n_z + n_w) \times 1}$ into $\mathcal{R}^{(n_z + n_w) \times 1}$, and that one can always backsolve for $\mathcal{P}$ from $\tilde{\mathcal{P}}$. For more complete details we refer to [BHV].

Below, in this section we shall say that a square rational matrix function W is <u>biproper</u> if W is analytic and invertible at infinity and we are in the continuous time case, or if W is analytic and invertible at zero and we are in the discrete time case. In addition we say that the square rational matrix function R is σ-outer if both R and R^{-1} are analytic on σ (where σ is the closed right half plane in the continuous time case and the closed unit disk in the discrete time case).

5.2 Stability of feedback systems and interpolation

In this section we delineate the connections between internal stability of a feedback system in Figure 5.1 and the theory of null-pole structure and interpolation developed in the previous sections. We assume that we are given a rational matrix function $\mathcal{P} = \begin{bmatrix} \mathcal{P}_{11} & \mathcal{P}_{12} \\ \mathcal{P}_{21} & \mathcal{P}_{22} \end{bmatrix}$ as in Section 5.1 with $\mathcal{P}_{21}$ surjective and $\mathcal{P}_{12}$ injective, and then define $\tilde{\mathcal{P}} = \begin{bmatrix} \tilde{\mathcal{P}}_{11} & \tilde{\mathcal{P}}_{12} & \vdots & \tilde{\mathcal{P}}_{13} \\ \tilde{\mathcal{P}}_{21} & \tilde{\mathcal{P}}_{22} & \vdots & \tilde{\mathcal{P}}_{23} \end{bmatrix}$ as in (5.6). We consider $\tilde{\mathcal{P}}$ as a block row matrix function $\left(\text{with } \tilde{\mathcal{P}}_1 = \begin{bmatrix} \tilde{\mathcal{P}}_{11} & \tilde{\mathcal{P}}_{12} \\ \tilde{\mathcal{P}}_{21} & \tilde{\mathcal{P}}_{22} \end{bmatrix} \text{ and } \tilde{\mathcal{P}}_2 = \begin{bmatrix} \tilde{\mathcal{P}}_{13} \\ \tilde{\mathcal{P}}_{23} \end{bmatrix}\right)$ and apply the notions of extended null-pole structure for block row matrix functions explained in Section 2. Define integers m, n, K, L by

$$m = n_z, n = n_w, K = n_z - n_u, L = n_w - n_y,$$

so that $\tilde{\mathcal{P}}$ has size $(m + n) \times (m - K + n)$ with $\tilde{\mathcal{P}}_{11}$ of size $m \times (m - K)$, $\tilde{\mathcal{P}}_{12}$ of size $m \times (n - L)$, etc. Let σ be the closed right half plane (continuous time case) or the closed unit disk (discrete time case), and let $(\tilde{\omega}, P, Q) = \left(C_\pi, A_\pi, A_\zeta, B_\zeta, \Gamma, P, Q\right)$ be an extended complete null-pole data set over σ for the block row matrix function $\tilde{\mathcal{P}}$ in the sense explained in Section 2. Then in fact $(\tilde{\omega}, P, Q)$ has the finer block structure

$$(\tilde{\omega}, P, Q) = \left(\begin{bmatrix} C_+ \\ C_- \end{bmatrix}, A_\pi, A_\zeta, [B_+ \ B_-], \Gamma, [P_+ \ P_-], \begin{bmatrix} Q_+ \\ Q_- \end{bmatrix}\right)$$

where $C_+, C_-, A_\pi, A_\zeta, B_+, B_-, \Gamma$ are matrices of respective sizes $m \times n_\pi, n \times n_\pi, n_\pi \times n_\pi, n_\zeta \times n_\zeta, n_\zeta \times m, n_\zeta \times n$, and $n_\zeta \times n_\pi$, and where P_+, P_-, Q_+, Q_- are matrix polynomials of respective sizes $K \times m$, $K \times n, m \times L, n \times L$. The following is the main result of this section.

THEOREM 5.1. *Let* $\mathcal{P} = \begin{bmatrix} \mathcal{P}_{11} & \mathcal{P}_{12} \\ \mathcal{P}_{21} & \mathcal{P}_{22} \end{bmatrix}$ *with* $\mathcal{P}_{12}$ *injective and* $\mathcal{P}_{21}$ *surjective be a rational matrix function describing the plant in Figure 5.1, and define the rational block row matrix function* $\tilde{\mathcal{P}} = \begin{bmatrix} \tilde{\mathcal{P}}_{11} & \tilde{\mathcal{P}}_{12} & \vdots & \tilde{\mathcal{P}}_{13} \\ \tilde{\mathcal{P}}_{21} & \tilde{\mathcal{P}}_{22} & \vdots & \tilde{\mathcal{P}}_{23} \end{bmatrix}$ *of size* $(m + n) \times (m - K + n)$ *as in (5.6). Then the following are equivalent:*

(i) $\mathcal{P}$ *is stabilizable*

(ii) *The extended null-pole subspace* $S := \tilde{\mathcal{P}}\left(\mathcal{R}^{(m-K+n-L) \times 1}(\sigma) \oplus \mathcal{R}^{L \times 1}\right)$ *of* $\tilde{\mathcal{P}}$ *satisfies:*

(a) $S \cap \left(\mathcal{R}^{m \times 1} \oplus 0\right) \subset \mathcal{R}^{m \times 1}(\sigma) \oplus 0$ *and*

(b) $P_{0 \oplus \mathcal{R}^{n \times 1}}\left(S \cap \mathcal{R}^{(m+n) \times 1}(\sigma)\right) = 0 \oplus \mathcal{R}^{n \times 1}(\sigma)$ *where* $P_{0 \oplus \mathcal{R}^{n \times 1}}\left(\begin{bmatrix} a \\ b \end{bmatrix}\right) = \begin{bmatrix} 0 \\ b \end{bmatrix}$.

(iii) *If* $(\tilde{\omega}, P, Q) = \left(\begin{bmatrix} C_+ \\ C_- \end{bmatrix}, A_\pi, A_\zeta, [B_+ \ B_-], \Gamma, [P_+ \ P_-], \begin{bmatrix} Q_+ \\ Q_- \end{bmatrix}\right)$ *is an extended complete set of null-pole data over* σ *for the block row matrix function* $\tilde{\mathcal{P}}$, *then*

$$\omega = \left(C_+, C_-, A_\pi, A_\zeta, B_+, B_-, \Gamma, P_+, P_-, Q_+, Q_- \right)$$

is a σ-admissible interpolation data set (see (3.12)).

Moreover, if $\mathcal{P}$ is stabilizable, the following holds true:

(1) A rational matrix function K stabilizes $\mathcal{P}$ if and only if K has a σ-coprime factorization

$$K = N_K D_K^{-1}$$

with $N_K \in \mathcal{R}^{(m-K) \times (n-L)}(\sigma)$ and with $D_K \in \mathcal{R}^{(n-L) \times (n-L)}(\sigma)$ σ-biproper such that

$$[\tilde{\mathcal{P}}_{21} N_K + \tilde{\mathcal{P}}_{22} D_K, \tilde{\mathcal{P}}_{23}]\left(\mathcal{R}^{(n-L) \times 1}(\sigma) \oplus \mathcal{R}^{L \times 1} \right) = [\tilde{\mathcal{P}}_{21}, \tilde{\mathcal{P}}_{22}, \tilde{\mathcal{P}}_{23}]\left(\mathcal{R}^{(m-K+n-L) \times 1}(\sigma) \oplus \mathcal{R}^{L \times 1} \right).$$

(2) An $m \times n$ rational matrix function W is the closed loop transfer function T_{zw} associated with a stabilizing compensator K for $\mathcal{P}$ if and only if W is analytic on σ and satisfies the interpolation conditions (3.1') - (3.5') associated with the σ-admissible interpolation data set ω, and $[0 \; I_{n-L} \; 0]\tilde{\mathcal{P}}^{-L} \begin{bmatrix} W \\ I \end{bmatrix} \Psi \begin{bmatrix} I_{n-L} \\ 0 \end{bmatrix}$ is biproper. Here $\tilde{\mathcal{P}}^{-L}$ is any left inverse of $\tilde{\mathcal{P}}$ and Ψ is any regular rational $n \times n$ matrix function such that

$$\Psi\left(\mathcal{R}^{(n-L) \times 1}(\{\lambda\}) \oplus \mathcal{R}^L \right) = \left[\tilde{\mathcal{P}}_{21} \; \tilde{\mathcal{P}}_{22} \; \tilde{\mathcal{P}}_{23} \right]\left(\mathcal{R}^{(m-K+n-L) \times 1}(\{\lambda\}) \oplus \mathcal{R}^{L \times 1} \right)$$

with

$$\lambda = \begin{cases} \infty & \text{for the continuous time case} \\ 0 & \text{for the discrete time case.} \end{cases}$$

As a corollary we obtain a solution of an inverse problem, namely: given an admissible interpolation data set, describe the plants $\mathcal{P}$ for which the closed loop transfer functions T_{zw} associated with internally stabilizing compensators are characterized by the prescribed set of interpolation conditions (3.1') - (3.5'). In particular such plants $\mathcal{P}$ always exist; thus interpolation and stabilization of feedback systems are equivalent.

COROLLARY 5.2. *Let $\omega = \left(C_+, C_-, A_\pi, A_\zeta, B_+, B_-, \Gamma, P_+, P_-, Q_+, Q_- \right)$ be a σ-admissible interpolation data set (where either $\sigma = \bar{\Pi}$ or $\sigma = \bar{\mathcal{D}}$) and let $\mathcal{P} = \begin{bmatrix} \mathcal{P}_{11} & \mathcal{P}_{12} \\ \mathcal{P}_{21} & \mathcal{P}_{22} \end{bmatrix}$ be a rational matrix function describing a plant as in Section 6.1. Then the following are equivalent:*

(i) The proper rational matrix function K stabilizes $\mathcal{P}$ if and only if $W := \mathcal{P}_{11} + \mathcal{P}_{12} K (I - \mathcal{P}_{22} K)^{-1} \mathcal{P}_{21}$ is stable and satisfies the interpolation conditions (3.1') - (3.5') associated with ω

(ii) *The transform*

$$\tilde{\mathcal{P}} = \begin{bmatrix} \mathcal{P}_{11} - \mathcal{P}_{11}\begin{bmatrix} \mathcal{P}_{21} \\ \mathcal{P}_{21}^0 \end{bmatrix}^{-1}\begin{bmatrix} \mathcal{P}_{22} \\ \mathcal{P}_{22}^0 \end{bmatrix} & \mathcal{P}_{11}\begin{bmatrix} \mathcal{P}_{21} \\ \mathcal{P}_{21}^0 \end{bmatrix}^{-1} \\ -\begin{bmatrix} \mathcal{P}_{21} \\ \mathcal{P}_{21}^0 \end{bmatrix}^{-1}\begin{bmatrix} \mathcal{P}_{22} \\ \mathcal{P}_{22}^0 \end{bmatrix} & \begin{bmatrix} \mathcal{P}_{11} \\ \mathcal{P}_{11}^0 \end{bmatrix}^{-1} \end{bmatrix}$$

of $\mathcal{P}$ to the generalized chain formalism has

$$(\tilde{\omega}, P, Q) = \left(\begin{bmatrix} C_+ \\ C_- \end{bmatrix}, A_\pi, A_\zeta, [B_+ \ B_-], \Gamma, [P_+ \ P_-], \begin{bmatrix} Q_+ \\ Q_- \end{bmatrix} \right)$$

as an extended complete set of null-pole data over σ.

Corollary 5.2 follows immediately from Theorem 5.1 so we omit a formal proof.

PROOF OF THEOREM 5.1. The equivalence of (ii) and (iii) is the content of Theorem 4.3. Hence we need only verify (i) $\Longleftrightarrow$ (ii) together with the parametrization of stabilizing compensators (1) and of the associated closed loop transfer functions T_{zw} (2).

Let K be a given compensator. As explained in Section 5.1, K stabilizes $\mathcal{P}$ if and only if one can solve the system of equations (5.7) uniquely for stable z, u, y_1 in terms of any prespecified stable w, v_1, v_2. In more concrete form, this means: given any stable h_1, h_2, h_3 there must exist k_1, k_2, k_3, k_4 with k_1, k_2, k_3 stable and unique such that

$$(5.9) \qquad \begin{bmatrix} 0 \\ h_1 \\ h_2 - K h_3 \end{bmatrix} = \begin{bmatrix} k_1 - \tilde{\mathcal{P}}_{11}k_2 - \tilde{\mathcal{P}}_{12}k_3 - \tilde{\mathcal{P}}_{23}k_4 \\ -\tilde{\mathcal{P}}_{21}k_2 - \tilde{\mathcal{P}}_{22}k_3 - \tilde{\mathcal{P}}_{23}k_4 \\ k_2 - K k_3 \end{bmatrix}.$$

Note that $\begin{bmatrix} \tilde{\mathcal{P}}_{21} & \tilde{\mathcal{P}}_{12} & \tilde{\mathcal{P}}_{23} \\ -I & K & 0 \end{bmatrix}$ is square, so in fact, necessarily the pair $(h_1, h_2 - K h_3)$ must determine k_2, k_3, k_4 uniquely.

Now suppose that K with coprime factorization $K = N_K D_K^{-1}$ is a stabilizing compensator for $\mathcal{P}$. Then in particular we can solve (5.9) with $h_2 = 0, h_3 = 0$ and h_1 arbitrary. From $k_2 - K k_3 = 0$ we get that necessarily $k_2 = N_K g$, $k_3 = D_K g$ for some stable g. To simultaneously solve the second equation in (5.9), one must have such a stable g together with a k_4 (not necessarily stable) so that

$$-h_1 = \left(\tilde{\mathcal{P}}_{21}N_K + \tilde{\mathcal{P}}_{22}D_K \right)g + \tilde{\mathcal{P}}_{23}k_4.$$

From the first equation in (5.9), in addition we must have

$$\left(\tilde{\mathcal{P}}_{11}N_K + \tilde{\mathcal{P}}_{12}D_K \right)g + \tilde{\mathcal{P}}_{23}k_4 =: k_1 \in \mathcal{R}^{m \times 1}(\sigma).$$

As h_1 is an arbitrary element of $\mathcal{R}^{n \times 1}(\sigma)$, we have thus verified condition (ii-b) in the statement of the theorem.

To verify (ii-a), suppose that $k_2 \in \mathcal{R}^{(m-K) \times 1}(\sigma)$, $k_3 \in \mathcal{R}^{(n-L) \times 1}(\sigma)$, $k_4 \in \mathcal{R}^{L \times 1}$ are such that

$$\tilde{\mathcal{P}}_{21}k_2 + \tilde{\mathcal{P}}_{22}k_3 + \tilde{\mathcal{P}}_{23}k_4 = 0.$$

Then the system of equations

$$0 = -\tilde{\mathcal{P}}_{21}k_2 - \tilde{\mathcal{P}}_{22}k_3 - \tilde{\mathcal{P}}_{23}k_4$$

$$k_2 - Kk_3 = k_2 - Kk_3$$

uniquely determines k_2 and k_3, since well-posedness implies the regularity of the matrix function $\begin{bmatrix} \tilde{\mathcal{P}}_{21} & \tilde{\mathcal{P}}_{22} & \tilde{\mathcal{P}}_{23} \\ -I & K & 0 \end{bmatrix}$. Since K is stabilizing for $\mathcal{P}$, from the first equation in (5.9) we see that necessarily

$$\tilde{\mathcal{P}}_{11}k_2 + \tilde{\mathcal{P}}_{12}k_3 + \tilde{\mathcal{P}}_{13}k_4 =: k_1 \in \mathcal{R}^{n\times 1}(\sigma).$$

This verifies (ii-a).

Next we verify

$$[\tilde{\mathcal{P}}_{21}N_K + \tilde{\mathcal{P}}_{22}D_K \ \mathcal{P}_{23}]\left(\mathcal{R}^{(n-L)\times 1}(\sigma) \oplus \mathcal{R}^{L\times 1}\right)$$

$$\text{(5.10)} \qquad = [\tilde{\mathcal{P}}_{21} \ \tilde{\mathcal{P}}_{22} \ \tilde{\mathcal{P}}_{23}]\left(\mathcal{R}^{(m-K+n-L)\times 1}(\sigma) \oplus \mathcal{R}^{L\times 1}\right).$$

To see this, choose $h_1 = 0$ and let h_2 and h_3 be general stable functions in (5.9). From the last equation in (5.9), there must be a $g \in \mathcal{R}^{(n-L)\times 1}(\sigma)$ with $k_2 = h_2 + N_K g, k_3 = h_3 + D_K g$. Now the second equation in (5.9) says there is some choice of such a g and a $k_4 \in \mathcal{R}^{L\times 1}$ such that

$$(\tilde{\mathcal{P}}_{21}N_K + \tilde{\mathcal{P}}_{22}D_K)g = \tilde{\mathcal{P}}_{21}h_2 + \tilde{\mathcal{P}}_{22}h_3 + \tilde{\mathcal{P}}_{23}k_4.$$

This verfies (5.10).

Conversely, suppose S satisfies (ii-a) and (ii-b) and $K \in \mathcal{R}^{(m-K)\times(n-L)}$ has a coprime factorization $K = N_K D_K^{-1}$ such that the pair (N_K, D_K) satisfies (5.10). To show that K is stabilizing for $\mathcal{P}$ we need to verify that (5.9) is solvable for $k_1 \in \mathcal{R}^{m\times 1}(\sigma)$, $k_2 \in \mathcal{R}^{(m-K)\times 1}(\sigma)$, $k_3 \in \mathcal{R}^{(n-L)\times 1}(\sigma)$ and $k_4 \in \mathcal{R}^{L\times 1}$ for any $h_1 \in \mathcal{R}^{n\times 1}(\sigma)$, $h_2 \in \mathcal{R}^{(m-K)\times 1}(\sigma)$ and $h_3 \in \mathcal{R}^{(n-L)\times 1}(\sigma)$. By linearity it suffices to consider two special cases.

Case 1: $h_1 \in \mathcal{R}^{n\times 1}(\sigma), h_2 = 0, h_3 = 0$. By (ii-b) there are stable k_2, k_3 and a not necessarily stable k_4 so that $-h_1 = \tilde{\mathcal{P}}_{21}k_2 + \tilde{\mathcal{P}}_{22}k_3 + \tilde{\mathcal{P}}_{23}k_4$ and such that $k_1 := \tilde{\mathcal{P}}_{11}k_2 + \tilde{\mathcal{P}}_{12}k_3 + \tilde{\mathcal{P}}_{13}k_4$ is stable. Then (k_1, k_2, k_3, k_4) is the desired solution of (5.9).

Case 2: $h_1 = 0, h_2 \in \mathcal{R}^{(m-K)\times 1}(\sigma), h_3 \in \mathcal{R}^{(n-L)\times 1}(\sigma)$. By (5.10) we can find $g \in \mathcal{R}^{(n-L)\times 1}(\sigma)$ and $k_4 \in \mathcal{R}^{L\times 1}$ so that

$$-[\tilde{\mathcal{P}}_{21}h_2 + \tilde{\mathcal{P}}_{22}h_3] = (\tilde{\mathcal{P}}_{21}N_K + \tilde{\mathcal{P}}_{22}D_K)g + \tilde{\mathcal{P}}_{23}k_4.$$

Then $k_2 = h_2 + N_K g, k_3 = h_3 + D_K g, k_4$ give a solution of the last two equations in (5.9). From (ii-b) combined with (ii-a) we see that also $k_1 := \tilde{\mathcal{P}}_{11}k_2 + \tilde{\mathcal{P}}_{11}k_2 + \tilde{\mathcal{P}}_{23}k_4$ must be stable as well. Then $\left(k_1, k_2, k_3, k_4\right)$ gives the desired solution of (5.9) in this case. Hence $K = N_K D_K^{-1}$ is stabilizing for $\mathcal{P}$ as asserted.

To complete the proof of Theorem 5.1 it remains only to verify (2). But this follows from the characterization (1) of stabilizing compensators and the characterization of the range of the linear fractional map $(N_K, D_K) \to [\tilde{\mathcal{P}}_{11}N_K + \tilde{\mathcal{P}}_{12}D_K, \tilde{\mathcal{P}}_{13}][\tilde{\mathcal{P}}_{21}N_K + \tilde{\mathcal{P}}_{22}D_K, \tilde{\mathcal{P}}_{23}]^{-1}$ for stable pairs (N_K, D_K) satisfying (5.10) given by Theorem 4.1. The side condition that $[0 \ I_{n-L} \ 0]\tilde{\mathcal{P}}^{-L}\begin{bmatrix} W \\ I \end{bmatrix}\Psi$ $[I_{n-L} \ 0]$ be biproper is imposed to restrict (N_K, D_K) to pairs for which $K = N_K D_K^{-1}$ is well defined and proper.

References

[ABDS] D. Alpay, P. Bruinsma, A. Dijksma, H.S.V. de Snoo, Interpolation problems, extensions of symmetric operators and reproducing kernel spaces I, in *Topics in Matrix and Operator Theory* (ed. by H. Bart, I. Gohberg and M. A. Kaashoek), pp. 35-82, OT 50, Birkhäuser Verlag, Basel-Boston-Berlin, 1991.

[B] V. Belevitch, *Classical Network Theory*, Holden Day, San Francisco, 1968.

[BC] J. A. Ball and N. Cohen, *Sensitivity minimization in an H^∞ norm: parametrization of all suboptimal solutions*, Int. J. Control 46 (1987), 785-816.

[BCRR] J. A. Ball, N. Cohen, M. Rakowski and L. Rodman, Spectral data and minimal divisibility of nonregular meromorphic matrix functions, Technical Report 91.04, College of William & Mary, 1991.

[BGK] H. Bart, I. Gohberg and M. A. Kaashoek, *Minimal Factorization of Matrix and Operator Functions*, Birkhäuser, 1979

[BGR1] J. A. Ball, I. Gohberg and L. Rodman, Realization and interpolation of rational matrix Functions, in *Topics in Interpolation Theory of Rational Matrix Functions* (ed. I. Gohberg), pp. 1-72, OT 33, Birkhäuser Verlag, Basel Boston Berlin, 1988.

[BGR2] J. A. Ball, I. Gohberg and L. Rodman, Two-sided Lagrange-Sylvester interpolation problems for rational matrix functions, in *Proceeding Symposia in Pure Mathematics, Vol. 51*, (ed. W. B. Arveson and R. G. Douglas), pp. 17-83, Amer. Math. Soc., Providence, 1990.

[BGR3] J. A. Ball, I. Gohberg and L. Rodman, Minimal factorization of meromorphic matrix functions in terms of local data, *Integral Equations and Operator Theory*, 10 (1987), 309-348.

[BGR4] J. A. Ball, I. Gohberg and L. Rodman, *Interpolation of Rational Matrix Functions*, OT

45, Birkhäuser Verlag, Basel-Boston-Berlin, 1990.

[BGR5] J. A. Ball, I. Gohberg and L. Rodman, Sensitivity minimization and tangential Nevanlinna-Pick interpolation in contour integral form, in *Signal Processing Part II: Control Theory and Applications* (ed. F. A. Grünbaum et al), IMA Vol. in Math. and Appl. vol. 23, pp. 3-25, Springer-Verlag, New York, 1990.

[BGR6] J. A. Ball, I. Gohberg and L. Rodman, Tangential interpolation problems for rational matrix functions, in *Proceedings of Symposium in Applied Mathematics vol. 40*, pp. 59-86, Amer. Math. Soc., Providence, 1990.

[BH] J. A. Ball and J. W. Helton, A Beurling-Lax theorem for the Lie group U(m, n) which contains most classical interpolation, *J. Operator Theory*, 9, 1983, 107-142.

[BHV] J. A. Ball, J. W. Helton and M. Verma, A factorization principle for stabilization of linear control systems, *Int. J. of Robust and Nonlinear Control*, to appear.

[BR1] J. A. Ball and M. Rakowski, Minimal McMillan degree rational matrix functions with prescribed zero-pole structure, *Linear Algebra and its Applications*, 137/138 (1990), 325-349.

[BR2] J. A. Ball and M. Rakowski, Zero-pole structure of nonregular rational matrix functions, in *Extension and Interpolation of Linear Operators and Matrix Functions* (ed. by I. Gohberg), OT 47, pp. 137-193, Birkhäuser Verlag, Basel Boston Berlin.

[BR3] J. A. Ball and M. Rakowski, Null-pole subspaces of rectangular rational matrix functions, *Linear Algebra and its Applications*, 159 (1991), 81-120.

[BR4] J. A. Ball and M. Rakowski, Transfer functions with a given local zero pole structure, in *New Trends in Systems Theory*, (ed. G. Conte, A. M. Perdon and B. Wyman), pp. 81-88, Birkhäuser Verlag, Basel-Boston-Berlin, 1991.

[BR5] J. A. Ball and M. Rakowski, Interpolation by rational matrix functions and stability of feedback systems: the 2-block case, preprint.

[BRan] J. A. Ball and A. C. M. Ran, Local inverse spectral problems for rational matrix functions, *Integral Equations and Operator Theory*, 10 (1987), 349-415.

[CP] G. Conte, A. M. Perdon, On the causal factorization problem, *IEEE Transactions on Automatic Control*, AC-30 (1985), 811-813.

[D] H. Dym, J. contractive matrix functions, interpolation and displacement rank, Regional conference series in mathematics, 71, Amer. Math. Soc., Providence, R.I., 1989.

[DGKF] J. C. Doyle, K. Glover, P. P. Khargonekar and B. A. Francis, State-space solutions to standard H^2 and H^∞ control problems, *IEEE Trans. Auto. Control*, AC-34, (1989), 831-847.

[F] G. D. Forney, Jr., Minimal bases of rational vector spaces, with applications to multivariable linear systems, *SIAM Journal of Control*, 13 (1975), 493-520.

[FF] C. Foias and A. E. Frazho, *The Commutant Lifting Approach to Interpolation Problems*, Birkhäuser Verlag, Basel-Boston-Berlin, 1990.

[Fr] B. A. Francis, *A Course in H^∞ Control Theory*, Springer Verlag, New York, 1987.

[GK] I. Gohberg and M. A. Kaashoek, An inverse spectral problem for rational matrix functions and minimal divisibility, *Integral Equations and Operator Theory*, 10 (1987), 437-465.

[Hu] Y. S. Hung, H^∞ interpolation of rational matrices, *Int. J. Control*, 48 (1988), 1659-1713.

[K] T. Kailath, *Linear Systems*, Prentice-Hall, Englewood Cliffs, N. J., 1980.

[Ki] H. Kimura, Directional interpolation approach to H^∞-optimization and robust stabilization, *IEEE Trans. Auto. Control*, AC-32 (1987), 1085-1093.

[M] A. F. Monna, *Analyse non-archimédienne*, Springer, Verlag, Berlin Heidelberg New York, 1970.

[McFG] D. C. McFarlane and K. Glover, *Robust Controller Design Using Normalized Coprime Factor Plant Descriptions*, Lecture Notes in Control and Information Sciences Vol. 138, Springer-Verlag, New York, 1990.

[NF] B. Sz.-Nagy and C. Foias, Harmonic Analysis of Operators on Hilbert Space, *American Elsevier*, New York, 1970.

[R] M. Rakowski, Generalized Pseudoinverses of Matrix Valued Functions, *Int. Equations and Operator Theory*, 14 (1991), 564-585.

[YBL] D. C. Youla, J. Bongiorno and Y. Lu, Single loop stabilization of linear multivariable dynamic plants, *Automatica*, 10 (1974), 151-173.

[YJB] D. C. Youla, H. A. Jabr and J. J. Bongiorno, Modern Wiener-Hopf design of optimal controllers: I and II, *IEEE Trans. Auto. Control*, AC-291 (1977), 3-13.

[V] M. Vidyasager, *Control Systems Synthesis: A Factorization Approach*, MIT Press, Cambridge, Mass., 1985.

Department of Mathematics   Department of Mathematics
Virginia Tech   Southwestern Oklahoma State University
Blacksburg, VA 24061   Weatherford, OK 73096

MSC: Primary 47A57, Secondary 93B52, 93B36

Operator Theory:
Advances and Applications, Vol. 59
© 1992 Birkhäuser Verlag Basel

MATRICIAL COUPLING AND EQUIVALENCE AFTER EXTENSION

H. Bart and V.E. Tsekanovskii

The purpose of this paper is to clarify the notions of matricial coupling and equivalence after extension. Matricial coupling and equivalence after extension are relationships that may or may not exist between bounded linear operators. It is known that matricial coupling implies equivalence after extension. The starting point here is the observation that the converse is also true: *Matricial coupling and equivalence after extension amount to the same.* For special cases (such as, for instance, Fredholm operators) necessary and sufficient conditions for matricial coupling are given in terms of null spaces and ranges. For matrices, the issue of matricial coupling is considered as a completion problem.

1   Introduction

Let T and S be bounded linear operators acting between (complex) Banach spaces. We say that T and S are *matricially coupled* if they can be embedded into 2×2 operator matrices that are each others inverse in the following way

$$\begin{pmatrix} T & T_2 \\ T_1 & T_0 \end{pmatrix}^{-1} = \begin{pmatrix} S_0 & S_1 \\ S_2 & S \end{pmatrix}. \tag{1}$$

This notion was introduced and employed in [7]. In the addendum to [7], connections with earlier work by A. Devinatz and M. Shinbrot [18] and by S. Levin [38] are explained (cf. [46]). For a recent account on matricial coupling, see the monograph [19]. Concrete examples of matricial coupling, involving important classes of operators, can be found in [2], [7], [8], [10], [19], [21], [24], [26], [33] and [45].

The operators T and S are called *equivalent after extension* if there exist Banach spaces Z and W such that $T \oplus I_Z$ and $S \oplus I_W$ are equivalent operators. This means that there exist invertible bounded linear operators E and F such that

$$\begin{pmatrix} T & 0 \\ 0 & I_Z \end{pmatrix} = E \begin{pmatrix} S & 0 \\ 0 & I_W \end{pmatrix} F. \tag{2}$$

Two basic references in this context are [22] and [23]. The general background of these papers is the study of analytic operator functions. Thus the relevant issue in [22] and [23] is *analytic* equivalence after extension, i.e., the situation where the operators T, S, E and F in (2) depend analytically on a complex parameter. Other early references with the same background are [14], [16], [30], [31], [32], [34], [41] and [42]. For a recent application of analytic equivalence after extension involving unbounded operators, see [35]. Ordinary analytic equivalence (without extension) plays a prominent role in [1], [29] and [37]. More references, also to publications not dealing with operator functions but with single operators, will be given in Section 3.

Evidently, operators that are equivalent after extension have many features in common. Although this is less obvious, the same conclusion holds for operators that are matricially coupled. The reason behind this is that matricial coupling implies equivalence after extension. For details see [7] and [19], Section III.4.

The main point of the present paper is the observation that not only does matricial coupling imply equivalence after extension, in fact the two concepts amount to the same. The proof involves the construction of a coupling relation (1) out of an equivalence relation of the type (2). This is the main issue in Section 2. Section 3 contains examples. Two examples are directly taken from the literature; in the other three, known material is brought into the context of matricial coupling. Along the way, we give additional references.

In Section 4, we specialize to generalized invertible operators. For such operators, matricial coupling is characterized in terms of null spaces and ranges. An example is given to show that the invertibility condition is essential. Things are further worked out for finite rank operators, Fredholm operators and matrices. For matrices, one has the following simple result: If T is an $m_T \times n_T$ matrix and S is an $m_S \times n_S$ matrix, then T and S are matricially coupled if and only if

$$\operatorname{rank} T - \operatorname{rank} S = m_T - m_S = n_T - n_S. \tag{3}$$

Section 4 ends with a discussion of matricial coupling of matrices viewed as a completion problem: Under the assumption that (3) is satisfied, construct matrices T_0, T_1, T_2, S_0, S_1 and S_2 of the appropriate sizes such that (1) is fulfilled. Extra details are provided for the case when T and S are selfadjoint.

A few remarks about notation and terminology. The letters $\mathcal{R}$ and $\mathcal{C}$ stand for the real line and the complex plane, respectively. All linear spaces are assumed to be complex. The identity operator on a linear space Z is denoted by I_Z, or simply I. By $\dim Z$ we mean the dimension of Z. For two Banach spaces X and Y, the notation $X \simeq Y$ is used to indicate that X and Y are isomorphic. This means that there exists an invertible bounded linear operator from X onto Y. If X is a Banach space and M is a closed subspace of X, then X/M stands for the quotient space of X over M. The dimension of X/M is called the codimension of M (in X) and written as $\operatorname{codim} M$. The null space and range of a linear operator T are denoted by $\ker T$ and $\operatorname{im} T$, respectively. The symbol $\oplus$ signals the operation of taking direct sums, not only of linear spaces, but also of operators and matrices.

Acknowledgement. The foundation for this paper was laid in May 1990 while the first author was visiting Donetsk (USSR) on the invitation of E.R. Tsekanovskii, the father of the second author.

2 Coupling versus equivalence

We begin by recalling the notion of matricial coupling (cf. [7] and [19]). Let X_1, X_2, Y_1 and Y_2 be (complex) Banach spaces. Two bounded linear operators $T : X_1 \longrightarrow X_2$ and $S : Y_1 \longrightarrow Y_2$ are said to be *matricially coupled* if they can be embedded into invertible 2×2 operator matrices

$$\begin{pmatrix} T & T_2 \\ T_1 & T_0 \end{pmatrix} : X_1 \oplus Y_2 \longrightarrow X_2 \oplus Y_1, \tag{4}$$

$$\begin{pmatrix} S_0 & S_1 \\ S_2 & S \end{pmatrix} : X_2 \oplus Y_1 \longrightarrow X_1 \oplus Y_2, \tag{5}$$

involving bounded linear operators only, such that

$$\begin{pmatrix} T & T_2 \\ T_1 & T_0 \end{pmatrix}^{-1} = \begin{pmatrix} S_0 & S_1 \\ S_2 & S \end{pmatrix}. \tag{6}$$

The identity (6) is then called a *coupling relation* for T and S, while the 2×2 operator matrices appearing in (4) and (5) are referred to as *coupling matrices*.

Next, let us formally give the definitions of equivalence and equivalence after extension. The operators $T : X_1 \longrightarrow X_2$ and $S : Y_1 \longrightarrow Y_2$ are called *equivalent*, written $T \sim S$, if there exist invertible bounded linear operators $V_1 : X_1 \longrightarrow Y_1$ and $V_2 : Y_2 \longrightarrow X_2$ such that $T = V_2 S V_1$. Generalizing this concept, we say that T and S are *equivalent after extension* if there exist Banach spaces Z and W such that $T \oplus I_Z \sim S \oplus I_W$. In this context, the spaces Z and W are sometimes referred to as *extension* spaces. Ordinary equivalence, of course, corresponds to the situation where these extension spaces can be chosen to be the trivial space.

From [7] it is known that matricial coupling implies equivalence after extension (cf. [19], Section III.4). Our main result here is that the converse is also true.

Theorem 1 *Let $T : X_1 \longrightarrow X_2$ and $S : Y_1 \longrightarrow Y_2$ be bounded linear operators acting between Banach spaces. Then T and S are matricially coupled if and only if T and S are equivalent after extension.*

Proof. Assume T and S are equivalent after extension, i.e., there exist Banach spaces Z and W such that $T \oplus I_Z$ and $S \oplus I_W$ are equivalent. Let

$$E = \begin{pmatrix} E_{11} & E_{12} \\ E_{21} & E_{22} \end{pmatrix} : Y_2 \oplus W \longrightarrow X_2 \oplus Z$$

and

$$F = \begin{pmatrix} F_{11} & F_{12} \\ F_{21} & F_{22} \end{pmatrix} : X_1 \oplus Z \longrightarrow Y_1 \oplus W$$

be invertible bounded linear operators such that $T \oplus I_Z = E(S \oplus I_W)F$, i.e.,

$$\begin{pmatrix} T & 0 \\ 0 & I_Z \end{pmatrix} = \begin{pmatrix} E_{11} & E_{12} \\ E_{21} & E_{22} \end{pmatrix} \begin{pmatrix} S & 0 \\ 0 & I_W \end{pmatrix} \begin{pmatrix} F_{11} & F_{12} \\ F_{21} & F_{22} \end{pmatrix}.$$

Write the inverses E^{-1} and F^{-1} as

$$E^{-1} = \begin{pmatrix} E_{11}^{(-1)} & E_{12}^{(-1)} \\ E_{21}^{(-1)} & E_{22}^{(-1)} \end{pmatrix} : X_2 \oplus Z \longrightarrow Y_2 \oplus W$$

and

$$F^{-1} = \begin{pmatrix} F_{11}^{(-1)} & F_{12}^{(-1)} \\ F_{21}^{(-1)} & F_{22}^{(-1)} \end{pmatrix} : Y_1 \oplus W \longrightarrow X_1 \oplus Z.$$

A straightforward computation, taking into account the identities implied by the above set up, shows that the operators

$$\begin{pmatrix} T & -E_{11} \\ F_{11} & F_{12}E_{21} \end{pmatrix} : X_1 \oplus Y_2 \longrightarrow X_2 \oplus Y_1$$

and

$$\begin{pmatrix} F_{12}^{(-1)} E_{21}^{(-1)} & F_{11}^{(-1)} \\ -E_{11}^{(-1)} & S \end{pmatrix} : X_2 \oplus Y_1 \longrightarrow X_1 \oplus Y_2$$

are invertible and each others inverse. Thus

$$\begin{pmatrix} T & -E_{11} \\ F_{11} & F_{12}E_{21} \end{pmatrix}^{-1} = \begin{pmatrix} F_{12}^{(-1)} E_{21}^{(-1)} & F_{11}^{(-1)} \\ -E_{11}^{(-1)} & S \end{pmatrix}$$

is a coupling relation for T and S. This proves the if part of the theorem.

For the proof of the if part, we could simply refer to [7] or [19]. For reasons of completeness and for later reference, we prefer however to give a brief indication of the argument.

Suppose T and S are matricially coupled with coupling relation (6). Following [7], we introduce

$$E = \begin{pmatrix} -T_2 & TS_0 \\ I_{Y_2} & S_2 \end{pmatrix} : Y_2 \oplus X_2 \longrightarrow X_2 \oplus Y_2,$$

$$F = \begin{pmatrix} T_1 & T_0 \\ T & T_2 \end{pmatrix} : X_1 \oplus Y_2 \longrightarrow Y_1 \oplus X_2.$$

Then E and F are invertible with inverses

$$E^{-1} = \begin{pmatrix} -S_2 & ST_0 \\ I_{X_2} & T_2 \end{pmatrix} : X_2 \oplus Y_2 \longrightarrow Y_2 \oplus X_2,$$

$$F^{-1} = \begin{pmatrix} S_1 & S_0 \\ S & S_2 \end{pmatrix} : Y_1 \oplus X_2 \longrightarrow X_1 \oplus Y_2.$$

A direct computation shows that $T \oplus I_{Y_2} = E(S \oplus I_{X_2})F$. Thus $T \oplus I_{Y_2}$ and $S \oplus I_{X_2}$ are equivalent. This completes the proof.

Of particular interest is the case when the operators T and S depend analytically on a complex parameter. Theorem 1 and its proof then lead to the conclusion that *analytic* matricial coupling amounts to the same as *analytic* equivalence after extension (cf. [7], Section I.1 and [19], Section III.4).

Another remark that can be made on the basis of the proof of Theorem 1 is the following. Suppose $T : X_1 \longrightarrow X_2$ and $S : Y_1 \longrightarrow Y_2$ are equivalent after extension, i.e., there exist Banach spaces Z and W such that $T \oplus I_Z \sim S \oplus I_W$. Then Z and W can be taken to be equal to Y_2 and X_2, respectively. Another possible choice is $Z = Y_1$ and $W = X_1$ (cf. [7], Section I.1). Thus, if the underlying spaces X_1, X_2, Y_1 and Y_2 belong to a certain class of Banach spaces (for instance separable Hilbert spaces), then the extension spaces Z and W can be taken in the same class. Roughly speaking, equivalence by extension, if at all possible, can always be achieved with "relatively small" or "relatively nice" extension spaces.

We conclude this section with some additional observations. But first we introduce a convenient notation. Let $T : X_1 \longrightarrow X_2$ and $S : Y_1 \longrightarrow Y_2$ be bounded linear operators acting between Banach spaces. We shall write $T \overset{*}{\sim} S$ when T and S are matricially coupled or, what amounts to the same, T and S are equivalent after extension.

The relation $\overset{*}{\sim}$ is reflexive, symmetric and transitive. This is obvious from the viewpoint of equivalence after extension. In terms of matricial coupling things are as follows. Reflexivity is seen from

$$\begin{pmatrix} T & -I_{X_2} \\ I_{X_1} & 0 \end{pmatrix}^{-1} = \begin{pmatrix} 0 & I_{X_1} \\ -I_{X_2} & T \end{pmatrix}.$$

Symmetry is evident from the fact that (6) can be rewritten as

$$\begin{pmatrix} S & S_2 \\ S_1 & S_0 \end{pmatrix}^{-1} = \begin{pmatrix} T_0 & T_1 \\ T_2 & T \end{pmatrix}.$$

Finally, if $T \overset{*}{\sim} S$ and $S \overset{*}{\sim} R$, with coupling relations

$$\begin{pmatrix} T & A_2 \\ A_1 & A_0 \end{pmatrix}^{-1} = \begin{pmatrix} B_0 & B_1 \\ B_2 & S \end{pmatrix}, \qquad \begin{pmatrix} S & C_2 \\ C_1 & C_0 \end{pmatrix}^{-1} = \begin{pmatrix} D_0 & D_1 \\ D_2 & R \end{pmatrix},$$

then $T \overset{*}{\sim} R$ with coupling relation

$$\begin{pmatrix} T & A_2 C_2 \\ -C_1 A_1 & C_0 - C_1 A_0 C_2 \end{pmatrix}^{-1} = \begin{pmatrix} B_0 - B_1 D_0 B_2 & -B_1 D_1 \\ D_2 B_2 & R \end{pmatrix}.$$

This can be verified by calculation.

The relation $\overset{*}{\sim}$ implies certain connections between the operators involved. Those that are most relevant in the present context are stated in the next proposition.

Proposition 1 *Let $T : X_1 \longrightarrow X_2$ and $S : Y_1 \longrightarrow Y_2$ be bounded linear operators, and assume $T \overset{*}{\sim} S$. Then $\ker T \simeq \ker S$. Also $\operatorname{im} T$ is closed if and only if $\operatorname{im} S$ is closed, and in that case $X_2/\operatorname{im} T \simeq Y_2/\operatorname{im} S$.*

All elements needed to establish the proposition can be found in [7], Section I.2 and [19], Section III.4. The details are easy to fill in and therefore omitted. We take the opportunity here to point out that there is a misprint in [7]. On the first line of [7], page 44, the symbol B_{21} should be replaced by A_{12}.

3 Examples

Interesting instances of matricial coupling can be found in the publications mentioned in the Introduction. These concern integral operators on a finite interval with semi-separable kernel, singular integral equations, Wiener-Hopf integral operators, block Toeplitz equations, etc.. Here we present five examples. In the first three, known material is brought into the context of matricial coupling. The fourth example can be seen as a special case of the Example given in [7], Section I.1, and the fifth summarizes the results of [7], Section IV.1 and [19], Section XIII.8.

Example 1 Suppose we have two scalar polynomials

$$a(\lambda) = \lambda^m + a_{m-1}\lambda^{m-1} + \ldots + a_1\lambda + a_0,$$

$$b(\lambda) = \lambda^m + b_{m-1}\lambda^{m-1} + \ldots + b_1\lambda + b_0.$$

The *resultant* (or *Sylvester matrix*) associated with a and b is the $2m \times 2m$ matrix

$$R = R(a,b) = \begin{pmatrix}
a_0 & a_1 & \cdots & a_{m-1} & 1 & 0 & \cdots & 0 \\
0 & a_0 & \cdots & a_{m-2} & a_{m-1} & 1 & & 0 \\
\vdots & & \ddots & & & & \ddots & \ddots \\
0 & \cdots & 0 & a_0 & & & a_{m-1} & 1 \\
b_0 & b_1 & \cdots & b_{m-1} & 1 & & & \\
0 & b_0 & \cdots & b_{m-2} & b_{m-1} & 1 & & \\
\vdots & & \ddots & & & & \ddots & \ddots \\
0 & \cdots & 0 & b_0 & & & b_{m-1} & 1
\end{pmatrix}.$$

The *Bezoutian* (or *Bezout matrix*) associated with a and b is the $m \times m$ matrix

$$B = B(a,b) = (b_{ij})_{i,j=1}^{m}$$

given by

$$\frac{a(\lambda)b(\mu) - a(\mu)b(\lambda)}{\lambda - \mu} = \sum_{i,j=1}^{m} b_{ij}\lambda^{i-1}\mu^{j-1}.$$

As is well-known, the matrices R and B provide information about the common zeros of a and b (see, e.g., [36], Section 13.3). Our aim here is to show that R and B are matricially coupled. Matrices are identified with linear operators in the usual way.

It is convenient to introduce the following auxiliary $m \times m$ matrices:

$$S(a) = \begin{pmatrix} a_1 & a_2 & \ldots & a_{m-1} & 1 \\ a_2 & a_3 & \ldots & 1 & 0 \\ \vdots & & & & \vdots \\ a_{m-1} & 1 & & & \\ 1 & 0 & \ldots & 0 & 0 \end{pmatrix},$$

$$T(a) = \begin{pmatrix} a_0 & a_1 & \ldots & a_{m-2} & a_{m-1} \\ 0 & a_0 & \ldots & a_{m-3} & a_{m-2} \\ \vdots & & \ddots & & \vdots \\ 0 & 0 & & a_0 & a_1 \\ 0 & 0 & \ldots & 0 & a_0 \end{pmatrix},$$

$$J = \begin{pmatrix} 0 & 0 & \ldots & 0 & 1 \\ 0 & 0 & & 1 & 0 \\ \vdots & & & \vdots & \\ 0 & 1 & & & 0 \\ 1 & 0 & \ldots & 0 & 0 \end{pmatrix}.$$

Observe that $R = R(a, b)$ can be written as

$$R = \begin{pmatrix} T(a) & JS(a) \\ T(b) & JS(b) \end{pmatrix}. \tag{7}$$

From [36], Section 13.3 we know that

$$S(a)T(b) - S(b)T(a) = B, \qquad S(a)JS(b) - S(b)JS(a) = 0.$$

Clearly $J^2 = I_m$, where I_m stands for the $m \times m$ identity matrix. A simple calculation now shows that

$$\begin{pmatrix} T(a) & JS(a) & 0 \\ T(b) & JS(b) & -S(a)^{-1} \\ I_m & 0 & 0 \end{pmatrix}^{-1} = \begin{pmatrix} 0 & 0 & I_m \\ S(a)^{-1}J & 0 & -S(a)^{-1}JT(a) \\ S(b) & -S(a) & B \end{pmatrix}.$$

In view of (7), this is a coupling relation for R and B.

By the results of Section 2, we have $R \oplus I_m \sim B \oplus I_{2m}$, i.e. R and B are equivalent by *two-sided* extension involving the $m \times m$ and $2m \times 2m$ identity matrix. In the present situation things can actually be done with *one-sided* extension. In fact, $R \sim B \oplus I_m$. Details can be found in [36], Section 13.3 (see also the discussion on finite rank operators in Section 4 below).

The equivalence after extension of the Bezoutian and the resultant already appears in [20]. For an analogous result for matrix polynomials, see [39]. It is also known that the Vandermonde matrix and the resultant for two matrix polynomials are equivalent after extension (cf. [25]).

Example 2 This example is inspired by [40], Section 3. Let $A : Z \longrightarrow W$, $B : X \longrightarrow Z$, $C : W \longrightarrow Y$, and $D : X \longrightarrow Y$ be bounded linear operators acting between Banach spaces. Then $D + CAB$ is a well-defined bounded linear operator from X into Y. Put

$$M = \begin{pmatrix} -I_W & 0 & A \\ C & D & 0 \\ 0 & B & -I_Z \end{pmatrix} : W \oplus X \oplus Z \longrightarrow W \oplus Y \oplus Z.$$

Then $D + CAB \overset{*}{\sim} M$ and the identity

$$\begin{pmatrix} D + CAB & -C & -I_Y & -CA \\ AB & -I_W & 0 & -A \\ I_X & 0 & 0 & 0 \\ B & 0 & 0 & -I_Z \end{pmatrix}^{-1} = \begin{pmatrix} 0 & 0 & I_X & 0 \\ 0 & -I_W & 0 & A \\ I_Y & C & D & 0 \\ 0 & 0 & 0 & -I_Z \end{pmatrix}$$

is a coupling relation for $D + CAB$ and M.

This coupling relation implies that

$$D + CAB \oplus I_{W \oplus Y \oplus Z} \sim M \oplus I_Y, \qquad D + CAB \oplus I_{W \oplus X \oplus Z} \sim M \oplus I_X.$$

Both equivalences involve two-sided extensions, but, as in Example 1, things can be done with one-sided extension. Indeed, it is not difficult to prove that M is equivalent to $D + CAB \oplus I_{W \oplus Z}$. The details are left to the reader.

Example 3 Consider the monic operator polynomial

$$L(\lambda) = \lambda^n I + \lambda^{n-1} A_{n-1} + \ldots + \lambda A_1 + A_0.$$

Here $A_0, \ldots, A_{n-1}$ are bounded linear operators acting on a Banach space X. Put

$$C_L = \begin{pmatrix} 0 & I & 0 & \ldots & \ldots & 0 \\ \vdots & 0 & I & 0 & \ldots & 0 \\ \vdots & \vdots & 0 & \ddots & \ddots & \vdots \\ \vdots & \vdots & \vdots & \ddots & \ddots & 0 \\ 0 & 0 & 0 & \ldots & 0 & I \\ -A_0 & -A_1 & -A_2 & \ldots & \ldots & -A_{n-1} \end{pmatrix} : X^n \longrightarrow X^n.$$

It is well-known that

$$L(\lambda) \oplus I_{X^{n-1}} \sim \lambda I - C_L$$

and, in fact, we have a case here of analytic equivalence after one-sided extension. Clearly it is also a case of *linearization* by (one-sided) extension. For details, see [3], [27], [28], [36] and [44].

Now, let us consider things from the point of view of matricial coupling. For $k = 0, \ldots, n-1$, put

$$L_k(\lambda) = -\left(\lambda^k I + \sum_{j=0}^{k-1} \lambda^j A_{n-k-j}\right),$$

so $L_0(\lambda) = -I$ in particular. Then we have the coupling relation

$$\begin{pmatrix} L(\lambda) & L_{n-1}(\lambda) & L_{n-2}(\lambda) & \ldots & L_2(\lambda) & L_1(\lambda) & L_0(\lambda) \\ I & 0 & 0 & \ldots & 0 & 0 & 0 \\ \lambda I & -I & 0 & \ldots & 0 & 0 & 0 \\ \lambda^2 I & -\lambda I & -I & & 0 & 0 & 0 \\ \vdots & \vdots & \vdots & & & & \vdots \\ -\lambda^{n-1} I & -\lambda^{n-2} I & -\lambda^{n-3} I & \ldots & -\lambda I & -I & 0 \end{pmatrix}^{-1} =$$

$$\begin{pmatrix} 0 & I & 0 & \ldots & 0 & 0 & 0 \\ 0 & \lambda I & -I & & 0 & 0 & 0 \\ 0 & 0 & \lambda I & & 0 & 0 & 0 \\ \vdots & \vdots & & & & & \\ 0 & 0 & 0 & \ldots & 0 & \lambda I & -I \\ -I & A_0 & A_1 & \ldots & A_{n-3} & A_{n-2} & \lambda I + A_{n-1} \end{pmatrix},$$

showing that $L(\lambda)$ and $\lambda I - C_L$ are matricially coupled. Note that this is a case of analytic matricial coupling.

Example 4 Let $A : X \longrightarrow X$, $B : Y \longrightarrow X$, $C : X \longrightarrow Y$ and $D : Y \longrightarrow Y$ be bounded linear operators between Banach spaces. For λ in the resolvent set $\rho(A)$ of A, we put

$$W(\lambda) = D + C(\lambda I_X - A)^{-1} B. \tag{8}$$

Assume that D is invertible and write $A^\times = A - BD^{-1}C$. It is well-known that

$$W(\lambda) \oplus I_X \sim (\lambda I_X - A^\times) \oplus I_Y \tag{9}$$

and, in fact, we have another case here of analytic equivalence after (two-sided) extension. For details and additional information, see [4], Section 2.4.

Considering things from the viewpoint of matricial coupling, we see that $W(\lambda) \stackrel{*}{\sim} \lambda I_X - A^\times$ with coupling relation

$$\begin{pmatrix} W(\lambda) & -C(\lambda I_X - A)^{-1} \\ -(\lambda I_X - A)^{-1} B & (\lambda I_X - A)^{-1} \end{pmatrix}^{-1} = \begin{pmatrix} D^{-1} & D^{-1}C \\ BD^{-1} & \lambda I_X - A^\times \end{pmatrix}.$$

Note that this is again a case of analytic matricial coupling.

An expression of the type (8) is called a *realization* for W. Under very mild conditions analytic operator functions admit such realizations. For instance, if the operator function W is analytic on a neighbourhood of ∞, then W admits a realization. For more information on this issue, see [4]. Whenever an operator function W can be written in the form (8) with invertible D, it admits a linearization by two-sided extension (9), and hence certain features of it can be studied by using spectral theoretical tools. Under additional (invertibility) conditions on B or C, even linearization by one-sided extension, i.e. analytic equivalence of the type

$$W(\lambda) \oplus I \sim \lambda I_X - A^\times,$$

can be achieved (cf. [4], Section 2.4; see also [31]).

Example 5 Let $K : L_p([0,\infty),Y) \longrightarrow L_p([0,\infty),Y)$ be the convolution integral operator given by

$$[K\varphi](t) = \int_0^\infty k(t-s)\varphi(s)ds.$$

Here Y is a (non-trivial) Banach space, $1 \leqq p \leqq \infty$ and k is a Bochner integrable kernel whose values are bounded linear operators on Y. The familiar Wiener-Hopf equation

$$\varphi(t) - \int_0^\infty k(t-s)\varphi(s)ds = f(t), \quad t \geqq 0$$

involving Y-valued L_p-functions φ and f can be written as $(I - K)\varphi = f$.

Assume that the so called symbol

$$W(\lambda) = I_Y - \int_{-\infty}^\infty e^{i\lambda t}k(t)dt$$

admits an analytic continuation to a neighbourhood in the Rieman sphere $\mathcal{C}_\infty$ of the extended real line $\mathcal{R}_\infty$. Then W admits a realization

$$W(\lambda) = I_Y + C(\lambda I_X - A)^{-1}B, \quad \lambda \in \mathcal{R} \subset \rho(A).$$

In case Y is finite dimensional, the (state) space X can be chosen to be finite dimensional if and only if W is rational. For details, see [4], [19] and [28].

Suppose, in addition, that W takes invertible values on $\mathcal{R}$. In view of (9), this means that the spectrum $\sigma(A^\times)$ of $A^\times = A - BC$ lies off the real line. Let P, respectively $P^\times$, be the Riesz projection corresponding to the part of $\sigma(A)$, respectively $\sigma(A^\times)$, lying in the upper half plane, and put $S = P^\times |_{\mathrm{im}P} : \mathrm{im}\, P \longrightarrow \mathrm{im}\, P^\times$. So S is the restriction of $P^\times$ to im P considered as an operator into im $P^\times$. Then

$$I - K \overset{*}{\sim} S. \tag{10}$$

For an explicit coupling relation between $I - K$ and S we refer to [7], Section IV.1 and [19], Section XIII.8. One of the (many) consequences of (10) is that $I - K$ is invertible if and only if $X = \mathrm{im}\, P \oplus \ker P^\times$. For additional information, generalizations and related material, see [2], [4], [5], [6], [7], [8], [9], [10], [11], [12], [19], [21], [24], [26], [33] and [45].

4   Special classes of operators

Proposition 1 in Section 2 gives rise to the following question:*Suppose $T : X_1 \longrightarrow X_2$ and $S : Y_1 \longrightarrow Y_2$ have closed range, $\ker T \simeq \ker S$ and $X_2/\mathrm{im}\, T \simeq Y_2/\mathrm{im}\, S$. Does it follow that $T \overset{*}{\sim} S$?* We shall see that without extra conditions on the operators involved, the answer to this question is negative (Example 6 below). But first we shall make clear that under additional invertibility assumptions it is postive.

Let X and Y be Banach spaces, and let $T : X \longrightarrow Y$ be a bounded linear operator. We say that T is *generalized invertible* if there exists a bounded linear operator $T^+ : Y \longrightarrow X$ such that $T = TT^+T$ and $T^+ = T^+TT^+$. In that case T^+ is called a *generalized inverse* of T. Special instances are left invertible operators having left inverses ($T^+T = I_X$) and right invertible operators having right inverses ($TT^+ = I_Y$). Note that T is generalized invertible if and only if there exists a bounded linear operator $T^\dagger : Y \longrightarrow X$ such that $T = TT^\dagger T$ (take $T^+ = T^\dagger TT^\dagger$ to get a generalized inverse of T). Also T is generalized invertible if and only if $\ker T$ is complemented in X and $\mathrm{im}\, T$ is complemented in Y (cf. [43]).

Theorem 2 *Let $T : X_1 \longrightarrow X_2$ and $S : Y_1 \longrightarrow Y_2$ be bounded linear operators acting between Banach spaces, and assume that T and S are generalized invertible. Then $T \overset{*}{\sim} S$ if and only if $\ker T \simeq \ker S$ and $X_2/\mathrm{im}\, T \simeq Y_2/\mathrm{im}\, S$.*

To place the result against its proper background, note that if two operators are matricially coupled (equivalent after extension) and one of them is generalized (left, right) invertible, then the other is generalized (left, right) invertible, too. For details, see [7], Section I.2 and [19], Section III.4.

The only if part of Theorem 2 is covered by Proposition 1. Note that generalized invertible operators have closed range. The if part of the theorem can be proved by establishing the following more detailed result. *Let $T^+ : X_2 \longrightarrow X_1$ and $S^+ : Y_2 \longrightarrow Y_1$ be generalized inverses of T and S, respectively. Then there exist bounded linear operators*

$$T_1 : \ X_1 \longrightarrow Y_1, \ \ T_2 : Y_2 \longrightarrow X_2,$$

$$S_1 : Y_1 \longrightarrow X_1, \ \ S_2 : \ X_2 \longrightarrow Y_2,$$

such that

$$\begin{pmatrix} T & T_2 \\ T_1 & S^+ \end{pmatrix}^{-1} = \begin{pmatrix} T^+ & S_1 \\ S_2 & S \end{pmatrix}$$

is a coupling relation for T and S. The argument is straightforward and uses the fact that, with respect to appropriate decompositions of the underlying spaces, the operators T, T^+, S and S^+ can be written in the form

$$T = \begin{pmatrix} T_0 & 0 \\ 0 & 0 \end{pmatrix}, \ \ T^+ = \begin{pmatrix} T_0^{-1} & 0 \\ 0 & 0 \end{pmatrix},$$

$$S = \begin{pmatrix} 0 & 0 \\ 0 & S_0 \end{pmatrix}, \ \ S^+ = \begin{pmatrix} 0 & 0 \\ 0 & S_0^{-1} \end{pmatrix},$$

with T_0 and S_0 invertible. A complete proof is given in [13].

The only if part of Theorem 2 is true even without the generalized invertibility condition on T and S (cf. Proposition 1). For the if part, things are different. Counterexamples are easy to construct when one allows the spaces X_1, X_2, Y_1 and Y_2 to be different. The following example is concerned with the case when all these spaces are the same. It is inspired by an example given by A. Pietsch (see [43], page 366). The example also provides a negative answer to the question raised at the beginning of this section.

Example 6 Let ℓ_∞ be the Banach space of all bounded complex sequences provided with the usual supremum norm, and let c_0 be the subspace of ℓ_∞ consisting of all sequences converging to zero. Then c_0 is a closed subspace of ℓ_∞, but c_0 is not complemented in ℓ_∞ (cf. [17]). Put $X = \ell_\infty \oplus c_0 \oplus (\ell_\infty/c_0)$, and introduce $T : X \longrightarrow X$ and $S : X \longrightarrow X$ by stipulating

$$T\left[(x_1, x_2, x_3, \cdots), (y_1, y_2, y_3, \cdots), \kappa(z_1, z_2, z_3, \cdots)\right] =$$

$$= \left[(x_2, x_4, x_6, \cdots), (0, 0, 0, \cdots), \kappa(x_1, 0, x_3, 0, x_5, \cdots)\right],$$

$$S\left[(x_1, x_2, x_3, \cdots), (y_1, y_2, y_3, \cdots), \kappa(z_1, z_2, z_3, \cdots)\right] =$$

$$= \left[(x_1, x_2, x_3, \cdots), (0, 0, 0, \cdots), \kappa(z_1, 0, z_3, 0, z_5, \cdots)\right],$$

where $\kappa : \ell_\infty \longrightarrow \ell_\infty/c_0$ is the canonical projection of ℓ_∞ onto ℓ_∞/c_0. Then T and S are well-defined bounded linear operators on X. Since S is idempotent, the range of S is closed. Also im $T =$ im S and so, in particular, $X/\text{im } T \simeq X/\text{im } S$. Analysis of the null spaces of T and S shows that ker $T \simeq$ ker S too. However, in spite of all of this, T and S are not matricially coupled. Indeed, the idempotent operator S is generalized invertible, but T is not, since ker T is not complemented in X.

A bounded linear operator T acting between Banach spaces is called a *finite rank operator* if dim im $T < \infty$. The number dim im T is then called the *rank* of T and denoted by rank T. Finite rank operators are generalized invertible. The following observations are pertinent to the topic of this paper. Details may be found in [13].

Let T and S be finite rank operators from a Banach space X into a Banach space Y. If rank $T =$ rank S, then $T \sim S$. The converse is also true, but completely trivial.

For Hilbert spaces, we have the following result. Let X and Y be infinite dimensional Hilbert spaces. Then $T \overset{*}{\sim} S$ for all finite rank operators from X into Y. This is immediate from Theorem 2.

Returning to the general situation, assume that $T : X_1 \longrightarrow X_2$ and $S : Y_1 \longrightarrow Y_2$ are finite rank operators between Banach spaces. Suppose $T \overset{*}{\sim} S$ and rank $T \geq$ rank S. Let H be a finite dimensional space with dim $H =$ rank $T-$ rank S. Then $T \sim S \oplus I_H$. Thus two finite rank operators that are equivalent after extension are equivalent after a *one-sided* extension involving a *finite dimensional* extension space. For other material on the reduction of extension spaces, see [15], Section 3.3 and [32], Section 5 (cf. also Theorem 3 below).

Extra details can also be obtained for Fredholm operators. Let X and Y be Banach spaces. A bounded linear operator $T : X \longrightarrow Y$ is called a *Fredholm operator* if ker A

has finite dimension and im S has finite codimension (in Y). Fredholm operators have closed range and are generalized invertible. If two operators are matricially coupled (equivalent after extension) and one of them is Fredholm, then the other is Fredholm too.

Theorem 3 *Let $T : X_1 \longrightarrow X_2$ and $S : Y_1 \longrightarrow Y_2$ be Fredholm operators between Banach spaces. Then $T \overset{*}{\sim} S$ if and only if*

$$\dim \ker T = \dim \ker S, \quad \operatorname{codim} \operatorname{im} T = \operatorname{codim} \operatorname{im} S. \tag{11}$$

Moreover, when $T \overset{*}{\sim} S$, and Z and W are Banach spaces, then $T \oplus I_Z \sim S \oplus I_W$ if and only if $X_1 \oplus Z \simeq Y_1 \oplus W$ and $X_2 \oplus Z \simeq Y_2 \oplus W$.

The second part of the theorem tells us, for the Fredholm case, what freedom there is in choosing the extension spaces. The crux of the matter is this: *Suppose T and S are Fredholm operators from a Banach space X into a Banach space Y. Then $T \sim S$ if and only if (11) is satisfied (i.e. $T \overset{*}{\sim} S$).*

It is a trivial matter to construct examples showing that the Fredholm condition in Theorem 3 is essential. For the (simple) proof of Theorem 3 and some related observations, see [13].

Finally, we specialize to (complex) matrices. As we shall see, some of the things discussed earlier can then be made more explicit. Matrices are identified with linear operators in the usual way. Theorem 3 immediately implies the following result. *Let A be an $m_A \times n_A$ matrix and let B be an $m_B \times n_B$ matrix. Then $A \overset{*}{\sim} B$ if and only if*

$$\operatorname{rank} A - \operatorname{rank} B = m_A - m_B = n_A - n_B. \tag{12}$$

In the remainder of this section, we shall consider matricial coupling of matrices as a completion problem.

The precise statement of the problem reads as follows. *Given an $m_A \times n_A$ matrix A and an $m_B \times n_B$ matrix B such that (12) holds, construct matrices A_0, A_1, A_2, B_0, B_1 and B_2 of the appropriate sizes such that the coupling relation*

$$\begin{pmatrix} A & A_2 \\ A_1 & A_0 \end{pmatrix}^{-1} = \begin{pmatrix} B_0 & B_1 \\ B_2 & B \end{pmatrix} \tag{13}$$

is satisfied. These appropriate sizes are: $n_B \times m_B$ for A_0, $n_B \times n_A$ for A_1, $m_A \times m_B$ for A_2, $n_A \times m_A$ for B_0, $n_A \times n_B$ for B_1, and $m_B \times m_A$ for B_2.

In order to facilitate the discussion, we introduce the following notations

$$r_A = \operatorname{rank} A, \quad r_B = \operatorname{rank} B,$$

$$p = m_A - r_A, \quad q = n_A - r_A.$$

We also let I_s stand for the $s \times s$ identity matrix and $O_{s,t}$ for the $s \times t$ zero matrix.

Now choose invertible matrices R_A, C_A, R_B and C_B (of the appropriate sizes) such that

$$R_A A C_A = I_{r_A} \oplus O_{p,q}, \quad R_B B C_B = O_{p,q} \oplus I_{r_B}. \tag{14}$$

Here it is used that $p = m_A - r_A = m_B - r_B$ and $q = n_A - r_A = n_B - r_B$. The matrices R_A and R_B correspond to row operations, the matrices C_A and C_B to column operations.

It is easy to verify that the matrices appearing in the right hand sides of the identities (14) are matricially coupled with coupling relation

$$\begin{pmatrix} I_{r_A} & O_{r_A,q} & O_{r_A,p} & O_{r_A,r_B} \\ O_{p,r_A} & O_{p,q} & I_p & O_{p,r_B} \\ O_{q,r_A} & I_q & O_{q,p} & O_{q,r_B} \\ O_{r_B,r_A} & O_{r_B,q} & O_{r_B,p} & I_{r_B} \end{pmatrix}^{-1} = \begin{pmatrix} I_{r_A} & O_{r_A,p} & O_{r_A,q} & O_{r_A,r_B} \\ O_{q,r_A} & O_{q,p} & I_q & O_{q,r_B} \\ O_{p,r_A} & I_p & O_{p,q} & O_{p,r_B} \\ O_{r_B,r_A} & O_{r_B,p} & O_{r_B,q} & I_{r_B} \end{pmatrix}.$$

From this one can obtain a coupling relation for A and B by multiplying from the left and the right by

$$\begin{pmatrix} C_A & 0 \\ 0 & R_B^{-1} \end{pmatrix}, \quad \begin{pmatrix} R_A & 0 \\ 0 & C_B^{-1} \end{pmatrix},$$

respectively. In fact, this gives a coupling relation (13) with, for instance,

$$A_0 = C_B \begin{pmatrix} O_{q,r_B} \\ I_{r_B} \end{pmatrix} \begin{pmatrix} O_{r_B,p} & I_{r_B} \end{pmatrix} R_B,$$

$$A_1 = C_B \begin{pmatrix} I_q \\ O_{r_B,q} \end{pmatrix} \begin{pmatrix} O_{q,r_A} & I_q \end{pmatrix} C_A^{-1}.$$

Thus A_0 is the product of the matrix obtained from C_B by omitting the first $q = n_B - r_B$ columns and the matrix obtained from R_B by omitting the first $p = m_B - r_B$ rows. Similarly, A_1 is the product of the matrix resulting from C_B by omitting the last $n_B - q = r_B$ columns and the matrix resulting from C_A^{-1} by omitting the first $n_A - q = r_A$ rows. Analogous descriptions can be given of A_2, B_0, B_1 and B_2 (cf. [13]).

We conclude this section with a remark on the case when both A and B are selfadjoint. To fix notation, let A be a selfadjoint $m \times m$ matrix and let B be a selfadjoint $n \times n$ matrix. We assume that $r_A - r_B = m - n$, so $A \overset{*}{\sim} B$. Here, as before, $r_A =$ rank A and $r_B =$ rank B.

The selfadjoint matrices A and B can be brought into diagonal from with the help of unitary transformations. Using such diagonalizations instead of (14) and working along the same lines as above, one arrives at a coupling relation (13) such that the coupling matrices

$$\begin{pmatrix} A & A_2 \\ A_1 & A_0 \end{pmatrix}, \quad \begin{pmatrix} B_0 & B_1 \\ B_2 & B \end{pmatrix} \tag{15}$$

are selfadjoint. It is also possible to describe the spectra of these matrices: If $\alpha_1, \ldots, \alpha_{r_A}$ are the nonzero eigenvalues of A and $\beta_1, \ldots, \beta_{r_B}$ are the nonzero eigenvalues of B, then the eigenvalues of the first matrix in (15) are $\alpha_1, \ldots, \alpha_{r_A}, \beta_1^{-1}, \ldots, \beta_{r_B}^{-1}, -1, \ldots, -1, +1, \ldots, +1$, where both -1 and $+1$ are repeated $p(= m - r_A = n - r_B)$ times. To get these of the second matrix in (15), take reciprocals. We leave the details to the reader.

References

[1] Apostol, C.: On a spectral equivalence of operators, in: *Topics in Operator Theory (Constantin Apostol Memorial Issue)*, Operator Theory: Advances and Applications, Vol. 32, Birkhäuser, Basel, 1988, pp. 15-35.

[2] Bart, H.: Transfer functions and operator theory, *Linear Algebra Appl.* 84 (1986), 33-61.

[3] Bart, H., Gohberg, I., Kaashoek, M.A.: Operator polynomials as inverses of characteristic functions, *Integral Equations and Operator Theory* 1 (1978), 1-18.

[4] Bart, H., Gohberg, I., Kaashoek, M.A.: *Minimal Factorization of Matrix and Operator Functions*, Operator Theory: Advances and Applications, Vol. 1, Birkhäuser, Basel, 1979.

[5] Bart, H., Gohberg, I., Kaashoek, M.A.: Wiener-Hopf integral equations, Toeplitz matrices and linear systems, in: *Toeplitz Centennial*, Operator Theory: Advances and Applications, Vol. 4, Birkhäuser, Basel, 1982, pp. 85-135.

[6] Bart, H., Gohberg, I., Kaashoek, M.A.: Convolution equations and linear systems, *Integral Equations and Operator Theory* 5 (1982), 283-340.

[7] Bart, H., Gohberg, I., Kaashoek, M.A.: The coupling method for solving integral equations, in: *Topics in Operator Theory and Networks, the Rehovot Workshop* (Dym, H. and Gohberg, I., eds.), Operator Theory: Advances and Applications, Vol. 12, Birkhäuser, Basel, 1984, pp. 39-73.
Addendum: *Integral Equations and Operator Theory* 8 (1985), 890-891.

[8] Bart, H., Gohberg, I., Kaashoek, M.A.: Fredholm theory of Wiener-Hopf equations in terms of realization of their symbols, *Integral Equations and Operator Theory* 8 (1985), 590-613.

[9] Bart, H., Gohberg, I., Kaashoek, M.A.: Wiener-Hopf factorization, inverse Fourier transforms and exponentially dichotomous operators, *J. Funct. Analysis* 68 (1986), 1-42.

[10] Bart, H., Gohberg, I., Kaashoek, M.A.: Wiener-Hopf equations with symbols analytic in a strip, in: *Constructive Methods of Wiener-Hopf factorization* (Gohberg, I. and Kaashoek, M.A., eds.), Operator Theory: Advances and Applications, Vol. 21, Birkhäuser, Basel, 1986, pp. 39-74.

[11] Bart, H., Gohberg, I., Kaashoek, M.A.: The state space method in analysis, in: *Proceedings ICIAM 87, Paris-La Vilette* (Burgh, A.H.P. van der and Mattheij, R.M.M., eds.), Reidel, 1987, pp. 1-16.

[12] Bart, H., Kroon L.G.: An indicator for Wiener-Hopf integral equations with invertible analytic symbol, *Integral Equations and Operator Theory* 6 (1983), 1-20.
Addendum: *Integral Equations and Operator Theory* 6 (1983), 903-904.

[13] Bart, H., Tsekanovskii, V.E.: *Matricial coupling and equivalence after extension*, Report 9170 A, Econometric Institute, Erasmus University, Rotterdam 1991.

[14] Boer, H. den: Linearization of operator functions on arbitrary open sets, *Integral Equations and Operator Theory* 1 (1978), 19-27.

[15] Boer, H. den: *Block diagonalization of Matrix Functions*, Ph. D. Thesis, Vrije Universiteit, Amsterdam, 1981.

[16] Boer, H. den, Thijsse, G. Ph. A.: Semi-stability of sums of partial multiplicities under additive perturbation, *Integral Equations and Operator Theory* 3 (1980), 23-42.

[17] Day, M.M.: *Normed Linear Spaces*, 3rd ed., Springer, New York, 1973.

[18] Devinatz, A., Shinbrot, M.: General Wiener-Hopf operators, *Trans. A.M.S.* 145 (1969), 467-494.

[19] Gohberg, I., Goldberg, S., Kaashoek, M.A.: *Classes of Linear Operators, Vol. I*, Operator Theory: Advances and Applications, Vol. 49, Birkhäuser, Basel, 1990.

[20] Gohberg,I., Heinig, G.: The resultant matrix and its generalizations, I. The resultant operator for matrix polynomials, *Acta Sc. Math.* 37 (1975), 41-61 (Russian).

[21] Gohberg, I., Kaashoek, M.A.: Time varying linear systems with boundary conditions and integral equations, I, The transfer operator and its properties, *Integral Equations and Operator Theory* 7 (1984), 325-391.

[22] Gohberg, I., Kaashoek, M.A., Lay, D.C.: Spectral classification of operators and operator functions, *Bull. Amer. Math. Soc.* 82 (1976), 587-589.

[23] Gohberg, I., Kaashoek, M.A., Lay, D.C.: Equivalence, linearization and decompositions of holomorphic operator functions, *J. Funct. Anal.* 28 (1978), 102-144.

[24] Gohberg, I., Kaashoek, M.A., Lerer, L., Rodman, L.: On Toeplitz and Wiener-Hopf operators with contour-wise rational matrix and operator symbols, in: *Constructive Methods of Wiener-Hopf factorization* (Gohberg, I. and Kaashoek, M.A., eds.), Operator Theory: Advances and Applications, Vol. 21, Birkhäuser, Basel, 1986, 75-125.

[25] Gohberg, I., Kaashoek, M.A., Lerer, L., Rodman, L.: Common multiples and common divisors of matrix polynomials, II. Vandermonde and resultant matrices, *Lin. Multilin. Alg.* 12 (1982), 159-203.

[26] Gohberg, I., Kaashoek, M.A., Schagen, F. van: Non-compact integral operators with semi-separable kernels and their discrete analogues: Inversion and Fredholm properties, *Integral Equations and Operator Theory* 7 (1984), 642-703.

[27] Gohberg, I., Lancaster, P., Rodman, L.,: *Matrix Polynomials*, Academic Press, New York, 1982.

[28] Gohberg, I., Lancaster, P., Rodman, L.,: *Invariant Subspaces of Matrices with Applications*, J. Wiley and Sons, New York, 1986.

[29] Gohberg, I.C., Sigal E.I.: An operator generalization of the logarithmic residue theorem and the theorem of Rouché, *Mat. Sbornik* 84 (126) (1971), 607-629 (Russian); English. Transl., *Math. USSR Sbornik* 13 (1971), 603-625.

[30] Heinig, G.: Über ein kontinuierliches Analogon der Begleitmatrix eines Polynoms und die Linearisierung einiger Klassen holomorpher Operatorfunktionen, *Beiträge Anal.* No. 13 (1979), 111-126.

[31] Heinig, G.: Linearisierung und Realisierung holomorpher Operatorfunktionen, *Wissenschaftliche Zeitschrift der Technischen Hochschule Karl-Marx-Stadt*, XXII, H. 5 (1980), 453-459.

[32] Kaashoek, M.A., Mee, C.V.M. van der, Rodman, L.: Analytic operator functions with compact spectrum. I. Spectral nodes, linearization and equivalence, *Integral Equations and Operator Theory* 4 (1981), 504-547.

[33] Kaashoek, M.A., Schermer, J.N.M.: Inversion of convolution equations on a finite interval and realization triples, *Integral Equations and Operator Theory* 13 (1990), 76-103.

[34] Kaashoek, M.A., Ven, M.P.A. van de: A linearization for operator polynomials with coefficients in certain operator ideals, *Annali Mat. Pura Appl.* (IV) 15 (1980), 329-336.

[35] Kaashoek, M.A., Verduyn Lunel, S.M.: *Characteristic Matrices and Spectral Properties of Evolutionary Systems*, IMA Preprint Series no. 707, University of Minnesota, 1990.

[36] Lancaster, P., Tismenetsky, M.: *The theory of matrices, Second Edition with Applications*, Academic Press, Orlando, Fl., 1985.

[37] Leiterer, J.: Local and global equivalence of meromorphic operator functions, I, *Math. Nachr.* 83 (1978), 7-29; II, *Math. Nachr.* 84 (1978), 145-170.

[38] Levin, S.: On invertibility of finite sections of Toeplitz matrices, *Appl. Anal.* 13 (1982), 173-184.

[39] Lerer, L., Tismenetsky, M.: The Bezoutian and the eigenvalue-separation problem for matrix polynomials, *Integral Equations and Operator Theory* 5 (1982), 386-445.

[40] Linnemann, A., Fixed modes in parametrized systems, *Int. J. Control* 38 (1983), 319-335.

[41] Mee, C.V.M. van der: *Realization and linearization*, Rapport 109, Wiskundig Seminarium der Vrije Universiteit, Amsterdam, 1979.

[42] Mitiagin, B.: Linearization of holomorphic operator functions, I, II., *Integral Equations and Operator Theory* 1 (1978), 114-361 and 226-249.

[43] Pietsch, A. Zur Theorie der σ-Transformationen in lokalkonvexen Vektorräumen, *Math. Nachr.* 21 (1960), 347-369.

[44] Rodman, L.: *An Introduction to Operator Polynomials*, Operator Theory: Advances and Applications, Vol. 38, Birkhäuser, 1989.

[45] Roozemond, L.: *Systems of Non-normal and First Kind Wiener-Hopf Equations*, Ph.D. Thesis, Vrije Universiteit, Amsterdam, 1987.

[46] Speck, F.-O.: *General Wiener-Hopf Factorization Methods*, Pitman, Boston, 1985.

H. Bart
Econometric Institute Rotterdam
Erasmus University
P.O. Box 1738
3000 DR Rotterdam, The Netherlands

V.E. Tsekanovskii
18 Capen Blvd
Buffalo, N.Y. 14214
USA

MSC: Primary 47A05, Secondary 47A20

Operator Theory:
Advances and Applications, Vol. 59
© 1992 Birkhäuser Verlag Basel

OPERATOR MEANS AND THE RELATIVE OPERATOR ENTROPY

Jun Ichi Fujii

The notion of operator monotone functions was introduced by Löwner and that of operator concave functions by Kraus who is his student. Operator means were introduced by Ando and the general theory of them was established by Kubo and Ando himself. By their theory, a nonnegative operator monotone function is now considered as a variation of an operator mean. However this theory does not include the logarithm and the entropy function which are operator monotone and often used in information theory. These functions are operator concave and satisfy Jensen's inequality. So, considering operator means from the historical viewpoint, we shall introduce the relative operator entropy by generalizing the Kubo-Ando theory. Though its definition is derived from the Kubo-Ando theory of operator means, it can be constructed also in some ways. The relative operator entropy has of course some entropy-like properties.

1. INTRODUCTION

The theory of operator means is initiated by T.Ando [2] and established by F.Kubo and himself [20] in connection with Löwner's theory for operator functions [19,21]. Roughly speaking, there are two origins of this theory. One is the parallel sum for positive semidefinite matrices which was discussed by W.N.Anderson et al [1,9] in electrical network theory. The other is the geometric mean for positive sesquilinear forms which was discussed by W.Pusz and S.L.Woronowicz [23].

On the other hand, H.Umegaki [26] introduced the relative entropy for states on an operator algebra and it was developed by A.Uhlmann [25] by making use of the method of Pusz and Woronowicz. M.Nakamura and Umegaki [22] also introduced the operator entropy for positive operators and showed that the entropy function $\eta(x) = -x \log x$ is operator concave. So E.Kamei and the author [11] defined the *relative operator entropy* $S(A|B)$ for positive operators as a generalization of the Kubo-Ando theory of operator means. For positive invertible operators A and B, it is defined by

$$S(A|B) = A^{1/2} \log(A^{-1/2} B A^{-1/2}) A^{1/2}.$$

In this note, we show the relative operator entropy has some properties like operator means and the relative entropy and it can be constructed in some ways which its origins reflect. As an application of $S(A|B)$ to operator algebras, we see the relation between the relative operator entropy and Jones' index.

2. ORIGINS OF OPERATOR MEANS

First we see the parallel sum $A : B$ which corresponds with the impedance matrix of the parallel connection in electrical network theory. It was introduced first for positive semidefinite matrices in Anderson and Duffin [1] and second for positive operators on a Hilbert space in Fillmore and Williams [9]. It is characterized by

$$\langle A : Bx, x \rangle \equiv \inf \{ \langle Ay, y \rangle + \langle Bz, z \rangle \mid y + z = x \}.$$

If A and B are invertible, then

$$A : B = (A^{-1} + B^{-1})^{-1} = A(A + B)^{-1} B$$
$$= A^{1/2}(1 + A^{1/2} B^{-1} A^{1/2})^{-1} A^{1/2} = B^{1/2}(1 + B^{1/2} A^{-1} B^{1/2})^{-1} B^{1/2}.$$

In the theory of operator algebra, Pusz and Woronowicz [23] introduced the *geometrical mean* $\sqrt{\varphi\psi}$ for positive sesquilinear forms φ, ψ on a vector space V: Put

$$\mathcal{N} \equiv \{ x \in V \mid \varphi(x, x) + \psi(x, x) = 0 \}.$$

For a quotient map, $V \to V/\mathcal{N}$, $x \mapsto \tilde{x}$, define an inner product on $V/\mathcal{N}$ by

$$< \tilde{x}, \tilde{y} > \equiv \varphi(x, y) + \psi(x, y).$$

Then we have a Hilbert space $\mathcal{H}$ as the completion of $V/\mathcal{N}$ and the *derivatives* A, B on $\mathcal{H}$ by

$$< A\tilde{x}, \tilde{y} > = \varphi(x, y) \qquad \text{and} \qquad < B\tilde{x}, \tilde{y} > = \psi(x, y).$$

Since A and B commutes by $A + B = 1$, we can define a positive sesquilinear form $\sqrt{\varphi\psi}$ by

$$\sqrt{\varphi\psi}(x, y) \equiv < A^{1/2}B^{1/2}\tilde{x}, \tilde{y} > .$$

Then they showed that its definition does not depend on representations:

THEOREM(Pusz-Woronowicz). *If there exists a map, $x \mapsto \hat{x}$, onto a dense set of a Hilbert space H with commuting derivatives C and D*

$$\varphi(x, x) = < C\hat{x}|\hat{x} >_H, \qquad \psi(x, x) = < D\hat{x}|\hat{x} >_H,$$

then

$$\sqrt{\varphi\psi}(x, y) = < C^{1/2}D^{1/2}\hat{x} \,|\, \hat{y} >_H .$$

More generally, if $f(t, s)$ is a suitable (homogeneous) function (see also [24]), then one can define $f(\varphi, \psi)$ by

$$f(\varphi, \psi)(x, y) = < f(C, D)\hat{x} \,|\, \hat{y} >_H .$$

3. OPERATOR MEANS AND OPERATOR MONOTONE FUNCTIONS

Seeing these objects, Ando [2] introduced some operator means of positive operators on a Hilbert space:

$$\text{geometric mean}: \qquad AgB \equiv \max\left\{ X \geq 0 \,\middle|\, \begin{pmatrix} A & X \\ X & B \end{pmatrix} \geq 0 \right\},$$

$$\text{harmonic mean}: \qquad AhB \equiv \max\left\{ X \geq 0 \,\middle|\, \begin{pmatrix} 2A & 0 \\ 0 & 2B \end{pmatrix} \geq \begin{pmatrix} X & X \\ X & X \end{pmatrix} \right\}.$$

As a matter of fact, we have $AhB = 2A : B$ and $< AgBx, y > = \sqrt{< A\cdot, \cdot >< B\cdot, \cdot >}(x, y)$. Like numerical case, the following inequalities hold:

$$AhB \leq AgB \leq AaB \equiv \frac{A + B}{2}.$$

In their common properties, note the following inequality called the *transformer one*:

$$T^*(AmB)T \leq T^*AT \; m \; T^*BT.$$

If T is invertible then the equality holds in the above. Indeed, we have

$$T^*(AmB)T \leq T^*ATmT^*BT = T^*T^{*-1}(T^*ATmT^*BT)T^{-1}T \leq T^*(AmB)T.$$

In particular,

$$AmB = A^{1/2}(1 \; m \; A^{-1/2}BA^{-1/2})A^{1/2} = B^{1/2}(B^{-1/2}AB^{-1/2} \; m \; 1)B^{1/2}$$

for invertible A and B.

On the other hand, Löwner [21] defined the notion of operator monotone functions. A real-valued (continuous) function f on an interval I in the real line is called *operator monotone on I* and denoted by $f \in OM(I)$ if

$$A \leq B \quad \text{implies} \quad f(A) \leq f(B)$$

for all selfadjoint operators A, B with $\sigma(A), \sigma(B) \subset I$.

Then, for an open interval I, a function f is monotone-increasing analytic function and characterized in some ways:

1. Every *Löwner matrix* is positive semi-definite:
$$\left(\frac{f(t_i) - f(s_j)}{t_i - s_j} \right) \geq 0 \quad \text{for} \quad s_1 < t_1 < s_2 < t_2 < ... < s_n < t_n \in I.$$

2. There esists an analytic continuation $\hat{f}$ of f to the upper half plane and $\text{Im } \hat{f}(z) > 0$ for $\text{Im } z > 0$.

3. f has a suitable integral representation, see also [2,5].

Now we see the general operator means due to Kubo and Ando [20]. A binary operation m among positive operators on a Hilbert space is called an *operator mean* if it satisfies the following four axioms:

$$\textbf{monotonousness:} \qquad A \leq C, \; B \leq D \Longrightarrow AmB \leq CmD,$$
$$\textbf{lower continuity:} \qquad A_n \downarrow A, \; B_n \downarrow B \Longrightarrow A_nmB_n \downarrow AmB,$$

transformer inequality: $\qquad T^*(AmB)T \leq T^*ATmT^*BT, \quad$ and

normalization: $\qquad\qquad\qquad AmA = A.$

A nonnormalized operator mean is called a *connection*. For invertible A, we have

$$(1) \qquad\qquad AmB = A^{1/2} f_m(A^{-1/2}BA^{-1/2})A^{1/2}$$

and $f_m(x) = 1mx$ is operator monotone on $[0,\infty)$. (Note that $f_m(x)$ is a scalar since $f_m(x)$ commutes with all unitary operators by the transformer 'equality'.) By making use of an integral representation of operator monotone functions, we have a positive Radon measure μ_m on $[0,\infty]$ with

$$(2) \qquad\qquad AmB = aA + bB + \int_{(0,\infty)} (tA) : B\frac{1+t}{t}d\mu_m(t)$$

where $a = f_m(0) = \mu_m(\{0\})$ and $b = \inf t f_m(1/t) = \mu_m(\{\infty\})$. So the heart of the Kubo-Ando theory might be the following isomorphisms among them:

THEOREM (Kubo-Ando). *Maps $m \mapsto f_m$ and $m \mapsto \mu_m$ defined by* (1) *and* (2) *give affine order-isomorphisms from the connections to the nonnegative continuous operator monotone functions on $[0,\infty)$ and the positive Radon measures on $[0,\infty]$. If m is an operator mean, then $f_m(1) = 1$ and μ is a probability measure.*

Here f_m (resp. μ_m) is called the *representing function* (resp. *measure*) for m.

4. OPERATOR CONCAVE FUNCTIONS AND JENSEN'S INEQUALITY

Like operator monotone functions, a real-valued (continuous) function F on I is called *operator concave on I* and denoted by $F \in OC(I)$ if

$$F(tA + (1-t)B) \geq tF(A) + (1-t)F(B) \qquad (0 \leq t \leq 1)$$

for all selfadjoint operators A, B with $\sigma(A), \sigma(B) \subset I$. (If $-F$ is operator concave, then F is called *operator convex*.) Then, for an open interval I, a function F is concave analytic function and characterized by (see [5])

$$F_{[a]}(x) \equiv -\frac{F(x) - F(a)}{x - a} \in OM(I) \qquad (a \in I).$$

Typical examples of operator concave function is the logarithm and the *entropy function* $\eta(x) \equiv -x \log x$. In fact, Nakamura and Umegaki [22] proved the operator concavity of η and introduced the *operator entropy*

$$H(A) \equiv -A \log A \geq 0$$

for positive contraction A in $B(H)$ (see also Davis [7]).

In the Kubo-Ando theory, the following functions are operator concave: $f(x) = 1mx$, $f^\circ(x) = xm1 \in OC[0, \infty)$ and $F_m(x) \equiv xm(1 - x) \in OC[0, 1]$. Moreover, F_m gives an bridge between $OC[0, 1]^+$ and $OM(0, \infty)^+$ via operator means, see [10]:

THEOREM 4.1. *A map $m \mapsto F_m$ defines an affine order-isomorphism from the connections to nonnegative operator concave functions on $[0, 1]$.*

One of the outstanding properties of operator concave functions is so-called Jensen's inequality. For a unital completely positive map Φ on an operator algebra and a positive operator A, Davis [6] showed $\quad \Phi(F(A)) \leq F(\Phi(A)) \quad$ for an operator concave function F. By Stinespring's theorem, a completely positive map is essentially a map $X \mapsto C^*XC$. For a nonnegative function f, note that $f \in OM[0, \infty)$ if and only if $f \in OC[0, \infty)$ cf. [16]. So Jensen's inequality by Hansen [15] is

$$C^* f(A)C \leq f(C^* AC) \quad \text{for} \quad \|C\| \leq 1, A \geq 0.$$

For nonnegative $f \in OM[0, \infty)$, there exists a connection m; $f(x) = 1mx$. Then, the transformer inequality implies

$$C^* f(A)C = C^*(1mA)C \leq C^* C \, m \, C^* AC \leq 1mC^* AC = f(C^* AC).$$

Hansen and Pedersen [16] gave equivalent conditions that Jensen's inequality holds:

THEOREM(Hansen-Pedersen). *For a continuous real function F on $[0, a)$, the followings are equivalent: For $0 \le A, B < a$,*

(1) $C^* F(A)C \le F(C^* AC)$ *for* $\|C\| \le 1$,

(2) $F \in OC[0, a)$ *and* $F(0) \ge 0$,

(3) $PF(A)P \le F(PAP)$ *for every projection P,*

(4) $C^* F(A)C + D^* F(B)D \le F(C^* AC + D^* BD)$ *for* $C^* C + D^* D \le 1$.

In the Kubo-Ando theory, for $f(x) = 1mx \in OM(0, \infty)^+$, the transpose is $f^\circ(x) = xm1 = xf(1/x)$. Adopting this definition for $f \in OM(0, \infty)$, we have $f^\circ(x) = -x \log x = \eta(x)$ for $f(x) = \log x$. In general, the transpose of $f \in OM(0, \infty)$ is just a function satisfying the above equivalent conditions (see [2,14,16]):

THEOREM 4.2. $f(x) \in OM(0, \infty)$ *if and only if $f^\circ(x) \in OC[0, \infty)$ and $f^\circ(0) \ge 0$.*

This theorem suggests that one can generalize the Kubo-Ando theory dealing with $OM(0, \infty)^+$, see [14].

5. RELATIVE OPERATOR ENTROPY

Now we introduce the *relative operator entropy* $S(A|B)$ for positive operators A and B on a Hilbert space. If A and B is invertible, then it is defined as

$$S(A|B) \equiv A^{1/2} \log(A^{-1/2} B A^{-1/2}) A^{1/2}$$
$$= B^{1/2} \eta(B^{-1/2} A B^{-1/2}) B^{1/2}.$$

The above formula shows that $S(A|B)$ can be defined as a bounded operator if B is invertible. Moreover, $S(A|B + \varepsilon)$ is monotone decreasing as $\varepsilon \downarrow 0$ by $\log x \in OM(0, \infty)$. So, even if B is not invertible, we can define $S(A|B)$ by

$$(3) \qquad\qquad S(A|B) \equiv \text{s-}\lim_{\varepsilon \downarrow 0} S(A|B + \varepsilon)$$

if the limit exists. Here one of the existence conditions is (see [14]):

THEOREM 5.1. *The strong limit in (3) exists if and only if there exists c with*

$$c \leq tB - (\log t)A \quad (t > 1).$$

Under the existence, the following properties like operator means hold:

      right monotonousness: $B \leq C \Longrightarrow S(A|B) \leq S(A|C),$

      right lower continuity: $B_n \downarrow B \Longrightarrow S(A|B_n) \downarrow S(A|B),$

      transformer inequality: $T^*S(A|B)T \leq S(T^*AT|T^*BT).$

Conversely, if an operator function $S'(A|B)$ satisfies the above axioms, then there exist $f \in OM(0, \infty)$ and $F \in OC[0, \infty)$ with $F(0) \geq 0$ such that

$$S'(A|B) = A^{1/2}f(A^{-1/2}BA^{-1/2})A^{1/2} = B^{1/2}F(B^{-1/2}AB^{-1/2})B^{1/2}$$

for invertible $A, B \geq 0$, so that the class of such functions S' is a generalization of that of operator means or connections, see [14]. In addition, the relative operator entropy has entropy-like properties, e.g.:

      subadditivity: $S(A + B|C + D) \geq S(A|C) + S(C|D),$

      joint concavity: $S(A|B) \geq tS(A_1|B_1) + (1 - t)S(A_2|B_2)$

if $A = tA_1 + (1 - t)A_2$ and $B = tB_1 + (1 - t)B_2$ for $0 \leq t \leq 1$.

      informational monotonity: $\Phi(S(A|B)) \leq S(\Phi(A)|\Phi(B))$

for a normal positive linear map Φ from a W*-algebra containing A and B to a suitable W*-algebra such that $\Phi(1)$ is invertible. In particular, we have

      Peierls-Bogoliubov inequality: $\varphi(S(A|B)) \leq \varphi(A)(\log \varphi(A) - \log \varphi(B))$

for a normal positive linear functional φ on a W*-algebra containing A and B.

      Now we apply $S(A|B)$ to operator algebras. Let E be the conditional expectation of a type II_1 factor $\mathcal{M}$ onto a subfactor $\mathcal{N}$, define a *maximal entropy* $S(\mathcal{N})$ as

$$S(\mathcal{N}) \equiv \sup\{\|S(A|E(A))\| \mid A \in \mathcal{M}_1^+\}.$$

Then, for Jones' index $[\mathcal{M} : \mathcal{N}]$, we have (see [13])

THEOREM 5.2. *Let $\mathcal{N}$ be a subfactor of a type II_1 factor $\mathcal{M}$. Then,*

$$S(\mathcal{N}) = \log[\mathcal{M} : \mathcal{N}].$$

Here we recall the *relative entropy* $S(\varphi|\psi)$ for states φ, ψ on an operator algebra, cf., [27]. Derived from the Kullback-Leibler information (divergence):

$$\sum_{k=1}^{n} p_k \log \frac{p_k}{q_k} \qquad \text{for probability vectors } p, q,$$

Umegaki [26] introduced the relative entropy $S(\varphi|\psi)$ for states φ, ψ on a semi-finite von Neumann algebra, which is defined as

$$S(\varphi|\psi) = \tau(A \log A - A \log B)$$

where A and B are density operators of φ and ψ respectively, i.e.,

$$\varphi(X) = \tau(AX) \quad \text{and} \quad \psi(X) = \tau(BX).$$

Araki [3] generalized it by making use of the Tomita-Takesaki theory, Uhlmann [25] by the quadratic interpolation and Pusz-Woronowicz [24] by their functional calculus. These generalizations are all equivalent. The constructions of the last two entropies are based on the Pusz-Woronowicz calculus: Put positive sesquilinear forms $\Phi(X, Y) = \varphi(X^*Y)$ and $\Psi(X, Y) = \psi(XY^*)$, then

$$S_{PW}(\varphi|\psi) \equiv (\Phi \log \frac{\Phi}{\Psi})(1, 1),$$
$$S_U(\varphi|\psi) \equiv -\liminf_{t\downarrow 0} \frac{(\Phi^{1-t}\Psi^t) - \Phi}{t}(1, 1).$$

According to these definition, we see some constructions of $S(A|B)$. Making use of the fact $(x^t - 1)/t \downarrow \log x$ as $t \downarrow 0$, we have (see [11,12])

$$\textbf{Uhlmann type:} \qquad S(A|B) = \operatorname*{s-lim}_{t\downarrow 0} \frac{A g_t B - A}{t}.$$

where g_t is the operator mean satisfying $1 g_t x = x^t$. Note that this formula gives an approximation of $S(A|B)$. Putting $\Phi(x, y) = \ <Ax, y>\ $ and $\Psi(x, y) = \ <Bx, y>\ $, we have

Pusz-Woronoiwicz type: $\quad < S(A|B)x, y > = -(\Phi \log \frac{\Phi}{\Psi})(x, y).$

In [18], S.Izumino discussed quotient of operators by making use of Douglas' majorization theorem [8] and the parallel sum, which is considered as a space-free version of the Pusz-Woronowicz method. By making use of this, we also construct $S(A|B)$: Let $R = (A + B)^{1/2}$. Then, there exist X and Y with $XR = A^{1/2}$ and $YR = B^{1/2}$, which are uniquely determined by kernel conditions

$$\ker R \subset \ker X \cap \ker Y.$$

Here $X^*X + Y^*Y$ is the projection onto $\overline{\operatorname{ran} R}$ and X^*X commutes with Y^*Y. Then, for $F(x) = S(x|1-x) = -x\log(x/(1-x))$, we have

Izumino type: $\qquad S(A|B) = R(F(X^*X))R.$

Recently, Hiai discussed a bridge between the relative entropy and the relative operator entropy. Note that if the density operators A and B commute, then

$$S(\varphi|\psi)(= S_{PW}(\varphi|\psi) = S_U(\varphi|\psi)) = -\tau(S(A|B)).$$

Hiai and Petz [17] pointed out that the last term

$$S_{BS}(\varphi|\psi) \equiv -\tau(S(A|B))$$

had already been discussed by Belavkin and Staszewski [4]. Hiai and Petz showed that

$$S(\varphi|\psi) \geq S_{BS}(\varphi|\psi)$$

for states on a finite dimensional C*-algebra. Furthermore, Hiai informed us by private communication that it also holds for states defined by trace class operators.

REFERENCES

[1] W.N.Anderson and R.J.Duffin: Series and parallel addition of matrices, J. Math. Anal. Appl., **26**(1969), 576-594.

[2] T.Ando: Topics on operator inequalities, Hokkaido Univ. Lecture Note, 1978.

[3] H.Araki: Relative entropy of states of von Neumann algebras, Publ. RIMS, Kyoto Univ., **11** (1976), 809-833.

[4] V.P.Belavkin and P.Staszewski: C*-algebraic generalization of relative entropy and entropy, Ann. Inst. H. Poincaré Sect. A.**37**(1982), 51-58.

[5] J.Bendat and S.Sherman: Monotone and convex operator functions, Trans. Amer. Math. Soc., **79** (1955), 58-71.

[6] C.Davis: A Schwarz inequality for convex operator functions, Proc. Amer. Math. Soc., **8**(1957), 42-44.

[7] C.Davis: Operator-valued entropy of a quantum mechanical measurement, Proc. Jap. Acad., **37**(1961), 533-538.

[8] R.G.Douglas: On majorization, factorization and range inclusion of operators in Hilbert space, Proc. Amer. Math. Soc., **17**(1966), 413-416.

[9] P.A.Fillmore and J.P.Williams: On operator ranges, Adv. in Math., **7**(1971), 254-281.

[10] J.I.Fujii: Operator concave functions and means of positive linear functionals, Math. Japon., **25** (1980),453-461.

[11] J.I.Fujii and E.Kamei: Relative operator entropy in noncommutative information theory, Math. Japon., **34** (1989), 341-348.

[12] J.I.Fujii and E.Kamei: Uhlmann's interpolational method for operator means. Math. Japon., **34** (1989), 541-547.

[13] J.I.Fujii and Y.Seo: Jones' index and the relative operator entropy, Math. Japon., **34**(1989), 349-351.

[14] J.I.Fujii, M.Fujii and Y.Seo: An extension of the Kubo-Ando theory: Solidarities, Math. Japon, **35**(1990), 387-396.

[15] F.Hansen: An operator inequality, Math. Ann., **246**(1980), 249-250.

[16] F.Hansen and G.K.Pedersen: Jensen's Inequality for operators and Löwner's theorem, Math. Ann., **258**(1982), 229-241.

[17] F.Hiai and D.Petz: The proper formula for relative entropy and its asymptotics in quantum probability, Preprint.

[18] S.Izumino: Quotients of bounded operators, Proc. Amer. Math. Soc., **106**(1989), 427-435.

[19] F.Kraus: Über konvexe Matrixfunctionen, Math. Z., **41**(1936), 18-42.

[20] F.Kubo and T.Ando: Means of positive linear operators, Math. Ann., **248** (1980) 205-224.

[21] K.Löwner: Über monotone Matrixfunctionen, Math. Z., **38**(1934), 177-216.

[22] M.Nakamura and H.Umegaki: A note on the entropy for operator algebras, Proc. Jap. Acad., **37** (1961), 149-154.

[23] W.Pusz and S.L.Woronowicz: Functional calculus for sesquilinear forms and the purification map, Rep. on Math. Phys., **8** (1975), 159-170.

[24] W.Pusz and S.L.Woronowicz: Form convex functions and the WYDL and other inequalities, Let. in Math. Phys., **2**(1978), 505-512.

[25] A.Uhlmann: Relative entropy and the Wigner-Yanase-Dyson-Lieb concavity in an interpolation theory, Commun. Math. Phys., **54** (1977), 22-32.

[26] H.Umegaki: Conditional expectation in an operator algebra IV, Kodai Math. Sem. Rep. **14** (1962), 59-85.

[27] H.Umegaki and M.Ohya: Entropies in Quantum Theory (in Japanese), Kyoritsu, Tokyo (1984).

Department of Arts and Sciences

(Information Science),

Osaka Kyoiku University,

Kasiwara Osaka 582 Japan

MSC 1991: Primary 94A17, 47A63

Secondary 45B15, 47A60

Operator Theory:
Advances and Applications, Vol. 59
© 1992 Birkhäuser Verlag Basel

AN APPLICATION OF FURUTA'S INEQUALITY TO ANDO'S THEOREM

Masatoshi Fujii*, Takayuki Furuta ** and Eizaburo Kamei ***

Several authors have given mean theoretic considerations to Furuta's inequality which is an extension of Löwner-Heinz inequality. Ando discussed it on the geometric mean. In this note, Furuta's inequality is applied to a generalization of Ando's theorem.

1. INTRODUCTION.

Following after Furuta's inequality, several operator inequalities have been presented in [1,3,4,5,6,9,12]. We now think that they have suggested us a new progress of Furuta's inequality. In particular, Ando's result in [1] has been inspiring us this possibility, cf. [3,5,6]. Here we state Furuta's inequality [7] which is the starting point in our discussion.

FURUTA'S INEQUALITY. *Let A and B be positive operators acting on a Hilbert space. If $A \geq B \geq 0$, then*

$$(1) \qquad (B^r A^p B^r)^{1/q} \geq B^{(p+2r)/q}$$

and

$$(2) \qquad A^{(p+2r)/q} \geq (A^r B^p A^r)^{1/q}$$

for all $p, r \geq 0$ and $q \geq 1$ with $(1+2r)q \geq p+2r$.

If we take $p = 2r$ and $q = 2$ in (2), then we have

$$(3) \qquad A^p \geq (A^{p/2} B^p A^{p/2})^{1/2}$$

for all $p \geq 0$. From the viewpoint of this, Ando [1] showed that for a pair of selfadjoint operators A and B, $A \geq B$ if and only if the exponential version of (3) holds, i.e.,

$$e^{pA} \geq (e^{pA/2} e^{pB} e^{pA/2})^{1/2}$$

for all $p \geq 0$. So we pay attention to the exponential order due to Hansen [10] defined by $e^A \geq e^B$, and introduce an order among positive invertible operators which is just opposite to the exponential one. That is, $A \gg B$ means $\log A \geq \log B$. We call it the chaotic order because $\log A$ might be regarded as degree of the chaos of A. Thus Ando's result in [1] is rephrased as follows :

THEOREM A. *Let A and B be positive invertible operators. Then the following conditions are equivalent :*

(a) $A \gg B$.

(b) The following inequality holds for all $p \geq 0$;

$$(3) \qquad A^p \geq (A^{p/2} B^p A^{p/2})^{1/2}, \ \ i.e., \ A^{-p} \ g \ B^p \leq I.$$

(c) The operator function $G(p) = A^{-p} \ g \ B^p$ is monotone decreasing for $p \geq 0$, where g is the geometric mean.

In this note, we first propose the following operator inequalities like Furuta's one, which are improvements of the results in the preceding note [6] :

THEOREM 1. *Let A and B be positive invertible operators. If $A \gg B$, then*

$$(4) \qquad (B^r A^p B^r)^{2r/(p+2r)} \geq B^{2r}$$

and

$$(5) \qquad A^{2r} \geq (A^r B^p A^r)^{2r/(p+2r)}$$

for all $p, r \geq 0$.

This is a nice application of Furuta's inequality and implies the monotonity of an operator function discussed in [3] (see the next section), which is nothing but an extension of Theorem A by Ando. As a consequence, we also obtain Furuta's inequality in case of $2rq \geq p + 2r$ under the chaotic order.

2. OPERATOR FUNCTIONS.

Means of operators established by Kubo and Ando [13] fit right in with our plan as in [4,5,6]. A binary operation m among positive operators is called a mean if m is upper-continuous and it satisfies the monotonity and the transformer inequality

$$T^*(A \ m \ B)T \leq T^* A T \ m \ T^* B T$$

for all T. We note that if T is invertible, then it is replaced by the equality

$$T^*(A \ m \ B)T \ = \ T^* A T \ m \ T^* B T.$$

Now, by the principal result in [13], there is a unique mean m_s corresponding to the operator monotone function x^s for $0 \leq s \leq 1$;

$$1 \ m_s \ x \ = \ x^s$$

for $x \geq 0$. Particularly the mean $g = m_{1/2}$ is called the geometric one as in the case of scalars. In the below, we denote $m_{(1+s)/(p+s)}$ by $m_{(p,s)}$ for all $p \geq 1$ and $s \geq 0$. Here we can state our recent result in [3], which is a nice application of Furuta's inequality.

THEOREM B. *If $A \geq B \geq 0$, then*

$$(6) \qquad M(p,r) = B^{-2r} \ m_{(p,2r)} \ A^p$$

is a monotone increasing function, that is,

$$M(p+t, r+s) \geq M(p,r)$$

for $p \geq 1$ and $r, s, t \geq 0$.

On the other hand, we have attempted mean theoretic approach to Furuta's inequality in [2,8,11,12]. It is expressed as

$$(7) \qquad M(p,r) = B^{-2r} \ m_{(p,2r)} \ A^p \geq B$$

and equivalently

$$(7') \qquad N(p,r) = A^{-2r} \ m_{(p,2r)} \ B^p \leq A$$

under the assumption $A \geq B$, $p \geq 1$ and $r \geq 0$. However the argument in [12], [9] and [3] might say that the key point of Furuta's inequality should be seen as

$$(8) \qquad M(p,r) = B^{-2r} \ m_{(p,2r)} \ A^p \geq A$$

under the same assumption.

Concluding this section, we state that (4) and (5) are rephrased to

(4')
$$B^{-2r}\, m_{2r/(p+2r)}\, A^p \geq 1$$

and

(5')
$$A^{-2r}\, m_{2r/(p+2r)}\, B^p \leq 1$$

respectively. If we take $p = 2r$ in (5'), then it is just (3) in Theorem A.

3. FURUTA'S TYPE INEQUALITIES.

In this section, we prove Furuta's type inequalities (4) and (5) in Theorem 1. We will use Ando's result (3). Moreover we need the following lemma on a mean m_s, cf. [9].

LEMMA 2.   *Let C and D be positive invertible operators and $0 \leq s \leq 1$. Then*

(a) $C\, m_s\, D = D\, m_{1-s}\, C$,

(b) $D\, m_s\, C = (D^{-1}\, m_s\, C^{-1})^{-1}$,

and consequently

(c) $C\, m_s\, D = (D^{-1}\, m_{1-s}\, C^{-1})^{-1}$.

PROOF.   The function $f_m(x) = 1\, m\, x$ is called the representing function for a mean m and the map : $m \to f_m$ is an affine isomorphism of the means onto the operator monotone functions by [13]. Therefore it suffices to check that

$$x\, m_s\, 1 = 1\, m_{1-s}\, x \quad \text{and} \quad 1\, m_s\, x = (1\, m_s\, x^{-1})^{-1}$$

for $x > 0$. Actually we have easily

$$x\, m_s\, 1 = x(1\, m_s\, x^{-1}) = x^{1-s} = 1\, m_{1-s}\, x$$

by the transformer 'equality', and

$$1\, m_s\, x = x^s = (x^{-s})^{-1} = (1\, m_s\, x^{-1})^{-1}.$$

PROOF OF THEOREM 1.   Assume that $A \gg B$. Then it follows from Theorem A that

$$C = A^p \geq (A^{p/2} B^p A^{p/2})^{1/2} = D.$$

Suppose that $2r \geq p \geq 0$ and take $t \geq 0$ with $2r = p(1 + 2t)$. Furuta's inequality (7') ensures that

$$C \geq C^{-2t}\, m_{(2,2t)}\, D^2.$$

In other words,

$$C^{1+2t} \geq (C^t D^2 C^t)^{(1+2t)/(2+2t)}$$

and so

$$A^{2r} \geq (A^r B^p A^r)^{2r/(p+2r)}.$$

Next we have to show the case where $p \geq 2r \geq 0$. Since $B^{-1} \gg A^{-1}$, Theorem A also implies that

$$B^{-2r} \geq (B^{-r} A^{-2r} B^{-r})^{1/2},$$

that is,

$$(B^r A^{2r} B^r)^{1/2} \geq B^{2r}.$$

Again applying Furuta's inequality (7) to this, we have

$$B^{-4rt} \ m_{(2,2t)} \ B^r A^{2r} B^r \geq B^{2r},$$

and so

$$B^{-2r(1+2t)} \ m_{(2,2t)} \ A^{2r} \geq I.$$

If we choose $t \geq 0$ with $p = 2r(1 + 2t)$ since $p \geq 2r \geq 0$, then $1 - (1 + 2t)/(2 + 2t) = 2r/(p + 2r)$ and consequently it is equivalent to

$$A^{-2r} \ m_{2r/(p+2r)} \ B^p \leq I$$

by Lemma 2 (c). This completes the proof.

The following corollary of Theorem 1 plays an important role in the next section.

COROLLARY 3. *If $A \gg B$, then*

(9) $$(B^r A^p B^r)^{s/(p+2r)} \geq B^s$$

for $p \geq 0$ and $2r \geq s \geq 0$, and

(10) $$A^s \geq (A^{p/2} B^{2r} A^{p/2})^{s/(p+2r)}$$

for all $r \geq 0$ and $p \geq s \geq 0$.

4. AN APPLICATION TO ANDO'S THEOREM.

Finally we discuss a generalization of Theorem A by Ando [1]. Such an attempt has been done by [3], cf. also [9]. The purpose of this section is to complete it.

A modification of Theorem B might be considered as in [3]. Let us define

$$m_{(p,s,t)} = m_{(t+s)/(p+s)}$$

for $p \geq t \geq 0$ and $s \geq 0$. Clearly $m_{(p,s,1)} = m_{(p,s)}$.

THEOREM 4.   *If $A \gg B$, then for a given $t \geq 0$*

$$M_t(p, r) = B^{-2r}\, m_{(p, 2r, t)}\, A^p$$

is monotone increasing for $p \geq t$ and $r \geq 0$.

PROOF.   First of all, we prove that for a fixed $r > 0$, $M_t(p+s, r) \geq M_t(p, r)$ for $p \geq s > 0$. Putting $m = m_{(p+s, 2r, t)}$, it follows from (10) that

$$
\begin{aligned}
M_t(p + s, r) &= B^{-2r}\, m\, A^{p+s} \\
&= A^{p/2}(A^{-p/2}B^{-2r}A^{-p/2}\, m\, A^s)A^{p/2} \\
&\geq A^{p/2}((A^{p/2}B^{2r}A^{p/2})^{-1}\, m\, (A^{p/2}B^{2r}A^{p/2})^{s/(p+2r)})A^{p/2} \\
&= A^{p/2}(A^{-p/2}B^{-2r}A^{-p/2})^{(p-t)/(p+2r)}A^{p/2} \\
&= A^p\, m_{(p-t)/(p+2r)}\, B^{-2r} \\
&= B^{-2r}\, m_{(p, 2r, t)}\, A^p.
\end{aligned}
$$

The last equality is implied by Lemma 2 (a).

Next we show the monotonity on r. Putting $m = m_{(p, 2r+s, t)}$ for $2r \geq s \geq 0$, it follows from (9) that

$$
\begin{aligned}
M_t(p, r + s/2) &= B^{-r}(B^{-s}\, m\, B^r A^p B^r)B^{-r} \\
&\geq B^{-r}((B^r A^p B^r)^{-s/(p+2r)}\, m\, B^r A^p B^r)B^{-r} \\
&= B^{-r}(B^r A^p B^r)^{(t+2r)/(p+2r)}B^{-r} \\
&= M_t(p, r).
\end{aligned}
$$

As a result, Theorem A has the following generalization.

THEOREM 5.   *For positive invertible operators A and B, the following conditions are equivalent :*

(a) $A \gg B$.

(b) For each fixed $t \geq 0$, $M_t(p, r) \geq A^t$ for $r \geq 0$ and $p \geq t$.

(c) For each fixed $t \geq 0$, $M_t(p, r)$ is a monotone increasing function for $r \geq 0$ and $p \geq t$.

Finally, we mention that Furuta's inequality is extended to the following in the sense of (8). Actually, if we take $t = 1$ in (b) of Theorem 5, then we have :

COROLLARY 6.   *If $A \gg B$, then (8) holds, that is,*

$$(8) \qquad\qquad M(p, r) = B^{-2r}\, m_{(p, 2r)}\, A^p \geq A.$$

REFERENCES

[1] T.Ando, *On some operator inequalities*, Math.Ann., **279** (1987), 157-159.

[2] M. Fujii, *Furuta's inequality and its mean theoretic approach*, J. Operator Theory, **23** (1990), 67-72.

[3] M.Fujii, T.Furuta and E.Kamei, *Operator functions associated with Furuta's inequality*, Linear Alg. its Appl., **149** (1991), 91-96.

[4] M.Fujii and E.Kamei, *Furuta's inequality for the chaotic order*, Math.Japon., **36** (1991), 603-606.

[5] M.Fujii and E.Kamei, *Furuta's inequality for the chaotic order, II*, Math. Japon., **36** (1991), 717-722.

[6] M.Fujii and E.Kamei, *Furuta's inequality and a generalization of Ando's theorem*, Proc. Amer. Math. Soc., in press.

[7] T.Furuta, $A \geq B \geq 0$ *assures* $(B^r A^p B^r)^{1/q} \geq B^{(p+2r)/q}$ *for* $r \geq 0, p \geq 0, q \geq 1$ *with* $(1 + 2r)q \geq p + 2r$, Proc.Amer.Math.Soc., **101** (1987), 85-88.

[8] T.Furuta, *A proof via operator means of an order preserving inequality*, Linear Alg. its Appl., **113** (1989), 129-130.

[9] T.Furuta, *Two operator functions with monotone property*, Proc.Amer.Math.Soc., **111** (1991), 511-516.

[10] F.Hansen, *Selfadjoint means and operator monotone functions*, Math.Ann., **256** (1981), 29-35.

[11] E. Kamei, *Furuta's inequality via operator mean*, Math.Japon., **33** (1988), 737-739.

[12] E. Kamei, *A satellite to Furuta's inequality*, Math.Japon., **33** (1988), 883-886.

[13] F.Kubo and T.Ando, *Means of positive linear operators*, Math.Ann., **246** (1980), 205-224.

* Department of Mathematics, Osaka Kyoiku University, Tennoji, Osaka 543, Japan

** Department of Applied Mathematics, Faculty of Science, Science University of Tokyo, Kagurazaka, Shinjuku, Tokyo 162, Japan

*** Momodani Senior Highschool, Ikuno, Osaka 544, Japan

MSC 1991: Primary 47A63

    Secondary 47B15

Operator Theory:
Advances and Applications, Vol. 59
© 1992 Birkhäuser Verlag Basel

APPLICATIONS OF ORDER PRESERVING OPERATOR INEQUALITIES

TAKAYUKI FURUTA

$A \geq B \geq 0$ assures $(B^r A^p B^r)^{1/q} \geq B^{(p+2r)/q}$ for $r \geq 0$, $p \geq 0$, $q \geq 1$ with $(1 + 2r)q \geq (p + 2r)$. This is Furuta's inequality. In this paper, we show that Furuta's inequality can be applied to estimate the value of the relative operator entropy and also this inequality can be applied to extend Ando's result.

§0.   INTRODUCTION

An operator means a bounded linear operator on a complex Hilbert space. In this paper, a capital letter means an operator. An operator T is said to be positive if $(Tx, x) \geq 0$ for all x in a Hilbert space. We recall the following famous inequality ; if $A \geq B \geq 0$, then $A^\alpha \geq B^\alpha$ for each $\alpha \in [0, 1]$. This inequality is called the Löwner-Heinz theorem discovered in [14] and [12]. Moreover nice operator algebraic proof was shown in [16]. Closely related to this inequality, it is well known that $A \geq B \geq 0$ does not always ensure $A^p \geq B^p$ for $p > 1$ in general. As an extension of this Löwner-Heinz theorem, we established Furuta's inequality in [7] as follows; if $A \geq B \geq 0$, then for each $r \geq 0$,

$$(B^r A^p B^r)^{1/q} \geq B^{(p+2r)/q}$$

and

$$A^{(p+2r)/q} \geq (A^r B^p A^r)^{1/q}$$

hold for each p and q such that $p \geq 0, q \geq 1$ and $(1 + 2r)q \geq p + 2r$. We remark that Furuta's inequality yields the Löwner-Heinz theorem when we put $r = 0$. Also we remark that although $A^p \geq B^p$ for any $p > 1$ does not always hold even if $A \geq B \geq 0$, Furuta's inequality asserts that $f(A^p) \geq f(B^p)$ and $g(A^p) \geq g(B^p)$ hold under the suitable conditions where $f(X) = (B^r X B^r)^{1/q}$ and $g(Y) = (A^r Y A^r)^{1/q}$. Alternative proofs of Furuta's inequality are given in [4][8][9] and [13]. The relative operator entropy for positive invertible operators A and B is defined in [2] by

$$S(A \mid B) = A^{1/2}(log A^{-1/2} B A^{-1/2}) A^{1/2}.$$

In [11], we showed that Furuta's inequality could be applied to estimate the value of this relative operator entropy $S(A \mid B)$. For example, let A, B and C be positive invertible operators. Then $log C \geq log A \geq log B$ holds if and only if

$$S(A^{-r} \mid C^p) \geq S(A^{-r} \mid A^p) \geq S(A^{-r} \mid B^p)$$

holds for all $p \geq 0$ and all $r \geq 0$. In particular $log C \geq log A^{-1} \geq log B$ ensures $S(A \mid C) \geq -2A log A \geq S(A \mid B)$ for positive invertible operators A, B and C. In this paper, we shall attempt to extend this result by using Furuta's inequality.

In [11], we showed an elementary proof of the following result which is an extension of Ando's one [1]. Let A and B be selfadjoint operators. Then $A \geq B$ holds if and only if for a fixed $t \geq 0$,

$$Fe(p,r) = e^{-rB}(e^{rB}e^{pA}e^{rB})^{(t+2r)/(p+2r)}e^{-rB}$$

is an increasing function of both p and r for $p \geq t$ and $r \geq 0$. In this paper, also by using Furuta's inequality we shall attempt to extend this result.

§1.   APPLICATION TO THE RELATIVE OPERATOR ENTROPY

We shall show that Furuta's inequality can be applied to estimate the value of the relative operator entropy in this section. Recently in [2], the relative operator entropy $S(A \mid B)$ is defined by

$$S(A \mid B) = A^{1/2}(log A^{-1/2}BA^{-1/2})A^{1/2}$$

for positive invertible operators A and B. We remark that $S(A \mid I) = -A log A$ is the usual operator entropy. This relative operator entropy $S(A \mid B)$ can be considered as an extension of the entropy considered by Nakamura and Umegaki [15] and the relative entropy by Umegaki [17].

THEOREM 1. *Let A and B be positive invertible operators. Then the following assertions are mutually equivalent.*

(I) $log A \geq log B$.

(II$_0$) $A^p \geq (A^{p/2}B^p A^{p/2})^{1/2}$ *for all $p \geq 0$.*

(II$_1$) $A^p \geq (A^{p/2}B^s A^{p/2})^{p/(p+s)}$ *for all $p \geq 0$ and all $s \geq 0$.*

(II_2) $A^p \geq (A^{p/2}B^{s_0}A^{p/2})^{p/(p+s_0)}$ *for a fixed positive number s_0 and for all p such that $p \in [0, p_0]$, where p_0 is a fixed positive number.*

(II_3) $A^{p_0} \geq (A^{p_0/2}B^s A^{p_0/2})^{p_0/(p_0+s)}$ *for a fixed positive number p_0 and for all s such that $s \in [0, s_0]$, where s_0 is a fixed positive number.*

(III_1) $log A^{p+s} \geq log(A^{p/2}B^s A^{p/2})$ *for all $p \geq 0$ and all $s \geq 0$.*

(III_2) $log A^{p+s_0} \geq log(A^{p/2}B^{s_0}A^{p/2})$ *for a fixed positive number s_0 and for all p such that $p \in [0, p_0]$, where p_0 is a fixed positive number.*

THEOREM 2. *Let A and B be positive invertible operators. Then the following assertions are mutually equivalent.*

(I) $log C \geq log A \geq log B$

(II_0) $(A^{p/2}C^p A^{p/2})^{1/2} \geq A^p \geq (A^{p/2}B^p A^{p/2})^{1/2}$ *for all $p \geq 0$.*

(II_1) $(A^{p/2}C^s A^{p/2})^{p/(p+s)} \geq A^p \geq (A^{p/2}B^s A^{p/2})^{p/(p+s)}$ *for all $p \geq 0$ and all $s \geq 0$.*

(II_2) $(A^{p/2}C^{s_0}A^{p/2})^{p/(p+s_0)} \geq A^p \geq (A^{p/2}B^{s_0}A^{p/2})^{p/(p+s_0)}$ *for a fixed positive number s_0 and for all p such that $p \in [0, p_0]$, where p_0 is a fixed positive number.*

(II_3) $(A^{p_0/2}C^s A^{p_0/2})^{p_0/(p_0+s)} \geq A^{p_0} \geq (A^{p_0/2}B^s A^{p_0/2})^{p_0/(p_0+s)}$ *for a fixed positive number p_0 and for all s such that $s \in [0, s_0]$, where s_0 is a fixed positive number.*

(III_1) $log(A^{p/2}C^s A^{p/2}) \geq log A^{p+s} \geq log(A^{p/2}B^s A^{p/2})$ *for all $p \geq 0$ and all $s \geq 0$.*

(III_2) $log(A^{p/2}C^{s_0}A^{p/2}) \geq log A^{p+s_0} \geq log(A^{p/2}B^{s_0}A^{p/2})$ *for a fixed positive number s_0 and for all p such that $p \in [0, p_0]$, where p_0 is a fixed positive number.*

(IV_1) $S(A^{-p} \mid C^s) \geq S(A^{-p} \mid A^s) \geq S(A^{-p} \mid B^s)$ *for all $p \geq 0$ and all $s \geq 0$.*

(IV_2) $S(A^{-p} \mid C^{s_0}) \geq S(A^{-p} \mid A^{s_0}) \geq S(A^{-p} \mid B^{s_0})$ *for a fixed positive number s_0 and for all p such that $p \in [0, p_0]$, where p_0 is a fixed positive number.*

COROLLARY 1 [11]. *Let A, B and C be positive invertible operators. If $log C \geq log A^{-1} \geq log B$, then $S(A \mid C) \geq -2A log A \geq S(A \mid B)$.*

In order to give proofs to Theorem 1 and Theorem 2, we need the following Furuta's inequaliy in [7].

THEOREM A (Furuta's inequality). *Let A and B be positive operators on a Hilbert space. If $A \geq B \geq 0$, then*

(i) $$(B^r A^p B^r)^{(1+2r)/(p+2r)} \geq B^{1+2r}$$

and

(ii) $$A^{1+2r} \geq (A^r B^p A^r)^{(1+2r)/(p+2r)}$$

hold for all $p \geq 1$ and $r \geq 0$.

LEMMA 1 [10]. *Let A and B be invertible positive operators.   For any real number r,*

$$(BAB)^r = BA^{1/2}(A^{1/2}B^2 A^{1/2})^{r-1}A^{1/2}B.$$

LEMMA 2. *Let A and B be positive invertible operators. Then for any $p, s \geq 0$, the following assertions are mutually equivalent.*

(i) $$A^p \geq (A^{p/2} B^s A^{p/2})^{p/(s+p)},$$

(ii) $$(B^{s/2} A^p B^{s/2})^{s/(s+p)} \geq B^s.$$

Proof of Lemma 2. Assume (i). Then by Lemma 1,

$$A^p \geq (A^{p/2} B^s A^{p/2})^{p/(s+p)}$$
$$= A^{p/2} B^{s/2}(B^{s/2} A^p B^{s/2})^{-s/(s+p)} B^{s/2} A^{p/2},$$

that is,

$$B^{-s} \geq (B^{s/2} A^p B^{s/2})^{-s/(s+p)}$$

holds. Taking inverses proves (ii). Conversely, we have (i) from (ii) by the same way.

Proof of Theorem 1. (I) $\Longleftrightarrow$ (II$_0$) is shown in [1]. (II$_1$) $\Longrightarrow$ (II$_0$) is obvious by putting $s=p$ in (II$_1$). We show (II$_0$) $\Longrightarrow$ (II$_1$). Assume (II$_0$) ; $A^s \geq (A^{s/2} B^s A^{s/2})^{1/2}$ for all $s \geq 0$. Then by (ii) of Theorem A, we have the following inequality (1)

(1) $$A^{s(1+2t)} \geq \{A^{st}(A^{s/2} B^s A^{s/2})^{m/2} A^{st}\}^{(1+2t)/(m+2t)} \quad \text{for } m \geq 1 \text{ and } t \geq 0.$$

Putting $m = 2$ in (1), we have

(2) $$A^{s(1+2t)} \geq \{A^{s(t+1/2)} B^s A^{s(t+1/2)}\}^{(1+2t)/(2+2t)} \quad \text{for } t \geq 0.$$

Put $p = s(1 + 2t)$ in (2). Then $(1 + 2t)/(2 + 2t) = p/(s + p)$, so we have

(3) $$A^p \geq (A^{p/2} B^s A^{p/2})^{p/(s+p)} \text{ for all } p \text{ and } s \text{ such that } p \geq s \geq 0,$$

because $p = s(1 + 2t) \geq s$.

On the other hand (II$_0$) is equivalent to the following (4) by Lemma 2,

$$(4) \qquad (B^{p/2}A^pB^{p/2})^{1/2} \geq B^p \text{ for all } p \geq 0.$$

Then appling (i) of Theorem A to (4), we have the following (5)

$$(5) \qquad \{B^{pu}(B^{p/2}A^pB^{p/2})^{m/2}B^{pu}\}^{(1+2u)/(m+2u)} \geq B^{p(1+2u)} \qquad \text{for } m \geq 1 \text{ and } u \geq 0.$$

Put $m = 2$ in (5). Then we have

$$(6) \qquad \{B^{p(u+1/2)}A^pB^{p(u+1/2)}\}^{(1+2u)/(2+2u)} \geq B^{p(1+2u)} \qquad \text{for } u \geq 0.$$

Put $s = p(1 + 2u)$ in (6). Then $(1 + 2u)/(2 + 2u) = s/(p + s)$, so we have

$$(7) \qquad (B^{s/2}A^pB^{s/2})^{s/(s+p)} \geq B^s \text{ for all p and s such that } s \geq p \geq 0,$$

because $s = p(1 + 2u) \geq p$. (7) is equivalent to the following (8) by Lemma 2

$$(8) \qquad A^p \geq (A^{p/2}B^sA^{p/2})^{p/(s+p)} \text{ for all p and s such that } s \geq p \geq 0.$$

Hence the proof of (II$_1$) is complete by (3) and (8).

(II$_1$) $\implies$ (II$_2$) and (II$_1$) $\implies$ (II$_3$) are obvious since (II$_2$) and (II$_3$) are both special cases of (II$_1$). (II$_1$) $\implies$ (III$_1$) and (II$_2$) $\implies$ (III$_2$) are obtained by taking logarithm of both sides of (II$_1$) and (II$_2$) respectively since $logt$ is an operator monotone function. (III$_1$) $\implies$ (III$_2$) is obvious since (III$_2$) is a special case of (III$_1$). We show (III$_2$) $\implies$ (I). Letting $p = 0$ in (III$_2$), we have $s_0 log A \geq s_0 log B$, that is, (I). Finally we show (II$_3$) $\implies$ (I). Assume (II$_3$). Then by Lemma 2, (II$_3$) is equivalent to the following (9)

$$(9) \qquad (B^{s/2}A^{p_0}B^{s/2})^{s/(p_0+s)} \geq B^s$$

holds for a fixed positive number p_0 and for all s such that $s \in [0, s_0]$, where s_0 is a fixed positive number. Taking logarithm of both sides of (9) since $logt$ is an operator monotone function, we get

$$log(B^{s/2}A^{p_0}B^{s/2}) \geq (p_0 + s)logB.$$

Letting $s = 0$, we have $p_0 log A \geq p_0 log B$, that is, (I). Hence the proof of Theorem 1 is complete.

We remark that equivalence relation among (I), (II$_0$) and (II$_1$) is shown in [11].

Proof of Theorem 2. (I) $\iff$ (II$_1$). The hypothesis $log C \geq log A$ in (I), that is, $log A^{-1} \geq log C^{-1}$ is equivalent to

$$A^{-p} \geq (A^{-p/2}C^{-s}A^{-p/2})^{p/(p+s)} \text{ for all } p \geq 0 \text{ and all } s \geq 0$$

by (I) and (II$_1$) of Theorem 1. Taking inverses implies

$$(A^{p/2}C^s A^{p/2})^{p/(p+s)} \geq A^p \text{ for all } p \geq 0 \text{ and all } s \geq 0$$

and the rest of (II$_1$) is already shown by (I) and (II$_1$) of Theorem 1. For the proof (II$_1$) $\Longrightarrow$ (I) , we have only to trace the reverse implication in the proof (I) $\Longrightarrow$ (II$_1$). So the proof of (I) $\Longleftrightarrow$ (II$_1$) is complete. By the same method as in the proof of Theorem 1 and together wih the same technique as in the proof (I) $\Longleftrightarrow$ (II$_1$) in theorem 2, we can easily obtain the equivalence relation among (I), (II$_0$), (II$_1$), (II$_2$), (II$_3$), (III$_1$) and (III$_2$).

(III$_1$) $\Longleftrightarrow$ (IV$_1$). (III$_1$) is equivalent to the following inequalities

$$A^{-p/2}log(A^{p/2}C^s A^{p/2})A^{-p/2} \geq A^{-p/2}log(A^{p+s})A^{-p/2}$$

$$\geq A^{-p/2}log(A^{p/2}B^s A^{p/2})A^{-p/2}$$

for all $p \geq 0$ and all $s \geq 0$, equivalently

$$S(A^{-p} \mid C^s) \geq S(A^{-p} \mid A^s) \geq S(A^{-p} \mid B^s)$$

for all $p \geq 0$ and all $s \geq 0$, which is just (IV$_1$).

(IV$_1$) $\Longrightarrow$ (IV$_2$) is obvious since (IV$_2$) is a special case of (IV$_1$).

(IV$_2$) $\Longrightarrow$ (I). Put $p = 0$ in (IV$_2$). Then we have

$$logC \geq logA \geq logB$$

since s_0 is a fixed positive number.

Hence the proof of Theorem 2 is complete.

Proof of Corollary 1. Corollary 1 easily follows by Theorem 2.

§2. APPLICATION TO SOME EXTENDED RESULT OF ANDO'S ONE

Ando [1] shows the following excellent result.

THEOREM B [1]. *Let A and B be selfadjoint operators. Then the following assertions are mutually equivalent.*

(i) $A \geq B$

(ii) $e^{-rA/2}(e^{rA/2}e^{rB}e^{rA/2})^{1/2}e^{-rA/2} \leq 1 \text{ for all } r \geq 0$

(iii) $e^{-rA/2}(e^{rA/2}e^{rB}e^{rA/2})^{1/2}e^{-rA/2}$ *is a decreasing function of* $r \geq 0$.

As an extended result of Theorem B, we shall show the following Theorem 3 which includes Theorem 1 as a special case when $t = 0$.

THEOREM 3. *Let A and B be positive invertible operators, then the following assertions are mutually equivalent.*

(I) $logA \geq logB$.

(II_0) *For any fixed* $t \geq 0$,

$F(p,r) = B^{-r}(B^r A^p B^r)^{(t+2r)/(p+2r)} B^{-r}$ *is an increasing function of both* $p \geq t$

and $r \geq 0$.

(II_1) *For any fixed* $t \geq 0$, $r_0 > 0$, *and* $p_0 \geq 0$,

$F(p,r_0) = B^{-r_0}(B^{r_0} A^p B^{r_0})^{(t+2r_0)/(p+2r_0)} B^{-r_0}$ *is an increasing function of* p

such that $p \in [0, p_0]$ *for* $p \geq t$.

(II_2) *For any fixed* $t \geq 0$, $r_0 > 0$, *and* $p_0 \geq t$,

$F(p_0,r) = B^{-r}(B^r A^{p_0} B^r)^{(t+2r)/(p_0+2r)} B^{-r}$ *is an increasing function of* r *such that*

$r \in [0, r_0]$.

(III_0) *For any fixed* $t \geq 0$,

$G(p,r) = A^{-r}(A^r B^p A^r)^{(t+2r)/(p+2r)} A^{-r}$ *is a decreasing function of both* $p \geq t$ *and*

$r \geq 0$.

(III_1) *For any fixed* $t \geq 0$, *and* $r_0 > 0$, *and* $p_0 \geq 0$,

$G(p,r_0) = A^{-r_0}(A^{r_0} B^p A^{r_0})^{(t+2r_0)/(p+2r_0)} A^{-r_0}$ *is a decreasing function of* p *such*

that $p \in [0, p_0]$ *for* $p \geq t$.

(III_2) *For any fixed* $t \geq 0$, $r_0 > 0$, *and* $p_0 \geq t$,

$G(p_0,r) = A^{-r}(A^r B^{p_0} A^r)^{(t+2r)/(p_0+2r)} A^{-r}$ *is a decreasing function of* r *such that*

$r \in [0, r_0]$.

(IV) *For any fixed* $t \geq 0$, *and* $r \geq 0$,

$log(B^r A^p B^r)^{(t+2r)/(p+2r)}$ *is an increasing function of* p *for* $p \geq t$.

(V) *For any fixed $t \geq 0$, and $r \geq 0$,*

$log(A^r B^p A^r)^{(t+2r)/(p+2r)}$ *is a decreasing function of p for $p \geq t$.*

Proof of Theorem 3. (I) $\Longrightarrow$ (II$_0$). Assume (I). First of all, we cite (10) by (I) and (II$_1$) of Theorem 1

$$(10) \qquad A^p \geq (A^{p/2} B^{2r} A^{p/2})^{p/(p+2r)} \text{ for all } p \geq 0 \text{ and all } r \geq 0.$$

Moreover (10) ensures the following (11) by the Löwner-Heinz theorem

$$(11) \qquad A^s \geq (A^{p/2} B^{2r} A^{p/2})^{s/(p+2r)} \text{ for all } p \geq s \geq 0 \text{ and all } r \geq 0.$$

Then by (11), we have

$$(B^r A^p B^r)^{(p+s+2r)/(p+2r)} = B^r A^{p/2} (A^{p/2} B^{2r} A^{p/2})^{s/(p+2r)} A^{p/2} B^r \qquad \text{by Lemma 1}$$

$$\leq B^r A^{p/2} A^s A^{p/2} B^r \qquad \text{by (11)}$$

$$= B^r A^{p+s} B^r.$$

So the following (12) and (13) hold for each $r \geq 0$ and each $p \geq s \geq 0$,

$$(12) \qquad B^r A^{p+s} B^r \geq (B^r A^p B^r)^{(p+s+2r)/(p+2r)}$$

and

$$(13) \qquad (A^r B^p A^r)^{(p+s+2r)/(p+2r)} \geq A^r B^{p+s} A^r.$$

(13) is an immediate consequence of (12) because $log B^{-1} \geq log A^{-1}$ ensures that

$$(A^{-r} B^{-p} A^{-r})^{(p+s+2r)/(p+2r)} \leq A^{-r} B^{-(p+s)} A^{-r}$$

holds for each $r \geq 0$ and for each p and s such that $p \geq s \geq 0$. Taking inverses gives (13). As $(t+2r)/(p+s+2r) \in [0,1]$ since $p \geq t \geq 0$, (12) ensures the following inequality by the Löwner-Heinz theorem

$$(B^r A^{p+s} B^r)^{(t+2r)/(p+s+2r)} \geq (B^r A^p B^r)^{(t+2r)/(p+2r)},$$

which implies the following results for a fixed $t \geq 0$ and $r \geq 0$

$$(14) \qquad (B^r A^p B^r)^{(t+2r)/(p+2r)} \text{ is an increasing function of } p \geq t,$$

and

$$(15) \qquad (A^r B^p A^r)^{(t+2r)/(p+2r)} \text{ is a decreasing function of } p \geq t,$$

because (15) is easily obtained by (13) and its proof is the same way as in the proof of (14) from (12).

Next we show the following inquality (16)

(16) $(A^{p/2}B^{2r}A^{p/2})^{(t-p)/(2r+p)} \geq (A^{p/2}B^{2s}A^{p/2})^{(t-p)/(2s+p)}$ for $r \geq s \geq 0$.

By (15), we have

(17) $(A^{p/2}B^{2r}A^{p/2})^{(t_1+p)/(2r+p)} \leq (A^{p/2}B^{2s}A^{p/2})^{(t_1+p)/(2s+p)}$ for $2r \geq 2s \geq t_1 \geq 0$.

Put $\alpha = (p-t)/(p+t_1) \in [0,1]$ since $p \geq t \geq 0$ and $t_1 \geq 0$. By the Löwner-Heinz theorem, taking α as exponents of both sides of (17) and moreover taking inverses of these both sides, we have (16). Therefore for $r \geq s \geq 0$,

$$
\begin{aligned}
F(p,r) &= B^{-r}(B^r A^p B^r)^{(t+2r)/(p+2r)} B^{-r} \\
&= A^{p/2}(A^{p/2}B^{2r}A^{p/2})^{(t-p)/(p+2r)} A^{p/2} &&\text{by Lemma 1} \\
&\geq A^{p/2}(A^{p/2}B^{2s}A^{p/2})^{(t-p)/(p+2s)} A^{p/2} &&\text{by (16)} \\
&= B^{-s}(B^s A^p B^s)^{(t+2s)/(p+2s)} B^{-s} &&\text{by Lemma 1} \\
&= F(p,s),
\end{aligned}
$$

so we have (II$_0$) since $F(p,r)$ is an increasing function of $p \geq t$ by (14). So the proof of (I) $\Longrightarrow$ (II$_0$) is complete.

(II$_0$) $\Longrightarrow$ (II$_1$) and (II$_0$) $\Longrightarrow$ (II$_2$) are obvious since both (II$_1$) and (II$_2$) are special cases of (II$_0$).

(II$_1$) $\Longrightarrow$ (I) . Assume (II$_1$). Then $F(p,r_0) \geq F(0,r_0)$ with $t = 0$, that is,

$$
B^{-r_0}(B^{r_0} A^p B^{r_0})^{2r_0/(p+2r_0)} B^{-r_0} \geq B^{-r_0} B^{2r_0} B^{-r_0} = I,
$$

equivalently,

$$
A^{p/2}(A^{p/2}B^{2r_0}A^{p/2})^{-p/(p+2r_0)} A^{p/2} \geq I \qquad\text{by Lemma 1}
$$

namely

(18) $$ A^p \geq (A^{p/2}B^{2r_0}A^{p/2})^{p/(p+2r_0)} $$

holds for all p such that $p \in [0,p_0]$ and a fixed $r_0 > 0$. Taking logarithm of both sides of (18) since $logt$ is an operator monotone function, we have

$$
(p + 2r_0)logA \geq log(A^{p/2}B^{2r_0}A^{p/2}).
$$

Letting $p \longrightarrow 0$, we have $logA \geq logB$ since r_0 is a fixed positive number.

(II$_2$) $\Longrightarrow$ (I). Assume (II$_2$). Then $F(p_0,r) \geq F(p_0,0)$ with $t = 0$, that is,

$$B^{-r}(B^r A^{p_0} B^r)^{2r/(p_0+2r)} B^{-r} \geq I,$$

equivalently,

$$(19) \qquad\qquad (B^r A^{p_0} B^r)^{2r/(p_0+2r)} \geq B^{2r}$$

for all r such that $r \in [0, r_0]$ and a fixed $p_0 > 0$. Taking logarithm of both sides of (19) since $logt$ is an operator monotone function, we have

$$log(B^r A^{p_0} B^r) \geq (p_0 + 2r)logB.$$

Letting $r \longrightarrow 0$, we have $logA \geq logB$ since p_0 is a fixed positive number.

(I) $\Longrightarrow$ (III$_0$). This is in the same way as (I) $\Longrightarrow$ (II$_0$).

(III$_0$) $\Longrightarrow$ (III$_1$) and (III$_0$) $\Longrightarrow$ (III$_2$) are obvious since both (III$_1$) and (III$_2$) are special cases of (III$_0$). (III$_1$) $\Longrightarrow$ (I) and (III$_2$) $\Longrightarrow$ (I) are obtained by the same ways as (II$_1$) $\Longrightarrow$ (I) and (II$_2$) $\Longrightarrow$ (I) respectively. (II$_0$) $\Longrightarrow$ (IV) and (III$_0$) $\Longrightarrow$ (V) are both trivial since $logt$ is an operator monotone function.

(IV) $\Longrightarrow$ (I). Assume (IV) with $t = 0$. Then

$$log(B^r A^p B^r)^{2r/(p+2r)} \geq logB^{2r},$$

that is,

$$log(B^r A^p B^r) \geq (p + 2r)logB.$$

Letting $r \longrightarrow 0$ and $p = 1$, we have $logA \geq logB$.

(V) $\Longrightarrow$ (I). This is in the same way as (IV) $\Longrightarrow$ (I).

Hence the proof of Theorem 3 is complete.

We remark that the equivalence relation between (I) and (II$_0$) has been shown in [6] as an extension of [5,Theorem 1].

I would like to express my sincere appreciation to Professor T. Ando for inviting me to WOTCA at Sapporo and his hospitality to me during this Conference which has been held and has been excellently organized during June 11-14, 1991.

I would like to express my cordial thanks to the referee for reading carefully the first version and for giving to me useful and nice comments.

References

[1] T.Ando, On some operator inequality, Math. Ann.,**279**(1987),157-159.

[2] J.I.Fujii and E.Kamei, Relative operator entropy in noncommutative information theory, Math. Japon.,**34**(1989),341-348.

[3] J.I. Fijii and E.Kamei, Uhlmann's interpolational method for operator means, Math. Japon.,**34**(1989),541-547.

[4] M.Fujii, Furuta's inequality and its mean theoretic approach, J. of Operator Theory,**23**(1990),67-72.

[5] M.Fujii, T.Furuta and E.Kamei, Operator functions associated with Furuta's inequality, Linear Alg. and Its Appl., **149**(1991),91-96.

[6] M.Fujii, T.Furuta and E.Kamei, An application of Furuta's inequality to Ando's theorem, preprint.

[7] T.Furuta: $A \geq B \geq 0$ assures $(B^r A^p B^r)^{1/q} \geq B^{(p+2r)/q}$ for $r \geq 0, p \geq 0, q \geq 1$ with $(1 + 2r)q \geq (p + 2r)$. Proc. Amer. Math. Soc., **101**(1987),85-88.

[8] T.Furuta, A proof via operator means of an order preserving inequality, Linear Alg. and Its Appl.,**113**(1989),129-130.

[9] T.Furuta, Elementary proof of an order preserving inequality, Proc. Japan Acad.,**65**(1989),126.

[10] T.Furuta, Two operator functions with monotone property, Proc. Amer. Math. Soc.,**111**(1991),511-516.

[11] T.Furuta, Furuta's inequality and its application to the relative operator entropy, to appear in J. of Operator Theory.

[12] E.Heinz, Beiträgze zur Stŕungstheorie der Spektralzerlegung, Math. Ann.,**123**(1951),415-438.

[13] E.Kamei, A satellite to Furuta's inequality, Math. Japon,**33**(1988),883-886.

[14] K.Löwner, Über monotone Matrixfunktion, Math. Z., **38**(1934),177-216.

[15] M.Nakamura and H.Umegaki, A note on the entropy for operator algebras, Proc. Japan Acad.,**37**(1961),149-154.

[16] G.K.Pedersen, Some operator monotone functios, Proc. Amer. Math. Soc.,**36**(1972),309-310.

[17] H.Umegaki, Conditional expectation in operator algebra **IV**, (entropy and information), Kadai Math. Sem. Rep.,**14**(1962),59-85

Department of Applied Mathematics
Faculty of Science
Science University of Tokyo
1-3 Kagurazaka, Shinjuku
Tokyo 162
Japan

MSC 1991: Primary 47A63

Operator Theory:
Advances and Applications, Vol. 59
© 1992 Birkhäuser Verlag Basel

THE BAND EXTENSION ON THE REAL LINE AS A LIMIT OF
DISCRETE BAND EXTENSIONS, I. THE MAIN LIMIT THEOREM

I. Gohberg and M.A. Kaashoek

In this paper it is proved that the band extension on the real line (viewed as a convolution operator) may be obtained as a limit in the operator norm of block Laurent operators of which the symbols are band extensions of appropriate discrete approximations of the given data.

0. INTRODUCTION

Let k be an $m \times m$ matrix function with entries in $L_2([-\tau, \tau])$. An $m \times m$ matrix function f with entries in $L_1(\mathbf{R}) \cap L_2(\mathbf{R})$ is called a *positive extension* of k if

   (a) $f(t) = k(t)$ for $-\tau \leq t \leq \tau$,

   (b) $I - \widehat{f}(\lambda)$ is a positive definite matrix for each $\lambda \in \mathbf{R}$.

Here $\widehat{f}$ denotes the Fourier transform of f. If (b) is fulfilled, then

$$(0.1) \qquad (I - \widehat{f}(\lambda))^{-1} = I - \widehat{\gamma}(\lambda), \ \lambda \in \mathbf{R}$$

where γ is again an $m \times m$ matrix function with entries in $L_1(\mathbf{R}) \cap L_2(\mathbf{R})$. A positive extension f of k is called a *band extension* if the function γ in (0.1) has the following additional property:

   (c) $\gamma(t) = 0$ a.e. on $\mathbf{R}\backslash[-\tau, \tau]$.

It is known (see [7]) that the band extension may also be characterized as the unique positive extension f of k that maximizes the *entropy integral* $\mathcal{E}(f)$, where

$$(0.2) \qquad \mathcal{E}(f) = \lim_{\varepsilon \downarrow 0} \frac{1}{2\pi} \int_{-\infty}^{\infty} \frac{\log \det(I - \widehat{f}(\lambda))}{\varepsilon^2 \lambda^2 + 1} \, d\lambda.$$

The main aim of the present paper is to establish the above mentioned maximum entropy characterization of the band extension by reducing it to the corresponding result for the discrete case, which concerns Fourier series on the unit circle with operator coefficients. Our reduction is based on partitioning of operators and does not use the usual discretization of the given k. Let us remark here that the maximum entropy principle for matrix and operator functions on the unit circle is well-understood and may be derived as a corollary of the abstract maximum entropy principle appearing in the general framework of the band method ([12]). However, for the continuous case there are different entropy formulas ([2], [3], [5], [7], see also [6], [16]), and the maximum entropy principle does not seem to follow from the abstract analogue in the band method (see [12] for an example).

This paper consists of two parts. In the present first part we show that the band extension on the line may be obtained from discrete band extensions on the circle by a limit in an appropriate norm. In this limit procedure the first step is to replace the given $m \times m$ matrix function k by a trigonometric polynomial with operator coefficients, namely

$$(0.3) \qquad I - \sum_{\nu=-(n-1)}^{n-1} z^\nu K_\nu^{(n)}.$$

Here n is a positive integer and $K_\nu^{(n)}$ is the operator on $L_2^m([0, \frac{1}{n}\tau])$ defined by

$$(K_\nu^{(n)}\varphi)(t) = \int_0^{\frac{1}{n}\tau} k(t - s + \frac{\nu}{n}\tau)\varphi(s)\, ds, \qquad 0 \leq t \leq \frac{1}{n}\tau.$$

Note that the trigonometric polynomial (0.3) uses the information about the given data on $-\tau + \frac{1}{n}\tau \leq t \leq \tau$, but not on $-\tau \leq t \leq -\tau + \frac{1}{n}\tau$. The next step is to build the band extension for the trigonometric polynomial (0.3), that is, we build an operator function

$$B^{(n)}(z) = \sum_{\nu=-\infty}^{\infty} z^\nu B_\nu^{(n)}$$

with the following poperties

1) $B_\nu^{(n)} = K_\nu^{(n)}, \quad \nu = -(n-1), \ldots, n-1,$

2) $I - B^{(n)}(z)$ is a positive definite operator on $L_2^m([0, \frac{1}{n}\tau])$ for each $|z| = 1,$

3) $(I - B^{(n)}(z))^{-1}$ is of the form $\sum_{\nu=-(n-1)}^{n-1} z^\nu C_\nu^{(n)}$.

From the construction it follows that the operators $B_\nu^{(n)}$ are Hilbert-Schmidt operators on $L_2^m([0, \frac{1}{n}\tau])$ and one can form a bounded linear operator $\widehat{B}^{(n)}$ on $L_2^m(\mathbf{R})$ such that

$$(\widehat{B}^{(n)}\psi)(t) = (B_{i-j}^n \psi(\cdot + \frac{j}{n}\tau))(t - \frac{i}{n}\tau), \quad \frac{i}{n}\tau \leq t \leq \frac{i+1}{n}\tau, \quad \psi \in L_2^m([\frac{j}{n}\tau, \frac{j+1}{n}\tau]).$$

The main result is that the sequence of operators $\widehat{B}^{(1)}$, $\widehat{B}^{(2)}, \ldots$ converges in the operator norm to L_b, where b is a band extension of k and L_b stands for the convolution operator on $L_2^m(\mathbf{R})$ defined by

$$(L_b\varphi)(t) = \int_{-\infty}^{\infty} b(t - s)\varphi(s)\,ds, \qquad t \in \mathbf{R}.$$

We prove this convergence also for some norms that are stronger than the operator norm.

This part of the paper is split into three sections (not counting the present introduction). The first section contains preliminary material and is of preparatory character. In the second section we recall the construction of the band extension both for the continuous and the discrete case. The main theorem and its proof appear in the third section. As a by-product we get the connection between discrete and continuous orthogonal polynomials for the positive definite case.

A few words about notation. An identity operator is denoted by I; from the context it should be clear on which space it acts. By S_2 we denote the class of Hilbert-Schmidt operators, where the operators are assumed to act on a Hilbert space. For an invertible Hilbert space operator A we let A^{-*} denote the adjoint of A^{-1}. The spectral norm of a matrix M is denoted by $\|M\|$, i.e., $\|M\|$ is the largest singular value of M.

I. PRELIMANARIES AND PREPARATIONS

I.1 Operator Wiener algebra and block Laurent operators. Let H be a separable Hilbert space. By definition (cf., Section II.1 in [11]) the *operator Wiener algebra* $\mathcal{W}(H;\mathrm{T})$ consists of all operator-valued functions G on the unit circle T such that

$$(1.1) \qquad G(z) = \sum_{\nu=-\infty}^{\infty} z^\nu G_\nu, \qquad z \in \mathrm{T},$$

where each G_ν is a bounded linear operator on H and

$$(1.2) \qquad \sum_{\nu=-\infty}^{\infty} \|G_\nu\| < \infty.$$

As usual, G_ν is called the ν-th *Fourier coefficient* of the function G. With the usual multiplication of operator-valued functions $\mathcal{W}(H;\mathrm{T})$ is a unital algebra, the unit being given by the function $E(z) = I$ of each $z \in \mathrm{T}$. Also, on $\mathcal{W}(H;\mathrm{T})$ there is a natural involution $*$, namely, for G as in (1.3) we have

$$(1.3) \qquad G^*(z) = \sum_{\nu=-\infty}^{\infty} z_\nu G_{-\nu}^* = G(z)^*, \qquad z \in \mathrm{T}.$$

Each $G \in \mathcal{W}(H;\mathrm{T})$ defines in a canonical way a bounded linear operator on $\oplus_{-\infty}^{\infty} H$. Here $\oplus_{-\infty}^{\infty} H$ denotes the Hilbert space of all square summable sequences $\mathbf{x} = (x_i)_{i=-\infty}^{\infty}$ with elements in H. Inner product and norm on $\oplus_{-\infty}^{\infty} H$ are given by

$$\langle \mathbf{x}, \mathbf{y} \rangle = \sum_{j=-\infty}^{\infty} \langle x_j, y_j \rangle, \qquad \|\mathbf{x}\| = \Big(\sum_{j=-\infty}^{\infty} \|x_j\|^2 \Big)^{\frac{1}{2}}.$$

Now, let $G \in \mathcal{W}(H;\mathrm{T})$ be given by (1.1), and define an operator M on $\oplus_{-\infty}^{\infty} H$ by setting

$$(1.4) \qquad (M\mathbf{x})_i = \sum_{j=-\infty}^{\infty} G_{i-j} x_j, \qquad i = 0, \pm 1, \pm 2, \ldots.$$

Then M is a well-defined bounded linear operator on $\oplus_{-\infty}^{\infty} H$. Instead of (1.4) we shall write $M = (G_{i-j})_{i,j=-\infty}^{\infty}$. We call M the *H-block Laurent operator with symbol G*.

I.2 The Wiener algebra over the Hilbert-Schmidt operators. Let H be a separable Hilbert space. By $\mathcal{W}(S_2, H;\mathrm{T})$ we denote the set of all operator-valued functions F on the unit circle such that

$$(2.1) \qquad F(z) = \sum_{\nu=-\infty}^{\infty} z^\nu F_\nu, \qquad z \in \mathrm{T},$$

where F_ν is a Hilbert-Schmidt operator on H for each ν and

$$(2.2\mathrm{a}) \qquad \|\|F\|\| := \sum_{\nu=-\infty}^{\infty} \|F_\nu\| < \infty,$$

$$(2.2\mathrm{b}) \qquad \|\|F\|\|_2 := \Big(\sum_{\nu=-\infty}^{\infty} \|F_\nu\|_2^2 \Big)^{\frac{1}{2}} < \infty.$$

Here $\| \cdot \|_2$ denotes the Hilbert-Schmidt norm (cf., Section VIII.2 in [8]). With the usual multiplication of operator-valued functions $\mathcal{W}(S_2, H; T)$ is an algebra (see Proposition 2.1 below). We shall refer to $\mathcal{W}(S_2, H; T)$ as the *Wiener algebra on* T *over the Hilbert-Schmidt operators on* H. From (2.2a) we see that $\mathcal{W}(S_2, H; T)$ is contained in $\mathcal{W}(H; T)$.

PROPOSITION 2.1. *The Wiener algebra on* T *over the Hilbert-Schmidt operators on* H *is a 2-sided ideal in the operator Wiener algebra* $\mathcal{W}(H; T)$ *and is closed under the involution* *.

PROOF. If A is a Hilbert-Schmidt operator, then (see [8], Section VIII.2) the same is true for A^* and $\|A^*\|_2 = \|A\|_2$. From these facts it follows that $\mathcal{W}(S_2, H; T)$ is closed under the involution *. To prove that $\mathcal{W}(S_2, H; T)$ is a 2-sided ideal in $\mathcal{W}(H; T)$, take $F \in \mathcal{W}(S_2, H; T)$ and $G \in \mathcal{W}(H; T)$. Let F be as in (2.1) and G as in (1.1). Then

$$(GF)(z) = G(z)F(z) = \sum_{\nu=-\infty}^{\infty} z^\nu S_\nu, \qquad z \in T,$$

where

$$(2.3) \qquad S_\nu = \sum_{k=-\infty}^{\infty} G_{\nu-k} F_k.$$

Note that $G_{\nu-k} F_k$ is a Hilbert-Schmidt operator and

$$\|G_{\nu-k} F_k\|_2 \le \|G_{\nu-k}\| \|F_k\|_2.$$

The sequence $(\|F_k\|_2)$ is bounded. Thus we can use (1.2) to show that the right hand side of (2.3) converges in the Hilbert-Schmidt norm. So S_ν is a Hilbert-Schmidt operator and

$$\|S_\nu\|_2 \le \sum_{k=-\infty}^{\infty} \|G_{\nu-k}\| \|F_k\|_2.$$

But then

$$(2.4) \qquad \sum_{\nu=-\infty}^{\infty} \|S_\nu\|_2^2 \le \gamma^2 \Big(\sum_{k=-\infty}^{\infty} \|F_k\|_2^2 \Big) < \infty,$$

where γ is equal to the left hand side of (1.2). Hence (2.2b) holds for GF in place of F. Since G and F are both in $\mathcal{W}(H; T)$, the inequality (2.2a) also holds for GF in place of F. Thus

$GF \in \mathcal{W}(S_2, H; \mathrm{T})$. In a similar way (or by duality) one proves that $FG \in \mathcal{W}(S_2, H; \mathrm{T})$. ∎

COROLLARY 2.2. *Let* $G(\cdot) = I - F(\cdot)$, *where* $F \in \mathcal{W}(S_2, H; \mathrm{T})$, *and assume that* $G(z)$ *is an invertible operator on* H *for each* $z \in \mathrm{T}$. *Then*

$$(2.5) \qquad (I - F(\cdot))^{-1} - I \in \mathcal{W}(S_2, H; \mathrm{T}).$$

PROOF. Let R be the operator-valued function defined by the left hand side of (2.5). By the Bochner-Phillips theorem, ([4], Theorem 1) we have $R \in \mathcal{W}(H; \mathrm{T})$. Now $(I - R(z))(I - F(z)) = I$ for each $z \in \mathrm{T}$, and hence $R + F - RF = 0$. Since $F \in \mathcal{W}(S_2, H; \mathrm{T})$, Proposition 2.1 implies that the same holds true for RF. Thus $F - RF \in \mathcal{W}(S_2, H; \mathrm{T})$, and (2.5) is proved. ∎

PROPOSITION 2.3. *The algebra* $\mathcal{W}(S_2, H; \mathrm{T})$ *is a Banach algebra with respect to the norm*

$$(2.6) \qquad \|F\|_{\mathcal{W}} := \max\{|||F|||, |||F|||_2\}.$$

PROOF. Recall that $S_2(H)$, the ideal of all Hilbert-Schmidt operators on H, is a Hilbert space with respect to the Hilbert-Schmidt norm. It follows that the space

$$(2.7) \qquad \{F \mid F(z) = \sum_{\nu=-\infty}^{\infty} z^{\nu} F_{\nu} \quad (z \in \mathrm{T}), \; F_{\nu} \in S_2(H) \quad (\nu \in \mathrm{Z}), \; |||F|||_2 < \infty\}$$

is a Hilbert space with respect to the norm $||| \cdot |||_2$. We also know that $\mathcal{W}(H; \mathrm{T})$ is a Banach space with respect to the norm $||| \cdot |||$. Now, let $(F(n))_{n=1}^{\infty}$ be a Cauchy sequence in $\mathcal{W}(S_2, H; \mathrm{T})$ with respect to the norm $\| \cdot \|_{\mathcal{W}}$. Then the sequence $(F(n))_{n=1}^{\infty}$ has a limit in $\mathcal{W}(H; \mathrm{T})$ and in the space (2.7). Obviously, both limits are equal. So, there exists $F \in \mathcal{W}(S_2, H; \mathrm{T})$ such that $\|F(n) - F\|_{\mathcal{W}}$ goes to zero if $n \to \infty$. Hence $\mathcal{W}(S_2, H; \mathrm{T})$ is complete with respect to the norm $\| \cdot \|_{\mathcal{W}}$.

It remains to prove that $\| \cdot \|_{\mathcal{W}}$ is an algebra norm. Take F and G in $\mathcal{W}(S_2, H; \mathrm{T})$. Then $G \in \mathcal{W}(H; \mathrm{T})$, and the arguments used in the proof of Proposition 2.1 (cf., formula (2.4)) imply that

$$(2.8) \qquad |||GF|||_2 \leq |||G||| \cdot |||F|||_2 \leq \|G\|_{\mathcal{W}} \|F\|_{\mathcal{W}}.$$

Also, $F \in \mathcal{W}(H;T)$ and hence

$$(2.9) \qquad |||GF||| \leq |||G||| \cdot |||F||| \leq \|G\|_{\mathcal{W}}\|F\|_{\mathcal{W}}.$$

Therefore $\|GF\|_{\mathcal{W}} \leq \|G\|_{\mathcal{W}}\|F\|_{\mathcal{W}}.$ ∎

I.3 Block partitioning. By $L_2^m(\mathbf{R})$ and $L_2^m([0,\sigma])$ we denote the Hilbert spaces of square integrable $\mathbf{C}^m$-valued functions on $\mathbf{R}$ and $[0,\sigma]$, respectively. Fix $\sigma > 0$. Often it will be convenient to identify operators on $L_2^m(\mathbf{R})$ with operators acting on $\oplus_{-\infty}^{\infty} L_2^m([0,\sigma])$, the Hilbert space of all square summable sequences with entries in $L_2^m([0,\sigma])$. To make this identification explicit we need a few auxiliary operators, namely:

$$(3.1) \qquad \eta_j : L_2^m([0,\sigma]) \to L_2^m(\mathbf{R}), \quad (\eta_j\varphi)(t) = \begin{cases} \varphi(t - \sigma j), & \sigma j \leq t \leq \sigma(j+1), \\ 0, & \text{otherwise}, \end{cases}$$

and

$$(3.2) \qquad \rho_i : L_2^m(\mathbf{R}) \to L_2^m([0,\sigma]), \qquad (\rho_i\psi)(t) = \psi(t + \sigma i), \quad 0 \leq t \leq \sigma.$$

Here j and i are arbitrary integers. We also need:

$$(3.3) \qquad J_\sigma : L_2^m(\mathbf{R}) \to \oplus_{-\infty}^{\infty} L_2^m([0,\sigma]), \qquad J_\sigma\psi = (\rho_i\psi)_{i=-\infty}^{\infty}.$$

The map J_σ is a unitary operator. We call J_σ the σ-partitioner of $L_2^m(\mathbf{R})$.

Now, let $T : L_2^m(\mathbf{R}) \to L_2^m(\mathbf{R})$ be a bounded linear operator. By definition, the σ-block partitioning of T is the double infinite operator matrix $(T_{ij})_{i,j=-\infty}^{\infty}$ whose (i,j)-th entry is given by

$$(3.4) \qquad T_{ij} = \rho_i T \eta_j.$$

Note that the canonical action of the operator matrix $(T_{ij})_{i,j=-\infty}^{\infty}$ on $\oplus_{-\infty}^{\infty} L_2^m([0,\sigma])$ is precisely the operator $J_\sigma T J_\sigma^{-1}$.

Let $\tau = n\sigma$, where n is a positive integer. With some obvious changes in notation one defines the σ-block partitioning of a bounded linear operator T on $L_2^m([0,\tau])$ to be the $n \times n$ operator matrix

$$(3.5) \qquad \begin{pmatrix} T_{11} & \cdots & T_{1n} \\ \vdots & & \vdots \\ T_{n1} & \cdots & T_{nn} \end{pmatrix},$$

where $T_{ij} = \rho_i S \eta_j$, $\quad i,j = 1,\ldots,n$. The operators η_j and ρ_i are defined as in (3.1) and (3.2), except that in these definitions one has to replace $L_2^m(\mathbf{R})$ by $L_2^m([0,\tau])$. The operator matrix in (3.5) is considered to be an operator on $\oplus_1^n L_2^m([0,\sigma])$, the Hilbert space direct sum of n copies of $L_2^m([0,\sigma])$.

I.4 Convolution operators. Let f be an $m \times m$ matrix function with entries in $L_1(\mathbf{R})$. The convolution operator on $L_2^m(\mathbf{R})$ associated with f will be denoted by L_f. Thus

$$(4.1) \qquad (L_f \varphi)(t) = \int_{-\infty}^{\infty} f(t - s)\varphi(s)\,ds, \qquad t \in \mathbf{R}.$$

We shall refer to L_f as the *convolution operator with kernel function f.*

PROPOSITION 4.1. *Let f be an $m \times m$ matrix function with entries in $L_1(\mathbf{R}) \cap L_2(\mathbf{R})$. Then the σ-block partitioning of the convolution operator L_f is a $L_2^m([0,\sigma])$-block Laurent operator, $F = (F_{i-j})_{i,j=-\infty}^{\infty}$, with symbol in the Wiener algebra on T over the Hilbert-Schmidt operators on $L_2^m([0,\sigma])$ and*

$$(4.2) \qquad \sum_{\nu=-\infty}^{\infty} \|F_\nu\| \leq 2 \int_{-\infty}^{\infty} \|f(t)\|\,dt,$$

$$(4.3) \qquad \sum_{\nu=-\infty}^{\infty} \|F_\nu\|_2^2 = \sigma \int_{-\infty}^{\infty} \|f(t)\|^2\,dt,$$

The proof of Proposition 4.1 is the same as that of Lemma I.5.1 in [9] and, therefore, it is omitted. Here we only note that for $\nu = 0, \pm 1, \pm 2, \ldots$ the operator F_ν in Proposition 4.1 is the Hilbert-Schmidt operator on $L_2^m([0,\sigma])$ defined by

$$(4.4) \qquad (F_\nu \varphi)(t) = \int_0^{\sigma} f(t - s + \nu\sigma)\varphi(s)\,ds, \qquad 0 \leq t \leq \sigma.$$

I.5 The algebra $\mathcal{B}(\tau)$. Throughout this section $\tau > 0$ is a fixed positive number. By $\mathcal{B}(\tau)$ we denote the set of bounded linear operators T on $L_2^m(\mathbf{R})$ such that the τ-block partitioning of T is a $L_2^m([0,\tau])$-block Laurent operator with symbol in $\mathcal{W}(S_2, L_2^m([0,\tau]); T)$. Take $T \in \mathcal{B}(\tau)$, and let $(T_{i-j})_{i,j=-\infty}^{\infty}$ be its τ-block partitioning. We define:

$$(5.1a) \qquad \||T\|| := \sum_{j=-\infty}^{\infty} \|T_j\| < \infty,$$

$$(5.1b) \qquad \||T\||_2 := \Big(\sum_{j=-\infty}^{\infty} \|T_j\|_2^2 \Big)^{\frac{1}{2}} < \infty,$$

and

$$\text{(5.2)} \qquad \|T\|_{\mathcal{B}(\tau)} := \max\{|||T|||, |||T|||_2\}.$$

Formula (5.1a) implies that $\| \cdot \|_{\mathcal{B}(\tau)}$ is stronger than the usual operator norm, i.e.,

$$\text{(5.3)} \qquad \|T\| \leq \|T\|_{\mathcal{B}(\tau)}, \qquad T \in \mathcal{B}(\tau).$$

Given $T \in \mathcal{B}(\tau)$, let F_T denote the symbol of the τ-block partitioning of T. Note that $\|T\|_{\mathcal{B}(\tau)} = \|F\|_{\mathcal{W}}$, where $\| \cdot \|_{\mathcal{W}}$ denotes the norm on $\mathcal{W}(S_2, L_2^m([0,\sigma];\mathrm{T})$ introduced in subsection I.2. So we know from Proposition I.2.3 that $\mathcal{B}(\tau)$ endowed with the norm $\| \cdot \|_{\mathcal{B}(\tau)}$ is a Banach algebra. In fact, $\mathcal{B}(\tau)$ and $\mathcal{W}(S_2, L_2^m([0,\sigma];\mathrm{T})$ are algebraically and isometrically isomorphic, the isomorphism being given by the map $T \mapsto F_T$. For later purposes we mention the following two inequalities (which follow from (2.8)):

$$\text{(5.4a)} \qquad |||TS|||_2 \leq |||T||| \cdot |||S|||_2, \quad T, S \in \mathcal{B}(\tau),$$

$$\text{(5.4b)} \qquad |||TS|||_2 \leq |||T|||_2|||S|||, \quad T, S \in \mathcal{B}(\tau).$$

Each $T \in \mathcal{B}(\tau)$ can be represented in the form

$$\text{(5.5)} \qquad (T\varphi)(t) = \int_{-\infty}^{\infty} \alpha(t,s)\varphi(s)\,ds, \qquad t \in \mathbf{R},$$

where $\alpha(t+\tau, s+\tau) = \alpha(t,s)$ a.e. on $\mathbf{R} \times \mathbf{R}$ and

$$\text{(5.6)} \qquad \int_0^\tau \int_{-\infty}^{\infty} \|\alpha(t,s)\|^2\,ds\,dt < \infty.$$

Formula (5.5) holds true for each $\varphi \in L_2^m(\mathbf{R})$ with compact support. To obtain the representation (5.5), let $\alpha_\nu(\cdot,\cdot)$ be the kernel function of the Hilbert-Schmidt integral operator T_ν (appearing in the τ-block partitioning of T), and put

$$\alpha(t,s) = \alpha_{i-j}(t - i\tau, s - j\tau), \qquad i\tau \leq t \leq (i+1)\tau, \quad j\tau \leq s \leq (j+1)\tau.$$

Then α has the desired periodicity , formula (5.6) holds true because of (5.1b), and we have the representation (5.5) because $(T_{i-j})_{i,j=-\infty}^{\infty}$ is the τ-block partitioning of T. We shall refer to α in (5.5) as the *kernel function* of T. For T as in (5.5) we have

$$\text{(5.7a)} \qquad |||T|||_2 = \Big(\int_0^\tau \int_{-\infty}^{\infty} \|\alpha(t,s)\|^2\,ds\,dt \Big)^{\frac{1}{2}},$$

$$\text{(5.7b)} \qquad |||T|||_2 = \Big(\int_0^\tau \int_{-\infty}^{\infty} \|\alpha(t,s)\|^2\,dt\,ds \Big)^{\frac{1}{2}}.$$

PROPOSITION 5.1. *Let $T \in \mathcal{B}(\tau)$. Put $\sigma = \frac{1}{n}\tau$, and let $(T_{i,j}^{(n)})_{i,j=-\infty}^{\infty}$ be the σ-block partitioning of T. Then*

$$(5.8a) \qquad \||T\||_2 = \Big(\sum_{j=-\infty}^{\infty} \sum_{i=0}^{n-1} \|T_{ij}^{(n)}\|_2^2 \Big)^{\frac{1}{2}},$$

$$(5.8b) \qquad \||T\||_2 = \Big(\sum_{i=-\infty}^{\infty} \sum_{j=0}^{n-1} \|T_{ij}^{(n)}\|_2^2 \Big)^{\frac{1}{2}}.$$

PROOF. Let α be the kernel function of T. We claim that

$$(T_{ij}^{(n)}\varphi)(t) = \int_0^{\sigma} \alpha_{(i,j)}(t,s)\varphi(s)\,ds, \qquad 0 \le t \le \sigma, \quad 0 \le s \le \sigma,$$

where $\alpha_{i,j}(t,s) = \alpha(t + i\sigma, s + j\sigma)$. Indeed, for $0 \le t \le \sigma$ we have (cf., Section I.3):

$$(\rho_i T \eta_j \varphi(t) = \int_{-\infty}^{\infty} \alpha(t + i\sigma, s)(\eta_j \varphi)(s)\,ds = \int_{\sigma j}^{\sigma(j+1)} \alpha(t + i\sigma, s)\varphi(s - j\sigma)\,ds$$

$$= \int_0^{\sigma} \alpha(t + i\sigma, s + j\sigma)\varphi(s)\,ds\ ,$$

which shows that $T_{ij}^{(n)}$ has the desired form. Now, by (5.7a)

$$
\begin{aligned}
\||T\||_2^2 &= \sum_{r=-\infty}^{\infty} \Big(\int_0^{\tau} \int_{r\tau}^{(r+1)\tau} \|\alpha(t,s)\|^2\,ds\,dt \\
&= \sum_{r=-\infty}^{\infty} \sum_{i=0}^{n-1} \sum_{j=0}^{n-1} \int_{i\sigma}^{(i+1)\sigma} \int_{r\tau+j\sigma}^{r\tau+(j+1)\sigma} \|\alpha(t,s)\|^2\,ds\,dt \\
&= \sum_{r=-\infty}^{\infty} \sum_{i=0}^{n-1} \sum_{j=0}^{n-1} \int_0^{\sigma} \int_0^{\sigma} \|\alpha_{i,rn+j}(t,s)\|^2\,ds\,dt \\
&= \sum_{r=-\infty}^{\infty} \sum_{i=0}^{n-1} \sum_{j=0}^{n-1} \|T_{i,rn+j}^{(n)}\|_2^2 \\
&= \sum_{j=-\infty}^{\infty} \sum_{i=0}^{n-1} \|T_{ij}^{(n)}\|_2^2,
\end{aligned}
$$

which proves (5.8a). In a similar, by (5.7b) in place of (5.7a), one derives (5.8b). ∎

PROPOSITION 5.2. *If $\sigma = \frac{1}{n}\tau$, then $\mathcal{B}(\sigma) \subset \mathcal{B}(\tau)$.*

PROOF. Take $T \in \mathcal{B}(\sigma)$, and let $(S_{i-j})_{i,j=-\infty}^{\infty}$ be the σ-block partitioning of T. For each $\nu \in \mathbb{Z}$ let T_ν be the operator on $L_2^m([0,\tau])$ whose σ-block partitioning is equal to the following block Toeplitz matrix

$$\begin{pmatrix} S_{\nu n} & \cdots & S_{(\nu-1)n+1} \\ \vdots & \ddots & \vdots \\ S_{(\nu+1)n-1} & \cdots & S_{\nu n} \end{pmatrix}.$$

Then $(T_{i-j})_{i,j=-\infty}^{\infty}$ is the τ-block partitioning of T. From these connections it is clear that $T \in \mathcal{B}(\tau)$. Moreover, we see that

$$(5.9) \qquad \|T\|_{\mathcal{B}(\tau)} \leq n\|T\|_{\mathcal{B}(\sigma)}.$$

∎

II. BAND EXTENSIONS

II.1 The circle case. Consider the following trigonometric operator polynomial:

$$(1.1) \qquad I - \sum_{\nu=-(N-1)}^{N-1} z^\nu F_\nu.$$

The coefficients $F_{-(N-1)}, \ldots, F_{N-1}$ are assumed to be bounded linear operators on the separable Hilbert space H. We shall call (1.1) a *discrete operator band*. To explain the word "band", note the H-block Laurent operator with symbol (1.1) is a double infinite banded matrix of which all diagonals are zero except the $2N - 1$ diagonals located symmetrically around the main diagonal (the main diagonal included).

An operator-valued function $I - B(\cdot) \in \mathcal{W}(H; \mathrm{T})$ is called a *band extension* of the discrete operator band (1.1) if

(i) for $|j| \leq N - 1$ the j-th Fourier coefficient of $B(\cdot)$ is equal to F_j,

(ii) $I - B(z)$ is a positive definite operator for each $z \in \mathrm{T}$,

(iii) for $|j| \geq N$ the j-th Fourier coefficient of $(I - B(\cdot))^{-1}$ is equal to zero.

If only (i) and (ii) are fulfilled, then $I - B(\cdot)$ is called a *positive extension* of (1.1). From [11], Section II.1, we know that a band extension exists if and only if the operator $I - \Lambda$, with

$$(1.2) \qquad \Lambda = \begin{pmatrix} F_0 & F_{-1} & \cdots & F_{-(N-1)} \\ F_1 & F_0 & \cdots & F_{-(N-2)} \\ \vdots & \vdots & \ddots & \vdots \\ F_{N-1} & F_{N-2} & \cdots & F_0 \end{pmatrix},$$

is a positive definite operator on H^N, the Hilbert space direct sum of N copies of H. Furthermore, in that case (see [11], Theorem II.1.2) there is precisely one band extension which may be obtained in the following way. Let $X_0, X_1, \ldots, X_{N-1}$ be given by

$$(1.3) \qquad \begin{pmatrix} X_0 \\ X_1 \\ \vdots \\ X_{N-1} \end{pmatrix} = (I - \Lambda)^{-1} \begin{pmatrix} F_0 \\ F_1 \\ \vdots \\ F_{N-1} \end{pmatrix},$$

and put $X(z) = \sum_{j=0}^{N-1} z^j X_j$. Then $I + X(z)$ is an invertible operator for each $|z| \le 1$ and the operator function $I - B(\cdot)$,

$$(1.4) \qquad I - B(z) := (I + X(z))^{-*}(I + X_0)(I + X(z))^{-1}, \qquad z \in \mathrm{T},$$

is the band extension of (1.1).

PROPOSITION 1.1. *Let* $I - B(\cdot)$ *be the band extension of the discrete operator band* (1.1). *If* F_j *is a Hilbert-Schmidt operator for each* $|j| \le N - 1$, *then* $B(\cdot)$ *is in the Wiener algebra on* T *over the Hilbert-Schmidt operators on* H.

PROOF. Let Λ be as in (1.2). By our hypotheses, $I - \Lambda$ is invertible. Put $\Lambda^\times = I - (I - \Lambda)^{-1}$, and write $\Lambda^\times$ as an $N \times N$ operator matrix with entries acting on H:

$$\Lambda^\times = \begin{pmatrix} \Lambda_{00}^\times & \cdots & \Lambda_{0,N-1}^\times \\ \vdots & & \vdots \\ \Lambda_{N-1,0}^\times & \cdots & \Lambda_{N-1,N-1}^\times \end{pmatrix}.$$

Let $X_0, \ldots, X_{N-1}$ be the operators defined by (1.3). Then

$$X_i = F_i - \sum_{j=0}^{N-1} \Lambda_{ij}^\times F_j, \qquad i = 0, \ldots, N-1,$$

and thus, by the ideal property of the Hilbert-Schmidt operators, $X_0, \ldots, X_{N-1}$ are in S_2. Put $X(z) = \sum_{j=0}^{N-1} z^j X_j$. We already know that $I + X(z)$ is invertible for each $z \in \mathrm{T}$. Since

$X(\cdot) \in \mathcal{W}(S_2, H; T)$, we can apply Corollary I.2.2 to show that the same holds true for $I - (I + X(\cdot))^{-1}$. Next use (1.4) and Proposition I.2.1 to obtain that $B(\cdot) \in \mathcal{W}(S_2, H; T)$.

∎

There is an alternative way (cf., Theorem II.1.2 in [11]) to construct the band extension, namely replace (1.3) by

$$(1.5) \qquad \begin{pmatrix} Y_{-(N-1)} \\ \vdots \\ Y_{-1} \\ Y_0 \end{pmatrix} = (I - \Lambda)^{-1} \begin{pmatrix} F_{-(N-1)} \\ \vdots \\ F_{-1} \\ F_0 \end{pmatrix},$$

and $X(z)$ by $Y(z) = \sum_{j=-(N-1)}^{0} z^j Y_j$. Also one may describe all positive extensions $I - F(\cdot)$ of the band (1.1) with $F(\cdot)$ in $\mathcal{W}(S_2, H; T)$ by a linear fractional map of which the coefficients are determined by $X(\cdot)$ and $Y(\cdot)$. In fact, Theorems II.1.1 and II.1.3 in [11] remain valid if in these theorems the operator Wiener algebra $\mathcal{W}(H; T)$ is replaced by the algebra

$$\widetilde{\mathcal{W}}(S_2, H; T) = \{\alpha E + F \mid \alpha \in \mathbb{C}, \ F \in \mathcal{W}(S_2, H; T)\}.$$

Here E is the unit defined by $E(z) = I$ for each $z \in T$.

II.2 The real line case. Throughout this section $\tau > 0$ is a fixed positive number and k is an $m \times m$ matrix function whose entries are in $L_1([-\tau, \tau])$. An $m \times m$ matrix function b with entries in $L_1(\mathbf{R})$ is said to be a *band extension* of k if

(i) $b(t) = k(t)$ for $-\tau \le t \le \tau$,

(ii) $I - L_b$ is positive definite,

(iii) the kernel function of $I - (I - L_b)^{-1}$ is zero almost everywhere on $\mathbf{R} \backslash [-\tau, \tau]$.

If only (i) and (ii) are fulfilled, then b is called a *positive extension* of k. Recall that L_b denotes the convolution operator with kernel function b (see Section I.4). Condition (ii) implies that $I - L_b$ is invertible. In that case $I - (I - L_b)^{-1}$ is again a convolution operator with kernel function γ, say. Condition (iii) requires that the support of γ is contained in the set $[-\tau, \tau]$.

With the given function k we associate the following integral operator on

$L_2^m([0,\tau])$:

$$(2.1) \qquad (K\varphi)(t) = \int_0^\tau k(t-s)\varphi(s)\,ds, \qquad 0 \le t \le \tau.$$

From [7] we know that a band extension of k exists if and only if the operator $I - K$ is a positive definite operator on $L_2^m([0,\tau])$. Furthermore, in that case (see [7]) there is precisely one band extension which one may construct in the following manner. Let x be the solution of the equation:

$$(2.2) \qquad x(t) - \int_0^\tau k(t-s)x(s)\,ds = k(t), \qquad 0 \le t \le \tau,$$

and put $x(t) = 0$ for $t \in \mathbf{R}\backslash[0,\tau]$. Then x is an $m \times m$ matrix function with entries in $L_1(\mathbf{R})$, the operator $I + L_x$ is invertible and $(I + L_x)^{-1} = I + L_{x^\times}$, where $x^\times$ is an $m \times m$ matrix with entries in $L_1(\mathbf{R})$ such that $x^\times(t) = 0$ for $t \in \mathbf{R}\backslash[0,\infty)$. With x determined in this way one obtains the band extension b of k as the kernel function of the convolution operator L_b defined by

$$(2.3) \qquad I - L_b := (I + L_x)^{-*}(I + L_x)^{-1}.$$

PROPOSITION 2.1. *Let b be the band extension of k. If the entries of k are in $L_2([-\tau,\tau])$, then b has entries in $L_1(\mathbf{R}) \cap L_2(\mathbf{R})$.*

PROOF. By our hypotheses, $I - K$ is an invertible operator on $L_2^m([0,\tau])$ and the right hand side of (2.2) is a function in $L_2^m([0,\tau])$. Thus $x \in L_2^m([0,\tau])$. Since $x(t) = 0$ for t outside $[0,\tau]$, we conclude that the entries of x are in $L_1(\mathbf{R}) \cap L_2(\mathbf{R})$. Let x_{ij} be the (i,j)-th entry of x, and let $x_{ij}^\times$ be the (i,j)-th entry of $x^\times$, where $x^\times$ is determined by

$$L_{x^\times} = (I + L_x)^{-1} - I.$$

From the latter identity it follows that

$$(2.4) \qquad x_{ij} + x_{ij}^\times + \sum_{\nu=1}^m x_{i\nu} * x_{\nu j}^\times = 0, \qquad i,j = 1,\ldots,m.$$

Here $*$ denotes the convolution product in $L_1(\mathbf{R})$. Now, recall that for $f \in L_2(\mathbf{R})$ and $g \in L_1(\mathbf{R})$, the convolution product $f * g \in L_2(\mathbf{R})$. Thus $L_1(\mathbf{R}) \cap L_2(\mathbf{R})$ is an ideal in

$L_1(\mathbf{R})$ with respect to the convolution product. So we see from (2.4) that the entries of $x^\times$ are also in $L_1(\mathbf{R}) \cap L_2(\mathbf{R})$. Put $x^\#(t) = x^\times(-t)^*$ for $t \in \mathbf{R}$. Then, by (2.3), we have

$$b = -(x^\# + x^\times + x^\# * x^\times),$$

and hence the entries of b are in $L_1(\mathbf{R}) \cap L_2(\mathbf{R})$. $\blacksquare$

There is an alternative way (cf., [7]) to construct the band extension b, namely, replace L_x by L_y, where y is the solution of

$$y(t) - \int_{-\tau}^0 k(t-s)y(s)\,ds = k(t), \qquad -\tau \le t \le 0$$

and $y(t) = 0$ for $t \in \mathbf{R}\backslash[-\tau, 0]$. Also (cf., [6] and [10]), all positive extensions f of k such that the entries of f are in $L_1(\mathbf{R}) \cap L_2(\mathbf{R})$ may be described by a linear fractional map of which the coefficients are determined by the functions x and y. In fact, the extension theorems in Section I.4 of [10] remain valid if in these theorems the role of $L_1(\mathbf{R})$ is taken over by $L_1(\mathbf{R}) \cap L_2(\mathbf{R})$.

III CONTINUOUS VERSUS DISCRETE

III.1 Main theorem. Throughout this section $\tau > 0$ is a fixed positive number and k is a given $m \times m$ matrix function with entries in $L_2([-\tau, \tau])$. By K we denote the operator on $L_2^m([0, \tau])$ defined by

$$(1.1) \qquad (K\varphi)(t) = \int_0^\tau k(t-s)\varphi(s)\,ds, \qquad 0 \le t \le \tau,$$

and we assume that $I - K$ is positive definite.

Put $\sigma = \frac{1}{n}\tau$, with n a positive integer, and consider the following discrete operator band:

$$(1.2) \qquad I - \sum_{\nu=-(n-1)}^{n-1} z^\nu K_\nu^{(n)},$$

where for $\nu = -(n-1), \ldots, n-1$ the operator $K_\nu^{(n)}$ is the operator on $L_2^m([0, \sigma])$ defined by

$$(1.3) \qquad (K_\nu^{(n)}\varphi)(t) = \int_0^\sigma k(t-s+\nu\sigma)\varphi(s)\,ds, \qquad 0 \le t \le \sigma.$$

Put

$$(1.4) \qquad \Lambda(n) = \begin{pmatrix} K_0^{(n)} & K_{-1}^{(n)} & \cdots & K_{-(n-1)}^{(n)} \\ K_1^{(n)} & K_0^{(n)} & \cdots & K_{-(n-2)}^{(n)} \\ \vdots & \vdots & \ddots & \vdots \\ K_{n-1}^{(n)} & K_{n-2}^{(n)} & \cdots & K_0^{(n)} \end{pmatrix} .$$

Note that $\Lambda(n)$ and the operator K in (1.1) are unitarily equivalent. In fact, $\Lambda(n)$ is the σ-block partitioning of the operator K. It follows that $I - \Lambda(n)$ is a positive definite operator.

The positive definiteness of $I - K$ implies (see Section II.2) that k has a band extension on the real line, and (by Proposition II.2.1) the entries of b are in $L_1(\mathbf{R}) \cap L_2(\mathbf{R})$. Similarly, since $I - \Lambda(n)$ is positive definite, the discrete operator band (1.2) has a band extension $I - B^{(n)}(\cdot)$ in $\mathcal{W}(L_2^m([0,\sigma]);\mathrm{T})$, and, by Proposition II.1.1, the function $B^{(n)}(\cdot)$ belongs to the Wiener algebra on T over the Hilbert-Schmidt operators on $L_2^m([0,\sigma])$.

THEOREM 1.1. *Let L_b be the convolution operator on $L_2^m(\mathbf{R})$ whose kernel function is the band extension b of k, and for $n = 1, 2, \ldots$ let $\widehat{B}^{(n)}$ be the operator on $L_2^m(\mathbf{R})$ whose σ-block partitioning $(\sigma = \frac{1}{n}\tau)$ is the $L_2^m([0,\sigma])$-block Laurent operator with symbol $B^{(n)}(\cdot)$, where $I - B^{(n)}(\cdot)$ is the band extension of the discrete operator band (1.2). Then $L_b, \widehat{B}^{(1)}, \widehat{B}^{(2)}, \ldots$ are in $\mathcal{B}(\tau)$, and*

$$(1.5) \qquad \lim_{n \to \infty} \|L_b - \widehat{B}^{(n)}\|_{\mathcal{B}(\tau)} = 0 .$$

Recall that $\mathcal{B}(\tau)$ is the algebra of all bounded linear operators on $L_2^m(\mathbf{R})$ such that the τ-block partitioning of T is the $L_2^m([0,\tau])$-block Laurent operator with symbol in $\mathcal{W}(S_2, L_2^m([0,\tau]);\mathrm{T})$ (see Section I.5). From Proposition I.4.1 we know that $L_b \in \mathcal{B}(\tau)$. The operators $\widehat{B}^{(1)}, \widehat{B}^{(2)}, \ldots$ are in $\mathcal{B}(\tau)$ because of Proposition I.5.2. The main result is the limit formula (1.5), which we shall prove in the third subsection.

III.2 Main auxiliary theorems. Let $\tau > 0$ be a fixed positive number. Recall (see Section I.5) that each operator $T \in \mathcal{B}(\tau)$ may be represented as an integral operator. The expression $T = T_\alpha$ means that α is the kernel function of T, i.e., T is represented as in formula (I.5.5). We say that the *support* of α belongs to the set V

(notation: supp $\alpha \subset V$) if $\alpha(t,s) = 0$ almost everywhere on the complement of V. In $\mathcal{B}(\tau)$ we consider the following subspaces:

$$\mathcal{B}_1 = \{T \in \mathcal{B}(\tau) \mid T = T_\alpha, \quad \text{supp } \alpha \subset \{(t,s) \mid t - s \geq \tau\}\},$$

$$\mathcal{B}_2 = \{T \in \mathcal{B}(\tau) \mid T = T_\alpha, \quad \text{supp } \alpha \subset \{(t,s) \mid 0 \leq t - s \leq \tau\}\},$$

$$\mathcal{B}_3 = \{T \in \mathcal{B}(\tau) \mid T = T_\alpha, \quad \text{supp } \alpha \subset \{(t,s) \mid -\tau \leq t - s \leq 0\}\},$$

$$\mathcal{B}_4 = \{T \in \mathcal{B}(\tau) \mid T = T_\alpha, \quad \text{supp } \alpha \subset \{(t,s) \mid t - s \leq -\tau\}\}.$$

Obviously, we have the following direct sum decomposition:

$$(2.1) \qquad\qquad \mathcal{B}(\tau) = \mathcal{B}_1 \dotplus \mathcal{B}_2 \dotplus \mathcal{B}_3 \dotplus \mathcal{B}_4.$$

The corresponding projections are denoted by Q_1, Q_2, Q_3 and Q_4. So, for example, Q_2 is the projection of $\mathcal{B}(\tau)$ onto $\mathcal{B}_2$ along the other spaces in the decomposition (2.1).

Recall (see Section I.5) that the norm $\|\cdot\|_{\mathcal{B}(\tau)}$ on $\mathcal{B}(\tau)$ is defined by

$$(2.2) \qquad\qquad \|T\|_{\mathcal{B}(\tau)} = \max\{|||T|||, |||T|||_2\},$$

where $|||\cdot|||$ and $|||\cdot|||_2$ are given by (I.5.1a) and (I.5.1b), respectively. From the expression for $||| \ |||_2$ in formula (I.5.7a) it is clear that

$$(2.3) \qquad\qquad |||Q_\nu T|||_2 \leq |||T|||_2 \qquad (\nu = 1, 2, 3, 4).$$

LEMMA 2.1. *For $T \in \mathcal{B}_2 \dotplus \mathcal{B}_3$ we have*

$$(2.4) \qquad\qquad |||T||| \leq \sqrt{3}|||T|||_2.$$

PROOF. Let $(T_{i-j})_{i,j=-\infty}^{\infty}$ be the τ-block partitioning of T. Since $T \in \mathcal{B}_2 \dotplus \mathcal{B}_3$, the τ-block partitioning of T is tridiagonal, i.e., $T_\nu = 0$ for $|\nu| \geq 2$. Now use that $\|A\| \leq \|A\|_2$ for any Hilbert-Schmidt operator A. It follows that

$$|||T||| = \|T_{-1}\| + \|T_0\| + \|T_1\|$$

$$\leq \|T_{-1}\|_2 + \|T_0\|_2 + \|T_1\|_2$$

$$\leq \sqrt{3}(\|T_{-1}\|_2^2 + \|T_0\|_2^2 + \|T_1\|_2^2)^{\frac{1}{2}},$$

which yields (2.4). $\blacksquare$

From (2.3) and Lemma 2.1 one sees that the projections Q_2 and Q_3 are bounded linear operators on $\mathcal{B}(\tau)$ with respect to the norm $\|\cdot\|_{\mathcal{B}(\tau)}$.

Next, let $\sigma = \frac{1}{n}\tau$. We also need the following subspaces:

$$\mathcal{B}_1(n) = \{T \in \mathcal{B}(\tau) \mid \rho_i T \eta_j = 0 \text{ for } (i,j) \notin \{i - j \geq n\}\},$$

$$\mathcal{B}_2^0(n) = \{T \in \mathcal{B}(\tau) \mid \rho_i T \eta_j = 0 \text{ for } (i,j) \notin \{0 < i - j \leq n - 1\}\},$$

$$\mathcal{B}_d(n) = \{T \in \mathcal{B}(\tau) \mid \rho_i T \eta_j = 0 \text{ for } i \neq j\}\},$$

$$\mathcal{B}_3^0(n) = \{T \in \mathcal{B}(\tau) \mid \rho_i T \eta_j = 0 \text{ for } (i,j) \notin \{0 > i - j \geq -(n - 1)\}\},$$

$$\mathcal{B}_4(n) = \{T \in \mathcal{B}(\tau) \mid \rho_i T \eta_j = 0 \text{ for } (i,j) \notin \{i - j \leq -n\}\}.$$

Here ρ_i and η_j are as in the first paragrapf of Section I.3; in other words $\rho_i T \eta_j$ is the (i,j)-th entry in the σ-block partitioning of T. We have the following direct sum decomposition:

$$(2.5) \qquad \mathcal{B}(\tau) = \mathcal{B}_1(n) \dot{+} \mathcal{B}_2^0(n) \dot{+} \mathcal{B}_d(n) \dot{+} \mathcal{B}_3^0(n) \dot{+} \mathcal{B}_4(n).$$

The corresponding projections are denoted by $Q_1(n), Q_2^0(n), Q_d(n), Q_3^0(n)$ and $Q_4(n)$. We set

$$(2.6) \qquad Q_2(n) := Q_d(n) + Q_2^0(n), \quad Q_3(n) := Q_d(n) + Q_3^0(n).$$

The following inclusions are valid:

$$(2.7a) \qquad \mathcal{B}_2^0(n) \subset \mathcal{B}_2, \quad \mathcal{B}_3^0(n) \subset \mathcal{B}_3,$$

$$(2.7b) \qquad \mathcal{B}_2^0(n) \dot{+} \mathcal{B}_d(n) \dot{+} \mathcal{B}_3^0(n) \subset \mathcal{B}_2 \dot{+} \mathcal{B}_3.$$

By the expression for $\|||\cdot|||_2$ in formula (I.5.7a) we have

$$(2.8) \qquad \||| Q_\nu(n) |||_2 \leq \||| T |||_2 \qquad (\nu = 1, 2, 3, 4, d)$$

for each $T \in \mathcal{B}(\tau)$. From (2.8), the inclusion (2.7b) and Lemma 2.1 we conclude that the projections $Q_2(n), Q_3(n)$ and $Q_d(n)$ are bounded linear operators on $\mathcal{B}(\tau)$ with respect to the norm $\|\cdot\|_{\mathcal{B}(\tau)}$.

LEMMA 2.2. *For $T \in \mathcal{B}(\tau)$, we have*

(2.9a)
$$\lim_{n \to \infty} |||Q_d(n)T|||_2 = 0,$$

(2.9b)
$$\lim_{n \to \infty} |||Q_2 T - Q_2(n)T|||_2 = 0,$$

(2.9c)
$$\lim_{n \to \infty} |||Q_3 T - Q_3(n)T|||_2 = 0.$$

PROOF. We shall prove (2.9b); the other two limits are established in an analogous way. For $R \in \mathcal{B}(\tau)$ we let R_{ij} denote the (i, j)-th entry in the σ-block partitioning of R. Here $\sigma = \frac{1}{n}\tau$. We have

$$(Q_2(n)T)_{ij} = \begin{cases} T_{ij}, & 0 \le i - j \le n - 1, \\ 0, & \text{otherwise.} \end{cases}$$

Furthermore,

$$(Q_2 T)_{ij} = (Q_2(n)T)_{ij}, \text{ for } j \ne i \text{ and } j \ne i - n.$$

So, by Proposition I.5.1,

$$|||Q_2 T - Q_2(n)T|||_2^2 = \sum_{i=0}^{n-1} \{\|(Q_2 T - T)_{ii}\|_2^2 + \|(Q_2 T)_{i,i-n}\|_2^2\}.$$

Next observe that

$$\|(Q_2 T - T)_{ij}\|_2 \le \|T_{ij}\|_2, \quad \|(Q_2 T)_{ij}\| \le \|T_{ij}\|_2,$$

and thus

(2.10)
$$|||Q_2 T - Q_2(n)T|||_2^2 \le \sum_{i=0}^{n-1} \{\|T_{ii}\|_2^2 + \|T_{i,i-n}\|_2^2\} = \int_\Omega \|\alpha(t, s)\|^2 \, dt \, ds.$$

Here α is the kernel function of T and $\Omega = \Omega' \cup \Omega''$, where

$$\Omega' = \cup_{i=0}^{n-1}\{(t, s) \mid i\sigma \le s \le (i+1)\sigma, \quad i\sigma \le t \le (i+1)\sigma\},$$

$$\Omega'' = \cup_{i=0}^{n-1}\{(t, s) \mid i\sigma \le s \le (i+1)\sigma, \quad (n+i)\sigma \le t \le (n+i+1)\sigma\}.$$

The total Lebesgue measure of Ω is equal to $2\tau^2 n^{-1}$ (and hence goes to zero if $n \to \infty$). Since $\|\alpha(\cdot, \cdot)\|^2$ is integrable over each compact subset of $\mathbf{R} \times \mathbf{R}$, Lebesgue's dominated convergence theorem implies that the last term in the right hand side of (2.10) goes to zero, which proves (2.9b). ∎

Let k be an $m \times m$ matrix function with entries in $L_2([-\tau, \tau])$. Put

$$(2.11) \qquad \widetilde{k}(t) = \begin{cases} k(t), & -\tau \leq t \leq \tau, \\ 0, & \text{otherwise.} \end{cases}$$

and let $\widetilde{K}$ be the convolution operator on $L_2^m(\mathbf{R})$ with kernel function $\widetilde{k}$; in other words, $\widetilde{K} = L_{\widetilde{k}}$. From Proposition I.4.1 we know that $\widetilde{K} \in \mathcal{B}(\tau)$. Since $\widetilde{k}(t) = 0$ for $t \notin [-\tau, \tau]$, we have

$$(2.12) \qquad \widetilde{K} = (Q_2 + Q_3)\widetilde{K}.$$

Next, put $\sigma = \frac{1}{n}\tau$, and let $K(n)$ be the operator on $L_2^m(\mathbf{R})$ whose σ-block partitioning $(K_{ij}(n))_{i,j=-\infty}^{\infty}$ is given by

$$(2.13) \qquad K_{ij}(n) = \begin{cases} K_{i-j}^{(n)}, & |i - j| \leq n - 1, \\ 0, & \text{otherwise,} \end{cases}$$

where, for $\nu = -(n-1), \ldots, n-1$,

$$(2.14) \qquad (K_\nu^{(n)}\varphi)(t) = \int_0^{\sigma} k(t - s + \nu\sigma)\varphi(s)\, ds, \qquad 0 \leq t \leq \sigma.$$

In other words:

$$(2.15) \qquad K(n) = (Q_2^0(n) + Q_d(n) + Q_3^0(n))\widetilde{K}.$$

Now, use (2.12), (2.15) and apply Lemmas 2.1 and 2.2. It follows that

$$(2.16) \qquad \lim_{n \to \infty} \|\widetilde{K} - K(n)\|_{\mathcal{B}(\tau)} = 0.$$

PROPOSITION 2.3. *The operators*

$$(2.17) \qquad S : \mathcal{B}(\tau) \to \mathcal{B}(\tau), \quad S(X) := X - Q_2(\widetilde{K}X),$$

$$(2.18) \qquad S_n : \mathcal{B}(\tau) \to \mathcal{B}(\tau), \quad S_n(X) := X - Q_2(n)(K(n)X),$$

are bounded linear operators on the Banach space $\mathcal{B}(\tau)$, and in the operator norm:

$$(2.19) \qquad \lim_{n \to \infty} \|S - S_n\| = 0.$$

PROOF. We already know that Q_2 and $Q_2(n)$ are bounded as linear operators on $\mathcal{B}(\tau)$. Since $\mathcal{B}(\tau)$ is a Banach algebra, it follows that the operators S and S_n are bounded operators on $\mathcal{B}(\tau)$. To prove (2.19) it suffices to show that for each $X \in \mathcal{B}(\tau)$ we have

$$(2.20) \qquad \|Q_2(\widetilde{K}X) - Q_2(n)(K(n)X)\|_{\mathcal{B}(\tau)} = \gamma(n)\|X\|_{\mathcal{B}(\tau)},$$

where $\gamma(n)$ is a constant depending on n only such that $\gamma(n) \to 0$ if $n \to \infty$.

Recall that $\sigma = \frac{1}{n}\tau$. In what follows we write T_{ij} for the (i,j)-th entry in the σ-block partitioning of $T \in \mathcal{B}(\tau)$. As was shown in the proof of Lemma 2.2, for each $T \in \mathcal{B}(\tau)$, we have

$$\||Q_2T - Q_2(n)T\||_2^2 \leq \sum_{i=0}^{n-1}\{\|T_{ii}\|_2^2 + \|T_{i,i-n}\|_2^2\}.$$

Let us apply this result to $T = K(n)X$. Note (see (2.13)) that

$$(K(n)X)_{ij} = \sum_{\nu=-\infty}^{\infty}(K(n))_{i\nu}X_{\nu j} = \sum_{\nu=-n+1+i}^{n-1+i}K_{i-\nu}^{(n)}X_{\nu j}.$$

It follows that

$$\|(K(n)X)_{ii}\|_2^2 \leq \left(\sum_{\nu=-n+1+i}^{n-1+i}\|K_{i-\nu}^{(n)}\|_2\|X_{\nu i}\|_2\right)^2$$

$$\leq \left(\sum_{\nu=-n+1+i}^{n-1+i}\|K_{i-\nu}^{(n)}\|_2^2\right)\left(\sum_{\nu=-n+1+i}^{n-1+i}\|X_{\nu i}\|_2^2\right)$$

$$= \left(\sum_{j=-n+1}^{n-1}\|K_j^{(n)}\|_2^2\right)\left(\sum_{\nu=-n+1+i}^{n-1+i}\|X_{\nu i}\|_2^2\right).$$

In a similar way one shows that

$$\|(K(n)X)_{i,i-n}\|_2^2 \leq \left(\sum_{j=-n+1}^{n-1}\|K_j^{(n)}\|_2^2\right)\left(\sum_{\nu=-n+1+i}^{n-1+i}\|X_{\nu,i-n}\|_2^2\right).$$

From Proposition I.4.1 (applied to $\widetilde{k}$) we see that

$$\sum_{j=-n+1}^{n-1}\|K_j^{(n)}\|_2^2 \leq \sigma\int_{-\tau}^{\tau}\|k(t)\|^2\,dt.$$

Furthermore,

$$\sum_{i=0}^{n-1}\sum_{\nu=-n+1+i}^{n-1+i}\|X_{\nu i}\|_2^2 \le \sum_{i=0}^{n-1}\sum_{\nu=-\infty}^{\infty}\|X_{\nu i}\|_2^2 = \|\|X\|\|_2^2,$$

because of Proposition I.5.1. Simalarly,

$$\sum_{i=0}^{n-1}\sum_{\nu=-n+1+i}^{n-1+i}\|X_{\nu,i-n}\|_2^2 \le \|\|X\|\|_2^2.$$

So we have proved that

$$(2.21) \qquad \|\|Q_2(K(n)X) - Q_2(n)(K(n)X)\|\|_2^2 \le \frac{2\tau}{n}\Big(\int_{-\tau}^{\tau}\|k(t)\|^2\,dt\Big)\|\|X\|\|_2^2.$$

Next, by (2.3) and inequality (I.5.4b),

$$\|\|Q_2(\widetilde{K}X) - Q_2(K(n)X)\|\|_2 \le \|\|\widetilde{K}X - K(n)X\|\|_2$$
$$\le \|\|\widetilde{K} - K(n)\|\|_2\|\|X\|\|,$$

and thus

$$(2.22) \qquad \|\|Q_2(\widetilde{K}X) - Q_2(K(n)X)\|\|_2 \le \|\widetilde{K} - K(n)\|_{\mathcal{B}(\tau)}\|\|X\|\|.$$

From (2.21) and (2.22) we see that

$$\|\|Q_2(\widetilde{K}X) - Q_2(n)(K(n)X)\|\|_2 \le \Big\{\sqrt{\frac{2\pi}{n}}\Big(\int_{-\tau}^{\tau}\|k(t)\|^2\,dt\Big)^{\frac{1}{2}} + \|\widetilde{K} - K(n)\|_{\mathcal{B}(\tau)}\Big\}\|X\|_{\mathcal{B}(\tau)}.$$

Note that $Q_2(\widetilde{K}X) - Q_2(n)(K(n)X) \in \mathcal{B}_2\dotplus\mathcal{B}_3$, because of (2.7b), and thus we can apply Lemma 2.1 to show that

$$\|Q_2(\widetilde{K}X) - Q_2(n)(K(n)X)\|_{\mathcal{B}(\tau)} \le \sqrt{3}\|\|Q_2(\widetilde{K}X) - Q_2(n)(K(n)X)\|\|_2.$$

It follows that (2.20) holds with

$$0 \le \gamma(n) \le \sqrt{\frac{6\tau}{n}}\Big(\int_{-\tau}^{\tau}\|k(t)\|^2\,dt\Big)^{\frac{1}{2}} + \sqrt{3}\|\widetilde{K} - K(n)\|_{\mathcal{B}(\tau)},$$

and hence $\gamma(n) \to 0 \quad (n \to \infty)$ by (2.16). $\blacksquare$

Let K be the integral operator on $L_2^m([0,\tau])$ defined by

$$(2.23) \qquad (K\varphi)(t) - \int_0^{\tau} k(t-s)\varphi(s)\,ds, \qquad 0 \le t \le \tau.$$

PROPOSITION 2.4. *If the operator $I - K$ is invertible, then the operator S on $\mathcal{B}(\tau)$ defined by (2.17) is also invertible.*

PROOF. The proof of the invertibility of S is split into two parts. First we consider S on $\mathcal{B}_2$.

Part (a). Note that $S\mathcal{B}_2 \subset \mathcal{B}_2$. Define $\widehat{S} : \mathcal{B}_2 \to \mathcal{B}_2$ to be the restriction of S to $\mathcal{B}_2$. Let $Y \in \mathcal{B}_2$, and let y be the kernel function of Y. We know that y has its support in the set

$$\Delta_2 = \{(t,s) \in \mathbf{R}^2 \mid s \leq t \leq \tau + s\}.$$

Assume that $X \in \mathcal{B}_2$ and $\widehat{S}(X) = Y$. Let x be the kernel function of X, and let f be the kernel function of $F := \widetilde{K}X$. Since k has its support in $[-\tau, \tau]$ and x has its support in Δ_2, we see that

$$f(t,s) = \int_s^{\tau+s} k(t-u)x(u,s)\,du, \qquad s \leq t \leq \tau + s.$$

Thus the identity $\widehat{S}(X) = Y$ implies that

$$(2.24) \qquad x(t,s) - \int_s^{\tau+s} k(t-u)x(u,s)\,du = y(t,s), \ \text{ a.e. on } \Delta_2.$$

The last identity may be rewritten as:

$$(2.25) \qquad x(t+s,s) - \int_0^{\tau} k(t-u)x(u+s,s)\,du = y(t+s,s), \ \text{ a.e. on } \Delta_2.$$

Put

$$(2.26a) \qquad \widetilde{x}(t,s) = x(t+s,s), \quad 0 \leq t \leq \tau, \quad 0 \leq s \leq \tau,$$

$$(2.26b) \qquad \widetilde{y}(t,s) = y(t+s,s), \quad 0 \leq t \leq \tau, \quad 0 \leq s \leq \tau,$$

The functions $\widetilde{x}$ and $\widetilde{y}$ are square integrable on $[0,\tau] \times [0,\tau]$. Let $\widetilde{X}$ and $\widetilde{Y}$ be the integral operators on $L_2^m([0,\tau])$ with kernel functions $\widetilde{x}$ and $\widetilde{y}$, respectively. Then (2.25) implies that

$$(2.27) \qquad (I - K)\widetilde{X} = \widetilde{Y}.$$

Since $I - K$ is invertible, it follows that $\widetilde{X}$ is uniquely determined by $\widetilde{Y}$, and therefore X is uniquely determined by Y. In other words, $\widehat{S}$ is one-one.

The above arguments also show how to solve the equation $\widehat{S}(X) = Y$ for a given Y in $\mathcal{B}_2$. Indeed, let y be the kernel function of Y, and define $\widetilde{y}$ by (2.26b). Then $\widetilde{y}$ is square integrable on $[0, \tau] \times [0, \tau]$. Let $\widetilde{Y}$ be the corresponding Hilbert-Schmidt integral operator on $L_2^m([0, \tau])$, and put

$$(2.28) \qquad\qquad \widetilde{X} := (I - K)^{-1}\widetilde{Y}.$$

We know that $(I - K)^{-1} - I$ is a Hilbert-Schmidt operator on $L_2^m([0, \tau])$, and hence the same is true for $\widetilde{X}$. Let $\widetilde{x}$ be the kernel function of $\widetilde{X}$, and define $x : \mathbf{R}^2 \to \mathbb{C}$ by (2.26a) and the following rules:

$$x(t, s) = 0, \quad (t, s) \notin \Delta_2,$$

$$x(t + \tau, s + \tau) = x(t, s), \quad (t, s) \in \mathbf{R}^2.$$

Then x is uniquely determined by $\widetilde{x}$, and there is a unique $X \in \mathcal{B}_2$ such that x is the kernel function of X. From (2.28) and the definition of x it follows that (2.24) is fulfilled, and hence $\widehat{S}(X) = Y$. We have proved that $\widehat{S}$ is one-one and onto.

Part (b). Take $Y \in \mathcal{B}(\tau)$, and let X be the unique solution in $\mathcal{B}_2$ of the equation

$$X - Q_2(\widetilde{K}X) = Q_2(Y) + Q_2\{\widetilde{K}(I - Q_2)Y\}.$$

The result of Part (a) guarantees that such an operator X exists. Now define,

$$Z := (I - Q_2)Y + X.$$

Then a simple computation shows that $S(Z) = Y$. Indeed,

$$\begin{aligned}
S(Z) &= Z - Q_2(\widetilde{K}Z) \\
&= (I - Q_2)Y + X - Q_2\{\widetilde{K}(I - Q_2)Y\} - Q_2(\widetilde{K}X) \\
&= (I - Q_2)Y + Q_2(Y) = Y.
\end{aligned}$$

It remains to prove that the equation $S(Z) = Y$ has no other solution in $\mathcal{B}(\tau)$. To do this, assume that $S(Z) = 0$. Then $(I - Q_2)Z = 0$ and $Q_2(Z) \in \mathcal{B}_2$ satisfies the equation $X - Q_2(\widetilde{K}X) = 0$. Thus, by the result of Part (a), we have $Q_2(Z) = 0$, and therefore $Z = 0$. ∎

III.3 Proof of the limit formula (1.5). In this subsection we prove formula (1.5). We continue to use the notations introduced in the previous subsection. Let S and S_n be the operators defined by (2.17) and (2.18), respectively. Let K be as in (2.23). According to our hypotheses the operator $I - K$ is positive definite. In particular, $I - K$ is invertible, and so (by Proposition 2.4) the operator S is invertible. Proposition 2.3 implies that for n sufficiently large S_n is invertible and

$$(3.1) \qquad \lim_{n \to \infty} \|S^{-1} - S_n^{-1}\| = 0.$$

Recall from Section II.2 that

$$(3.2) \qquad I - L_b = (I + L_x)^{-*}(I + L_x)^{-1},$$

where x is the (unique) solution of the equation (II.2.2). In operator form (II.2.2) can be rephrased as

$$(3.3) \qquad S(L_x) = Q_2\widetilde{K}.$$

From Section II.1 we know that

$$(3.4) \qquad I - \widehat{B}^{(n)} = (I + X(n))^{-*}(I + Q_d(n)X(n))(I + X(n))^{-1}.$$

where $X(n) \in \mathcal{B}_2(n)$ and satisfies the equation

$$(3.5) \qquad S_n(X(n)) = Q_2(n)K(n).$$

Indeed, (3.5) is just the analogue of (II.1.3) for the case considered here. Now apply (3.1). It follows that

$$\|L_x - X(n)\|_{\mathcal{B}(\tau)} \le \|S^{-1} - S_n^{-1}\| \cdot \|Q_2\widetilde{K}\|_{\mathcal{B}(\tau)} + \|S_n^{-1}\| \cdot \|Q_2\widetilde{K} - Q_2(n)K(n)\|_{\mathcal{B}(\tau)}.$$

The right hand side of the above inequality tends to zero if $n \to \infty$ (by (3.1) and (2.20)). Thus

$$(3.6) \qquad \lim_{n \to \infty} \|L_x - X(n)\|_{\mathcal{B}(\tau)} = 0.$$

Next, note that

$$(3.7) \qquad \|Q_d(n)X(n)\|_{\mathcal{B}(\tau)} \leq \|Q_d(n)L_x\|_{\mathcal{B}(\tau)} + \sqrt{3}\|L_x - X(n)\|_{\mathcal{B}(\tau)}.$$

Here we used that for each $T \in \mathcal{B}(\tau)$

$$\|Q_d(n)T\|_{\mathcal{B}(\tau)} \leq \sqrt{3}\|\|Q_d(n)T\|\|_2 \leq \sqrt{3}\|\|T\|\|_2 \leq \sqrt{3}\|T\|_{\mathcal{B}(\tau)},$$

because of (2.4) and the inequality (2.8) for $Q_d(n)$. By (2.9a) and (2.4) we have $\|Q_d(n)L_x\|_{\mathcal{B}(\tau)} \to 0$ if $n \to \infty$. Thus (3.6) yields

$$\|Q_d(n)X(n)\|_{\mathcal{B}(\tau)} \to 0, \qquad (n \to \infty),$$

and therefore

$$(3.8) \qquad \lim_{n \to \infty} \|I - (I + Q_d(n)X(n))^{-1}\|_{\mathcal{B}(\tau)} = 0.$$

From (3.6), (3.8) and the identities (3.2) and (3.4) we conclude that

$$\|(I - L_b)^{-1} - (I - \widehat{B}^{(n)})^{-1}\|_{\mathcal{B}(\tau)} \to 0 \quad (n \to \infty),$$

and thus $\|L_b - \widehat{B}^{(n)}\|_{\mathcal{B}(\tau)} \to 0 \quad (n \to \infty).$ ∎

III.4 Discrete and continuous orthogonal functions. Let k be an $m \times m$ matrix function with entries in $L_2([-\tau, \tau])$, and let K be the operator on $L_2^m([0, \tau])$ defined by

$$(K\varphi)(t) = \int_0^\tau k(t - s)\varphi(s)\, ds, \qquad 0 \leq t \leq \tau.$$

In what follows we assume that $I - K$ is invertible. The corresponding resolvent kernel is denoted by $\gamma(t, s)$, i.e.,

$$(4.1) \qquad ((I - K)^{-1}\varphi)(t) = \varphi(t) + \int_0^\tau \gamma(t, s)\varphi(s)\, ds, \qquad 0 \leq t \leq \tau.$$

By definition (see [14], [15]) the (first kind) *orthogonal function* associated with k is the entire $m \times m$ matrix function $D(\cdot)$ defined by

$$(4.2) \qquad D(\lambda) = e^{i\tau\lambda}(I + \int_0^\tau \gamma(t, 0)e^{-i\lambda t}\, dt), \qquad \lambda \in \mathbf{C}.$$

Next, let n be an arbitrary positive integer, and let

$$(4.3) \qquad \begin{pmatrix} I - K_0^{(n)} & -K_{-1}^{(n)} & \cdots & -K_{-(n-1)}^{(n)} \\ -K_1^{(n)} & I - K_0^{(n)} & \cdots & -K_{-(n-2)}^{(n)} \\ \vdots & \vdots & \ddots & \vdots \\ -K_{n-1}^{(n)} & -K_{n-2}^{(n)} & \cdots & I - K_0^{(n)} \end{pmatrix}$$

be the $\frac{\tau}{n}$-block partitioning of $I - K$. In particular, $K_\nu^{(n)}$ $(\nu = -(n-1), \ldots, n-1)$ is the operator on $L_2^m([0, \frac{1}{n}\tau])$ defined by

$$(K_\nu^{(n)}\varphi)(t) = \int_0^{\frac{\tau}{n}} k(t - s + \frac{\nu}{n}\tau)\varphi(s)\,ds, \qquad 0 \le t \le \frac{\tau}{n}.$$

Since $I - K$ is invertible, the $n \times n$ operator matrix in (4.3) is invertible. By definition (see [1], [13]), the *orthogonal polynomial* associated with (4.3) is the operator polynomial $P_n(z)$ given by

$$(4.4) \qquad P_n(z) = z^{n-1}(I + X_0) + z^{n-2}X_1 + \ldots + X_{n-1},$$

where

$$(4.5) \qquad \begin{pmatrix} I + X_0 \\ X_1 \\ \vdots \\ X_{n-1} \end{pmatrix} = \begin{pmatrix} I - K_0^{(n)} & -K_{-1}^{(n)} & \cdots & -K_{-(n-1)}^{(n)} \\ -K_1^{(n)} & I - K_0^{(n)} & \cdots & -K_{-(n-2)}^{(n)} \\ \vdots & \vdots & \ddots & \vdots \\ -K_{n-1}^{(n)} & -K_{n-2}^{(n)} & \cdots & I - K_0^{(n)} \end{pmatrix} \begin{pmatrix} I \\ 0 \\ \vdots \\ 0 \end{pmatrix}.$$

We shall see that for $n \to \infty$ the n-th orthogonal polynomial P_n converges in a certain sense to the orthogonal function D associated with k. To make this statement precise, let T_n be the operator on $L_2^m(\mathbf{R})$ whose $\frac{\tau}{n}$-block partitioning is the $L_2^m([0, \frac{1}{n}\tau])$-block Laurent operator with symbol $\sum_{j=0}^{n-1} z^j X_j$. Since k has its entries in $L_2([-\tau, \tau])$, the operators $K_\nu^{(n)}$ $(\nu = -(n-1), \ldots, n-1)$ are Hilbert-Schmidt operators. We may rewrite (4.5) as

$$(4.6) \qquad \begin{pmatrix} X_0 \\ X_1 \\ \vdots \\ X_{n-1} \end{pmatrix} = \begin{pmatrix} I - K_0^{(n)} & -K_{-1}^{(n)} & \cdots & -K_{-(n-1)}^{(n)} \\ -K_1^{(n)} & I - K_0^{(n)} & \cdots & -K_{-(n-2)}^{(n)} \\ \vdots & \vdots & \ddots & \vdots \\ -K_{n-1}^{(n)} & -K_{n-2}^{(n)} & \cdots & I - K_0^{(n)} \end{pmatrix} \begin{pmatrix} K_0^{(n)} \\ K_1^{(n)} \\ \vdots \\ K_{n-1}^{(n)} \end{pmatrix}.$$

It follows (by the same arguments as used in the proof of Proposition II.1.1) that $X_0, \ldots, X_{n-1}$ are Hilbert-Schmidt operators, and thus (cf., Section I.5) the operators T_n belong to the algebra $\mathcal{B}(\frac{1}{n}\tau)$. Let x_n be the kernel function of T_n. From (4.6) we see that x_n is given by

$$(4.7\text{a}) \quad x_n(t,s) = k(t-s) + \int_0^\tau \gamma(t,u)k(u-s)\,du, \qquad 0 \le t \le \tau, \quad 0 \le s \le \frac{1}{n}\tau,$$

$$(4.7\text{b}) \quad x_n(t,s) = 0, \quad t \in \mathbf{R}\backslash[0,\tau], \qquad 0 \le s \le \frac{1}{n}\tau,$$

$$(4.7\text{c}) \quad x_n(t+\frac{1}{n}\tau, s+\frac{1}{n}\tau) = x_n(t,s), \qquad (t,s) \in \mathbf{R}.$$

Now, put

$$x(t) = \begin{cases} \gamma(t,0), & 0 \le t \le \tau, \\ 0, & t \in \mathbf{R}\backslash[0,\tau]. \end{cases}$$

Then we see from (4.1) that

$$(4.8) \qquad x(t) = k(t) + \int_0^\tau \gamma(t,u)k(u)\,du, \qquad 0 \le t \le \tau,$$

and hence the formulas (4.7a) - (4.7c) suggest that for $n \to \infty$ the function $x_n(t,s)$ converges to $x(t-s)$. The precise result is the following.

PROPOSITION 4.1. *Let* $T, T_1, T_2, \ldots$ *be the integral operators on* $L_2^m(\mathbf{R})$ *with kernel functions* $x(t-s), x_1(t,s), x_2(t,s), \ldots$, *respectively. Then*

$$(4.9) \qquad \|T - T(n)\|_{\mathcal{B}(\tau)} \to 0 \qquad (n \to \infty).$$

PROOF. Note that $T = L_x$. Since x has its entries in $L_1(\mathbf{R}) \cap L_1(\mathbf{R})$, we know (see Proposition I.4.1) that $T \in \mathcal{B}(\tau)$. Proposition I.5.2 implies that $T_n \in \mathcal{B}(\tau)$ for $n \ge 1$. Next we use notations and results from the previous section. From the definition of x it follows that

$$(4.10) \qquad S(T) = Q_2\widetilde{K}.$$

Furthermore, because of (3.6), we have

$$(4.11) \qquad S_n(T_n) = Q_2(n)K(n).$$

But then, by the arguments used in the proof of the limit formula (1.5) (see Section III.3), we obtain (4.9). ∎

REFERENCES

[1]   A. Atzmon, N-th orthogonal operator polynomials, in: *Orthogonal matrix-valued polynomials and applications*, OT 34, Birkhäuser Verlag, Basel, 1988; pp. 47-63.

[2]   D.Z. Arov and M.G. Krein, Problems of search of the minimum of entropy in indeterminate extension problems, *Funct. Anal. Appl.* 15 (1981), 123-126.

[3]   J.J. Benedetto, A quantative maximum entropy theorem fro the real line, *Integral Equations and Operator Theory* 10 (1987), 761-779.

[4]   S. Bochner and R.S. Phillips, Absolutely convergent Fourier expansions for non commutative rings, *Annals of Mathematics* 43 (1942) 409-418.

[5]   J. Chover, On normalized entropy and the extensions of a positive definite function. *J. Math. Mech.* 10 (1961), 927-945.

[6]   H. Dym, *J contractive matrix functions, reproducing kernel Hilbert spaces and interpolation*, CBMS 71, Amer. Math. Soc., Providence RI, 1989.

[7]   H. Dym and I Gohberg, On an extension problem, generalized Fourier analysis and an entropy formula, *Integral Equations Operator Theory*, 3 (1980), 143-215.

[8]   I. Gohberg, S. Goldberg and M.A. Kaashoek, *Classes of Linear Operators*, Volume I, Birkhäuser Verlag, Basel, 1990.

[9]   I. Gohberg and M.A. Kaashoek, Asymptotic formulas of Szegö-Kac-Achiezer type, *Asymptotic Analysis*, 5 (1992), 187-220.

[10]   I. Gohberg, M.A. Kaashoek and H.J. Woerdeman, The band method for positive and contractive extension problems, *J. Operator Theory*, 22 (1989), 109-105.

[11]   I. Gohberg, M.A. Kaashoek and H.J. Woerdeman, The band method for positive and contractive extension problems: An alternative version and new applications, *Integral Equations Operator Theory* 12 (1989), 343-382.

[12]   I. Gohberg, M.A. Kaashoek and H.J. Woerdeman, A maximum entropy priciple in the general framework of the band method, *J. Funct, Anal. Anal.* 95 (1991), 231-254.

[13]   I. Gohberg and L. Lerer, Matrix generalizations of M.G. Krein theorems oon orthogonal polynomials, in: *Orthogonal matrix-valued polynomials and applications*, OT 34, Birkhäuser Verlag, Basel, 1988; pp. 137-202.

[14]   M.G. Krein, Continuous analogues of propositions about orthogonal polynomials on the unit circle (Russian), *Dokl. Akad. Nauk USSR*, 105:4 (1955), 637-640.

[15]   M.G. Krein and H. Langer, On some continuation problems which are closely related to the theory of operators in spaces Π_κ. IV, *J. Operator Theory* 13 (1985), 299-417.

[16]   D. Mustafa and K. Glover, *Minimum entropy H_∞ control*, Lecture Notes in Control and Information Sciences 146, Springer Verlag, Berlin, 1990.

I. Gohberg   M.A. Kaashoek
Raymond and Beverly Sackler   Faculteit Wiskunde en Informatica
Faculty of Exact Sciences   Vrije Universiteit
School of Mathematical Sciences   De Boelelaan 1081a
Tel-Aviv University   1081 HV Amsterdam
Ramat-Aviv, Israel   The Netherlands

MSC: Primary 47A57, Secondary 45A10, 47A20

Operator Theory:
Advances and Applications, Vol. 59
© 1992 Birkhäuser Verlag Basel

INTERPOLATING SEQUENCES IN THE MAXIMAL IDEAL SPACE OF H^∞ II

Keiji Izuchi

The objective of this paper is to study interpolating sequences in $M(H^\infty)$ the maximal ideal space of H^∞. Using the pseudohyperbolic distance, L. Carleson gave a characterization of interpolating sequences in D the open unit disk. But its condition does not characterize interpolating sequences in $M(H^\infty)$. Carleson's condition has several equivalent conditions in D. It is studied the relations between these conditions and interpolating sequences in $M(H^\infty)$.

1. INTRODUCTION

Let H^∞ be the Banach algebra of bounded analytic functions on the open unit disc D with the supremum norm. We denote by $M(H^\infty)$ the maximal ideal space of H^∞ with the weak-*topology. We identify a function in H^∞ with its Gelfand transform. We may consider that $D \subset M(H^\infty)$, and by the corona theorem D is dense in $M(H^\infty)$. The maximal ideal space $M(L^\infty)$ of $L^\infty(\partial D)$ may be considered as a subset of $M(H^\infty)$. For points x and y in $M(H^\infty)$, we define

$$\rho(x,y) = \sup\{|h(x)|;\, h(y) = 0,\, h \in H^\infty,\, \|h\| \le 1\}.$$

When z and w are points in D, $\rho(z,w) = |z - w| / |1 - \bar{z}w|$. For a point x in $M(H^\infty)$, we put

$$P(x) = \{y \in M(H^\infty);\, \rho(y,x) < 1\},$$

which is called a Gleason part. If $P(x) = \{x\}$, x is called a trivial point. Every point in $M(L^\infty)$ is trivial. If $P(x) \neq \{x\}$, then x and $P(x)$ are called nontrivial. In this case, there is a one to one continuous map L_x from D onto $P(x)$ such that $x = L_x(0)$ and $H^\infty \circ L_x \subset H^\infty$. Moreover if L_x is a homeomorphism, $P(x)$ is called a homeomorphic part. $P(0) = D$ is a typical homeomorphic part. We put

$$G = \{x \in M(H^\infty);\, x \text{ is not nontrivial}\}.$$

Then G is an open subset of $M(H^\infty)$ (see [6]). For a sequence $\{z_n\}_n$ in D with $\Sigma_{n=1}^{\infty} 1 - |z_n| < \infty$, a function

$$b(z) = \prod_{n=1}^{\infty} \frac{-\bar{z}_n}{|z_n|} \frac{z - z_n}{1 - \bar{z}_n z}, \qquad z \in D$$

is called a Blaschke product with zeros $\{z_n\}_n$. For a positive singular measure μ on ∂D, a function

$$S[\mu](z) = \exp\left(-\int \frac{e^{i\theta} + z}{e^{i\theta} - z} \, d\mu(\theta)\right)$$

is called a singular inner function. For a bounded measurable function F on ∂D such that $\log|F|$ is integrable, a function

$$O[\log|F|](z) = \exp\left(\int \frac{e^{i\theta} + z}{e^{i\theta} - z} \log\left|F(e^{i\theta})\right| \, d\theta/2\pi\right)$$

is called an outer function. These functions are contained in H^∞, and every function f in H^∞ is represented by $f = O[\log|f|]\,S[\mu]\,b$ except a constant factor (see [5]). Let $f_n = O[\log|f_n|]\,S[\mu_n]\,b_n$ be a function in H^∞ with $\|f_n\| \leq 1$. If $\Sigma_{n=1}^\infty \log|f_n|$ is integrable, $\Sigma_{n=1}^\infty \mu_n(\partial D) < \infty$ and $\Pi_{n=1}^\infty b_n$ is still a Blaschke product, then we denote by $\Pi_{n=1}^\infty f_n$ the function $O[\Sigma_{n=1}^\infty \log|f_n|]\,S[\Sigma_{n=1}^\infty \mu_n]\,\Pi_{n=1}^\infty b_n$ in H^∞.

A sequence $\{x_n\}_n$ in $M(H^\infty)$ is called *interpolating* if for every sequence $\{a_n\}_n$ of bounded complex numbers there is a function f in H^∞ such that $f(x_n) = a_n$ for all n. An interpolating sequence $\{z_n\}_n$ in D is characterized by Carleson [2] as follows;

$$\inf_k \prod_{n:n\neq k} \left| \frac{z_k - z_n}{1 - \bar{z}_n z_k} \right| > 0.$$

Let

$$\psi_n(z) = \frac{-\bar{z}_n}{|z_n|} \frac{z - z_n}{1 - \bar{z}_n z}$$

be a Blaschke factor. Then $\psi_n(z_n) = 0$, $\|\psi_n\| = 1$ and

$$\inf_k \prod_{n:n\neq k} \rho(z_n, z_k) = \inf_k \left| \left(\prod_{n:n\neq k} \psi_n \right)(z_k) \right|.$$

From this observation, we can consider similar conditions for sequences $\{x_n\}_n$ in $M(H^\infty)$ as follows:

(A_1) $$\inf_k \prod_{n:n\neq k} \rho(x_n, x_k) > 0.$$

(A_2) There is a sequence $\{g_n\}_n$ in H^∞ such that

$$\|g_n\| \leq 1, \; g_n(x_k) = 0 \text{ if } k \neq n, \text{ and } \inf_n |g_n(x_n)| > 0.$$

(A_3) There is a sequence $\{f_n\}_n$ in H^∞ such that

$$\|f_n\| \leq 1, \; f_n(x_n) = 0, \; \prod_{n=1}^\infty f_n \in H^\infty, \text{ and } \inf_k \left| \left(\prod_{n:n\neq k} f_n \right)(x_k) \right| > 0.$$

In this paper, we shall study what kind of relations are there between interpolating sequences and conditions $(A_i), i = 1, 2, 3$. First, we have

FACT 1 (Proposition 1). $(A_3) \Rightarrow (A_2) \Rightarrow (A_1)$.

By the open mapping theorem,

FACT 2. *If $\{x_n\}_n$ is interpolating, $\{x_n\}_n$ satisfies (A_2).*

In Theorem 1, we shall prove that the converse assertion of Fact 2 is true when each point x_n is trivial. A sequence $\{x_n\}_n$ in $M(H^\infty)$ is called strongly discrete if there is a sequence of disjoint open subsets $\{U_n\}_n$ of $M(H^\infty)$ such that $x_n \in U_n$. Then we have

FACT 3. *If $\{x_n\}_n$ is interpolating, $\{x_n\}_n$ is strongly discrete.*

A Blaschke product is called interpolating if its zero sequence is interpolating. For a function f in H^∞, we put

$$Z(f) = \{x \in M(H^\infty);\ f(x) = 0\}.$$

Then we have

FACT 4 (see [8]). *Let $\{x_n\}_n$ be a strongly discrete sequence. If $\{x_n\}_n \subset Z(b)$ for some interpolating Blaschke product b, then $\{x_n\}_n$ is interpolating.*

In [4], Gorkin, Lingenberg and Mortini proved that if $\{x_n\}_n$ is a sequence in a homeomorphic part and satisfies condition (A_1), then there is an interpolating Blaschke product b such that $\{x_n\}_n \subset Z(b)$. In [7], the author got the same conclusion when $\{x_n\}_n$ is a sequence in a homeomorphic part and satisfies condition (A_2). Also in [8], the author studied a sequence whose elements are contained in distinct parts. In Theorem 2, we shall prove that if $\{x_n\}_n$ is a sequence in G, then $\{x_n\}_n$ satisfies condition (A_3) if and only if $\{x_n\}_n$ is strongly discrete and there is an interpolating Blaschke product b such that $\{x_n\}_n \subset Z(b)$. And in Theorem 3, we show that there is a strongly discrete sequence $\{x_n\}_n$ in a nontrivial part such that $\{x_n\}_n$ satisfies condition (A_1) but $\{x_n\}_n$ is not interpolating.

2. CONDITION (A_2)

In this section, the following lemma plays an important role.

LEMMA 1 [9, Theorem 3.1]. *Let b be a Blaschke product. Then there are Blaschke products b_1 and b_2 such that $b = b_1 b_2$ and $b_1(x) = b_2(x) = 0$ for every point x in $M(H^\infty)$ with $b_{|P(x)} = 0$.*

First we prove

PROPOSITION 1. $(A_3) \Rightarrow (A_2) \Rightarrow (A_1)$.

PROOF. The implication $(A_3) \Rightarrow (A_2)$ is trivial. To prove $(A_2) \Rightarrow (A_1)$, it is sufficient to prove that if $f \in H^\infty$ satisfies $\|f\| \le 1$ and $f(x_n) = 0$ for every n, then

$$\prod_{n=1}^{\infty} \rho(x, x_n) \ge |f(x)| \quad \text{for } x \in M(H^\infty).$$

Let $f = O[\log|f|] S[\mu] b$ be a canonical factorization. We devide $\{x_n\}_n$ into three parts:

$$\{\zeta_j\}_j = \{x_n; O[\log|f|] S[\mu]_{|P(x_n)} = 0\};$$
$$\{\xi_j\}_j = \{x_n; x_n \notin \{\zeta_j\}_j \text{ and } b_{|P(x_n)} = 0\};$$
$$\{\lambda_j\}_j = \{x_n; f \text{ does not vanish indentically on } P(x_n)\}.$$

Let k be an arbitrary positive integer. We put

$$f_k = O[k^{-1} \log|f|] S[k^{-1}\mu].$$

Then $f_k \in H^\infty, \|f_k\| \le 1$ and $O[\log|f|] S[\mu] = (f_k)^k$, hence $f_k(\zeta_j) = 0$ for every j. Therefore

$$(1) \qquad \prod_{j=1}^{k} \rho(x, \zeta_j) \ge |f_k(x)|^k = |(O[\log|f|] S[\mu])(x)|.$$

By [6, Theorem 5.3], there are interpolating Blaschke products $b_1, b_2, \ldots, b_k$ such that $b = (\Pi_{j=1}^{k} b_j) B$ and $b_j(\lambda_j) = 0$ for $j = 1, 2, \ldots, k$ for some Blaschke product B. Then we have

$$(2) \qquad \prod_{j=1}^{k} \rho(x, \lambda_j) \ge \prod_{j=1}^{k} |b_j(x)| = \left|(\prod_{j=1}^{k} b_j)(x)\right|.$$

We apply Lemma 1 for B succeedingly. Then we have a factorization $B = \Pi_{j=1}^{k} B_j$ such that $B_j(\xi_i) = 0$ for every i. Hence

$$\prod_{j=1}^{k} \rho(x, \xi_j) \ge \left|(\prod_{j=1}^{k} B_j)(x)\right| = |B(x)|.$$

By (2), we have

$$\prod_{j=1}^{k} \rho(x, \xi_j) \rho(x, \lambda_j) \ge |b(x)|.$$

Therefore by (1),

$$\prod_{j=1}^{k} \rho(x, \zeta_j) \rho(x, \xi_j) \rho(x, \lambda_j) \ge |f(x)|.$$

Letting $k \to \infty$, we have $\Pi_{n=1}^{\infty} \rho(x, x_n) \ge |f(x)|$.

Now we prove the following theorem. The idea of this proof comes from Axler and Gorkin [1, Theorem 3].

THEOREM 1. *Let $\{x_n\}_n$ be a sequence of trivial points in $M(H^\infty)$. Then $\{x_n\}_n$ is interpolating if and only if $\{x_n\}_n$ satisfies condition (A_2).*

PROOF. Suppose that $\{x_n\}_n$ satisfies condition (A_2). Then there is a sequence $\{F_n\}_n$ in H^∞ such that $\|F_n\| \leq 1$,

$$(1) \qquad\qquad F_n(x_k) \;=\; 0 \qquad \text{if } k \neq n, \text{ and}$$

$$(2) \qquad\qquad F_n(x_n) \;\neq\; 0 \qquad \text{for every } n.$$

We may assume that $F_n(x_n) > 0$ for every n. Let $\{\epsilon_n\}_n$ be a sequence of positive numbers such that

$$(3) \qquad\qquad \prod_{n=1}^{\infty} (1 + \epsilon_n) \; < \; \infty.$$

Take $0 < \delta_n < 1$ such that

$$\rho(1 - \delta_n, -1 + \delta_n) \;=\; \rho(0, F_n(x_n)).$$

Here we note that if $F_n(x_n)$ closes to 1 then δ_n closes to 0. Let $F_n = O_n S_n B_n$ be a canonical factorization. For a positive integer k, we apply Lemma 1 for B_n k-times succeedingly. As a result, we have a factorization $B_n = B_{n_1} B_{n_2} \cdots B_{n_k}$. Since x_j is a trivial point by our starting hypothsis, if $B_n(x_j) = 0$ for some j then $B_{n_1}(x_j) = \ldots = B_{n_k}(x_j) = 0$. Here we may assume that

$$\left| B_{n_1}(x_n) \right| \leq \left| B_{n_2}(x_n) \right| \leq \; \cdots \; \leq \left| B_{n_k}(x_n) \right|.$$

Then by (2),

$$\left| (O_n^{1/k} S_n^{1/k} B_{n_k})(x_n) \right| \geq F_n(x_n)^{1/k} > F_n(x_n),$$

hence $\left| (O_n^{1/k} S_n^{1/k} B_{n_k})(x_n) \right| \to 1$ $(k \to \infty)$. By (1), we have that $(O_n^{1/k} S_n^{1/k} B_{n_k})(x_j) = 0$ for $j \neq n$. Hence we can consider $O_n^{1/k} S_n^{1/k} B_{n_k}$ instead of F_n, and so that we may assume moreover that

$$(4) \qquad\qquad \delta_n \; < \; 1 - 1/\sqrt{1 + 2\epsilon_n}.$$

Let

$$b_n(z) \;=\; \frac{z - (1 - \delta_n)}{1 - (1 - \delta_n)z}, \qquad z \in D.$$

Then $b_n(0) = -1 + \delta_n$ and $b_n(F_n(x_n)) = 1 - \delta_n$. Let

$$h_n(z) \;=\; b_n \circ F_n(z)/(1 - \delta_n), \qquad z \in D.$$

Then $h_n \in H^\infty$,

$$(5) \qquad\qquad \|h_n\| \; \leq \; \sqrt{1 + 2\epsilon_n} \qquad \text{by (4), and}$$

$$(6) \qquad\qquad h_n(x_n) = 1, \text{ and } h_n(x_k) = -1 \quad \text{for every } k \text{ with } k \neq n.$$

Now let

$$f_n = ((1 + h_n)/2)^2 \quad \text{and} \quad g_n = ((1 - h_n)/2)^2.$$

Then f_n and g_n are contained in H^∞, and by (5)

$$|f_n| + |g_n| = (1 + |h_n|^2)/2 < 1 + \epsilon_n.$$

Moreover by (6),

$$(7) \qquad\qquad f_n(x_n) = 1, \text{ and } f_n(x_k) = 0 \quad \text{for } k \neq n;$$

$$(8) \qquad\qquad g_n(x_n) = 0, \text{ and } g_n(x_k) = 1 \quad \text{for } k \neq n.$$

Here let $G_n = f_n g_1 g_2 \cdots g_{n-1}$. Then by [1, Lemma 2] and (3),

$$(9) \qquad\qquad \sum_{n=1}^{\infty} |G_n| \leq \prod_{n=1}^{\infty} (1 + \epsilon_n) < \infty \qquad \text{on } D.$$

Now we show that $\{x_n\}_n$ is interpolating. Let $\{a_n\}_n$ be an arbitrary bounded sequence. Define

$$G = \sum_{n=1}^{\infty} a_n G_n.$$

By (9), we have $G \in H^\infty$, and

$$\begin{aligned}
G(x_k) &= \sum_{n=1}^{k-1} a_n G_n(x_k) + a_k G_k(x_k) + \left(\sum_{n=k+1}^{\infty} a_n G_n \right)(x_k) \\
&= a_k + \left(\sum_{n=k+1}^{\infty} a_n G_n \right)(x_k) \qquad \text{by (7) and (8).}
\end{aligned}$$

Here $\sum_{n=k+1}^{\infty} a_n f_n g_{k+1} g_{k+2} \cdots g_{n-1}$ is a function in H^∞ and

$$\begin{aligned}
\left(\sum_{n=k+1}^{\infty} a_n G_n \right)(x_k) &= (g_1 g_2 \cdots g_k)(x_k) \left(\sum_{n=k+1}^{\infty} a_n f_n g_{k+1} \cdots g_{n-1} \right)(x_k) \\
&= 0 \qquad \text{by (8).}
\end{aligned}$$

Hence $G(x_k) = a_k$ for every k. This implies that $\{x_n\}_n$ is interpolating.

The converse is already mentioned as Fact 2.

In the first part of the above proof, we show that if $\{x_n\}_n$ is a sequence of trivial points, then there exists a sequence $\{F_n\}_n$ in H^∞ such that

$$\|F_n\| \leq 1, \ F_n(x_k) = 0 \qquad \text{if } k \neq n, \text{ and}$$

$|F_n(x_n)|$ closes to 1 sufficiently for every n.

In the rest of the above proof, we prove that for a given sequence $\{x_n\}_n$ in $M(H^\infty)$, x_n needs not to be trivial, if there is $\{F_n\}_n$ in H^∞ satisfying the above conditions, then $\{x_n\}_n$ is interpolating. We note that this proof works for the spaces H^∞ on the other domains.

PROPOSITION 2. *Let H^∞ be the space of bounded analytic functions on the domain Ω in C^n. Let $\{x_n\}_n$ be a sequence in $M(H^\infty)$. If for every ϵ with $0 < \epsilon < 1$ there is a sequence $\{f_n\}_n$ in H^∞ such that $\|f_n\| \le 1$, $f_n(x_k) = 0$ for $k \ne n$ and $|f_n(x_n)| > \epsilon$ for every n, then $\{x_n\}_n$ is an interpolating sequence.*

We have the following problem.

PROBLEM 1. Let $\{x_n\}_n$ be a sequence of trivial points.

(1)   If $\{x_n\}_n$ is strongly discrete, is it interpolating ?

(2)   If $\{x_n\}_n$ satisfies condition (A_2), does it satisfy (A_3) ?

We note that Hoffman proved in his unpublished note that if $\{x_n\}_n$ is a strongly discrete sequence in $M(L^\infty)$, then $\{x_n\}_n$ is interpolating.

3. CONDITION (A_3)

In this section, we prove the following theorem.

THEOREM 2. *Let $\{x_n\}_n$ be a sequence in G. Then $\{x_n\}_n$ satisfies condition (A_3) if and only if $\{x_n\}_n$ is strongly discrete and there exists an interpolating Blaschke product b such that $\{x_n\}_n \subset Z(b)$.*

To prove this, we need some lemmas. For a subset E of $M(H^\infty)$, we denote by $cl\, E$ the closure of E in $M(H^\infty)$.

LEMMA 2 [5, p. 205]. *Let b be an interpolating Blaschke product with zeros $\{z_n\}_n$. Then $Z(b) = cl\,\{z_n\}_n$.*

LEMMA 3 [6, p. 101]. *If b is an interpolating Blaschke product, then $Z(b) \subset G$. Conversely, for a point x in G there is an interpolating Blaschke product b such that $x \in Z(b)$.*

For an interpolating Blaschke product b with zeros $\{z_n\}_n$, put

$$\delta(b) \;=\; \inf_k\; \prod_{n:n\ne k}\; \rho(z_n, z_k).$$

LEMMA 4 [6, p. 82]. *Let b be an interpolating Blaschke product and let x be a point in $M(H^\infty)$ with $b(x) = 0$. Then for $0 < \sigma < 1$ there is a Blaschke subproduct B of b such that $B(x) = 0$ and $\delta(B) > \sigma$.*

LEMMA 5 [3, p. 287]. *For $0 < \delta < 1$, there exists a positive constant $K(\delta)$ satisfying the following condition; let b be an interpolating Blaschke product with zeros $\{z_n\}_n$ such that $\delta(b) > \delta$. Then for every sequence $\{a_n\}_n$ of complex numbers with $|a_n| \le 1$, there is a function h in H^∞ such that $h(z_n) = a_n$ for all n and $\|h\| < K(\delta)$.*

PROOF OF THEOREM 2. First suppose that $\{x_n\}_n$ is strongly discrete and $\{x_n\}_n \subset Z(b)$ for some interpolating Blaschke product b. Take a sequence $\{U_n\}_n$ of disjoint open subsets of $M(H^\infty)$ such that $x_n \in U_n$ for every n. Let $\{z_k\}_k$ be the zeros of b in D. For each n, let b_n be the Blaschke product with zeros $\{z_k\}_k \cap U_n$. Since $\{U_n\}_n$ is a sequence of disjoint subsets, $\Pi_{n=1}^\infty b_n$ is a subproduct of b. By Lemma 2, $x_n \in cl\,\{z_k\}_k$. Hence $x_n \in cl\,(\{z_k\}_k \cap U_n)$, so that $b_n(x_n) = 0$. We also have

$$\left| \left(\prod_{j:j\neq n} b_j \right)(z) \right| \geq \inf_i \prod_{j:j\neq i} \rho(z_j, z_i) = \delta(b) > 0$$

for every $z \in \{z_k\}_k \cap U_n$. Hence

$$\inf_n \left| \left(\prod_{j:j\neq n} b_j \right)(x_n) \right| \geq \delta(b).$$

Thus $\{x_n\}_n$ satisfies condition (A_3).

Next suppose that $\{x_n\}_n \subset G$ and $\{x_n\}_n$ satisfies condition (A_3). Then there is a sequence $\{f_n\}_n$ in H^∞ such that $\|f_n\| \leq 1, \Pi_{n=1}^\infty f_n \in H^\infty$, and

$$(1) \qquad f_n(x_n) = 0 \text{ and } \inf_k \left| \left(\prod_{n:n\neq k} f_n(x_k) \right) \right| > \delta$$

for some $\delta > 0$. By considering $\{c_n f_n\}_n$ with $0 < c_n < 1$ and $\Pi_{n=1}^\infty c_n > 0$, we may assume that

$$(2) \qquad \|f_n\| < 1 \qquad \text{for every } n.$$

By (1), $\{x_k\}_k$ is strongly discrete, hence we can take a sequence $\{V_k\}_k$ of disjoint open subsets of $M(H^\infty)$ such that $x_k \in V_k$ and

$$(3) \qquad \inf \left\{ \left| \left(\prod_{n:n\neq k} f_n \right)(w) \right| ; w \in V_k \right\} > \delta.$$

Let $K(\delta)$ be a positive constant which is given in Lemma 5 associated with $\delta > 0$. By (2), there is a sequence $\{\epsilon_n\}_n$ of positive numbers such that

$$(4) \qquad \sum_{n=1}^\infty \epsilon_n < \delta,$$

$$(5) \qquad K(\delta)\,\epsilon_n < \delta/2 \qquad \text{for every } n, \text{ and}$$

$$(6) \qquad \|f_n\| + K(\delta)\,\epsilon_n < 1 \qquad \text{for every } n.$$

Since $x_n \in G$, by Lemmas 2 and 3 there is an interpolating Blaschke product b_n with zeros $\{w_{n,j}\}_j$ such that $b_n(x_n) = 0$ and

$$(7) \qquad \{w_{n,j}\}_j \subset V_n \cap D.$$

By Lemma 4, we may assume that

$$(8) \qquad\qquad \delta(b_n) > \delta \qquad \text{for every } n.$$

Since $f_n(x_n) = 0$, we may assume moreover that

$$(9) \qquad\qquad |f_n(w_{n,j})| < \epsilon_n \qquad \text{for every } j.$$

By considering tails of sequences $\{w_{n,j}\}_j$ for $n = 1, 2, \ldots$, we may assume that $\Sigma_{n,j} 1 - |w_{n,j}| < \infty$, that is, $\Pi_{n=1}^{\infty} b_n$ is a Blaschke product. By (3) and (7), for $n \neq k$ we have

$$|f_n(w_{k,j})| > \delta \qquad \text{for every } j.$$

Hence by (5),

$$|f_n(w_{k,j})| - K(\delta)\,\epsilon_n > \delta/2 \qquad \text{for every } j.$$

Let take $c > 0$ such that

$$c(x - 1) < \log x \qquad \text{for } \delta/2 < x < 1.$$

Then

$$(10) \qquad c\left(|f_n(w_{k,j})| - K(\delta)\,\epsilon_n - 1\right) < \log\left(|f_n(w_{k,j})| - K(\delta)\,\epsilon_n\right)$$

for $n \neq k$ and $j = 1, 2, \ldots$. Since $1 - x < -\log x$ for $0 < x < 1$, we have

$$(11) \qquad 0 < 1 - |f_n(w_{k,j})| < -\log|f_n(w_{k,j})|$$

for $n \neq k$ and $j = 1, 2, \ldots$.

Now we shall prove that $b = \Pi_{n=1}^{\infty} b_n$ is an interpolating Blaschke product. By (8) and (9), there is a function g_n in H^{∞} such that $g_n(w_{n,j}) = f_n(w_{n,j})$ for every j and

$$(12) \qquad\qquad \|g_n\| < K(\delta)\,\epsilon_n.$$

Then $f_n = g_n + b_n h_n$ for some h_n in H^{∞}. By (6) and (12), $\|h_n\| < 1$. Hence

$$(13) \qquad\qquad |f_n - g_n| < |b_n| \qquad \text{on } D.$$

Here we have the following for every j and k,

$$
\begin{aligned}
\left|\Big(\Pi_{n:n\neq k} b_n\Big)(w_{k,j})\right| \;&=\; \Pi_{n:n\neq k} |b_n(w_{k,j})| \\
&>\; \Pi_{n:n\neq k} |f_n(w_{k,j}) - g_n(w_{k,j})| && \text{by (13)} \\
&>\; \Pi_{n:n\neq k} \big(|f_n(w_{k,j})| - K(\delta)\,\epsilon_n\big) && \text{by (12)} \\
&=\; \exp\Big\{\Sigma_{n:n\neq k} \log\big(|f_n(w_{k,j})| - K(\delta)\,\epsilon_n\big)\Big\}
\end{aligned}
$$

$$
\begin{aligned}
&> \; \exp\Big\{ \sum_{n:n\neq k} c\,(|f_n(w_{k,j})| - K(\delta)\,\epsilon_n - 1)\Big\} && \text{by (10)} \\
&= \; \exp\Big[-c\,\big(\sum_{n:n\neq k} 1 - |f_n(w_{k,j})|\big)\Big]\,\exp\big(-c\,K(\delta)\sum_{n:n\neq k}\epsilon_n\big) \\
&> \; \exp\Big[-c\sum_{n:n\neq k} -\log|f_n(w_{k,j})|\Big]\,\exp\big(-c\,K(\delta)\,\delta\big) && \text{by (4) and (11)} \\
&= \; \big(\prod_{n:n\neq k}|f_n(w_{k,j})|\big)^c\,\exp\big(-c\,K(\delta)\,\delta\big) \\
&> \; \delta^c\exp\big(-c\,K(\delta)\,\delta\big) && \text{by (3) and (7).}
\end{aligned}
$$

Therefore we have

$$
\begin{aligned}
\delta(b) &= \inf_{k,j}\;\prod_{(n,i):(n,i)\neq(k,j)} \rho(w_{n,i}, w_{k,j}) \\
&= \inf_{k,j}\;\Big|\big(\prod_{n:n\neq k} b_n\big)(w_{k,j})\Big|\prod_{i:i\neq j}\rho(w_{k,i}, w_{k,j}) \\
&> \; \delta^c\exp\big(-c\,K(\delta)\,\delta\big)\,\inf_k \delta(b_k) \\
&> \; \delta^{c+1}\exp\big(-c\,K(\delta)\,\delta\big) && \text{by (8).}
\end{aligned}
$$

Thus b is an interpolating Blaschke product. Since $b_n(x_n) = 0$, $b(x_n) = (\prod_{n=1}^{\infty} b_n)(x_n) = 0$. This completes the proof.

In [8], the author actually proved the following.

PROPOSITION 3. *Let $\{x_n\}_n$ be a sequence in G such that $P(x_n)\cap cl\{x_k\}_{k\neq n}$ $= \emptyset$ for every n. If $\{x_n\}_n$ satisfies condition (A_2), then $\{x_n\}_n$ satisfies condition (A_3).*

If $\{x_n\}_n$ satisfies a more stronger topological condition, then we can get the same conclusion without condition(A_2).

PROPOSITION 4. *Let $\{x_n\}_n$ be a sequence in G. If $cl\,P(x_n)\cap cl\,(\cup_{k:k\neq n} P(x_k))$ $= \emptyset$ for every n, then $\{x_n\}_n$ satisfies condition (A_3).*

To prove this, we use the following lemma.

LEMMA 6 [8, Lemma 8]. *Let $x \in G$ and let E be a closed subset of $M(H^\infty)$ with $P(x) \cap E = \emptyset$. Then for $0 < \epsilon < 1$, there is an interpolating Blaschke product b such that $b(x) = 0$ and $|b| > \epsilon$ on E.*

PROOF OF PROPOSITION 4. By our assumption, there is a sequence $\{U_n\}_n$ of disjoint open subsets of $M(H^\infty)$ such that $P(x_n) \subset U_n$ for every n. Let $\{\epsilon_n\}_n$ be a sequence of positive numbers such that $0 < \epsilon_n < 1$ and $\prod_{n=1}^{\infty}\epsilon_n > 0$. By Lemma 6, there is an interpolating Blaschke product b_n such that $b_n(x_n) = 0$ and $|b_n| > \epsilon_n$ on $M(H^\infty)\setminus U_n$. By considering tails of $b_n, n = 1, 2, \ldots$, we may assume that $\prod_{n=1}^{\infty} b_n \in H^\infty$. Since $U_n \subset M(H^\infty)\setminus U_k$ for $k \neq n$, we have

$$
\Big|\prod_{k:k\neq n} b_k\Big| > \prod_{k:k\neq n}\epsilon_k \qquad \text{on } D\cap U_n \text{ for every } n.
$$

Since x_n is contained in $cl\,(D \cap U_n)$,

$$\left|(\prod_{k:k \neq n} b_k)(x_n)\right| > \prod_{k=1}^{\infty} \epsilon_k \qquad \text{for every } n.$$

Hence $\{x_n\}_n$ satisfies condition (A_3).

In [10], Lingenberg proved that if E is a closed subset of $M(H^{\infty})$ such that $E \subset G$ and $H^{\infty}_{|E} = C(E)$, the space of continuous functions on E, then there is an interpolating Blaschke product b such that $E \subset Z(b)$. Here we have the following problem.

PRPOBLEM 2. If $\{x_n\}_n$ is an interpolating sequence in G, is $cl\,\{x_n\}_n \subset G$ true ?

If the answer of this problem is affirmative, we have that if $\{x_n\}_n$ is interpolating and $\{x_n\}_n \subset G$, then there exists an interpolating Blaschke product b such that $\{x_n\}_n \subset Z(b)$. We have anothor problem relating to Problem 2.

PROBLEM 3. Let $\{x_n\}_n$ be a sequence in G. If $\{x_n\}_n$ satisfies condition (A_2), does $\{x_n\}_n$ satisfy condition (A_3) ?

By Theorem 2 and Lemma 3, it is not difficult to see that if Problem 3 is true then Problem 2 is true. We end this section with the following problem.

PROBLEM 4. Let $\{x_n\}_n$ be a sequence in G. If $\{x_n\}_n$ satisfies condition (A_2), is $\{x_n\}_n$ interpolating ?

4. CONDITON (A_1)

In [6, p. 109], Hoffman gave an example of a nontrivial part which is not a homeomorphic part. We use his example to prove the following theorem.

THEOREM 3. *There exists a sequence $\{x_n\}_n$ satisfing the following conditions.*

(i)   *$\{x_n\}_n$ is contained in a nontrivial part.*
(ii)   *$\{x_n\}_n$ is strongly discrete.*
(iii)   *$\{x_n\}_n$ satisfies condition (A_1).*
(iv)   *$\{x_n\}_n$ is not interpolating.*

PROOF. We work on in the right half plane C_+. Then $S = \{1 + ni\}_n$ is an interpolating sequence for $H^{\infty}(C_+)$. Let b be an interpolating Blaschke product with these zeros. Let the integers operate on S by translation vertically. That gives a group homeomorphism of $cl\,S$;

$$h_k : cl\,S \longrightarrow cl\,S$$

$$h_k(1 + ni) = 1 + (n + k)i.$$

Let K be a closed subset of $cl\, S \setminus S$ which is invariant under h_1 and which is minimal with that property (among closed sets). Let $m \in K$. The sequence

$$m_k = h_k(m), \qquad k = 1, 2, \ldots$$

is invariant under h_1. Therefore

$$(1) \qquad\qquad K = cl\,\{m_k\}_{k \geq N} \qquad \text{for every } N.$$

Let L_m be the Hoffman map from C_+ onto $P(m)$. Then $L_m(1) = m$ and $L_m(1 + ik) = m_k$. Hence by (1), $P(m)$ is not a homeomorphic part. Let $x_n = L_m(1 + 1/n + in)$ for $n = 1, 2, \ldots$. Then

$$(2) \qquad\qquad x_n \in P(m) \qquad \text{for every } n.$$

We note that

$$(3) \qquad\qquad \rho(1 + 1/n + in, 1 + in) \longrightarrow 0 \;(n \to \infty).$$

Since $\{1 + in\}_n$ is an interpolating sequence in C_+, $\{1 + 1/n + in\}_n$ is also interpolating. Hence $\{1 + 1/n + in\}_n$ satisfies condition(A_1). Since L_m preserves ρ-distance [6, p. 103],

$$(4) \qquad\qquad \{x_n\}_n \text{ satisfies condition } (A_1).$$

Since $b = 0$ on K, $b(x_n) \to 0\,(n \to \infty)$. But we have $b(x_n) \neq 0$. Hence

$$(5) \qquad\qquad \{x_n\}_n \text{ is strongly discrete.}$$

To prove that $\{x_n\}_n$ is not interpolating, it is sufficient to prove that $\{x_n\}_n$ does not satisfy (A_2). Suppose that there exists g_n in H^∞ such that $\|g_n\| \leq 1, g_n(x_k) = 0$ for $k \neq n$, and $g_n(x_n) \neq 0$. By (3), we have $\rho(m_k, x_k) \to 0\,(k \to \infty)$. Hence by (1), $g_n = 0$ on K. Therefore

$$|g_n(x_n)| \leq \rho(x_n, m_n) \longrightarrow 0 \;(n \to \infty)$$

This implies that $\{x_n\}_n$ does not satisfy condition (A_2).

REFERENCES

[1] S. Axler and P. Gorkin, Sequences in the maximal ideal space of H^∞, Proc. Amer. Math. Soc. 108(1990), 731-740.

[2] L. Carleson, An interpolation problem for bounded analytic functions, Amer. J. Math. 80(1958), 921-930.

[3] J. Garnett, Bounded analytic functions, Academic Press, New York and London, 1981.

[4] P. Gorkin, H. -M. Lingenberg and R. Mortini, Homeomorphic disks in the spectrum of H^∞, Indiana Univ. Math. J. 39(1990), 961-983.

[5] K. Hoffman, Banach spaces of analytic functions, Prentice Hall, Englewood Cliffs, New Jersey, 1962.

[6] K. Hoffman, Bounded analytic functions and Gleason parts, Ann. of Math. 86(1967), 74-111.

[7] K. Izuchi, Interpolating sequences in a homeomorphic part of H^∞, Proc. Amer. Math. Soc. 111(1991), 1057-1065.

[8] K. Izuchi, Interpolating sequences in the maximal ideal space of H^∞, J. Math. Soc. Japan 43(1991),721-731.

[9] K. Izuchi, Factorization of Blaschke products, to appear in Michigan Math. J.

[10] H. -M. Lingenberg, Interpolation sets in the maximal ideal space of H^∞, Michigan Math. J. 39(1992), 53-63.

Department of Mathematics
Kanagawa University
Yokohama 221, JAPAM

MSC 1991: Primary 30D50, 46J15

Operator Theory:
Advances and Applications, Vol. 59
© 1992 Birkhäuser Verlag Basel

OPERATOR MATRICES WITH CHORDAL INVERSE PATTERNS*

Charles R. Johnson[1] and Michael Lundquist

We consider invertible operator matrices whose conformally partitioned *inverses* have 0 blocks in positions corresponding to a chordal graph. In this event, we describe a) block entry formulae that express certain blocks (in particular, those corresponding to 0 blocks in the inverse) in terms of others, under a regularity condition, and b) in the Hermitian case, a formula for the inertia in terms of inertias of certain key blocks.

INTRODUCTION

For Hilbert spaces $\mathcal{H}_i, i = 1, \cdots, n$, let $\mathcal{H}$ be the Hilbert space defined by $\mathcal{H} = \mathcal{H}_1 \oplus \cdots \oplus \mathcal{H}_n$. Suppose, further, that $A : \mathcal{H} \to \mathcal{H}$ is a linear operator in matrix form, partitioned as

$$A = \begin{bmatrix} A_{11} & A_{12} & \ldots & A_{1n} \\ A_{21} & & & \\ \vdots & & & \vdots \\ A_{n1} & & \ldots & A_{nn} \end{bmatrix},$$

in which $A_{ij} : \mathcal{H}_j \to \mathcal{H}_i$, $i, j = 1, \cdots, n$. (We refer to such an A as an *operator matrix*.) We assume throughout that A is invertible and that $A^{-1} = B = [B_{ij}]$ is partitioned conformably. We are interested in the situation in which some of the blocks B_{ij} happen to be zero. In this event we present (1) some relations among blocks of A (under a further regularity condition) and (2) a formula for the inertia of A, in terms of that of certain principal submatrices, when A is Hermitian. For this purpose we define an undirected graph $G = G(B)$ on vertex set $N \equiv \{1, \cdots, n\}$ as follows: there is an edge $\{i, j\}, i \neq j$, in $G(B)$ unless both B_{ij} and B_{ji} are 0.

An undirected graph G is called *chordal* if no subgraph induced by 4 or more vertices is a cycle. Note that if $G(B)$ is not complete, then there are chordal graphs

*This manuscript was prepared while both authors were visitors at the Institute for Mathematics and its Applications, Minneapolis, Minnesota.

[1] The work of this author was supported by National Science Foundation grant DMS90-00839 and by Office of Naval Research contract N00014-90-J-1739.

G (that are also not complete) such that $G(B)$ is contained in G. Thus, if there is any symmetric sparsity in B, our results will apply (perhaps by ignoring the fact that some blocks are 0), even if $G(B)$ is not chordal.

A *clique* in an undirected graph G is a set of vertices whose vertex induced subgraph in G is complete (i.e. contains all possible edges $\{i,j\}, i \neq j$). A clique is *maximal* if it is not a proper subset of any other clique. Let $\mathcal{C} = \mathcal{C}(G) = \{\alpha_1, \cdots, \alpha_p\}$ be the collection of maximal cliques of the graph G. The *intersection graph* $\mathcal{G}$ of the maximal cliques is an undirected graph with vertex set $\mathcal{C}$ and an edge between α_i and α_j, $i \neq j$ if $\alpha_i \cap \alpha_j \neq \phi$. The graph G is connected and chordal if and only if $\mathcal{G}$ has a *spanning tree* $\mathcal{T}$ that satisfies the *intersection property*: $\alpha_i \cap \alpha_j \subseteq \alpha_k$ whenever α_k lies on the unique simple path in $\mathcal{T}$ from α_i to α_j. Such a tree $\mathcal{T}$ is called a *clique tree* for G and is generally not unique [2]. (See [3] for general background facts about chordal graphs.) Clique trees constitute an important tool for understanding the structure of a chordal graph. For example, for a pair of nonadjacent vertices u, v in G, a u, v *separator* is a set of vertices of G whose removal (along with all edges incident with them) leaves u and v in different connected components of the result. A u, v separator is called *minimal* if no proper subset of it is a u, v separator. A set of vertices is called a *minimal vertex separator* if it is a minimal u, v separator for some pair of vertices u, v. (Note that it is possible for a proper subset of a minimal vertex separator to also be a minimal vertex separator.) If α_i and α_j are adjacent cliques in a clique tree for a chordal graph G then $\alpha_i \cap \alpha_j$ is a minimal vertex separator for G. The collection of such intersections (including multiplicities) turns out to be independent of the clique tree and all minimal vertex separators for G occur among such intersections.

Given an n-by-n operator matrix $A = (A_{ij})$, we denote the operator submatrix lying in block rows $\alpha \subseteq N$ and block columns $\beta \subseteq N$ by $A[\alpha, \beta]$. When the submatrix is principal (i.e. $\beta = \alpha$), we abbreviate $A[\alpha, \alpha]$ to $A[\alpha]$.

We define the *inertia* of an Hermitian operator B on a Hilbert space $\mathcal{K}$ as follows. The triple $i(B) = (i_+(B), i_-(B), i_o(B))$ has components defined by

$i_+(B) \equiv$ the maximum dimension of an invariant subspace of B on which the quadratic form is positive.

$i_-(B) \equiv$ the maximum dimension of an invariant subspace of B on which the quadratic form is negative.

and

$i_o(B) \equiv$ the dimension of the kernel of B (ker B).

Each component of $i(B)$ may be a nonnegative integer, or ∞ in case the relevant dimension is not finite. We say that two Hermitian operators B_1 and B_2 on $\mathcal{K}$ are *congruent* if there is an invertible operator $C : \mathcal{K} \to \mathcal{K}$ such that

$$B_2 = C^* B_1 C.$$

According to the spectral theorem, if a bounded linear operator $A : \mathcal{H} \to \mathcal{H}$ is Hermitian, then A is unitarily congruent (similar) to a direct sum:

$$U^* A U = \begin{bmatrix} A_+ & 0 & 0 \\ 0 & A_- & 0 \\ 0 & 0 & 0 \end{bmatrix},$$

in which A_+ is positive definite and A_- is negative definite. As $i(A) = i(U^* A U)$, $i_+(A)$ is the "dimension" of the direct summand A_+, $i_-(A)$ the dimension of A_-, and $i_o(A)$ the dimension of the 0 direct summand, including the possibility of ∞ in each case. It is easily checked that the following three statements are then equivalent:

(i) A is congruent to $\begin{bmatrix} I & 0 & 0 \\ 0 & -I & 0 \\ 0 & 0 & 0 \end{bmatrix}$, in which the sizes of the diagonal blocks are $i_+(A)$, $i_-(A)$ and $i_0(A)$, respectively;

(ii) each of A_+ and A_- is invertible;

and

(iii) A has closed range.

We shall frequently need to make use of congruential representations of the form (i) and, so, assume throughout that each key principal submatrix (i.e. those corresponding to maximal cliques and minimal separators in the chordal graph G of the inverse of an invertible Hermitian matrix) has closed range. This may be a stronger assumption than is necessary for our formulae in section 3; so there is an open question here.

Chordal graphs have played a key role in the theory of positive definite completions of matrices and in determinantal formulae. For example, in [4] it was shown that if the undirected graph of the specified entries of a partial positive definite matrix (with specified diagonal) is chordal, then a positive definite completion exists. (See e.g. [6] for definitions and background.) Furthermore, if the graph of the specified entries is not chordal, then there is a partial positive definite matrix for which there is no positive definite completion. (These facts carry over in a natural way to operator matrices.) If there is a positive definite completion, then there is a unique determinant maximizing one that is characterized by having 0's in the *inverse* in all positions corresponding to originally unspecified entries. Thus, if the graph of the specified entries is chordal, then the ordinary (undirected) graph of the inverse of the determinant maximizer is (generically) the same chordal graph. (In the partial positive definite operator matrix case such a zeros in-the-inverse completion still exists when the data is chordal and is an open question otherwise.) This was one of the initial motivations for studying matrices with chordal inverse (nonzero) patterns. Other motivation includes the structure of inverses of banded matrices, and this is background for section 2.

If an invertible matrix A has an inverse pattern contained in a chordal graph G, then $\det A$ may be expressed in terms of certain key principal minors [1], as long as all relevant minors are nonzero:

$$\det A = \frac{\prod\limits_{\alpha \in \mathcal{C}} \det A[\alpha]}{\prod\limits_{\{\alpha,\beta\} \in \mathcal{E}} \det A[\alpha \cap \beta]}.$$

Here $\mathcal{C}$ is the collections of maximal cliques of G, and $\mathcal{T} = (\mathcal{C}, \mathcal{E})$ is a clique tree for G. Thus, the numerator is the product of principal minors associated with maximal cliques, while the denominator has those associated with minimal vertex separators (with proper multiplicities). There is no natural analog of this determinantal formula in the operator case, but the inertia formula presented in section 3 has a logarithmic resemblance to it.

2. ENTRY FORMULAE

Let $G = (N, E)$ be a chordal graph. We will say that an operator matrix $A = [A_{ij}]$ is *G-regular* if $A[\alpha]$ is invertible whenever $\alpha \subseteq V$ is either a maximal clique of G or a minimal vertex separator of G. In this section we will establish explicit formulae for some of the block entries of A when $G(A^{-1}) \subseteq G$. Specifically, those entries are the ones corresponding to edges that are *absent* from E (see Theorem 3).

LEMMA 1. *Let $A = [A_{ij}]$ be a 3-by-3 operator matrix, and assume that*

$$M_1 = \begin{bmatrix} A_{11} & A_{12} \\ A_{21} & A_{22} \end{bmatrix}, \quad M_2 = \begin{bmatrix} A_{22} & A_{23} \\ A_{32} & A_{33} \end{bmatrix} \text{ and } A_{22} \text{ are each invertible.}$$

Then $B = A^{-1}$ exists and satisfies $B_{13} = 0$ if and only if $A_{13} = A_{12} A_{22}^{-1} A_{23}$.

Proof. Let us compute the Schur complement of A_{22} in A:

(1)
$$\begin{bmatrix} I & -A_{12}A_{22}^{-1} & 0 \\ 0 & I & 0 \\ 0 & -A_{32}A_{22}^{-1} & I \end{bmatrix} \begin{bmatrix} A_{11} & A_{12} & A_{13} \\ A_{21} & A_{22} & A_{23} \\ A_{31} & A_{32} & A_{33} \end{bmatrix} \begin{bmatrix} I & 0 & 0 \\ -A_{22}^{-1}A_{21} & I & -A_{22}^{-1}A_{23} \\ 0 & 0 & I \end{bmatrix}$$

$$= \begin{bmatrix} A_{11} - A_{12}A_{22}^{-1}A_{21} & 0 & A_{13} - A_{12}A_{22}^{-1}A_{23} \\ 0 & A_{22} & 0 \\ A_{31} - A_{32}A_{22}^{-1}A_{21} & 0 & A_{33} - A_{32}A_{22}^{-1}A_{23} \end{bmatrix}.$$

If $B = A^{-1}$ exists, then

(2)
$$\begin{bmatrix} B_{11} & B_{13} \\ B_{31} & B_{33} \end{bmatrix} = \begin{bmatrix} A_{11} - A_{12}A_{22}^{-1}A_{21} & A_{13} - A_{12}A_{22}^{-1}A_{23} \\ A_{31} - A_{32}A_{22}^{-1}A_{21} & A_{33} - A_{32}A_{22}^{-1}A_{23} \end{bmatrix}^{-1},$$

and hence if $B_{13} = 0$, then necessarily we have $A_{13} = A_{12}A_{22}^{-1}A_{23}$. Conversely, if $A_{13} = A_{12}A_{22}^{-1}A_{23}$, then the (1,3) entry of the matrix on the right-hand side of (1) is zero. Note that $A_{11} - A_{12}A_{22}^{-1}A_{21}$ and $A_{33} - A_{32}A_{22}^{-1}A_{23}$ are invertible, because they are the Schur complements of A_{22} in M_1 and M_2. Hence A is invertible, and by (2) we have $B_{13} = 0$. $\square$

Under the conditions of the preceding lemma, if we would like $B_{31} = 0$ then we must also have $A_{31} = A_{32}A_{22}^{-1}A_{21}$. Notice that the graph of B in this case is a path:

$$G = \text{①——②——③}.$$

Suppose now that $A = [A_{ij}]$ is an invertible $n \times n$ operator matrix, that $1 < k \le m < n$, and that $A^{-1} = [B_{ij}]$ satisfies $B_{ij} = 0$ and $B_{ji} = 0$ whenever $i < k$ and $j > m$. In this case B has the block form

$$B = \begin{bmatrix} \widetilde{B}_{11} & \widetilde{B}_{12} & 0 \\ \widetilde{B}_{21} & \widetilde{B}_{22} & \widetilde{B}_{23} \\ 0 & \widetilde{B}_{32} & \widetilde{B}_{33} \end{bmatrix},$$

in which $\widetilde{B}_{11} = B[\{1,\ldots,k-1\}]$, $\widetilde{B}_{22} = B[\{k,\ldots,m\}]$ and $\widetilde{B}_{33} = B[\{m+1,\ldots,m\}]$. Let $A = [\tilde{A}_{ij}]$ be partitioned conformably. If in addition to the above conditions we also have that $A[\{1,\ldots,m\}]$, $A[\{k,\ldots,n\}]$ and $A[\{k,\ldots,m\}]$ are invertible, then we simply have the case covered in the preceding Lemma, and we may deduce that

$$A[\{1,\ldots,k-1\},\{m+1,\ldots,n\}]$$

$$= A[\{1,\ldots,k-1\},\{k,\ldots,m\}]\, A[\{k,\ldots,m\}]^{-1} A[\{k,\ldots,m\},\{m+1,\ldots,n\}],$$

with a similar formula holding for $A[\{m+1,\ldots,n\},\{1,\ldots,k-1\}]$. From this we may write explicit formulae for individual entries in A. For example, we may express any entry

A_{ij} for which $i < k$ and $j > m$ as

$$(3) \qquad A_{ij} = A[\{i\}, \{k, \ldots, m\}]\, A[\{k, \ldots, m\}]^{-1} A[\{k, \ldots, m\}, \{j\}].$$

There is an obvious similarity between this situation and that covered in Lemma 1, which one sees simply by looking at the block structure of A^{-1}. But there are also some similarities which may be observed by looking at graphs. In the block case we just considered, the graph $G(B)$ is a chordal graph consisting of exactly two maximal cliques, the sets $\alpha_1 = \{1, \ldots, m\}$ and $\alpha_2 = \{k, \ldots, n\}$. The intersection $\beta = \{k, \ldots, m\}$ of α_1 and α_2 is a minimal vertex separator of G (in fact, the only minimal vertex separator in this graph). The formula (3) may then be written

$$(4) \qquad A_{ij} = A[\{i\}, \beta] A[\beta]^{-1} A[\beta, \{j\}].$$

Note now in the 3-by-3 case that the equation $A_{13} = A_{12} A_{22}^{-1} A_{23}$ has the same form as (4) when we let $\beta = \{2\}$. In fact, since $\beta = \{2\}$ is a minimal separator of the vertices 1 and 3 in the graph

$$\textcircled{1}\!-\!\textcircled{2}\!-\!\textcircled{3}$$

we see that $\{2\}$ plays the same role in the 3-by-3 case as $\{k, \ldots, m\}$ does in the $n \times n$ case.

In Theorem 3 we will encounter expressions of the form

$$(5) \qquad A_{ij} = A[\{i\}, \beta_1]\, A[\beta_1]^{-1} A[\beta_1, \beta_2]\, A[\beta_2]^{-1} \cdots A[\beta_m]^{-1} A[\beta_m, \{j\}]$$

in which each β_k is a minimal vertex separator in a chordal graph. The sequence $(\beta_1, \ldots, \beta_m)$ is obtained by looking at a clique tree for the chordal graph, identifying a path $(\alpha_0, \alpha_1, \ldots, \alpha_m)$ in the tree, and setting $\beta_k = \alpha_{k-1} \cap \alpha_k$.

These expressions turn out to be the natural generalization of (4) to cases in which the graph of B is any chordal graph. In addition, the results of this section generalize results of [9] from the scalar case to the operator case.

LEMMA 2. *Let $A : \mathcal{H} \to \mathcal{H}$ be an invertible operator matrix, with $B = A^{-1}$. Let $G = (N, E)$ be the undirected graph of B. Let $\{i, j\} \notin E$ and let $\beta \subseteq N$ be any i, j separator for which $A[\beta]$ is invertible. Then*

$$A_{ij} = A[\{i\}, \beta] A[\beta]^{-1} A[\beta, \{j\}].$$

Proof. Without loss of generality we may assume that $\beta = \{k, \ldots, m\}$, with $k \leq m$, and that β separates any vertices r and s for which $r < k$ and $s > m$. Assuming then that $i < k$ and $j > m$, we may write B as

$$\begin{bmatrix} \widetilde{B}_{11} & \widetilde{B}_{12} & 0 \\ \widetilde{B}_{21} & \widetilde{B}_{22} & \widetilde{B}_{23} \\ 0 & \widetilde{B}_{22} & \widetilde{B}_{33} \end{bmatrix}.$$

The result now follows from Lemma 1 and the remarks that follow it. $\square$

If G is chordal and i and j are nonadjacent vertices then an i, j *clique path* will mean a path in any clique tree associated with G that joins a clique containing vertex i to a clique containing vertex j. One important property of any i, j clique path is that it will "contain" every minimal i, j separator in the following sense: If $(\alpha_0, \ldots, \alpha_m)$ is any i, j clique path, and if β is any minimal i, j separator then $\beta = \alpha_{k-1} \cap \alpha_k$ for some $k, 1 \leq k \leq m$. Another important property of an i, j clique path is that every set $\beta_k = \alpha_{k-1} \cap \alpha_k$, $1 \leq k \leq m$, is an i, j separator. It is *not* the case, however, that every β_k is a minimal i, j separator (see [9]).

THEOREM 3. *Let $G = (N, E)$ be a connected chordal graph, and let $A : \mathcal{H} \to \mathcal{H}$ be a G-regular operator matrix. then the following assertions are equivalent:*

(i) *A is invertible and $G(A^{-1}) \subseteq G$;*

(ii) *for every $\{i, j\} \notin E$ there exists a minimal i, j separator β such that*

$$A_{ij} = A[\{i\}, \beta] \, A[\beta]^{-1} A[\beta, \{j\}];$$

(iii) *for every $\{i, j\} \notin E$ and every minimal i, j separator β we have*

$$A_{ij} = A[\{i\}, \beta] \, A[\beta]^{-1} A[\beta, \{j\}];$$

(iv) *for every $\{i, j\} \notin E$, every i, j clique path $(\alpha_0, \alpha_1, \ldots, \alpha_m)$ and any $k, 1 \leq k \leq m$ we have*

$$A_{ij} = A[\{i\}, \beta_k] \, A[\beta_k]^{-1} A[\beta_k, \{j\}],$$

in which $\beta_k = \alpha_{k-1} \cap \alpha_k$;

and

(v) *for every $\{i, j\} \notin E$ and every i, j clique path $(\alpha_0, \alpha_1, \ldots \alpha_m)$ we have*

$$A_{ij} = A[\{i\}, \beta_1] \, A[\beta_1]^{-1} A[\beta_1, \beta_2] \cdots A[\beta_m]^{-1} A[\beta_m, \{j\}],$$

in which $\beta_k = \alpha_{k-1} \cap \alpha_k$.

Proof. We will establish the following implications:

$$(iv) \implies (iii) \implies (ii) \implies (iv);$$
$$(i) \iff (iv) \iff (v).$$

$(iv) \implies (iii)$ follows from the observation that every minimal i, j separator equals β_k for some k, $1 \le k \le m$.

$(iii) \implies (ii)$ is immediate.

For $(ii) \implies (iv)$, let $\{i, j\} \notin E$, and let $(\alpha_0, \alpha_1, \ldots, \alpha_m)$ be a shortest i, j clique path. We will induct on m. For $m = 1$ there is nothing to show, since in this case $\beta_1 = \alpha_0 \cap \alpha_1$ is the only minimal i, j separator. Now let $m \ge 2$, and suppose that (iv) holds for all nonadjacent pairs of vertices for which the shortest clique path has length less than m. Since every minimal i, j separator equals β_k for some k, we have, by (ii),

$$A_{ij} = A[\{i\}, \beta_k] A[\beta_k]^{-1} A[\beta_k, \{j\}]$$

for some k, $1 \le k \le m$. It will therefore suffice to show that for $k = 1, 2, \ldots, m-1$ we have

$$(6) \qquad A[\{i\}, \beta_k] A[\beta_k]^{-1} A[\beta_k, \{j\}] = A[\{i\}, \beta_{k+1}] A[\beta_{k+1}]^{-1} A[\beta_{k+1}, \{j\}]$$

Let us first observe that for $k = 1, \ldots, m-1$,

$$(7) \qquad A[\beta_k, \{j\}] = A[\beta_k, \beta_{k+1}] A[\beta_{k+1}]^{-1} A[\beta_{k+1}, \{j\}].$$

Indeed, suppose $r \in \beta_k$. Then $(\alpha_k, \alpha_{k+1}, \ldots, \alpha_m)$ is an r, j clique path of length $m - k$, and by the induction hypothesis we may write

$$A_{rj} = A[\{r\}, \beta_{k+1}] A[\beta_{k+1}]^{-1} A[\beta_{k+1}, \{j\}],$$

and equation (7) follows. A similar argument shows that for $k = 2, \ldots, m$ we have

$$(8) \qquad A[\{i\}, \beta_{k+1}] = A[\{i\}, \beta_k] A[\beta_k]^{-1} A[\beta_k, \beta_{k+1}].$$

By (7) and (8), both sides of (6) are equal to

$$A[\{i\}, \beta_k] A[\beta_k]^{-1} A[\beta_k, \beta_{k+1}] A[\beta_{k+1}]^{-1} A[\beta_{k+1}, \{j\}],$$

and hence (6) holds, as required.

$(i) \implies (iv)$ follows from Lemma 2.

For $(iv) \implies (i)$, let the maximal cliques of G be $\alpha_1, \alpha_2, \ldots, \alpha_p$, $p \geq 2$. We will induct on p. In case $p = 2$ then the result follows from Lemma 1, so let $p > 2$ and suppose that the implication holds whenever the maximal cliques number fewer than p. Let $\mathcal{T}$ be a clique tree associated with G, let $\{\alpha_k, \alpha_{k+1}\}$ be any edge of $\mathcal{T}$, and suppose the vertex sets of the two connected components of $\mathcal{T} - \{\alpha_k, \alpha_{k+1}\}$ are $\mathcal{C}_1 = \{\alpha_1, \ldots, \alpha_k\}$ and $\mathcal{C}_2 = \{\alpha_{k+1}, \ldots, \alpha_p\}$. Set $V_1 = \cup_{i=1}^{k}\alpha_i$ and $V_2 = \cup_{i=k+1}^{p}\alpha_i$. (Let G_V be the subgraph of G induced by the subset V of vertices.) Since induced subgraphs of a chordal graph are necessarily chordal, G_{V_1} and G_{V_2} are chordal graphs, and since (iv) holds for the matrix A, (iv) holds as well for $A[V_1]$ and $A[V_2]$. By the induction hypothesis, $A[V_1]$ and $A[V_2]$ are invertible. Note also that $V_1 \cap V_2 = \alpha_k \cap \alpha_{k+1}$, which follows from the intersection property. Since $A[V_1 \cap V_2]$ is invertible, we may now apply Lemma 1 to the matrix A (in which A_{11} is replaced by $A[V_1 \setminus V_2]$, A_{22} by $A[V_1 \cap V_2]$ and A_{33} by $A[V_2 \setminus V_1]$), and conclude that $A^{-1}[V_1 \setminus V_2, V_2 \setminus V_1] = 0$ and $A^{-1}[V_2 \setminus V_1, V_1 \setminus V_2] = 0$. In other words, if we set $B = A^{-1}$ then $B_{ij} = 0$ and $B_{ji} = 0$ whenever $i \in V_1 \setminus V_2$ and $j \in V_2 \setminus V_1$. Now if $\{i, j\} \notin E$ then α_k and α_{k+1} may be chosen (renumbering the α's if necessary) so that $i \in V_1 \setminus V_2$ and $j \in V_2 \setminus V_1$. Hence it must be that $B_{ij} = 0$ and $B_{ji} = 0$ whenever $\{i, j\} \notin E$.

For $(iv) \implies (v)$, let $\{i, j\} \notin E$, and let $(\alpha_0, \alpha_1, \ldots, \alpha_m)$ be any i, j clique path. First, we must observe that for any $k, 1 \leq k \leq m$,

$$(9) \qquad A[\beta_k, \{j\}] = A[\beta_k, \beta_{k+1}]A[\beta_{k+1}]^{-1}A[\beta_{k+1}, \{j\}].$$

Indeed, by assumption, for any $r \in \beta_k$ we have

$$A_{rj} = A[\{r\}, \beta_{k+1}]A[\beta_{k+1}]^{-1}A[\beta_{k+1}, \{j\}],$$

and (9) follows from this. By successively applying (9) we obtain

$$\begin{aligned}
A_{ij} &= A[\{i\}, \beta_1]A[\beta_1]^{-1}A[\beta_1, \{j\}] \\
&= A[\{i\}, \beta_1]A[\beta_1]^{-1}A[\beta_1, \beta_2]A[\beta_2]^{-1}A[\beta_2, \{j\}] \\
&\ \ \vdots \\
&= A[\{i\}, \beta_1]A[\beta_1]^{-1}A[\beta_1, \beta_2] \cdots A[\beta_{m-1}, \beta_m]A[\beta_m]^{-1}A[\beta_m, \{j\}],
\end{aligned}$$

as required.

For $(v) \implies (iv)$, let $\{i,j\} \notin E$, and let $(\alpha_0, \ldots, \alpha_m)$ be an i,j clique path. Let $r \in \alpha_{k-1}$, $1 < k \le m$. We may write [because of assumption (v)]

$$A_{ir} = A[\{i\}, \beta_1] A[\beta_1]^{-1} \cdots A[\beta_{k-1}]^{-1} A[\beta_{k-1}, \{r\}],$$

and because $r \in \beta_k \implies r \in \alpha_k$ we thus have

$$(10) \qquad A[\{i\}, \beta_k] = A[\{i\}, \beta_1] A[\beta_1]^{-1} \cdots A[\beta_{k-1}]^{-1} A[\beta_{k-1}, \beta_k].$$

It may be similarly shown that

$$(11) \qquad A[\beta_k, \{j\}] = A[\beta_k, \beta_{k+1}] A[\beta_{k+1}]^{-1} \cdots A[\beta_m]^{-1} A[\beta_m, \{j\}].$$

By using (10) and (11) we therefore obtain

$$\begin{aligned} A_{ij} &= A[\{i\}, \beta_1] \cdots A[\beta_{k-1}, \beta_k] A[\beta_k]^{-1} A[\beta_k, \beta_{k+1}] \cdots A[\beta_m, \{j\}] \\ &= A[\{i\}, \beta_k] A[\beta_k]^{-1} A[\beta_k, \{j\}], \end{aligned}$$

as required. $\square$

3. INERTIA FORMULA

In [8], it was shown that if $A \in M_n(\mathbb{C})$ is an invertible Hermitian matrix and if $G = G(A^{-1})$ is a chordal graph, then the inertia of A may be expressed in terms of the inertias of certain principal submatrices of A. Precisely, let $\mathcal{C}$ denote the collection of maximal cliques of G, and let $\mathcal{T} = (\mathcal{C}, \mathcal{E})$ be a clique tree associated with G. If $G(A^{-1}) = G$, then it turns out that

$$(11) \qquad i(A) = \sum_{\alpha \in \mathcal{C}} i(A[\alpha]) - \sum_{\{\alpha, \beta\} \in \mathcal{E}} i(A[\alpha \cap \beta]).$$

It is helpful to think of (11) as a generalization of the fact that if A^{-1} is block diagonal (meaning, of course, that A is block diagonal) then the inertia of A is simply the sum of the inertias of the diagonal blocks of A. To see what (11) tells us in a specific case, suppose that A^{-1} has a pentadiagonal nonzero-pattern, as in

$$A^{-1} \sim \begin{bmatrix} X & X & X & & & \\ X & X & X & X & & \\ X & X & X & X & X \\ & X & X & X & X \\ & & X & X & X \end{bmatrix}$$

The graph of A^{-1} is then

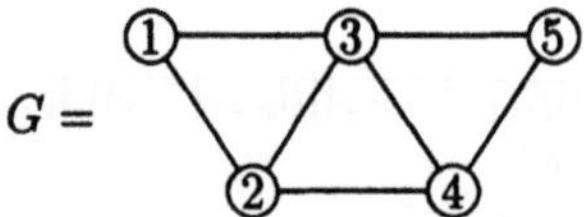

which is chordal. The maximal cliques of G are $\alpha_1 = \{1, 2, 3\}$, $\alpha_2 = \{2, 3, 4\}$ and $\alpha_3 = \{3, 4, 5\}$, and the clique tree associated with the graph G is

$$\textcircled{α_1} \!\!-\!\! \textcircled{α_2} \!\!-\!\! \textcircled{α_3}.$$

Equation (11) now tells us that the inertia of A is given by

$$i(A) = i(A[\{1,2,3\}]) + i(A[\{2,3,4\}]) + i(A[\{3,4,5\}])$$
$$- i(A[\{2,3\}]) - i(A[\{3,4\}]).$$

Thus, we may compute the inertia of A by adding the inertia of these submatrices:

$$\begin{bmatrix} X & X & X & X & X \\ X & X & X & X & X \\ X & X & X & X & X \\ X & X & X & X & X \\ X & X & X & X & X \end{bmatrix}$$

and subtracting the inertias of these:

$$\begin{bmatrix} X & X & X & X & X \\ X & X & X & X & X \\ X & X & X & X & X \\ X & X & X & X & X \\ X & X & X & X & X \end{bmatrix}$$

Our goal in this section is to generalize formula (11) to the case in which $A = [A_{ij}]$ is an invertible n-by-n Hermitian operator matrix. We will be concerned with the case in which one of the components of inertia is finite, so that in (11) we will replace i by i_+, i_- or i_o.

For a chordal graph $G = (N, E)$, we will say that an invertible n-by-n operator matrix A is *weakly G-regular* (or simply *weakly regular*) if for every maximal clique or minimal vertex separator α both $A[\alpha]$ and $A^{-1}[\alpha^c]$ have closed range.

LEMMA 4. *Let* $M : \mathcal{H}_1 \oplus \mathcal{H}_2 \to \mathcal{H}_1 \oplus \mathcal{H}_2$ *be represented by the 2-by-2 matrix*

$$M = \begin{bmatrix} A & B \\ C & D \end{bmatrix}.$$

Suppose that A is invertible, and that

$$M^{-1} = \begin{bmatrix} P & Q \\ R & S \end{bmatrix}.$$

Then $\dim \ker A = \dim \ker S$.

Proof. Let $x_1, x_2, \ldots, x_n$ be linearly independent elements of $\ker A$. Then for $1 \le k \le n$ we have

$$\begin{bmatrix} A & B \\ C & D \end{bmatrix} \begin{bmatrix} x_k \\ O \end{bmatrix} = \begin{bmatrix} O \\ y_k \end{bmatrix},$$

in which $y_k = Cx_k$, $k = 1, \ldots, n$. Since M is invertible it follows that $y_1, y_2, \ldots, y_n$ are linearly independent. Observe now that

$$\begin{bmatrix} P & Q \\ R & S \end{bmatrix} \begin{bmatrix} O \\ y_k \end{bmatrix} = \begin{bmatrix} x_k \\ O \end{bmatrix},$$

from which it follows that $y_k \in \ker S$. It follows now that $\dim \ker S \ge \dim \ker A$; by reversing the argument we find that $\dim \ker A \ge \dim \ker S$. Thus $\dim \ker A = \dim \ker S$. $\square$

LEMMA 5. *Let* $M : \mathcal{H}_1 \oplus \mathcal{H}_2 \to \mathcal{H}_1 \oplus \mathcal{H}_2$ *be Hermitian and invertible, and suppose that*

$$M = \begin{bmatrix} A & B \\ B^* & C \end{bmatrix}.$$

If $i_+(M) < \infty$, then $i_o(A) < \infty$.

Proof. Clearly $i_+(A) < \infty$, so let $n = i_+(A)$. Let H be an invertible operator for which $H^*AH = I_n \oplus -I \oplus O$, in which I_n denotes the identity operator on an n-dimensional subspace, and $-I$ and O are operators on spaces of respective dimensions $i_-(A)$ and $i_o(A)$. Then

$$\begin{bmatrix} H^* & O \\ O & I \end{bmatrix} \begin{bmatrix} A & B \\ B^* & C \end{bmatrix} \begin{bmatrix} H & 0 \\ O & I \end{bmatrix} = \begin{bmatrix} I_n & & & B_1 \\ & -I & & B_2 \\ & & O & B_3 \\ B_1^* & B_2^* & B_3^* & C \end{bmatrix}.$$

We may reduce this further by another congruence:

$$\begin{bmatrix} I_n & & & \\ & I & & \\ & & I & \\ -B_1^* & B_2^* & O & I \end{bmatrix} \begin{bmatrix} I_n & & & B_1 \\ & -I & & B_2 \\ & & O & B_3 \\ B_1^* & B_2^* & B_3^* & C \end{bmatrix} \begin{bmatrix} I_n & & & -B_1 \\ & I & & B_2 \\ & & I & O \\ & & & I \end{bmatrix}$$

$$= \begin{bmatrix} I_n & & & O \\ & -I & & O \\ & & O & B_3 \\ O & O & B_3 & S \end{bmatrix},$$

in which $S = C - B_1^* B_1 + B_2^* B_2$. Hence

$$i_+(M) = n + i_+\left(\begin{bmatrix} O & B_3 \\ B_3^* & S \end{bmatrix}\right),$$

and thus

$$i_+\left(\begin{bmatrix} O & B_3 \\ B_3^* & S \end{bmatrix}\right) < \infty.$$

But this implies that the zero block in this matrix must act on a space of finite dimension. Recalling that this dimension equals $i_o(A)$, we obtain the desired conclusion. $\square$

The following Lemma generalizes a result of [5] to operator matrices from the finite-dimensional case (see also [8]).

LEMMA 6. *Let* $M : \mathcal{H}_1 \oplus \mathcal{H}_2 \to \mathcal{H}_1 \oplus \mathcal{H}_2$ *be Hermitian and invertible, with*

$$M = \begin{bmatrix} A & B \\ B^* & C \end{bmatrix} \quad and \quad M^{-1} = \begin{bmatrix} P & Q \\ Q^* & R \end{bmatrix}.$$

If $i_+(M) < \infty$, *and if* A *and* R *both have closed range, then*

$$i_+(M) = i_+(A) + i_o(A) + i_+(R).$$

Proof. If $i_o(A) = 0$ then A is invertible and the result follows from the fact that R is the inverse of the Schur complement $C - B^* A^{-1} B$ and that $i_+(M) = i_+(A) + i_+(C - B^* A^{-1} B)$. Hence, suppose that $i_o(A) > 0$. Since $i_+(M) < \infty$ we have as well from Lemma 5 that $i_o(A) < \infty$.

Hence, let $n = i_o(A)$, and let us consider the special case in which $R = O$. Since we require, by Lemma 4, that $i_o(R) = i_o(A) = n$, R must act on an n-dimensional space. Hence we have

$$M^{-1} = \begin{bmatrix} P & Q \\ Q^* & O_n \end{bmatrix}$$

where O_n denotes the zero operator on n-dimensional Hilbert space. By an appropriately chosen congruence of the form $T_1 = H \oplus I$, we may reduce M to the form

$$M_1 = T_1^* M T_1 = \begin{bmatrix} I_k & & & B_1 \\ & -I & & B_2 \\ & & O_n & B_3 \\ B_1^* & B_2^* & B_3^* & C \end{bmatrix},$$

where $k = i_+(A)$. With

$$T_2 = \begin{bmatrix} I_k & & & -B_1 \\ & -I & & B_2 \\ & & I_n & O \\ & & & I \end{bmatrix},$$

we then have

$$M_2 = T_2^* M_1 T_2 = \begin{bmatrix} I_k & & & O \\ & -I & & O \\ & & O_n & B_3 \\ O & O & B_3^* & S \end{bmatrix},$$

in which $S = C - B_1^* B_1 + B_2^* B_2$. The matrix

$$\begin{bmatrix} O_n & B_3 \\ B_3^* & S \end{bmatrix}$$

is an invertible operator on a $2n$-by-$2n$ Hilbert space, and in this case its inertia must be $(n, n, 0)$. From the form of M_2 we see that we must have

$$\begin{aligned}
i_+(M) &= k + i_+\left(\begin{bmatrix} O_k & B_3 \\ B_3^* & S \end{bmatrix}\right) \\
&= k + n \\
&= i_+(A) + i_o(A).
\end{aligned}$$

Since $i_+(R) = 0$, this last expression equals $i_+(A) + i_o(A) + i_+(R)$.

Now let us consider the general case, in which we make no assumption concerning the dimension of the space on which R acts. Choose an invertible matrix of the form $T_1 = I \oplus H$ so that $T_1^* M^{-1} T_1$ has the form

$$M_1^{-1} = T_1^* M^{-1} T_1 = \begin{bmatrix} P & Q_1 & Q_2 & Q_3 \\ Q_1^* & I_\ell & & \\ Q_2^* & & -I & \\ Q_3^* & & & O_n \end{bmatrix},$$

in which $\ell = i_+(R)$ and $n = i_o(R)\ [= i_o(A)]$. Then with

$$T_2 = \begin{bmatrix} I & & & \\ -Q_1^* & I_\ell & & \\ Q_2^* & & I & \\ O & & & I_n \end{bmatrix},$$

we obtain

$$M_2^{-1} = T_2^* M_1^{-1} T_2 = \begin{bmatrix} S & O & O & Q_3 \\ O & I_\ell & & \\ O & & -I & \\ Q_3^* & & & O_n \end{bmatrix}.$$

From the form of M_2^{-1}, and by simple calculations, we find that $M_2 = T_2^{-1}T_1^{-1}M(T_1^{-1})^*(T_2^{-1})^*$ has the form

$$M_2 = \begin{bmatrix} A & O & O & B_2 \\ O & I_\ell & & \\ O & & -I & \\ B_2^* & & & C_2 \end{bmatrix}$$

for some operators B_2 and C_2. Hence we have

$$(12) \qquad i_+(M) = i_+(R) + i_+\left(\begin{bmatrix} A & B_2 \\ B_2^* & C_2 \end{bmatrix}\right).$$

Observe that

$$\begin{bmatrix} A & B_2 \\ B_2^* & C_2 \end{bmatrix}^{-1} = \begin{bmatrix} S & Q_3 \\ Q_3^* & O_n \end{bmatrix},$$

and thus by the special case we considered previously,

$$(13) \qquad i_+\left(\begin{bmatrix} A & B_2 \\ B_2^* & C_2 \end{bmatrix}\right) = i_+(A) + i_o(A).$$

Thus combining (12) and (13) we obtain

$$i_+(M) = i_+(A) + i_o(A) + i_+(R),$$

as required. $\square$

The following lemma will be used in the proof of the main result of this section. First, let $G = (V, E)$ be any connected chordal graph, and let $\mathcal{T} = (\mathcal{C}, \mathcal{E})$ be any clique tree associated with G. For any pair of maximal cliques α and β that are adjacent in $\mathcal{T}$, let $\mathcal{T}_\alpha$ and $\mathcal{T}_\beta$ be the subtrees of $\mathcal{T} - \{\alpha, \beta\}$ that contain, respectively, α and β, and let $\mathcal{C}_\alpha$ and $\mathcal{C}_\beta$ be the vertex sets of $\mathcal{T}_\alpha$ and $\mathcal{T}_\beta$. Define

$$V_{\alpha\backslash\beta} = \left(\bigcup_{\gamma \in \mathcal{C}_\alpha} \gamma\right) \backslash \beta,$$

with $V_{\beta\backslash\alpha}$ defined similarly.

LEMMA 7. [2] *Under the assumptions of the preceding paragraph, the following hold:*

(i) $V_{\alpha\backslash\beta} \cap V_{\beta\backslash\alpha} = \emptyset$;

(ii) $(\alpha \cap \beta)^c = V_{\alpha\backslash\beta} \cup V_{\beta\backslash\alpha}$;

and

(iii) α^c *is the disjoint union*

$$\alpha^c = \bigcup_{\beta \in \text{adj}\,\alpha} V_{\beta\setminus\alpha},$$

in which $\text{adj}\,\alpha = \{\beta \in \mathcal{C} : \{\alpha,\beta\} \in \mathcal{E}\}$.

We should note the following consequences of Lemma 7. Suppose $B = [B_{ij}]$ is a matrix satisfying $G(B) \subseteq G$, in which G is a chordal graph, and let $\mathcal{T}$ be a clique tree associated with G. If $\{\alpha, \beta\}$ is an edge of $\mathcal{T}$, then $B[(\alpha \cap \beta)^c]$ is essentially a direct sum of the matrices $B[V_{\alpha\setminus\beta}]$ and $B[V_{\beta\setminus\alpha}]$. The reason for this is that there are no edges between vertices in $V_{\alpha\setminus\beta}$ and vertices in $V_{\beta\setminus\alpha}$, and hence $B_{ij} = O$ whenever $i \in V_{\alpha\setminus\beta}$ and $j \in V_{\beta\setminus\alpha}$. Similarly, if α is any maximal clique of G then $B[\alpha^c]$ is essentially a direct sum matrices of the form $B[V_{\beta\setminus\alpha}]$ as β runs through all cliques that are adjacent in $\mathcal{T}$ to α.

LEMMA 8. *Let* $\mathcal{H} = \mathcal{H}_1 \oplus \cdots \oplus \mathcal{H}_n$, *let* $A: \mathcal{H} \to \mathcal{H}$ *be an invertible operator matrix, and let* $G = G(A^{-1})$ *be a connected chordal graph. If* $\mathcal{T} = (\mathcal{C}, \mathcal{E})$ *is any clique tree associated with* G, *then*

$$(14) \qquad \sum_{\alpha \in \mathcal{C}} \dim \ker A[\alpha] = \sum_{\{\alpha,\beta\} \in \mathcal{E}} \dim \ker A[\alpha \cap \beta].$$

Proof. Let us look first at the left-hand side of (14). by Lemma 4 and by Lemma 7 we have

$$(15) \qquad \begin{aligned} \sum_{\alpha \in \mathcal{C}} \dim \ker A[\alpha] &= \sum_{\alpha \in \mathcal{C}} \dim \ker A^{-1}[\alpha^c] \\ &= \sum_{\alpha \in \mathcal{C}} \sum_{\beta \in \text{adj}\,\alpha} \dim \ker A^{-1}[V_{\beta\setminus\alpha}]. \end{aligned}$$

On the other hand, by applying Lemmas 4 and 7 we may see that the right-hand side of (14) is

$$(16) \qquad \begin{aligned} \sum_{\{\alpha,\beta\} \in \mathcal{E}} \dim \ker A[\alpha \cap \beta] &= \sum_{\{\alpha,\beta\} \in \mathcal{E}} \dim \ker A^{-1}[(\alpha \cap \beta)^c] \\ &= \sum_{\{\alpha,\beta\} \in \mathcal{E}} \left(\dim \ker A^{-1}[V_{\beta\setminus\alpha}] + \dim \ker A^{-1}[V_{\alpha\setminus\beta}] \right). \end{aligned}$$

Observe that with every edge $\{\alpha, \beta\}$ of $\mathcal{T}$ we may associate exactly two terms in the right-most expression of (15), namely $\dim \ker A^{-1}[V_{\beta\setminus\alpha}]$ and $\dim \ker A^{-1}[V_{\alpha\setminus\beta}]$. But this just means that (15) and (16) contain all the same terms, and hence (14) is established. $\square$

THEOREM 9. *Let $G = (N, E)$ be a connected chordal graph, let $A = [A_{ij}]$ be an n-by-n weakly G-regular Hermitian operator matrix, and suppose that $G(A^{-1}) \subseteq G$. If $i_+(A) < \infty$, then for any clique tree $\mathfrak{T} = (\mathcal{C}, \mathcal{E})$ associated with G we have*

$$i_+(A) = \sum_{\alpha \in \mathcal{C}} i_+(A[\alpha]) - \sum_{\{\alpha, \beta\} \in \mathcal{E}} i_+(A[\alpha \cap \beta]).$$

Proof. Since $i_+(A) < \infty$, we must have $i_+(A[\alpha]) < \infty$ for any $\alpha \subseteq N$, and by Lemma 5 we know that $i_o(A[\alpha]) < \infty$ for any $\alpha \subseteq N$. By Lemma 6 we may write

$$
\begin{aligned}
(17) \quad & \sum_{\alpha \in \mathcal{C}} i_+(A[\alpha]) - \sum_{\{\alpha, \beta\} \in \mathcal{E}} i_+(A[\alpha \cap \beta]) \\
&= \sum_{\alpha \in \mathcal{C}} \left[i_+(A) - i_+(A^{-1}[\alpha^c]) - i_o(A[\alpha]) \right] \\
&\quad - \sum_{\{\alpha, \beta\} \in \mathcal{E}} \left[i_+(A) - i_+(A^{-1}[(\alpha \cap \beta)^c]) - i_o(A[\alpha \cap \beta]) \right] \\
&= \sum_{\alpha \in \mathcal{C}} i_+(A) - \sum_{\{\alpha, \beta\} \in \mathcal{E}} i_+(A) - \sum_{\alpha \in \mathcal{C}} i_+(A^{-1}[\alpha^c]) \\
&\quad + \sum_{\{\alpha, \beta\} \in \mathcal{E}} i_+(A^{-1}[(\alpha \cap \beta)^c]) - \sum_{\alpha \in \mathcal{C}} i_o(A[\alpha]) + \sum_{\{\alpha, \beta\} \in \mathcal{E}} i_o(A[\alpha \cap \beta]).
\end{aligned}
$$

The last two terms of the last expression in (17) cancel by Lemma 8, and the two middle terms cancel by an argument similar to that used in the proof of Lemma 8. Finally, since $\mathfrak{T}$ has exactly one more vertex than the number of edges, the right-hand side of (17) equals $i_+(A)$. This proves the theorem. $\square$

Of course, a similar statement is true for $i_-(A)$, and the corresponding statement for $i_o(A)$ is already contained in Lemma 8.

ACKNOWLEDGEMENT

The authors wish to thank M. Bakonyi and I. Spitkovski for helpful discussions of some operator theoretic background for the present paper.

REFERENCES

1.   W. Barrett and C.R. Johnson, *Determinantal Formulae for Matrices with Sparse Inverses*, Linear Algebra Appl. 56 (1984), pp. 73–88.

2.   W. Barrett, C.R. Johnson and M. Lundquist, *Determinantal Formulae for Matrix Completions Associated with Chordal Graphs*, Linear Algebra Appl. 121 (1989), pp. 265–289.

3.   M. Golumbic *Algorithmic Graph Theory and Perfect Graphs*, Academic Press, New York, 1980.

4.   R. Grone, C.R. Johnson, E. Sá and H. Wolkowicz, *Positive Definite Completions of Partial Hermitian Matrices*, Linear Algebra Appl. 58 (1984), pp. 109–124.

5.   E.V. Haynsworth, *Determination of the Inertia of a Partitioned Hermitian Matrix*, Linear Algebra Appl. 1 (1968), pp. 73–81.

6.   C.R. Johnson, *Matrix Completion Problems: A Survey*, Proceedings of Symposia in Applied Mathematics 40 (American Math. Soc.) (1990), pp. 171–198.

7.   C.R. Johnson and W. Barrett, *Spanning Tree Extensions of the Hadamard-Fischer Inequalities*, Linear Algebra Appl. 66 (1985), pp. 177–193.

8.   C.R. Johnson and M. Lundquist, *An Inertia Formula for Hermitian Matrices with Sparse Inverses*, Linear Algebra Appl., to appear.

9.   C.R. Johnson and M. Lundquist, *Matrices with Chordal Inverse Zero Patterns*, Linear and Multilinear Algebra, submitted.

Charles R. Johnson,
Department of Mathematics,
College of William and Mary,
Williamsburg, VA 23185,
U.S.A.

Michael Lundquist,
Department of Mathematics,
Brigham Young University,
Provo, Utah 84602,
U.S.A.

MSC: Primary 15A09, Secondary 15A21, 15A99, 47A20

Operator Theory:
Advances and Applications, Vol. 59
© 1992 Birkhäuser Verlag Basel

MODELS AND UNITARY EQUIVALENCE OF CYCLIC
SELFADJOINT OPERATORS IN PONTRJAGIN SPACES

P. Jonas, H. Langer, B. Textorius

It is shown that a cyclic selfadjoint operator in a Pontrjagin space is unitarily equivalent to the operator A_ϕ of multiplication by the independent variable in some space $\sqcap(\phi)$ generated by a "distribution" ϕ. Further, criteria for the unitary equivalence of two such operators $A_\phi, A_{\hat\phi}$ are given.

INTRODUCTION

It is well-known that a cyclic selfadjoint operator in a Hilbert space is unitarily equivalent to the operator of multiplication by the independent variable in a space $L_2(\sigma)$ with a positive measure σ. In the present paper we prove a corresponding result for a bounded cyclic selfadjoint operator A in a Pontrjagin space: It is shown that A is unitarily equivalent to the operator A_ϕ of multiplication by the independent variable in some space $\sqcap(\phi)$, generated by a "distribution" ϕ (which is a certain linear functional on a space of test functions, e.g. the polynomials in one complex variable). The class $\mathcal{F}$ of these "distributions" ϕ is introduced in Section 1.

We mention that, for an element $\phi \in \mathcal{F}$ there exists a finite exceptional set $s(\phi)$ such that ϕ restricted to $\mathbb{C}\backslash s(\phi)$ is a positive measure on $\mathbb{R}\backslash s(\phi)$ (in the notation of Section 1.3, $s(\phi) = s(\varphi) \cup \sigma_0(\psi)$, if $\phi = \varphi + \psi$ is the decomposition (1.5) of $\phi \in \mathcal{F}$). In the exceptional points, ϕ can be more complicated due to the presence of a finite number of negative squares of the inner product. In Section 2 the space $\sqcap(\phi)$ is defined and, by means of the integral representation of ϕ (see Lemma 1.2), a model of $\sqcap(\phi)$, which is an orthogonal sum of a Hilbert space $L_2(\phi)$ with some measure σ and a finite-dimensional space, is given. In Section 3 the operator A_ϕ of multiplication by the independent variable in $\sqcap(\phi)$ is introduced and represented as a matrix in the model space of Section 2. Thus it follows that each bounded cyclic selfadjoint operator in a Pontrjagin space is unitarily equivalent to such a matrix model. Naturally, this model is, in some sense, a finite-dimensional perturbation of the operator of multiplication by the independent variable in $L_2(\sigma)$.

In Section 4 conditions for the unitary equivalence of two operators $A_\phi, A_{\hat\phi}$ for $\phi, \hat\phi \in \mathcal{F}$ are given. For this equivalence it turns out to be necessary that the corresponding measures $\sigma, \hat\sigma$ are equivalent and, moreover, that the square root of the density $d\sigma/d\hat\sigma$ has "values" and, sometimes, also "derivatives" at the real exceptional points. This necessary condition for the unitary equivalence of A_ϕ and $A_{\tilde\phi}$ is, in fact, necessary and sufficient for the unitary equivalence of the spectral functions of A_ϕ and $A_{\tilde\phi}$. If $\kappa = 1$, also a necessary and sufficient condition for the unitary equivalence of A_ϕ and $A_{\tilde\phi}$ is given.

In this paper we restrict ourselves to a **bounded** selfadjoint operator in a Pontrjagin space. The case of a densely defined unbounded selfadjoint operator is only technically more complicated: The space of polynomials has to be replaced by another suitable set of test functions. Scalar and operator valued distributions have already played a role in the spectral theory of selfadjoint operators in Pontrjagin and Krein spaces e.g. in the papers [6], [7].

In order to find the model space $\sqcap(\phi)$ and the model operator A_ϕ we might also have started from the construction of the space $\sqcap(Q)$ and the operator A_Q used in [11] for a function $Q \in N_\kappa$ (with bounded spectrum). In this connection we mention that these functions of class N_κ can be considered as the Stieltjes transforms of the elements $\phi \in \mathcal{F}$, the functions of the class $\mathcal{P}_\kappa$ (see, e.g., [12]) with bounded spectrum are the Fourier transforms of these ϕ (compare also [8]). It seems to be interesting to study corresponding models for the case that A is not cyclic but has a finite number of generating vectors, which in a Pontrjagin space can always be supposed without loss of generality.

1. THE CLASS $\mathcal{F}$ OF LINEAR FUNCTIONALS

1.1. Distributions of the class $\mathcal{F}(\mathbb{R})$. By $\mathcal{F}(\mathbb{R})$ we denote the set of all distributions φ on $\mathbb{R}$ with compact support such that the following holds.

(a) φ is real, that is, φ has real values on real test functions.

(b) There exists a finite set $s(\varphi) \subset \mathbb{R}$ (the case $s(\varphi) = \emptyset$ is not excluded) such that φ restricted to $\mathbb{R}\backslash s(\varphi)$ is a (possibly unbounded) positive measure, and $s(\varphi)$ is the smallest set with this property. For $\alpha \in \mathbb{R}$, the set of all $\varphi \in \mathcal{F}(\mathbb{R})$ with $s(\varphi) = \{\alpha\}$ is denoted by $\mathcal{F}(\mathbb{R}, \alpha)$

Let $\varphi \in \mathcal{F}(\mathbb{R})$. Asumme that $n > 1, s(\varphi) = \{\alpha_1, \ldots, \alpha_n\}, -\infty < \alpha_1 < \ldots < < \alpha_n < \infty$, and let $t_i, i = 1, \ldots, n-1$, be real points with $\alpha_i < t_i < \alpha_{i+1}, i = 1, \ldots, n-1$, such that φ has no masses in the points t_i. We set $\triangle_1 := (-\infty, t_1], \triangle_i := (t_{i-1}, t_i], i = 2, \ldots, n-1, \triangle_n := (t_{n-1}, \infty)$. A system of intervals $\triangle_i, i = 1, \ldots, n$, with these properties is called a φ-**minimal decomposition** of $\mathbb{R}$. Let $\chi_{\triangle_i}$ be the characteristic function on $\mathbb{R}$ of $\triangle_i$. Then $\chi_{\triangle_i}\varphi \in \mathcal{F}(\mathbb{R}, \alpha_i), i = 1, \ldots, n$, and

$$\varphi \cdot f = \sum_{i=1}^{n} (\chi_{\triangle i}\varphi) \cdot f, \quad f \in \mathbb{C}^\infty(\mathbb{R}).$$

Here "$\cdot$" denotes the usual duality of distributions and $\mathbb{C}^\infty$ functions on $\mathbb{R}$.

If $\alpha \in \mathbb{R}$ and $\varphi \in \mathcal{F}(\mathbb{R}, \alpha)$, the order $\mu(\varphi)$ of φ is, as usual, the smallest $n \in \mathbb{N}_0 (= \mathbb{N} \cup \{0\})$ such that φ is the n-th derivative of a (signed) measure on $\mathbb{R}$. We denote by $\mu_0(\alpha; \varphi), (\mu_r(\alpha; \varphi), \mu_l(\alpha; \varphi))$ the smallest $n \in \mathbb{N}_0$ such that, for some measure τ

on $\mathbb{R}$, the n-th derivative $\tau^{(n)}$ of τ and φ coincide on $(-\infty, \alpha) \cup (\alpha, \infty)$ $((\alpha, \infty), (-\infty, \alpha),$ respectively). The numbers $\mu_0(\alpha; \varphi)$, $\mu_r(\alpha; \varphi)$, $\mu_l(\alpha; \varphi)$ are called **reduced order, right reduced order, left reduced order,** respectively, of φ at α. Evidently, we have $\mu_0(\alpha; \varphi) = \max\{\mu_r(\alpha; \varphi), \mu_l(\alpha; \varphi)\}$. Some more properties of these numbers are given in the following lemma (compare [6; Hilfsatz 1,2]).

LEMMA 1.1. If $\varphi \in \mathcal{F}(\mathbb{R}, \alpha)$, then the following statements hold.

(i) $\mu_r(\alpha; \varphi)$ $(\mu_l(\alpha; \varphi))$ coincides with the minimum of the numbers $n \in \mathbb{N}_0$ such that $(t - \alpha)^n \varphi$ is a bounded measure on (α, ∞) $((-\infty, \alpha),$ respectively).

(ii) $(t - \alpha)^{\mu(\varphi)} \varphi$ is a measure.

Here and in the sequel $\mathbf{t}$ denotes the function $f(t) \equiv t$.

PROOF. (i) Choose $a > |\alpha|$ with $\operatorname{supp} \varphi \subset (-a, a)$ and let n be such that $(t - \alpha)^n \varphi$ is a bounded measure on (α, a). If f is an element of

$$\mathcal{M} := \{f \in C_0^\infty(\mathbb{R}) : \operatorname{supp} f \subset (\alpha, a), \sup_{t \in (\alpha, a)} |f^{(n)}(t)| \leq 1\},$$

Taylor's formula implies $\sup\{|(t - \alpha)^{-n} f(t)| : t \in (\alpha, a)\} \leq 1$. It follows that

$$\sup\{|\varphi \cdot f| : f \in \mathcal{M}\} = \sup\{|(t - \alpha)^n \varphi \cdot (t - \alpha)^{-n} f| : f \in \mathcal{M}\} < \infty.$$

Then, by a standard argument of distribution theory, φ restricted to (α, a) is the n–th derivative of a bounded measure on (α, a), i.e. $\mu_r(\alpha; \varphi) \leq n$.

It remains to show that with $\mu_r := \mu_r(\alpha; \varphi)$, $(t - \alpha)^{\mu_r} \varphi | (\alpha, a)$ is a bounded measure. To this end choose a nonnegative $\beta \in C^\infty(\mathbb{R})$ equal to 0 on a neighbourhood of $(-\infty, 0]$ in $\mathbb{R}$ and equal to 1 on a neighbourhood of $[1, \infty)$ in $\mathbb{R}$, and set $\beta_k(t) := \beta(k(t-\alpha))$, $k \in \mathbb{N}$. A simple computation yields the uniform boundedness of the functions $((t - \alpha)^{\mu_r} \beta_k)^{(\mu_r)}$, $k \in \mathbb{N}$, on (α, a). Then, since there exists a measure φ_0 on $\mathbb{R}$ with compact support such that $\varphi_0^{(\mu_r)} = \varphi$ on (α, ∞), it follows that

$$(t - \alpha)^{\mu_r} \varphi \cdot \beta_k = \varphi \cdot (t - \alpha)^{\mu_r} \beta_k = (-1)^{\mu_r} \varphi_0 \cdot ((t - \alpha)^{\mu_r} \beta_k)^{(\mu_r)}.$$

The last expression is uniformly bounded with respect to k, hence $(t - \alpha)^{\mu_r} \varphi$ is a bounded measure on (α, a). The claim about $\mu_l(\alpha; \varphi)$ is proved analogously.

(ii) Evidently, $\mu := \mu(\varphi) \geq \mu_0(\alpha; \varphi)$. Then by (i) there exists a measure φ_0 on $\mathbb{R}$ with $\operatorname{supp} \varphi_0 \subset (-a, a)$ such that

$$(t - \alpha)^\mu \varphi = \varphi_0 + \sum_{l=1}^s a_l \delta_\alpha^{(l)},$$

where δ_α is the δ–measure concentrated in the point α. Let $l \in \mathbb{N}$ and define $f_{l,\epsilon}(t) := (t - \alpha + \epsilon)^l$ for $t \leq \alpha - \epsilon$, $f_{l,\epsilon}(t) = (t - \alpha - \epsilon)^l$ for $t \geq \alpha + \epsilon$, $f_{l,\epsilon}(t) = 0$ for $t \in (\alpha - \epsilon, \alpha + \epsilon)$. Then we have

$$(t - \alpha)^\mu \varphi \cdot (t - \alpha)^l = \lim_{\epsilon \to 0} (t - \alpha)^\mu \varphi \cdot (t - \alpha)^{-\mu} f_{l+\mu,\epsilon} =$$

$$= \lim_{\epsilon \to o} \varphi_0 \cdot (t - \alpha)^{-\mu} f_{l+\mu,\epsilon} = \varphi_0 \cdot (t - \alpha)^l.$$

It follows that $a_l = 0, l = 1, \ldots, s$, and (ii) is proved.

1.2. Integral representations of the distributions of $\mathcal{F}(\mathbb{R})$. The distributions of the classes $\mathcal{F}(\mathbb{R}, \alpha)$ can be represented by certain measures. Consider $\varphi \in \mathcal{F}(\mathbb{R}, \alpha)$ and set

$$k := \begin{cases} \frac{1}{2}\mu_0(\alpha; \varphi) & \text{if } \mu_0(\alpha; \varphi) \text{ is even,} \\ \frac{1}{2}(\mu_0(\alpha; \varphi) + 1) & \text{if } \mu_0(\alpha; \varphi) \text{ is odd.} \end{cases}$$

Then, by Lemma 1.1, the distribution $(t - \alpha)^{2k}\varphi$ is the restriction to $\mathbb{R}\backslash\{\alpha\}$ of a positive measure σ on $\mathbb{R}$ with compact support and $\sigma(\{\alpha\}) = 0$. If $k \geq 1$, then the function $(t - \alpha)^{-2}$ is not σ–integrable,

$$\int_{\mathbb{R}} (t - \alpha)^{-2} d\sigma(t) = \infty.$$

Indeed, otherwise $(t - \alpha)^{2k-2}\varphi$ would be a bounded measure on $\mathbb{R}\backslash\{\alpha\}$. Then, as $2k - 2 \geq \mu_0(\alpha; \varphi) - 1 \geq 0$, the distribution $(t - \alpha)^{\mu_0(\alpha;\varphi)-1}\varphi$ would be a bounded measure on $\mathbb{R}\backslash\{\alpha\}$, which in view of Lemma 1.1 is a contradiction to the minimality of $\mu_0(\alpha; \varphi)$.

By the definition of σ the distribution $(t - \alpha)^{2k}\varphi - \sigma$ is concentrated in the point α. If $k = 0$, define

$$c_i := (\varphi - \sigma) \cdot (t - \alpha)^i, \; i = 0, 1, \ldots,$$

if $k \geq 1$,

$$c_i = \begin{cases} \varphi \cdot (t - \alpha)^i & \text{for } i = 0, \ldots, 2k - 1 \\ ((t - \alpha)^{2k}\varphi - \sigma) \cdot (t - \alpha)^{i-2k} & \text{for } i = 2k, 2k + 1, \ldots \end{cases}$$

By Lemma 1.1, $c_i = 0$ if i is larger than the order of φ.

In the sequel for a function f with n derivatives at $t = \alpha$ we use the notation

$$f^{\{\alpha,0\}}(t) := f(t), \; f^{\{\alpha,n\}}(t) := f(t) - \sum_{i=0}^{n-1} i!^{-1}(t - \alpha)^i f^{(i)}(\alpha)$$

for $n \geq 1$, or, if α is clear from the context, shorter $f^{\{0\}}$ and $f^{\{n\}}$, respectively. Then, for every $f \in C^\infty(\mathbb{R})$,

$$\varphi \cdot f^{\{\alpha,2k\}} = (t - \alpha)^{2k}\varphi \cdot (t - \alpha)^{-2k} f^{\{\alpha,2k\}} =$$

$$(1.1) \qquad = \sigma \cdot (t - \alpha)^{-2k} f^{\{\alpha,2k\}} + ((t - \alpha)^{2k}\varphi - \sigma) \cdot (t - \alpha)^{-2k} f^{\{\alpha,2k\}}) =$$

$$= \sigma \cdot (t - \alpha)^{-2k} f^{\{\alpha, 2k\}} + \sum_{i \geq 2k} c_i \, i!^{-1} f^{(i)}(\alpha).$$

For $k \geq 1$ we have

$$(1.2) \qquad \varphi \cdot f = \varphi \cdot \left(\sum_{i=0}^{2k-1} i!^{-1}(t - \alpha)^i f^{(i)}(\alpha) + f^{\{\alpha, 2k\}} \right) =$$

$$= \sum_{i=0}^{2k-1} c_i i!^{-1} f^{(i)}(\alpha) + \varphi \cdot f^{\{\alpha, 2k\}}.$$

The relations (1.1) and (1.2) imply the first assertion of the following lemma.

LEMMA 1.2. If $\varphi \in \mathcal{F}(\mathbb{R}, \alpha)$, there exist numbers $k, l \in \mathbb{N}_0, c_0, c_1, \ldots, c_l \in \mathbb{R}, c_l \neq 0$ if $l \geq 1$, and a measure σ on $\mathbb{R}$ with compact support, $\sigma(\{\alpha\}) = 0$ and, if $k > 0, \int_{\mathbb{R}} (t - \alpha)^{-2} d\sigma(t) = \infty$, such that

$$(1.3) \qquad \varphi \cdot f = \int_{\mathbb{R}} f^{\{\alpha, 2k\}}(t) (t - \alpha)^{-2k} d\sigma(t) + \sum_{i=0}^{l} c_i \, i!^{-1} f^{(i)}(\alpha) \, (f \in C^\infty(\mathbb{R})).$$

The numbers $k, l, c_0, \ldots, c_l$ and the measure σ with the above properties are uniquely determined. Conversely, given $k, l, c_0, \ldots, c_l, \sigma$ with these properties then (1.3) defines a distribution $\varphi \in \mathcal{F}(\mathbb{R}, \alpha)$.

The uniqueness statement and the last assertion can be verified by an easy computation.

REMARK 1.3. If $\varphi \in \mathcal{F}(\mathbb{R}, \alpha)$ has the representation (1.3), then

$$\mu_0(\alpha; \varphi) = \begin{cases} 2k & \text{if } k = 0 \text{ or } k > 0 \text{ and } \int_{\mathbb{R}} |t - \alpha|^{-1} d\sigma(t) = \infty \\ 2k - 1 & \text{if } k > 0 \text{ and } \int_{\mathbb{R}} |t - \alpha|^{-1} d\sigma(t) < \infty. \end{cases}$$

The order $\mu(\varphi)$ of φ is the maximum of $\mu_0(\alpha; \varphi)$ and l. If

$$n := \max\{\nu \in \mathbb{N}_0 : 0 \leq \nu \leq 2k, \int_{-\infty}^{\alpha} |t - \alpha|^{-\nu} d\sigma < \infty\},$$

$$m := \max\{\mu \in \mathbb{N}_0 : 0 \leq \mu \leq 2k, \int_{\alpha}^{\infty} |t - \alpha|^{-\mu} d\sigma < \infty\},$$

then, according to Lemma 1.1,

$$\mu_l(\alpha;\varphi) = 2k - n, \ \mu_r(\alpha;\varphi) = 2k - m.$$

1.3. Linear functionals of the class $\mathcal{F}$. In the sequel we need a class of functionals which is slightly larger than $\mathcal{F}(\mathbb{R})$. Assume first $\beta_i \in \mathbb{C}^+ := \{z : \operatorname{Im} z > 0\}$, $\beta_i \neq \beta_k$ if $i \neq k, i, k = 1, \ldots, m, \nu_i \in \mathbb{N}, i = 1, \ldots, m$, and $d_{ij} \in \mathbb{C}, i = 1, \ldots, m; j = 0, 1, \ldots, \nu_i - 1$. We set $B := \{\beta_1, \ldots, \beta_m, \overline{\beta}_1, \ldots, \overline{\beta}_m\}$. For a function f which is locally holomorphic on B we define

$$(1.4) \qquad \psi(f) = \sum_{i=1}^{m} \sum_{j=0}^{\nu_i-1} (d_{ij}\, j!^{-1} f^{(j)}(\beta_i) + \overline{d_{ij}}\, j!^{-1} f^{(j)}(\overline{\beta}_i)).$$

The functional ψ is considered as a linear functional on the linear space $H(\mathbb{R} \cup B)$ of all locally holomorphic functions on $\mathbb{R} \cup B$. If for a function $f \in H(\mathbb{R} \cup B)$ we have $f(z) = \overline{f(\overline{z})}$ for all $z \in \mathbb{C}$ such that z and $\overline{z}$ are in the domain of f, then $\psi(f)$ is real.

We denote the set of all linear functionals ψ on $H(\mathbb{R} \cup B)$ of the form (1.4) by $\mathcal{F}(\mathbb{C}\backslash\mathbb{R}, B)$ and set $\mathcal{F}(\mathbb{C}\backslash\mathbb{R}, \emptyset) = \{0\}$. In a natural way $\mathcal{F}(\mathbb{R})$ is identified with a linear subspace of the algebraic dual space of $H(\mathbb{R} \cup B)$, and we define $\mathcal{F}(\mathbb{C}\backslash\mathbb{R}) := \cup_B \mathcal{F}(\mathbb{C}\backslash\mathbb{R}, B)$ and $\mathcal{F} := \cup_B (\mathcal{F}(\mathbb{R}) + \mathcal{F}(\mathbb{C}\backslash\mathbb{R}, B))$, where B runs through all finite $\mathbb{R}$–symmetric subsets of $\mathbb{C}\backslash\mathbb{R}$. If

$$(1.5) \qquad \phi = \varphi + \psi \in \mathcal{F}, \ \varphi \in \mathcal{F}(\mathbb{R}), \ \psi \in \cup_B \mathcal{F}(\mathbb{C}\backslash\mathbb{R}, B),$$

the minimal set B such that $\psi \in \mathcal{F}(\mathbb{C}\backslash\mathbb{R}, B)$ is denoted by $\sigma_0(\phi)$.

From well-known approximation results for holomorphic functions it follows that every $\phi \in \mathcal{F}$ is uniquely determined by its restriction to the linear space $\mathcal{P}$ of all polynomials.

2. THE PONTRJAGIN SPACE ASSOCIATED WITH $\phi \in \mathcal{F}$.

2.1. Completions. Let $(\mathcal{L}, [\cdot, \cdot])$ be an inner product space (see [3]), that is, $\mathcal{L}$ is a linear space equipped with the hermitian sesquilinear form $[\cdot, \cdot]$. By $\kappa_-(\mathcal{L})$ we denote the number of negative squares of $\mathcal{L}$ or of the inner product $[\cdot, \cdot]$, that is the supremum of the dimensions of all negative subspaces of $\mathcal{L}$ (for the notations see also [3], [2]). In what follows this number is always finite. Further, $\mathcal{L}^0 := \{x : [x, \mathcal{L}] = \{0\}\}$ is the isotropic subspace of $(\mathcal{L}, [\cdot, \cdot])$, and it is well–known that the factor space $(\mathcal{L}/\mathcal{L}^0, [\cdot, \cdot])$ admits a unique completion to a Pontrjagin space ([5, 2]), which is denoted by $(\mathcal{L}/\mathcal{L}^0)^\sim$.

LEMMA 2.1. Let $\mathcal{L}$ be a linear space, which is equipped with a nonnegative inner product $(\cdot, \cdot)$, and suppose that on $\mathcal{L}$ there are given linear functionals $v_1, \ldots, v_n$ such that no linear combination of the v_j's is bounded with respect to the seminorm $(\cdot, \cdot)^{\frac{1}{2}}$. Further, let $A = (\alpha_{jk})_1^n$ be a hermitian $n \times n$–matrix which is nonsingular and has κ negative (and

$n - \kappa$ positive) eigenvalues counted according to their multiplicities. Consider on $\mathcal{L}$ the inner product

$$[x,y] = (x,y) + \sum_{j,k=1}^{n} \alpha_{jk} v_k(x) \, \overline{v_j(y)} \quad (x,y \in \mathcal{L}).$$

Then this inner product has κ negative squares on $\mathcal{L}$. The completion of $(\mathcal{L}/\mathcal{L}^0, [\cdot,\cdot])$ is the Pontrjagin space $\mathcal{H} \oplus \mathcal{A}$ where $\mathcal{H}$ is the Hilbert space completion of $\hat{\mathcal{L}} := \mathcal{L}/\mathcal{L}_0, \mathcal{L}_0 :=$ $= \{x \in \mathcal{L} : (x,x) = 0\}$, and $\mathcal{A} := (\mathbb{C}^n, (A\cdot,\cdot)_{\mathbb{C}^n})$. More exactly, the mapping

$$\iota : \mathcal{L} \ni x \to (\hat{x}; v_1(x), \ldots, v_n(x))^\top$$

is an isometry of $(\mathcal{L}, [\cdot,\cdot])$ onto a dense subspace of $\mathcal{H} \oplus \mathcal{A}$, $\ker \iota = \mathcal{L}^0$.

PROOF. Evidently, the mapping ι is an isometry of $(\mathcal{L}, [\cdot,\cdot])$ into the π_κ-space $\mathcal{H} \oplus \mathcal{A}$. Hence $(\mathcal{L}, [\cdot,\cdot])$ has a finite number $(\leq \kappa)$ of negative squares. In order to prove that the range of ι is dense in $\mathcal{H} \oplus \mathcal{A}$ we show that for each $j_0 \in \{1,\ldots,n\}$ there exists a sequence $(y_\nu) \subset \mathcal{L}$ such that $(y_\nu, y_\nu) \to 0$, $v_j(y_\nu) \to 0$ if $j \in \{1,\ldots,n\}\backslash\{j_0\}$ and $v_{j_0}(y_\nu) \to 1, \nu \to \infty$. Indeed, if for each sequence (y_ν) the first two relations would imply $v_{j_0}(y_\nu) \to 0$, then v_{j_0} would be a continuous linear functional on $\mathcal{L}$ with respect to the semi-norm

$$\mathcal{L} \ni y \to (y,y)^{\frac{1}{2}} + \left(\sum_{\substack{j=1 \\ j \neq j_0}}^{n} |v_j(y)|^2 \right)^{\frac{1}{2}}.$$

This would imply a representation

$$v_{j_0}(y) = (y,\hat{e}) + \sum_{\substack{j=1 \\ j \neq j_0}}^{n} v_j(y) \hat{v}_j \quad (y \in \mathcal{L}),$$

with $\hat{e} \in \mathcal{H}$ and $\hat{v}_j \in \mathbb{C}$, which is impossible, as no nontrivial linear combination of the v_j's is a continuous linear functional on $(\mathcal{L}, (\cdot,\cdot)^{\frac{1}{2}})$.

The sequence (ιy_ν) converges to

$$f_{j_0} := (0; 0, \ldots, 0, 1, 0, \ldots, 0)^\top \in \mathcal{H} \oplus \mathcal{A},$$

where the 1 is at the j_0-th component. It follows that for arbitrary $x \in \mathcal{L}$ the element

$$\iota x - v_1(x) f_1 - \ldots - v_n(x) f_n = (\hat{x}; 0, \ldots, 0)^\top$$

is the limit in $\mathcal{H} \oplus \mathcal{A}$ of a sequence belonging to $\iota(\mathcal{L})$. Hence $\iota(\mathcal{L})$ is dense in $\mathcal{H} \oplus \mathcal{A}$.

The proof of the following lemma is straightforward and therefore left to the reader.

LEMMA 2.2. Let $(\mathcal{L}, [\cdot, \cdot])$ be an inner product space and

$$\mathcal{L} = \mathcal{L}_1[\dot{+}]\mathcal{L}_2[\dot{+}]\ldots[\dot{+}]\mathcal{L}_n$$

be a direct and orthogonal decomposition. Then for the isotropic parts it holds

$$\mathcal{L}^0 = \mathcal{L}_1^0[\dot{+}]\mathcal{L}_2^0[\dot{+}]\ldots[\dot{+}]\mathcal{L}_n^0,$$

and

$$\mathcal{L}/\mathcal{L}^0 = \mathcal{L}_1/\mathcal{L}_1^0[\dot{+}]\mathcal{L}_2/\mathcal{L}_2^0[\dot{+}]\ldots[\dot{+}]\mathcal{L}_n/\mathcal{L}_n^0.$$

If $\kappa_-(\mathcal{L}) < \infty$ then a sequence in $\mathcal{L}/\mathcal{L}^0$ is a Cauchy sequence if and only if for each $k = 1, 2, \ldots, n$ the corresponding sequence of projections in $\mathcal{L}_k/\mathcal{L}_k^0$ is a Cauchy sequence, moreover

$$(\mathcal{L}/\mathcal{L}^0)^{\sim} = (\mathcal{L}_1/\mathcal{L}_1^0)^{\sim}[\dot{+}](\mathcal{L}_2/\mathcal{L}_2^0)^{\sim}[\dot{+}]\ldots[\dot{+}](\mathcal{L}_n/\mathcal{L}_n^0)^{\sim}.$$

2.2. Inner products defined by the functionals of $\mathcal{F}$. Let $\phi \in \mathcal{F}$ be given and denote by $\mathcal{P}$ the linear space of all polynomials. On $\mathcal{P}$ an inner product $[\cdot, \cdot]_\phi$ is defined by the relation

$$[f, g]_\phi := \phi(f\bar{g}) \quad (f, g \in \mathcal{P}),$$

where $\bar{g}(z) := \overline{g(\bar{z})}$. This inner product can be extended by continuity to linear spaces which contain $\mathcal{P}$ as a proper subspace. To this end we observe the decomposition

$$\phi = \varphi + \psi, \quad \varphi \in \mathcal{F}(\mathbb{R}), \quad \psi \in \mathcal{F}(\mathbb{C}\backslash\mathbb{R}),$$

which implies

$$[f, g]_\phi = [f, g]_\varphi + [f, g]_\psi \quad (f, g \in \mathcal{P}).$$

Now $[\cdot, \cdot]_\phi$ can be extended by continuity to $C^\infty(\mathbb{R}) \times H(\sigma_0(\phi))$ or to $\mathcal{B}_2(\varphi) \times H(\sigma_0(\phi))$, where $\mathcal{B}_2(\varphi)$ is the linear space of all functions f on $\mathbb{R}$ such that

(i) f restricted to $\mathbb{R}\backslash s(\varphi)$ is φ– measurable and $\int_I |f|^2 d\varphi < \infty$ for each interval I such that $s(\varphi) \cap \bar{I} = \emptyset$;

(ii) for some $\delta_f > 0$, the restriction of f to $\{t \in \mathbb{R} : \text{dist}(s(\varphi), t) < \delta_f\}$ is a C^∞– function.

If $s(\varphi) \neq \emptyset, s(\varphi) = \{\alpha_1, \ldots, \alpha_n\}$, let $(\triangle_i)_{i=1}^n$ be a φ–minimal decomposition of $\mathbb{R}$ (see Section 1.1) such that α_i belongs to the interior of $\triangle_i$; if $s(\varphi) = \emptyset$ choose $n = 1$ and $\triangle_1 = \mathbb{R}$. Further, if $\psi \neq 0$, $\sigma_0(\psi) = \{\beta_1, \ldots, \beta_m, \bar{\beta}_1, \ldots, \bar{\beta}_m\}$, choose mutually disjoint neighbourhoods U_j of $\{\beta_j, \bar{\beta}_j\}, j = 1, 2, \ldots, m$, which do not intersect the real axis, $U := \bigcup_{j=1}^m U_j$. Define functions $\tilde{\chi}_{\triangle_i}$ and $\tilde{\chi}_j$ as follows:

$$\tilde{\chi}_{\triangle_i}(z) := \begin{cases} 1 & \text{if } z \in \triangle_i \\ 0 & \text{if } z \in (\mathbb{R}\backslash\triangle_i) \cup U, \end{cases} \quad \tilde{\chi}_j(z) := \begin{cases} 1 & \text{if } z \in U_j \\ 0 & \text{if } z \in (\bigcup_{k \neq j} U_k) \cup \mathbb{R}. \end{cases}$$

If $\mathcal{L}_k$ is the linear subspace of $\mathcal{B}_2(\varphi) \times H(\sigma_0(\phi))$ defined by

$$\mathcal{L}_i := \{\tilde{\chi}_{\Delta_i} f : f \in \mathcal{B}_2(\varphi) \times H(\sigma_0(\phi))\}, \quad i = 1, 2, \ldots, n,$$

$$\mathcal{L}_{n+j} := \{\tilde{\chi}_j f : f \in \mathcal{B}_2(\varphi) \times H(\sigma_0(\phi))\}, \quad j = 1, 2, \ldots, m,$$

it follows that

$$(2.1) \qquad \mathcal{B}_2(\varphi) \times H(\sigma_0(\phi)) = \mathcal{L}_1[\dot{+}]\mathcal{L}_2[\dot{+}]\ldots[\dot{+}]\mathcal{L}_{n+m},$$

where the sum is direct and $[\cdot,\cdot]_\phi$–orthogonal, and, if $\psi = 0$, the last m terms do not appear. If $s(\varphi) = \emptyset$, the space $\mathcal{L}_1$ is just the space of all functions on $\mathbb{R}$ which are square integrable with respect to the measure φ, equipped with the scalar product of $L_2(\varphi)$. In the following we study the other inner product spaces $(\mathcal{L}_j, [\cdot,\cdot]_\phi)$, that is we describe inner products $[\cdot,\cdot]_\varphi$ and $[\cdot,\cdot]_\psi$ for which $s(\varphi) = \{\alpha\}$ or $\sigma_0(\psi) = \{\beta,\overline{\beta}\}$.

The forms $[\cdot,\cdot]_\psi$, $\psi \in \mathcal{F}(\mathbb{C}\backslash\mathbb{R})$, $\sigma_0(\psi) = \{\beta,\overline{\beta}\}$ are easy to describe and well–known.

PROPOSITION 2.3. Let $\psi \in \mathcal{F}(\mathbb{C}\backslash\mathbb{R})$, $\sigma_0(\psi) = \{\beta,\overline{\beta}\}$, and

$$(2.2) \qquad \psi(p) = \sum_{j=0}^{\nu-1}(d_j \, j!^{-1}p^{(j)}(\beta) + \overline{d}_j \, j!^{-1}p^{(j)}(\overline{\beta})) \quad (p \in \mathcal{P}),$$

$d_j \in \mathbb{C}, j = 0, \ldots, \nu - 1; d_{\nu-1} \neq 0$. Then the mapping

$$\iota_\psi : \mathcal{P} \ni p \rightarrow (p(\beta), \ldots, (\nu-1)!^{-1}p^{(\nu-1)}(\beta); p(\overline{\beta}), \ldots, (\nu-1)!^{-1}p^{(\nu-1)}(\overline{\beta}))^\top$$

is an isometry of $(\mathcal{P}, [\cdot,\cdot]_\psi)$ onto the finite–dimensional nondegenerate inner product space $(\mathbb{C}^{2\nu}, (G\cdot,\cdot)_{\mathbb{C}^{2\nu}})$, where

$$G := \begin{bmatrix} 0 & \hat{G}^* \\ \hat{G} & 0 \end{bmatrix}, \quad \hat{G} := \begin{bmatrix} d_0 & d_1 & \ldots & d_{\nu-1} \\ d_1 & d_2 & \ldots & 0 \\ \vdots & \vdots & \ddots & \vdots \\ d_{\nu-1} & 0 & \ldots & 0 \end{bmatrix}$$

and, hence, induces an isometric isomorphism of $(\mathcal{P}/\mathcal{P}^0, [\cdot,\cdot]_\psi)$ onto $(\mathbb{C}^{2\nu}, (G\cdot,\cdot)_{\mathbb{C}^{2\nu}})$. We have

$$(2.3) \qquad \kappa_\psi := \kappa_-((\mathcal{P}, [\cdot,\cdot]_\psi)) = \nu.$$

PROOF. If $p, q \in \mathcal{P}$ it holds

$$[p, q]_\psi = \sum_{j=0}^{\nu-1} j!^{-1}(d_j \, (p\overline{q})^{(j)}(\beta) + \overline{d}_j(p\overline{q})^{(j)}(\overline{\beta}))$$

$$= \sum_{\mu=0}^{\nu-1}(p^{(\mu)}(\beta) \sum_{j=\mu}^{\nu-1} d_j\mu!^{-1}(j-\mu)!^{-1}\overline{q^{(j-\mu)}}(\beta)$$

$$+ p^{(\mu)}(\overline{\beta}) \sum_{j=\mu}^{\nu-1} \overline{d}_j\mu!^{-1}(j-\mu)!^{-1}\overline{q^{(j-\mu)}(\beta)})$$

which can be written as

$$[p, q]_\psi = (G \iota_\psi p, \iota_\psi q)_{\mathbb{C}^{2\nu}}.$$

The matrix G has ν positive and ν negative eigenvalues (counted according to their multiplicities).

Evidently, in Proposition 2.3 the linear space $\mathcal{P}$ can be replaced by the linear space $H := H(\{\beta, \overline{\beta}\})$ of all functions which are locally holomorphic on $\{\beta, \overline{\beta}\}$. The finite–dimensional space $(\mathcal{P}/\mathcal{P}^0, [\cdot, \cdot]_\psi)$, identified in a natural way with $(H/H^0, [\cdot, \cdot]_\psi)$, is denoted by $\sqcap(\psi)$.

The description of the forms $[\cdot, \cdot]_\varphi, \varphi \in \mathcal{F}(\mathbb{R}), s(\varphi) = \{\alpha\}$, is more complicated; it is given in the following section.

2.3. The spaces $\sqcap(\varphi)$. In this subsection we suppose that $\varphi \in \mathcal{F}(\mathbb{R})$ and that $s(\varphi)$ consists of just one point α. We start with the

LEMMA 2.4. Associate with $\varphi \in \mathcal{F}$ given as above the integers $k, l \in \mathbb{N}_0$ and the measure σ as in Lemma 1.2. Then no finite linear combination of the functionals

$$\mathcal{P} \ni p \to p_i = i!^{-1} p^{(i)}(\alpha), \quad i = 0, 1, \ldots,$$

and, if $k > 0$,

$$\mathcal{P} \ni p \to P_j' := \int_{\mathbb{R}} (t - \alpha)^{-(2k-j)} p^{\{2k-j\}}(t) \, d\sigma(t), \quad j = 0, \ldots, k-1,$$

is continuous with respect to the seminorm

$$\mathcal{P} \ni p \to \left(\int_{\mathbb{R}} (t - \alpha)^{-2k} |p^{\{k\}}(t)|^2 \, d\sigma(t) \right)^{\frac{1}{2}}.$$

PROOF. If the claim would not be true, there would exist an element $v \in L_2(\sigma)$ and numbers $\gamma_i, i = 0, \ldots, N, \, \delta_j, j = 0, \ldots, k-1$, such that $\sum_{i=0}^{N} |\gamma_i| + \sum_{j=0}^{k-1} |\delta_j| \neq 0$ and

$$(2.4) \qquad \sum_{i=1}^{N} \gamma_i p_i + \sum_{j=0}^{k-1} \delta_j P_j' = \int_{\mathbb{R}} (t - \alpha)^{-k} p^{\{k\}}(t) \overline{v(t)} \, d\sigma(t)$$

262   Jonas et al.

for $p \in \mathcal{P}$. For a polynomial p of degree $\leq k-1$ we have $p_i = 0$, $i = k, k+1, \ldots, p^{\{k\}} = 0$ and $P'_j = 0, j = 0, 1, \ldots, k-1$. Then (2.4) implies $\gamma_i = 0, i = 0, 1, \ldots, k-1$.

Consider now polynomials p such that $p_k = p_{k+1} = \ldots = p_r = 0$ with $r := \max\{N, 2k\}$. Then $p^{\{2k-j\}} = p^{\{k\}}, j = 0, 1, \ldots, k-1$, and (2.4) implies

$$\sum_{j=0}^{k-1} \delta_j \int_{\mathbb{R}} (t-\alpha)^{-k} p^{\{k\}}(t)(t-\alpha)^{-k+j} \, d\sigma(t) =$$

$$= \int_{\mathbb{R}} (t-\alpha)^{-k} p^{\{k\}}(t) \overline{v(t)} \, d\sigma(t).$$

The set of polynomials $(t-\alpha)^{-k} p^{\{k\}}(t)$ admitted here is the set of all polynomials whose derivatives vanish at $t = \alpha$ up to the certain order. As this set in dense in $L_2(\sigma)$, the relation (2.4) implies

$$\sum_{j=0}^{k-1} \delta_j (\mathbf{t}-\alpha)^{-k+j} = \overline{v} \in L_2(\sigma),$$

which, because of $\int_{\mathbb{R}} (t-\alpha)^{-2} d\sigma(t) = \infty$, yields $\delta_j = 0, j = 0, \ldots, k-1$. Finally, put $\tilde{p}(t) := (t-\alpha)^{-k} p^{\{k\}}(t)$. Then $\tilde{p}^{(n)}(\alpha) = (k+n)!^{-1} n! p^{(k+n)}(\alpha), n = 0, 1 \ldots$, and

$$\sum_{i=k}^{N} \gamma_i p_i = \sum_{n=0}^{N-k} \gamma_{k+n} n!^{-1} \tilde{p}^{(n)}(\alpha) = \int_{\mathbb{R}} \tilde{p}(t) \overline{v(t)} \, d\sigma(t).$$

As σ has no concentrated mass at α it follows that $\gamma_j = 0, j = k, \ldots, N$. This proves the lemma.

In order to formulate the next theorem we need some more notations. Again with $\varphi \in \mathcal{F}(\mathbb{R}), s(\varphi) = \{\alpha\}$ we associate $k, l \in \mathbb{N}_0$, the measure σ and also the numbers $c_0, \ldots, c_l \in \mathbb{R}$ as in Lemma 1.2, and put $r := \max\{0, l - 2k + 1\}$. If $p \in \mathcal{P}$, the numbers p_i and P'_j are defined as in Lemma 2.4. Now a linear mapping ι_φ from $\mathcal{P}$ into $L_2(\sigma) \oplus \mathbb{C}^{2k+r}$ is introduced as follows:

If $l \leq 2k - 1$ (that is $r = 0$):

$$(2.5) \qquad \iota_\varphi : p \to ((\mathbf{t}-\alpha)^{-k} p^{\{k\}}; p_0, \ldots, p_{k-1}; P_{k-1}, \ldots, P_0)^\top$$

where $P_j := P'_j + \sum_{i=k,\ldots,l; i+j \leq l} c_{i+j} p_i, \quad j = 0, \ldots, k-1;$

if $l \geq 2k$:

$$\iota_\varphi : p \to ((\mathbf{t}-\alpha)^{-k} p^{\{k\}}; p_0, \ldots, p_{k-1}; P_{k-1}, \ldots, P_0)^\top,$$

where now $P_j := P'_j + \sum\limits_{i=l-k+1,\ldots,l; i+j \leq l} c_{i+j} p_i,\ j = 0,\ldots,k-1.$

If $l \leq 2k - 1$ we denote by G the linear operator in $L_2(\sigma) \oplus \mathbb{C}^{2k}$ given by

$$(2.6) \qquad G = \left[\begin{array}{c|cccccc|ccc} I & & & 0 & & & & & 0 & \\ \hline & c_0 & c_1 & \cdots & & c_{k-1} & 0 & \cdots & & 1 \\ & c_1 & & & & \vdots & & & & \\ & & & & & c_l & & & & \\ 0 & \vdots & & & & 0 & & & & \\ & & & & & \vdots & & & & \\ & c_{k-1} & \cdots & c_l & 0 & \cdots & 0 & 1 & \cdots & 0 \\ \hline & 0 & & & & 1 & & & & \\ 0 & \vdots & & & & \vdots & & & & \\ & 1 & & & & 0 & & & & \end{array}\right] ;$$

if $l \geq 2k$, G is the linear operator

$$(2.7) \qquad G = \left[\begin{array}{c|cccc|cccc|ccc} I & & & 0 & & & & & & 0 & \cdots & 0 \\ \hline & c_0 & c_1 & \cdots & c_{k-1} & c_k & \cdots & c_{l-k} & & 0 & \cdots & 1 \\ 0 & c_1 & & & & & & & & & & \\ & \vdots & & & & & & & & & & \\ & c_{k-1} & & & & & & c_{l-1} & & 1 & \cdots & 0 \\ \hline & c_k & & & & & & c_l & & & & \\ 0 & \vdots & & & & & & \vdots & & & & \\ & c_{l-k} & \cdots & & c_{l-1} & c_l & \cdots & 0 & & & & \\ \hline & 0 & & & 1 & & & & & & & \\ 0 & \vdots & & & \vdots & & & & & & & \\ & 1 & & & 0 & & & & & & & \end{array}\right]$$

in $L_2(\sigma) \oplus \mathbb{C}^{2k+r}$.

THEOREM 2.5. Consider $\varphi \in \mathcal{F}(\mathbb{R})$ with $s(\varphi) = \{\alpha\}$, and let the numbers $k, l \in \mathbb{N}_0, c_0, \ldots, c_l \in \mathbb{R}, c_l \neq 0$ if $l > 0$, and the measure σ be as in Lemma 1.2. Then $(\mathcal{P}, [\cdot, \cdot]_\varphi)$ is an inner product space with κ_φ negative squares, where

$$(2.8) \qquad \kappa_\varphi := \max\left(k, \left[\frac{1}{2}(l+1)\right] + \delta\right)$$

with

$$\delta = \begin{cases} 1 & \text{if } l \text{ is even}, c_l < 0, l > 2k - 1, \\ 0 & \text{otherwise.} \end{cases}$$

The linear mapping ι_φ generates an isometric isomorphism of the completion

$$\sqcap(\varphi) := (\mathcal{P}/\mathcal{P}^0, [\cdot,\cdot]_\varphi)^\sim$$

of $(\mathcal{P}/\mathcal{P}^0, [\cdot,\cdot]_\varphi)$ onto the π_{κ_φ}–space

$$(L_2(\sigma) \oplus \mathbb{C}^{2k+r}, (G\cdot,\cdot)_{L_2(\sigma)\oplus\mathbb{C}^{2k+r}}).$$

Here "generates" means that ι_φ induces a linear mapping of $(\mathcal{P}/\mathcal{P}^0, [\cdot,\cdot]_\varphi)$ in $L_2(\sigma) \oplus \mathbb{C}^{2k+r}$ which extends by continuity to all of $\sqcap(\varphi)$.

PROOF. Assume first that $l \leq 2k-1$. Let $p, q \in \mathcal{P}$ and define q_i, Q'_i, Q_i analogously to p_i, P'_i, P_i. In view of the relation

$$(p\overline{q})^{\{2k\}}(t) = p^{\{k\}}(t)\overline{q^{\{k\}}(t)} + \sum_{j=0}^{k-1} (t-\alpha)^j \left(p_j\overline{q^{\{2k-j\}}(t)} + \overline{q}_j\, p^{\{2k-j\}}(t) \right)$$

it follows that

$$\begin{aligned}
[p,q]_\varphi &= \int_{\mathbb{R}} (t-\alpha)^{-2k} p^{\{k\}}(t)\overline{q^{\{k\}}(t)}\, d\sigma(t) + \sum_{j=0}^{k-1}(p_j\,\overline{Q'_j} + \overline{q}_j P'_j) + \sum_{i=0}^{l} c_i \sum_{\mu+\nu=i} p_\mu\overline{q}_\nu = \\
&= \int_{\mathbb{R}} (t-\alpha)^{-2k} p^{\{k\}}(t)\overline{q^{\{k\}}(t)}\, d\sigma(t) + \sum_{i=0}^{k-1} \overline{q}_i \sum_{j=0}^{k-1} c_{i+j}\, p_j \\
&\quad + \sum_{i=0}^{k-1} \overline{q}_i\left(\sum_{j=k}^{l} c_{i+j}p_j + P_i\right) + \sum_{j=0}^{k-1} p_j \left(\sum_{i=k}^{l} c_{i+j}\overline{q}_i + \overline{q}_j\right)P + \overline{Q'_j}) \\
&= \int_{\mathbb{R}} (t-\alpha)^{-2k} p^{\{k\}}(t)\overline{q^{\{k\}}(t)}\, d\sigma(t) + \sum_{i=0}^{k-1} \overline{q}_i \sum_{j=0}^{k-1} c_{i+j}p_j \\
&\quad + \sum_{i=0}^{k-1} \overline{q}_i P_i + \sum_{j=0}^{k-1} p_j \overline{Q}_j,
\end{aligned}$$

where we agree that $c_\nu = 0$ if $\nu > l$. This relation can be written as

$$[p,q]_\varphi = (G\iota_\varphi p, \iota_\varphi q)_{L_2(\sigma)\oplus\mathbb{C}^{2k}} \quad (p,q \in \mathcal{P})$$

with the Gram operator G as in (2.6).

Since $(Gx, x) = 0$ for all elements

$$x = (0; 0, \ldots, 0; \xi_0, \ldots, \xi_{k-1})^\top, \quad \xi_0, \ldots, \xi_{k-1} \in \mathbb{C},$$

e.g. the minimax characterization of the negative eigenvalues of G implies that G has k negative eigenvalues, counted according to their multiplicities. On the other hand, $\kappa_\varphi = k$ if $l \leq 2k - 1$ (see (2.8)).

According to Lemma 2.4, no linear combination of $p_0, \ldots, p_{k-1}, P_0, \ldots, P_{k-1}$ is continuous with respect to the seminorm

$$p \to \left(\int\limits_{\mathbb{R}} (t - \alpha)^{-2k} \, |p^{\{k\}}(t)|^2 d\sigma(t) \right)^{\frac{1}{2}},$$

and the claims of Theorem 2.5 in case $l \leq 2k - 1$ follow from Lemma 2.1.

Assume now that $l \geq 2k$. It follows as above that

$$[p, q]_\varphi = \int\limits_{\mathbb{R}} (t - \alpha)^{-2k} p^{\{k\}}(t) \overline{q^{\{k\}}(t)} \, d\sigma(t)$$

$$+ \sum_{i=0}^{l-k} \overline{q}_i \sum_{j=0}^{l-k} c_{i+j} \, p_j + \sum_{i=0}^{k-1} \overline{q}_i \left(\sum_{j=l-k+1}^{l} c_{i+j} \, p_j + P_i' \right)$$

$$+ \sum_{j=0}^{k-1} p_j \left(\sum_{i=l-k+1}^{l} c_{i+j} \, \overline{q}_i + \overline{Q_j'} \right).$$

This relation can be written as

$$[p, q]_\varphi = (G\iota_\varphi p, \iota_\varphi q)_{L_2(\sigma) \oplus \mathbb{C}^{2k+r}} \quad (p, q \in \mathcal{P})$$

with the Gram operator G in (2.7).

The number of negative eigenvalues of G is equal to $k + \frac{r}{2}$ if r is even and, if r is odd, $r = 2\rho + 1$, it is equal to $k + \rho + 1$ if $c_l < 0$ and $k + \rho$ if $c_l > 0$. This is just the value κ_φ (see (2.8)) in the case $l > 2k$. Now all assertions of Theorem 2.5 for the case $l > 2k$ follow as above from the Lemmas 2.1 and 2.4.

Evidently, in Theorem 2.5 the linear space $\mathcal{P}$ can be replaced by $C^\infty(\mathbb{R})$ or $\mathcal{B}_2(\varphi)$.

2.4. The spaces $\sqcap(\phi)$. Now let $\phi \in \mathcal{F}, \phi = \varphi + \psi$ with $\varphi \in \mathcal{F}(\mathbb{R}), \psi \in \mathcal{F}(\mathbb{C}\backslash\mathbb{R})$. The considerations in Subsection 2.2, in particular (2.1), and the fact that all the inner product spaces on the right hand side of this relation have a finite number of negative squares (see Proposition 2.3 and Theorem 2.5) imply that $(\mathcal{B}_2(\varphi) \times H(\sigma_0(\phi)), [\cdot, \cdot]_\phi)$ has a finite number of negative squares. We denote the completion of $(\mathcal{B}/\mathcal{B}^0, [\cdot, \cdot]_\phi)$ where $\mathcal{B} := \mathcal{B}_2(\varphi) \times H(\sigma_0(\phi))$, by $\sqcap(\phi)$. On account of Lemma 2.2 this space decomposes as follows:

$$(2.9) \qquad \sqcap(\phi) = \sqcap(\tilde{\chi}_{\Delta_1}\varphi)\,[\dotplus]\dots[\dotplus]\,\sqcap(\tilde{\chi}_{\Delta_n}\varphi)[\dotplus]\,\sqcap(\tilde{\chi}_1\psi)\,[\dotplus]\dots[\dotplus]\,\sqcap(\tilde{\chi}_m\psi).$$

Here $(\Delta_i)_{i=1}^n$ is again a φ–minimal decomposition of $\mathbb{R}$ and $\tilde{\chi}_{\Delta_1},\dots,\tilde{\chi}_{\Delta_n},\tilde{\chi}_1,\dots,\tilde{\chi}_m$ are defined as in 2.2. Models of the spaces $\sqcap(\tilde{\chi}_{\Delta_i}\varphi),\sqcap(\tilde{\chi}_j\psi),i=1,\dots,n,j=1,\dots,m,$ were given in Proposition 2.3 and Theorem 2.5, and a corresponding model of $\sqcap(\phi)$ is a direct orthogonal sum of models of these types. The number of negative squares of $\sqcap(\phi)$ is then, evidently, the sum of the numbers of negative squares of all the components on the right hand side of (2.9), which are given by (2.3) and (2.8), respectively.

Concluding this section we mention the following. The starting point of the above considerations was an inner product $[\cdot,\cdot]_\phi$ on the space $\mathcal{P}$, given by some $\phi \in \mathcal{F}$, and it turned out that this inner product has a finite number of negative squares. We could have started from any inner product $[\cdot,\cdot]$ on $\mathcal{P}$ with a finite number κ of negative squares. Then, up to a positive measure near ∞, the inner product $[\cdot,\cdot]$ is generated by some $\phi \in \mathcal{F}$. Namely, for each sufficiently large bounded open interval Δ there exist a measure σ_∞ on $\mathbb{R}\backslash\Delta$ and a $\phi \in \mathcal{F}$ which is zero on $\mathbb{R}\backslash\overline{\Delta}$ such that

$$(2.10) \qquad [p,q] = [p,q]_\phi + \int\limits_{\mathbb{R}\backslash\Delta} p(t)\,\overline{q(t)}\,d\sigma_\infty(t)\ (p,q \in \mathcal{P}).$$

Here the measure σ_∞ is such that

$$\int\limits_{\mathbb{R}\backslash\Delta} |t|^n d\sigma_\infty(t) < \infty$$

for all $n \in \mathbb{N}$.

This follows immediately from the results of [13]. Observe that the hermitian sesquilinear form $[\cdot,\cdot]$ on $\mathcal{P}$ determines a sequence (s_n) of "moments"

$$(2.11) \qquad s_n := [t^n, 1], n = 0, 1\dots,$$

which belongs to the class H_κ (see [13]). In general, even after the interval Δ has been chosen, neither ϕ nor the measure σ_∞ in (2.10) are uniquely determined by $[\cdot,\cdot]$. More exactly, they are uniquely determined (after Δ has been chosen and we agree that σ_∞ does not have point masses in the boundary points of Δ) if and only if the moment problem for the sequence (s_n) from (2.11) is determined.

3. MODELS FOR CYCLIC SELFADJOINT OPERATORS IN PONTRJAGIN SPACES

3.1. The operator of multiplication by the independent variable in $\sqcap(\phi)$. Let $\phi \in \mathcal{F}, \phi = \varphi + \psi$ with $\varphi \in \mathcal{F}(\mathbb{R}), \psi \in \mathcal{F}(\mathbb{C}\backslash\mathbb{R})$. On $(\mathcal{P},[\cdot,\cdot]_\phi)$ (or, what amounts to the same, on $\mathcal{B}_2(\varphi) \times H(\sigma_0(\phi))$) we consider the operator A_0 of multiplication with the independent variable

$$(A_0\, p)(z) := z\, p(z) \quad (p \in \mathcal{P}).$$

Evidently, $[A_0 p, q]_\phi = [p, A_0 q]_\phi (p, q \in \mathcal{P})$, hence A_0 generates a hermitian operator A_ϕ in $\sqcap(\phi)$. This operator A_ϕ is continuous. In order to see this, let μ be the order of the distribution φ and let $\triangle$ be an open interval with $\mathrm{supp}\,\varphi \subset \triangle$. Suppose again that ψ has the form (1.4), and set $\nu := \max\{\nu_i - 1 : i = 1, \dots, m\}$. Then the inner product $[\cdot, \cdot]_\phi$ and the operator A_0 are bounded with respect to the norm

$$\|p\| := \sup\left\{|p^{(i)}(t)| : 0 \le i \le \mu, t \in \triangle\right\} + \ \max\{|p^{(k)}(\beta)| : \beta \in \sigma_0(\phi), 0 \le k \le \nu\}$$

on $\mathcal{P}$. A result of M.G.Krein (see [9], [4]) implies that A_ϕ is bounded in $\sqcap(\phi)$ and, hence, can be extended by continuity to the whole space $\sqcap(\phi)$. The closure of A_ϕ, also denoted by A_ϕ, is called the operator of multiplication by the independent variable in $\sqcap(\phi)$.

Consider now a decomposition (2.9) of the space $\sqcap(\phi)$. The operator A_ϕ maps each component on the right–hand side of (2.9) into itself. This implies:

PROPOSITION 3.1. Under the above assumptions the operator A_ϕ in $\sqcap(\phi)$ is the direct orthogonal sum of the operators

$$A_{\tilde{\chi}_{\triangle_i}\varphi} \in \mathcal{L}(\sqcap(\tilde{\chi}_{\triangle_i}\varphi)), \ i = 1, 2, \dots, n, \ A_{\tilde{\chi}_j \psi} \in \mathcal{L}(\sqcap(\tilde{\chi}_j \psi)), \ j = 1, 2, \dots, m.$$

Therefore, in order to describe the operator A_ϕ, it is sufficient to describe the operators A_φ, $\varphi \in \mathcal{F}(\mathbb{R}; \alpha)$ and A_ψ, $\psi \in \mathcal{F}(\mathbb{C}\backslash\mathbb{R})$, $\sigma_0(\psi) = \{\beta, \overline{\beta}\}$. It is the aim of this subsection to find matrix representations of A_φ and A_ψ in the model spaces of Proposition 2.3 and Theorem 2.5. For the sake of simplicity these matrix representations of A_φ and A_ψ are denoted by the same symbols A_φ and A_ψ (although they are in fact $\iota A_\varphi \iota^{-1}$ with some isometric isomorphism ι).

THEOREM 3.2. Let $\varphi \in \mathcal{F}(\mathbb{R})$, $s(\varphi) = \{\alpha\}$, and suppose that $k, l, \sigma, c_0, \dots, c_l$ are associated with φ according to Lemma 1.2. Then in the space $L_2(\sigma) \oplus \mathbb{C}^{2k+r}$, equipped with the same inner product as in Theorem 2.5, the operator A_φ admits the following matrix representation:

$$(3.1) \quad \left[\begin{array}{c|ccccc|ccccc}
t\cdot & 0 & 0 & \dots & 0 & 1 & & & & & \\
\hline
 & \alpha & 0 & \dots & 0 & 0 & & & & & \\
 & 1 & \alpha & \dots & 0 & 0 & & & & & \\
 & \vdots & \ddots & \ddots & & \vdots & & & & & \\
 & 0 & 0 & \ddots & \alpha & 0 & & & & & \\
 & 0 & 0 & \dots & 1 & \alpha & & & & & \\
\hline
(\cdot, 1)_\sigma & 0 & 0 & \dots & 0 & c_{2k-1} & \alpha & 0 & \dots & 0 & 0 \\
0 & 0 & 0 & \dots & 0 & c_{2k-2} & 1 & \alpha & \dots & 0 & 0 \\
\vdots & \vdots & \vdots & & \vdots & \vdots & \vdots & \ddots & \ddots & & \vdots \\
0 & 0 & 0 & \dots & 0 & c_{k+1} & 0 & 0 & \dots & 1 & \alpha
\end{array}\right] \quad \text{if } l \le 2k - 1,$$

$$-k\ columns\ - \qquad -k\ columns-$$

$$(3.2) \quad
\left[
\begin{array}{c|cccc|cccc|cccc}
t\cdot & 0 & \dots & 0 & 1 & & & & & & & & \\
\hline
 & \alpha & \dots & 0 & 0 & & & & & & & & \\
 & 1 & & 0 & 0 & & & & & & & & \\
 & \vdots & \ddots & \ddots & \vdots & & & & & & & & \\
 & 0 & \dots & 1 & \alpha & & & & & & & & \\
\hline
 & 0 & \dots & 0 & 1 & \alpha & \dots & 0 & 0 & & & & \\
 & 0 & \dots & 0 & 0 & 1 & & 0 & 0 & & & & \\
 & \vdots & & \vdots & \vdots & \vdots & \ddots & \ddots & \vdots & & & & \\
 & 0 & \dots & 0 & 0 & 0 & \dots & 1 & \alpha & & & & \\
\hline
(\cdot,\mathbf{1})_\sigma & & & & & 0 & \dots & 0 & c_l & \alpha & \dots & 0 & 0 \\
0 & & & & & 0 & \dots & 0 & c_{l-1} & 1 & & 0 & 0 \\
\vdots & & & & & \vdots & & \vdots & \vdots & \vdots & \ddots & \ddots & \vdots \\
0 & & & & & 0 & \dots & 0 & c_{l-k+1} & 0 & \dots & 1 & \alpha
\end{array}
\right]
\quad \text{if } l \geq 2k-1,$$

$$-k\ columns-\qquad -r\ columns-\qquad -k\ columns-$$

In these matrices, all the nonindicated entries are zeros. The scalar product in $L_2(\sigma)$ is denoted by $(\cdot,\cdot)_\sigma$.

PROOF. We assume, for simplicity, $\alpha = 0$, and consider e.g. the case $l \leq 2k - 1$; if $l \geq 2k$ a similar reasoning applies. If $p \in \mathcal{P}$, with the mapping ι_φ introduced in (2.5), we write

$$\iota_\varphi(A_\varphi p) = (t^{-k}(tp)^{\{k\}};\bar{p}_0,\bar{p}_1,\dots,\bar{p}_{k-1};\bar{P}_{k-1},\dots,\bar{P}_1,\bar{P}_0)^\top$$

and express the components of this vector by those of $\iota_\varphi(p)$. Evidently,

$$\bar{p}_0 = 0, \quad \bar{p}_j = p_{j-1}, \quad j = 1,2,\dots,k-1.$$

Further,

$$(tp)^{\{n\}} = tp^{\{n-1\}} = tp^{\{n\}} + t^n(n-1)!^{-1}p^{(n-1)}(0) \quad (n \in \mathbb{N}),$$

and we find (with evident notation)

$$\tilde{P}_j' = \int t^{-(2k-j)}(tp)^{\{2k-j\}}(t)d\sigma(t)$$

$$= \int t^{-(2k-j-1)}p^{\{2k-j-1\}}(t)d\sigma(t) = P_{j+1}', \; j = 0,1,\ldots,k-1,$$

$$= \tilde{P}_0 = \sum_{j=k}^{l} c_j \tilde{P}_j + \tilde{P}_0' = \sum_{j=k}^{l} c_j p_{j-1} + P_1' = \sum_{j=k-1}^{l-1} c_{j+1} p_j + P_1$$

$$= c_k p_{k-1} + P_1,$$

$$\tilde{P}_1 = \sum_{j=k}^{l} c_{j+1} \tilde{p}_j + \tilde{P}_1' = \sum_{j=k}^{l-1} c_{j+1} p_{j-1} + P_2' = \sum_{j=k-1}^{l-1} c_{j+2} p_j + P_2'$$

$$= c_{k+1} p_{k-1} + P_2,$$

$$\vdots$$

$$\tilde{P}_{k-1} = \sum_{j=k}^{l} c_{j+k-1} \tilde{p}_j + \tilde{P}_{k-1}' = \sum_{j=k}^{l} c_{j+k-1} p_{j-1} + P_k'$$

$$= \sum_{j=k-1}^{l-1} c_{j+k} p_j + P_k' = c_{2k-1} p_{k-1} + P_k$$

where

$$P_k := \int t^{-k} p^{\{k\}}(t)d\sigma(t) = (t^{-k} p^{\{k\}}, 1)_\sigma.$$

It follows that A_φ has the matrix representation (3.1) with $\alpha = 0$.

The proof of the following theorem about the matrix representation of the operator $A_\psi, \psi \in \mathcal{F}(\mathbb{C}\backslash\mathbb{R})$, $\sigma_0(\psi) = \{\beta, \overline{\beta}\}$, is similar but much simpler and therefore left to the reader.

THEOREM 3.3. Let $\psi \in \mathcal{F}(\mathbb{C}\backslash\mathbb{R})$, $\sigma_0(\psi) = \{\beta, \overline{\beta}\}$ and suppose that ψ has the form (2.2). Then in the space $\mathbb{C}^{2\nu}$, equipped with the same inner product as in Proposition 2.3, the operator A_ψ admits the matrix representation

$$\left[\begin{array}{ccc|ccc} \beta & & 0 & & 0 & \\ 1 & \ddots & & & & \\ & \ddots & \ddots & & & \\ 0 & & 1 & \beta & & \\ \hline & & & \beta & & 0 \\ & & & 1 & \ddots & \\ & 0 & & & \ddots & \ddots \\ & & & 0 & & 1 & \overline{\beta} \end{array}\right]$$

3.2. Eigenspace, resolvent and spectral function of A_φ. In this section, if $\varphi \in \mathcal{F}(\mathbb{R}; \alpha)$, for the model operator A_φ in $L_2(\sigma) \oplus \mathbb{C}^{2k+r}$ from (3.1), (3.2) we study the algebraic eigenspace at α and the behaviour of its resolvent and spectral function near α. Without loss of generality we suppose that $\alpha = 0$.

First we mention that $\alpha = 0$ is an eigenvalue of A_φ with a nonpositive eigenvector and that A_φ has no other eigenvalues with this property. A maximal Jordan chain $x_0, x_1, \ldots, x_n$ of A_φ at $\alpha = 0$ is given as follows:

If $l \leq 2k - 1$, then $m = k - 1$ and

$$
\begin{aligned}
x_{k-1} &= (0; 0, \ldots, 0; 1, 0, \ldots, 0)^\top, \\
x_{k-2} &= (0; 0, \ldots, 0; 0, 1, \ldots 0)^\top, \\
&\;\;\vdots \\
x_0 &= (0; 0, \ldots, 0; 0, 0, \ldots, 1)^\top
\end{aligned}
$$

and the span of these elements is neutral. If $l \geq 2k$ then $m = l - k + 1$ and

$$
\begin{aligned}
x_{l-k} &= (0; 0, \ldots, 0; 1, 0, \ldots, 0; 0, \ldots, 0)^\top, \\
x_{l-k-1} &= (0; 0, \ldots, 0; 0, 1, \ldots, 0; 0, \ldots, 0)^\top, \\
&\;\;\vdots \\
x_k &= (0; 0, \ldots, 0; 0, 0, \ldots, 1; 0, \ldots, 0)^\top, \\
x_{k-1} &= (0; 0, \ldots, 0; 0, \ldots, 0; c_l, c_{l-1}, \ldots, c_{l-k+1})^\top, \\
&\;\;\vdots \\
x_0 &= (0; 0, \ldots, 0; 0, \ldots, 0; 0, \ldots, 0, c_l)^\top.
\end{aligned}
$$

In the second case it holds

$$
\left([x_i, x_j] \right)_{i,j=0}^{l-k} =
\begin{bmatrix}
0 & & & 0 & \\
\hline
& 0 & \cdots & 0 & c_l \\
0 & \vdots & \ddots & \ddots & \vdots \\
& 0 & & & c_{2k+1} \\
& c_l & \cdots & c_{2k+1} & c_{2k}
\end{bmatrix}
$$

where $[\cdot, \cdot] := (G\cdot, \cdot)_{L_2(\sigma) \oplus \mathbb{C}^{2k+r}}$ (see (2.7)). Hence the elements $x_0, \ldots, x_{k-1}$ span a neutral subspace, whereas on the span of $x_k, \ldots, x_{l-k}$ the inner product $[\cdot, \cdot]$ is nondegenerate.

Next we consider the matrix representation of the resolvent of A_φ. For the sake of simplicity we write down the matrix $(A_\varphi - zI)^{-1}$ if $k = 2$ and $l \leq 2k - 1 = 3$; its structure for arbitrary k and $l \leq 2k - 1$ will then be clear. This matrix is

$$\left[\begin{array}{c|cc|cc} (t-z)^{-1}\cdot & z^{-2}(t-z)^{-1} & z^{-1}(t-z)^{-1} & 0 & 0 \\ \hline 0 & -z^{-1} & 0 & 0 & 0 \\ 0 & -z^{-2} & -z^{-1} & 0 & 0 \\ \hline -z^{-1}(\cdot,(t-z)^{-1})_\sigma & b_{11}(z) & b_{12}(z) & -z^{-1} & 0 \\ -z^{-2}(\cdot,(t-z)^{-1})_\sigma & b_{21}(z) & b_{22}(z) & -z^{-2} & -z^{-1} \end{array}\right]$$

where

$$(b_{ij}(z))_1^2$$

$$= \left[\begin{array}{cc} -c_3 z^{-3} + z^{-3}\int (t-z)^{-1}d\sigma(t) & -c_3 z^{-2} + z^{-2}\int (t-z)^{-1}d\sigma(t) \\ -c_3 z^{-4} - c_2 z^{-3} + z^{-4}\int (t-z)^{-1}d\sigma(t) & -c_3 z^{-3} - c_2 z^{-2} + z^{-3}\int (t-z)^{-1}d\sigma(t) \end{array}\right]$$

The growth of $(A_\varphi - zI)^{-1}$ if z approaches zero along the imaginary axis or, more generally, nontangentially, is given by the term $b_{21}(z)$.

In the general case the growth of $(A_\varphi - zI)^{-1}$ is also determined by the entry b on the second place of the last row. If $l \le 2k - 1$, we have

$$b(z) = -c_k z^{-k-1} - c_{k+1} z^{-k-2} - \ldots - c_{2k-1} z^{-2k} + z^{-2k}\int (t-z)^{-1}d\sigma(t)$$

and, hence, for $z = iy$ and $y \downarrow 0$,

$$|y^{2k-1}b(iy)| \to \infty, \quad |y^{2k+1}b(iy)| \le c < \infty.$$

If $l \ge 2k$, we have

$$b(z) = -c_l z^{-l-1} - c_{l-1} z^{-l} - \ldots - c_{l-k+1} z^{-l+k-2} - z^{-2k}\int (t-z)^{-1}d\sigma(t).$$

This implies

$$y^{l+1}|b(iy)| \to |c_l| \text{ for } y \downarrow 0.$$

For sufficiently large exponents the powers of the model operator A_φ have a simple matrix form which is independent of the numbers $c_j, j = 0,\ldots,l$. If $l \ge 2k$, the operator A_φ^n with $n \ge k + r$ is given by

$$(3.3) \quad \left[\begin{array}{c|cccc|cc} t^n\cdot & t^{n-k} & t^{n-k-1} & \cdots & t^{n-1} & 0 & 0 \\ \hline 0 & & 0 & & & 0 & 0 \\ \hline 0 & & 0 & & & 0 & 0 \\ \hline (\cdot,t^{n-1})_\sigma & (1,t^{n-k-1})_\sigma & (1,t^{n-k})_\sigma & \cdots & (1,t^{n-2})_\sigma & & \\ (\cdot,t^{n-2})_\sigma & (1,t^{n-k-2})_\sigma & (1,t^{n-k-1})_\sigma & \cdots & (1,t^{n-3})_\sigma & & \\ \vdots & \vdots & \vdots & \ddots & \vdots & 0 & 0 \\ (\cdot,t^{n-k})_\sigma & (1,t^{n-2k})_\sigma & (1,t^{n-2k+1})_\sigma & \cdots & (1,t^{n-k-1})_\sigma & & \end{array}\right]$$

If we agree that in the case $l \leq 2k - 1$ the third row and column in (3.3) disappear, then (3.3) gives also the matrix representation of A_φ^n for $n \geq k, l \leq 2k - 1$.

Let E be the spectral function of A_φ and let $\triangle$ be an arbitrary interval with $0 \notin \overline{\triangle}$. Then making use of (3.3) and the fact that $E(\triangle)$ can be written as the strong limit of a sequence of polynomials of A_φ^n for odd n, or applying the Stieltjes-Livshic inversion formula to the model operator of the resolvent of A_φ, we obtain the following matrix representation for $E(\triangle)$:

$$
\left[
\begin{array}{c|cccc|c|c}
\chi_\triangle \cdot & t^{-k}\chi_\triangle & t^{-k+1}\chi_\triangle & \cdots & t^{-1}\chi_\triangle & 0 & 0 \\
\hline
0 & \multicolumn{4}{c|}{0} & 0 & 0 \\
\hline
0 & \multicolumn{4}{c|}{0} & 0 & 0 \\
\hline
(\cdot, \chi_\triangle t^{-1})_\sigma & (1, \chi_\triangle t^{-k-1})_\sigma & (1, \chi_\triangle t^{-k})_\sigma & & (1, \chi_\triangle t^{-2})_\sigma & & \\
(\cdot, \chi_\triangle t^{-2})_\sigma & (1, \chi_\triangle t^{-k-2})_\sigma & (1, \chi_\triangle t^{-k-1})_\sigma & & (1, \chi_\triangle t^{-3})_\sigma & & \\
\vdots & \vdots & \vdots & \ddots & \vdots & 0 & 0 \\
(\cdot, \chi_\triangle t^{-k})_\sigma & (1, \chi_\triangle t^{-2k})_\sigma & (1, \chi_\triangle t^{-2k+1})_\sigma & \cdots & (1, \chi_\triangle t^{-k-1})_\sigma & &
\end{array}
\right]
$$

Again, if $l \leq 2k - 1$, the third row and column disappear. It follows that

$$\|E(\triangle)\| = 0\left(\int_\triangle t^{-2k} d\sigma(t)\right),$$

if a boundary point of $\triangle$ approaches zero. The growth of $\|E(\triangle)\|$ is determined by k and independent of l. In particular, the point $\alpha = 0$ is a regular critical point of A_φ ([10], [14]) if and only if $k = 0$.

Evidently, by an appropriate choice of σ, k, l and $c_j, j = 0, \ldots, l$, examples for selfadjoint operators in Pontrjagin spaces with different growth properties for the resolvent and the spectral function can be constructed.

3.3. Cyclic selfadjoint operators in Pontrjagin spaces. Let A be a bounded cyclic selfadjoint operator in the Pontrjagin space $\sqcap$. Recall that A is called **cyclic**, if there exists a **generating element** $u \in \sqcap$ such that

$$\sqcap = c.l.s.\{A^j u : j = 0, 1, \ldots\}.$$

If $\phi \in \mathcal{F}$, then, evidently, the operator A_ϕ is cyclic in $\sqcap(\phi)$ with generating element **1** (or, more exactly, the element of $\sqcap(\phi)$ corresponding to **1**). The following theorem, which is the main result of this subsection, states that in this way we obtain all bounded cyclic selfadjoint operators in Pontrjagin spaces. In the following, "unitary" and "isometric" are always understood with respect to Pontrjagin space inner products.

THEOREM 3.4. Let A be a bounded cyclic selfadjoint operator in a Pontrjagin space $(\sqcap, [\cdot, \cdot])$ with generating element u. Then the linear functional

$$\phi : \mathcal{P} \ni p \to [p(A)u, u]$$

belongs to $\mathcal{F}$ and A is unitarily equivalent to the operator A_ϕ of multiplication by the independent variable in $\sqcap(\phi)$.

PROOF. If $p \in \mathcal{P}$, we have

$$(3.4) \qquad [p(A)u,u] = [p(A)E(\sigma(A)\cap \mathbb{R})u,u] + [p(A)E(\sigma(A)\backslash\mathbb{R})u,u]$$

where, for a spectral set σ of A, $E(\sigma)$ denotes the corresponding Riesz-Dunford projection. Further,

$$(3.5) \qquad\qquad [p(A)E(\sigma(A)\backslash\mathbb{R})u,u] =$$

$$= \sum_{\beta\in\sigma(A)\cap\mathbb{C}^+} ([p(A)E(\{\beta\})u,u] + [p(A)E(\{\bar\beta\})u,u])$$

and for $\beta \in \sigma(A)\backslash\mathbb{R}$ there exists a $\nu_\beta \in \mathbb{N}$ such that $(A-\beta I)^{\nu_\beta}E(\{\beta\}) = 0$. Hence

$$(3.6) \qquad [p(A)E(\{\beta\})u,u] = \sum_{\nu=0}^{\nu_\beta-1} \nu!^{-1}p^{(\nu)}(\beta)[(A-\beta I)^\nu E(\{\beta\})u,u] \ (p\in\mathcal{P}).$$

Moreover, for $\nu = 0,\ldots,\nu_\beta - 1$ we have

$$(3.7) \qquad\qquad [(A-\bar\beta I)^\nu E\{\bar\beta\})u,u] = \overline{[(A-\beta I)^\nu E(\{\beta\})u,u]}.$$

From (3.5), (3.6) and (3.7) it follows that the functional

$$p \to [p(A)E(\sigma(A)\backslash\mathbb{R})u,u]$$

belongs to $\mathcal{F}(\mathbb{C}\backslash\mathbb{R})$. Therefore, by (3.4), in order to prove that $\phi \in \mathcal{F}$ it is sufficient to show that the functional

$$\varphi : \mathcal{P} \ni p \to [p(A)u_0,u_0], \quad u_0 := E(\sigma(A)\cap\mathbb{R})u,$$

belongs to $\mathcal{F}(\mathbb{R})$.

Denote by p_0 a definitizing polynomial of $A_0 := A|\sqcap_0$, $\sqcap_0 := E(\sigma(A)\cap\mathbb{R})\sqcap$ with only real zeros, say $\alpha_1,\ldots,\alpha_n$ (mutually different) of orders $\mu_1,\ldots,\mu_n$, which is nonnegative on $\mathbb{R}$ (see, e.g., [14]). Let

$$(p_0(t))^{-1} = \sum_{i=1}^{n}\sum_{j=1}^{\mu_i} c_{ij}(t-\alpha_i)^{-j}, \quad t\in\mathbb{R}\backslash\{\alpha_1,\ldots,\alpha_n\}.$$

Then, for arbitrary $p \in \mathcal{P}$,

$$(3.8) \qquad\qquad p(t) = g(t;p) + p_0(t)h(t;p),$$

where

$$g\left(t;p\right) := p_0(t) \sum_{i=1}^{n} \sum_{j=1}^{\mu_i} c_{ij} \left(t - \alpha_i\right)^{-j}(p(\alpha_i) + \ldots + (j-1)!^{-1}p^{(j-1)}(\alpha_i)\left(t - \alpha_i\right)^{j-1}),$$

$$h\left(t;p\right) := \sum_{i=1}^{n} \sum_{j=1}^{\mu_i} c_{ij} \left(t - \alpha_i\right)^{-j} p^{\{\alpha_i, j\}}(t).$$

We choose a bounded open interval Δ which contains $\sigma(A_0)$, denote by μ the maximum of the μ_i, $i = 1, 2, \ldots, n$, and consider the set S of all polynomials p such that

$$\sup\{|p^{(k)}(t)| : 0 \le k \le \mu, t \in \Delta\} \le 1.$$

The polynomial $g\left(\cdot; p\right)$ depends on p only through the numbers $p^{(j)}(\alpha_i), i = 1, \ldots, n$, $j = 0, \ldots, \mu_i - 1$, therefore it holds

(3.9)
$$\sup\{\|[g\left(A_0;p\right)u_0, u_0]\| : p \in S\} < \infty.$$

As p_0 is a definitizing polynomial of A_0, the inner product $[p_0\left(A_0\right)\cdot, \cdot]$ is nonnegative in $\sqcap_0$. Evidently the operator A_0 is symmetric with respect to this inner product. From the result of M.G. Krein used already above (see [9]) it follows that A_0 induces a bounded selfadjoint operator in the Hilbert space which is generated in a canonical way from the inner product space $(\sqcap_0, [p_0\left(A_0\right)\cdot, \cdot])$. Moreover, the spectrum of this operator is contained in Δ. Therefore the functional

$$\mathcal{P} \ni q \to [p_0\left(A_0\right)q\left(A_0\right)u_0, u_0]$$

can be written as $\int_{\Delta} q\left(t\right)d\mu\left(t\right)$ with a measure μ supported on Δ. Taylor's formula implies that the polynomials $h\left(\cdot; p\right)$, $p \in S$, are uniformly bounded on Δ. Hence

$$\sup\{\|[p_0\left(A_0\right)h\left(A_0;p\right)u_0, u_0]\| : p \in S\} < \infty,$$

and from (3.8) and (3.9) it follows that

$$\sup\{\varphi\left(p\right) : p \in S\} < \infty.$$

This relation assures us that φ is a distribution (for a similar reasoning cf. [7; proof of Theorem 1]). Moreover, if $f \in C_0^{\infty}\left(\mathbb{R}\right)$ is nonnegative and $\alpha_i \notin$ supp f for all $i = 1, \ldots n$, then

$$\varphi \cdot f = p_0\varphi \cdot p_0^{-1}f \ge 0,$$

that is, $\varphi \in \mathcal{F}\left(\mathbb{R}\right)$.

In order to prove the unitary equivalence of A_ϕ and A, consider the isometric linear mapping U_0 from $(\mathcal{P}, [\cdot, \cdot]_\phi)$ onto a dense subspace of $\sqcap$ defined by

$$U_0\, p := p(A)u \quad (p \in \mathcal{P}).$$

Then $U_0(A_\phi p) = Ap(A)u$. Evidently, a polynomial p belongs to the isotropic subspace $\mathcal{P}^0$ of $(\mathcal{P}, [\cdot,\cdot]_\phi)$ if and only if $p(A)u = 0$. Then, if U_0' denotes the isometric bijection

$$p + \mathcal{P}^0 \to p(A)u$$

of $(\mathcal{P}/\mathcal{P}^0, [\cdot,\cdot]_\phi)$ into $\sqcap$, we have

$$U_0' A_\phi = A U_0' \text{ on } \mathcal{P}/\mathcal{P}^0.$$

The extension U of U_0' by continuity is an isometric isomorphism between $\sqcap(\phi)$ and $\sqcap$ satisfying

$$U A_\phi = AU,$$

and the theorem is proved.

UNITARY EQUIVALENCE OF CYCLIC SELFADJOINT
OPERATORS IN PONTRJAGIN SPACES

4.1. The general case. Recall that two selfadjoint operators $A, \hat{A}$ in Pontrjagin spaces $\sqcap$ and $\hat{\sqcap}$, respectively, are said to be unitarily equivalent if there exists a unitary operator U from $\sqcap$ onto $\hat{\sqcap}$ such that

$$(4.1) \qquad\qquad\qquad \hat{A} U = UA.$$

In this section we study the unitary equivalence of two cyclic selfadjoint operators $A, \hat{A}$ in Pontrjagin spaces $\sqcap$ and $\hat{\sqcap}$, respectively. More exactly, we fix generating elements $u, \hat{u}$ of A and $\hat{A}$, respectively. Then to A and $\hat{A}$ there correspond model operators $A_\phi, A_{\hat{\phi}}$ in spaces $\sqcap(\phi), \sqcap(\hat{\phi})$ for certain $\phi, \hat{\phi} \in \mathcal{F}$ (see Theorem 3.4), and we express the unitary equivalence of A_ϕ and $A_{\hat{\phi}}$ in terms of $\phi, \hat{\phi}$.

Evidently, if A and $\hat{A}$ are unitarly equivalent, then $\kappa_-(\sqcap) = \kappa_-(\hat{\sqcap})$, $\hat{E}(\triangle)U = UE(\triangle)$ for all admissible intervals $\triangle$ (where $E, \hat{E}$ denote the spectral functions of $A, \hat{A}$), $\sigma(A) = \sigma(\hat{A})$, $\sigma_p(A) = \sigma_p(\hat{A})$ and the algebraic eigenspaces of A and $\hat{A}$ corresponding to the same eigenvalue are isometric. As for a nonreal eigenvalue λ_0 of A and $\hat{A}$ the unitary equivalence of the algebraic eigenspaces means just that the lengths of the Jordan chains coincide (recall that the algebraic eigenspace of A at λ_0 consists of just one chain of finite length), we can suppose without loss of generality that A and $\hat{A}$ have only real spectrum and, after suitable decompositions of the spaces $\sqcap$ and $\hat{\sqcap}$, that A and $\hat{A}$ have just one eigenvalue α with a nonpositive eigenvector, and that $\alpha = 0$. So we are led to the following problem: Given $\varphi, \hat{\varphi} \in \mathcal{F}(\mathbb{R}; 0)$. Find necessary and sufficient conditions for the unitary equivalence of A_φ and $A_{\hat{\varphi}}$ in terms of φ and $\hat{\varphi}$. Here we suppose that $\varphi, \hat{\varphi}$ are given by their representations according to Lemma 1.2, that is,

$$\varphi \text{ is given by } k, l \in \mathbb{N}_0, \ c_0, \ldots, c_l \in \mathbb{R} \text{ and a measure } \sigma,$$

$$\hat{\varphi} \text{ is given by } \hat{k}, \hat{l} \in \mathbb{N}_0, \ \hat{c}_0, \ldots, \hat{c}_l \in \mathbb{R} \text{ and a measure } \hat{\sigma},$$

where we always assume that these data have all the properties mentioned in Lemma 1.2.

As $k\,(\hat{k})$ is the length of the isotropic part of the Jordan chain of $A_\varphi\,(A_{\hat\varphi})$ at 0, we have $k = \hat{k}$. Further, if $l \geq 2k$, then $l - k + 1$ is the maximal length of a Jordan chain of A_φ at 0, hence in this case $l = \hat{l}$. It follows that the sizes of the blocks of the matrices for A_φ and $A_{\hat\varphi}$ in (3.1) or (3.2) and for the Gram operators G_φ and $G_{\hat\varphi}$ from (2.6) or (2.7) coincide:

$$(4.2) \qquad G_\varphi = \begin{bmatrix} I & 0 & 0 & 0 \\ 0 & H_1 & H_3 & Z \\ 0 & H_3^* & H_2 & 0 \\ 0 & Z & 0 & 0 \end{bmatrix}, \quad A_\varphi = \begin{bmatrix} t\cdot & E & 0 & 0 \\ 0 & S_1 & 0 & 0 \\ 0 & J & S_2 & 0 \\ E' & D & C & S_1 \end{bmatrix}.$$

Here we agree that in case $r = 0$ the third row and column disappear. The blocks in (4.2) can be read off from (2.6), (2.7), (3.1), (3.2) with $\alpha = 0$; we only mention that $D = 0$ if $r > 0$. The corresponding blocks of $A_{\hat\varphi}$ are denoted by $\hat{H}_1, \hat{Z}$ etc. The operator U in (4.1) is partitioned in the same way: $U = (U_{ij})_1^4$.

As U maps the algebraic eigenspace of A_φ at 0 onto that of $A_{\hat\varphi}$ and these subspaces are given by the vectors with vanishing first and second block components, it follows that $U_{13} = U_{14} = U_{23} = U_{24} = 0$. Also the set of vectors with vanishing first, second and third block components is invariant under A_φ and $A_{\hat\varphi}$, hence $U_{34} = 0$.

Now we write the relation (4.1) for the block matrices of $A_\varphi, A_{\hat\varphi}, U$. Considering the components $\cdot 21$ on both sides it follows that $\hat{S}_1 U_{21} = U_{21}\,t\cdot$, hence $U_{21} = 0$ and, similarly, $U_{31} = 0$. Then (4.1) turns out to be equivalent to the following relations:

(4.3) $\hat{t}\cdot U_{11} = U_{11}\,t\cdot,$

(4.4) $\hat{t}\cdot U_{12} + \hat{E}U_{22} = U_{11}E + U_{12}S_1,$

(4.5) $\hat{S}_1 U_{22} = U_{22}S_1,$

(4.6) $\hat{J}U_{22} + \hat{S}_2 U_{32} = U_{32}S_1 + U_{33}J,$

(4.7) $\hat{S}_2 U_{33} = U_{33}S_2,$

(4.8) $\hat{E}'U_{11} + \hat{S}_1 U_{41} = U_{41}t\cdot + U_{44}E',$

(4.9) $\hat{E}'U_{12} + \hat{D}U_{22} + \hat{C}U_{32} + \hat{S}_1 U_{42} = U_{41}E + U_{42}S_1 + U_{43}J + U_{44}D,$

(4.10) $\hat{C}U_{33} + \hat{S}_1 U_{43} = U_{43}S_2 + U_{44}C,$

(4.11) $\hat{S}_1 U_{44} = U_{44}S_1.$

As U is isometric with respect to the inner products on $L_2(\sigma) \oplus \mathbb{C}^{2k+r}$ and $L_2(\hat\sigma) \oplus \mathbb{C}^{2k+r}$, generated by the Gram operators G_φ and $G_{\hat\varphi}$, respectively, we have also

(4.12) $U^* G_{\hat\varphi} U = G_\varphi,$

which is equivalent to

(4.13) $U_{11}^* U_{11} = I,$

(4.14) $U_{11}^* U_{12} + U_{41}^* Z U_{22} = 0,$

(4.15) $U_{12}^* U_{12} + U_{22}^* (\hat{H}_1 U_{22} + \hat{H}_3 U_{32} + Z U_{42}) + U_{32}^* (\hat{H}_3^* U_{22} + \hat{H}_2 U_{32}) + U_{42}^* Z U_{22} = H_1,$

(4.16) $U_{22}^* (\hat{H}_3 U_{33} + Z U_{43}) + U_{32}^* \hat{H}_2 U_{33} = H_3,$

(4.17) $U_{22}^* Z U_{44} = Z,$

(4.18) $U_{33}^* \hat{H}_2 U_{33} = H_2.$

The relations (4.3) and (4.13) imply that the operators of multiplication by the independent variable in $L_2(\sigma)$ and $L_2(\hat{\sigma})$ are unitarily equivalent. As is well-known this means the equivalence of the measures σ and $\hat{\sigma}$ and that U_{11} is the operator of multiplication by a $\hat{\sigma}$–measurable function $\hat{\gamma}$ such that $|\hat{\gamma}|^2 = d\sigma/d\hat{\sigma}$ ([1]).

Next we consider (4.5). It implies that U_{22} is a Toeplitz matrix of the form

$$U_{22} = \begin{bmatrix} \hat{u}_0 & 0 & \cdots & 0 & 0 \\ \hat{u}_1 & \hat{u}_0 & & 0 & 0 \\ \vdots & & \ddots & & \vdots \\ \hat{u}_{k-2} & \hat{u}_{k-3} & & \hat{u}_0 & 0 \\ \hat{u}_{k-1} & \hat{u}_{k-2} & \cdots & \hat{u}_1 & \hat{u}_0 \end{bmatrix}, u_j \in \mathbb{C}, \quad j = 0, \ldots, k-1.$$

Further, writing $U_{12} = (\hat{v}_k\ \hat{v}_{k-1} \ldots \hat{v}_1)$ with $\hat{v}_j \in L_2(\hat{\sigma}), j = 1, 2, \ldots, k$, the relation (4.4) is equivalent to

(4.19) $\hat{v}_1 = \mathbf{t}^{-1}(\hat{\gamma} - \hat{u}_0),\ \hat{v}_2 = \mathbf{t}^{-1}(\hat{v}_1 - \hat{u}_1), \ldots,\ \hat{v}_k = \mathbf{t}^{-1}(\hat{v}_{k-1} - \hat{u}_{k-1}).$

This implies

(4.20) $\hat{\gamma}(t) - (\hat{u}_0 + \hat{u}_1 t + \ldots + \hat{u}_{k-1} t^{k-1}) = t^k \hat{v}_k(t)$

$\hat{\sigma}$–a.e. That is, if $k > 0$ the function $\hat{\gamma}(\in L_2(\hat{\sigma}))$ has a well–defined "value" $\hat{u}_0$ at $t = 0$ and also "derivatives" up to the order $k-1$. That $\hat{u}_0, \hat{u}_1, \ldots, \hat{u}_{k-1}$ are uniquely determined by the relations (4.19) or by (4.20) follows from the condition $\int t^{-2} d\hat{\sigma}(t) = \infty$.

Similary, making use of (4.11) and (4.8) and putting $U_{41} = ((\cdot, v_1)_\sigma, \ldots, (\cdot, v_k)_\sigma))^\top$ and

$$U_{44} = \begin{bmatrix} \overline{u_0} & 0 & \cdots & 0 & 0 \\ \overline{u_1} & \overline{u_0} & \cdots & 0 & 0 \\ \vdots & \vdots & \ddots & \vdots & \vdots \\ \overline{u_{k-2}} & \overline{u_{k-3}} & \cdots & \overline{u_0} & 0 \\ \overline{u_{k-1}} & \overline{u_{k-2}} & \cdots & \overline{u_1} & \overline{u_0} \end{bmatrix}, u_j \in \mathbb{C}, \quad j = 0, \ldots, k-1,$$

we find

$$v_1 = \mathbf{t}^{-1}(\hat{\gamma}^{-1} - u_0),\ v_2 = \mathbf{t}^{-1}(v_1 - u_1), \ldots,\ v_k = \mathbf{t}^{-1}(v_{k-1} - u_{k-1})$$

and all these functions belong to $L_2(\sigma)$. It is easy to see that the relation (4.17) is now automatically satisfied. In particular, we have $\hat{u}_0 \neq 0$ and

$$\hat{u}_0 = u_0^{-1}.$$

Then by (4.20) there exists a polynomial p of order $\leq k-1$ with real coefficients and $p(0) = |\hat{u}_0|$ such that $\mathbf{t}^{-k}(|\hat{\gamma}| - p) \in L_2(\hat{\sigma})$. In particular, we have

$$(4.21) \qquad\qquad\qquad \mathbf{t}^{-1}(|\hat{\gamma}| - \hat{u}_0) \in L_2(\hat{\sigma}).$$

We summarize some of the above results to a necessary condition for unitary equivalence of A_φ and $A_{\hat{\varphi}}$. Here κ (and correspondingly $\hat{\kappa}$) are given by (2.8).

THEOREM 4.1. Let $\varphi, \hat{\varphi} \in \mathcal{F}(\mathbb{R}; 0)$ with representation according to Lemma 1.2. If the operators A_φ and $A_{\hat{\varphi}}$ are unitarily equivalent, the following statements hold:

(i) $\kappa = \hat{\kappa}$, $k = \hat{k}$, and, if $l \geq 2k$, $l = \hat{l}$.

(ii) The measures σ and $\hat{\sigma}$ are equivalent; if $k > 0$ there exists a polynomial p of order $\leq k-1$ with $p(0) \neq 0$ and real coefficients such that with $\hat{g} := (d\sigma/d\hat{\sigma})^{\frac{1}{2}}$ the function

$$\mathbf{t}^{-k}(\hat{g} - p)$$

belongs to $L_2(\hat{\sigma})$.

The meaning of the necessary conditions (i), (ii) in Theorem 4.1 is enlightened by the following result, which, in fact, contains Theorem 4.1.

THEOREM 4.2. Let $\varphi, \hat{\varphi}$ be as in Theorem 4.1 and denote by $E, \hat{E}$ the spectral functions of $A_\varphi, A_{\hat{\varphi}}$, respectively. Then the conditions (i), (ii) in Theorem 4.1 are necessary and sufficient for the unitary equivalence of the spectral functions E and $\hat{E}$.

PROOF. 1. Assume that E and $\hat{E}$ are unitarily equivalent. Then, evidently, $\kappa = \hat{\kappa}$. The number $k\,(\hat{k})$ coincides with dimension of the isotropic part of the closed linear span $\mathcal{L}_{(0)}\,(\hat{\mathcal{L}}_{(0)})$ of all ranges of $E(\Delta)\,(\hat{E}(\Delta))$, where Δ is an arbitrary interval with $0 \notin \overline{\Delta}$. Therefore $k = \hat{k}$. The orthogonal companion $\mathcal{L}_{(0)}^{[\perp]}\,(\hat{\mathcal{L}}_{(0)}^{[\perp]})$ of $\mathcal{L}_{(0)}\,(\hat{\mathcal{L}}_{(0)})$ is the algebraic eigenspace of $A_\varphi\,(A_{\hat{\varphi}})$ at 0. Then, in view of the results of Section 3.2, we have $l \geq 2k$ if and only if $\hat{l} \geq 2\hat{k}$, and in this case $l = \hat{l}$. Hence the condition (i) holds.

In order to prove (ii) set $n := 2(k+r)+1$ with $r = \max\{0, l-2k+1\}$. Then A_φ^{n-1} is nonnegative with respect to the Pontrjagin space inner product, and $A_\varphi^n = \int t^n dE(t)$ (cf., e.g., [7; 3.3]). A similar relation holds for $A_{\hat{\varphi}}^n$. Therefore the operators A_φ^n and $A_{\hat{\varphi}}^n$ are unitarily equivalent. Assume that $k > 0$ (if $k = 0$, a similar, much simpler reasoning applies). We write A_φ^n in a form similar to that of A_φ in (4.2) (see (3.3)): $t\cdot$ is replaced by $t^n\cdot$, $E = (\mathbf{t}^{n-k} \dots \mathbf{t}^{n-1})$, $S_1 = J = S_2 = C = 0$, $E' = ((\cdot, \mathbf{t}^{n-1})_\sigma \dots (\cdot, \mathbf{t}^{n-k})_\sigma)^\top$ and

$$D = \begin{bmatrix} (1,t^{n-k-1})_\sigma & (1,t^{n-k})_\sigma & \cdots & (1,t^{n-1})_\sigma \\ (1,t^{n-k-2})_\sigma & (1,t^{n-k-1})_\sigma & \cdots & (1,t^{n-2})_\sigma \\ \vdots & \vdots & \ddots & \vdots \\ (1,t^{n-2k})_\sigma & (1,t^{n-2k+1})_\sigma & \cdots & (1,t^{n-k-1})_\sigma \end{bmatrix}.$$

Let $U = (U_{ij})_1^4$ be a unitary operator from $(L_2(\sigma) \oplus \mathbb{C}^{2k+r}, (G_\varphi \cdot, \cdot)_{L_2(\sigma) \oplus \mathbb{C}^{2k+r}})$ onto $(L_2(\sigma) \oplus \mathbb{C}^{2k+r}, (G_{\hat{\varphi}} \cdot, \cdot)_{L_2(\hat{\sigma}) \oplus \mathbb{C}^{2k+r}})$ such that

(4.22) $\quad A_{\hat{\varphi}}^n U = U A_\varphi^n$

As above it follows that $U_{13} = U_{14} = U_{23} = U_{24} = U_{21} = U_{31} = 0$. The relation (4.22) is equivalent to the following:

(4.23) $\quad \hat{t}^n U_{11} = U_{11} t^n$,

(4.24) $\quad \hat{t}^n U_{12} + \hat{E} U_{22} = U_{11} E$,

(4.25) $\quad \hat{E}' U_{11} = U_{41} t^n + U_{44} E'$,

(4.26) $\quad \hat{E}' U_{12} + \hat{D} U_{22} = U_{41} E + U_{44} D$.

By (4.23) and (4.13) the measures σ and $\hat{\sigma}$ are equivalent, and U_{11} is the operator of multiplication by a $\hat{\sigma}$–measurable function $\hat{\gamma}$ such that $|\hat{\gamma}|^2 = d\sigma/d\hat{\sigma}$. We write $U_{12} = (\hat{v}_k \, \hat{v}_{k-1} \ldots \hat{v}_1)$, $U_{22} = (u_{ij})_0^{k-1}$. Then (4.24) is equivalent to

$$t^n \hat{v}_1 + t^{n-k} u_{0,k-1} + \ldots + t^{n-1} u_{k-1,k-1} = \hat{\gamma} t^{n-1},$$

(4.27) $\qquad\qquad t^n \hat{v}_2 + t^{n-k} u_{0,k-2} + \ldots + t^{n-1} u_{k-1,k-2} = \hat{\gamma} t^{n-2},$

$$\vdots$$

$$t^n \hat{v}_k + t^{n-k} u_{0,0} + \ldots + t^{n-1} u_{k-1,0} = \hat{\gamma} t^{n-k}.$$

In view of $\int t^{-2} d\hat{\sigma}(t) = \infty$ the first equation of (4.27) gives

$$\hat{v}_1 = t^{-1}(\hat{\gamma} - u_{k-1,k-1}), \; u_{0,k-1} = \ldots = u_{k-2,k-1} = 0.$$

The second equation gives

$$\hat{v}_2 = t^{-1}(\hat{v}_1 - u_{k-1,k-2}), \; u_{k-2,k-2} = u_{k-1,k-1},$$

$$u_{0,k-2} = \ldots = u_{k-3,k-2} = 0.$$

Pursuing this computation we obtain relations similar to (4.19) with $u_0 = u_{k-1,k-1}, \ldots, u_{k-1} = u_{k-1,0}$. Hence the condition (ii) is satisfied.

2. Assume now that (i) and (ii) are fulfilled. It is sufficient to prove that there exists a unitary operator U such that with the integer n defined above the relations (4.22) and (4.12) hold. Let first $k > 0$.

With $p(t) = \hat{u}_0 + \hat{u}_1 t + \ldots + \hat{u}_{k-1} t^{k-1}$ we associate functions $\hat{v}_1, \ldots, \hat{v}_k \in L_2(\hat{\sigma})$:

$$\hat{v}_k := \mathbf{t}^{-k}(\hat{g} - p), \quad \hat{v}_j := \mathbf{t}^{-j}(\hat{u}_j \mathbf{t}^j + \ldots + \hat{u}_{k-1} \mathbf{t}^{k-1} + \mathbf{t}^k \hat{v}_k), \quad j = 1, \ldots, k-1.$$

Then the operators $U_{11} = \hat{g} \cdot$, $U_{12} := (\hat{v}_k \ldots \hat{v}_1)$ and

$$U_{22} := \begin{bmatrix} \hat{u}_0 & 0 & \cdots & 0 \\ \hat{u}_1 & \hat{u}_0 & \cdots & 0 \\ \vdots & \vdots & \ddots & \vdots \\ \hat{u}_{k-1} & \hat{u}_{k-2} & \cdots & \hat{u}_0 \end{bmatrix}$$

satisfy the relations (4.13), (4.23) and (4.24).

From (ii) it follows that there exists a (uniquely determined) polynomial $q(t) = u_0 + u_1 t + \ldots + u_{k-1} t^{k-1}$ with real coefficients and $u_0 \neq 0$ such that the function

$$v_k := \mathbf{t}^{-k}(\hat{g}^{-1} - q)$$

belongs to $L_2(\sigma)$. It is easy to see that the coefficients of p and q satisfy the relations

$$(4.28) \qquad \hat{u}_0 u_0 = 1, \quad \sum_{i=0}^{j} \hat{u}_i u_{j-i} = 0, \quad j = 1, \ldots, k-1.$$

Define functions $v_1, \ldots, v_{k-1} \in L_2(\sigma)$ by

$$v_j := \mathbf{t}^{-j}(u_j \mathbf{t}^j + \ldots + u_{k-1} \mathbf{t}^{k-1} + \mathbf{t}^k v_k), \quad j = 1, \ldots, k-1.$$

Then with the operators $U_{41} := ((\cdot, v_1)_\sigma \ldots (\cdot, v_k)_\sigma)^\top$,

$$U_{44} := \begin{bmatrix} u_0 & 0 & \cdots & 0 \\ u_1 & u_0 & \cdots & 0 \\ \vdots & \vdots & \ddots & \vdots \\ u_{k-1} & u_{k-2} & \cdots & u_0 \end{bmatrix}$$

the relations (4.28) are equivalent to (4.17) and (4.14), (4.25) and (4.26) are satisfied.

In order to prove Theorem 4.2 for $k > 0$ it remains to find operators U_{32}, U_{33}, U_{42} and U_{43} which fulfil the relations (4.15), (4.16) and (4.18). Since by (i) the nondegenerate forms $(H_2\cdot,\cdot)$ and $(\hat{H}_2\cdot,\cdot)$ have the same signatures, there exists a matrix U_{33} which satisfies (4.18). We set $U_{32}=0$. Consider the equation (4.15):

$$U_{22}^* Z\, U_{42} + U_{42}^* Z\, U_{22} = H_1 - U_{12}^* U_{12} - U_{22}^* \hat{H}_1 U_{22} =: S.$$

Evidently, the operator $U_{42} := \frac{1}{2}(U_{22}^* Z)^{-1} S$ satisfies this relation. If we choose

$$U_{43} := (U_{22}^* Z)^{-1}(H_3 - U_{22}^* \hat{H}_3 U_{33}),$$

then the equation (4.16) is fulfilled.

If $k=0$, then the operator

$$U = \begin{bmatrix} U_{11} & 0 \\ 0 & U_{33} \end{bmatrix}$$

with U_{11} and U_{33} as defined above has the required properties. This completes the proof.

REMARK 4.3. It is easy to see that for $k=\hat{k}=0$ the conditions (i) and (ii) in Theorem 4.1 are also sufficient for the unitary equivalence of A_φ and $A_{\hat\varphi}$.

Necessary and sufficient conditions for the unitary equivalence of A_φ and $A_{\hat\varphi}$ can be given also under other additional assumptions. However, a complete treatment of the relations (4.3) - (4.11) and (4.13) - (4.18) seems to be complicated. In the following subsection we consider the case $\kappa=1$.

4.2. The case $\kappa=1$. Let $\varphi,\hat\varphi \in \mathcal{F}(\mathbb{R};0)$ be such that the numbers $\kappa,\hat\kappa$ given by (2.8) are one. By Remark 4.3 we can restrict ourselves to the case $k=\hat{k}=1$.

THEOREM 4.4. Let $\varphi,\hat\varphi \in \mathcal{F}(\mathbb{R},0)$, $\kappa=\hat\kappa=1$ (see (2.8)) and $k=\hat{k}=1$.

If $l=\hat{l}=2$, then the operators A_φ and $A_{\hat\varphi}$ are unitarily equivalent if and only if

(i) the measures σ and $\hat\sigma$ are equivalent,

(ii) with $\hat{g} := (d\sigma/d\hat\sigma)^{\frac{1}{2}}$ there exists a nonzero real number $\hat{g}_0$ such that the function $t \to t^{-1}(\hat{g}(t) - \hat{g}_0)$ belongs to $L_2(\hat\sigma)$,

(iii) $|\hat{g}_0|^2 \hat{c}_2 = c_2$.

If $l,\hat{l} \le 1$, then A_φ and $A_{\hat\varphi}$ are unitarily equivalent if and only if condition (i) and the following condition are satisfied.

(iv) There exists a complex function $\hat\gamma \in L_2(\hat\sigma)$ with $|\hat\gamma|^2 = d\sigma/d\hat\sigma$ and a nonzero number $\hat\gamma_0$ such that the function $t \to t^{-1}(\hat\gamma(t) - \hat\gamma_0)$ belongs to $L_2(\hat\sigma)$ and

$$(4.29) \qquad \hat{c}_1 \hat\gamma_0 + \int t^{-1}(\hat\gamma(t) - \hat\gamma_0)\, d\hat\sigma(t) = c_1 \overline{\hat\gamma_0^{-1}} + \int t^{-1} \overline{(\hat\gamma(t)^{-1} - \hat\gamma_0^{-1})}\, d\sigma(t).$$

PROOF. If $k = 1$ and $r \leq 1$, then for the blocks in (4.2) we have $S_1 = S_2 = \hat{S}_1 = \hat{S}_2 = 0$ and (if $r = 1$) $J = \hat{J} = 1$. Consider first the case $l = 2$, that is $r = 1$ (or $c_2 > 0$). As above the relations (4.3), (4.4) and (4.13) imply (i) and (ii), and (4.5) - (4.11) become

(4.30) $\quad U_{22} = U_{33}.$

(4.31) $\quad (U_{11}\cdot, \mathbf{1})_{\hat{\sigma}} = U_{41}t \cdot + U_{44}(\cdot, \mathbf{1})_{\sigma},$

(4.32) $\quad (U_{12}\cdot, \mathbf{1})_{\hat{\sigma}} + \hat{c}_2 U_{32} = U_{41}\mathbf{1} + U_{43},$

(4.33) $\quad \hat{c}_2 U_{33} = U_{44}c_2.$

Further, (4.14) - (4.18) are equivalent to

(4.34) $\quad U_{11}^* U_{12} + U_{41}^* U_{22} = 0,$

(4.35) $\quad U_{12}^* U_{12} + U_{22}^* \hat{c}_0 U_{22} + U_{32}^* \hat{c}_1 U_{22} + U_{42}^* U_{22} + U_{22}^* \hat{c}_1 U_{32} + U_{32}^* \hat{c}_2 U_{32} + U_{22}^* U_{42} = c_0,$

(4.36) $\quad U_{22}^* \hat{c}_1 U_{33} + U_{32}^* \hat{c}_2 U_{33} + U_{22}^* U_{43} = c_1,$

(4.37) $\quad U_{22}^* U_{44} = 1,$

(4.38) $\quad U_{33}^* \hat{c}_2 U_{33} = c_2.$

The relations (4.30) and (4.38) give $U_{22}^* U_{22} \hat{c}_2 = c_2$. As above (see (4.21)) we find $U_{22}^* U_{22} = |\hat{g}_0|^2$ which proves (iii).

Assume now that $l = \hat{l} = 2$ and the conditions (i), (ii) and (iii) hold. Then, if $U_{11} := \hat{g}\cdot, U_{12} := \mathbf{t}^{-1}(\hat{g} - \hat{g}_0), U_{22} = U_{33} := \hat{g}_0, U_{41} := (\cdot, v_1)_{\sigma}$ with $v_1 := \mathbf{t}^{-1}(\hat{g}^{-1} - \hat{g}_0^{-1}), U_{44} := \hat{g}_0^{-1}$ the conditions (4.3), (4.4), (4.13) and (4.30), (4.31), (4.33), (4.34), (4.37) and (4.38) are fulfilled.

There remain (4.32), (4.35) and (4.36) to be satisfied which give three equations for the three numbers U_{32}, U_{42} and U_{43}. The relations (4.32) and (4.36) lead to the following system of equations for (real) U_{32}, U_{43}:

$$\int U_{12}d\hat{\sigma} + \hat{c}_2 U_{32} = \int v_1 d\sigma + U_{43},$$

$$U_{22}\hat{c}_1 U_{33} + U_{32}\hat{c}_2 U_{33} + U_{22}U_{43} = c_1.$$

Its determinant of the coefficients of U_{32}, U_{43} is

$$det \begin{bmatrix} \hat{c}_2 & -1 \\ \hat{c}U_{33} & U_{22} \end{bmatrix} = 2\hat{c}_2 \hat{g}_0 \neq 0,$$

hence U_{32} and U_{43} are uniquely determined. Finally U_{42} follows from (4.35), and A_{φ} and $A_{\hat{\varphi}}$ are unitarily equivalent.

If $r = 0$ (or $c_2 = 0$), in the block matrices in (4.2) the third rows and columns disappear and $D = c_1, \hat{D} = \hat{c}_1$. Assume that A_{φ} and $A_{\hat{\varphi}}$ are unitarily equivalent. Then, as above, the relations (4.3) − (4.5), (4.13) and (4.17) give (i) and the first part of (iv); and in view of (4.8) and (4.17) the relation (4.9) is equivalent to (4.29).

Let now (i) and (iv) be fulfilled. Then if $U_{11} := \hat{\gamma}., U_{12} := \mathbf{t}^{-1}(\hat{\gamma} - \hat{\gamma}_0), U_{22} = \hat{\gamma}_0,$ $U_{41} := (\cdot, v_1)_\sigma$ with $v_1 = \mathbf{t}^{-1}(\hat{\gamma}^{-1} - \hat{\gamma}_0^{-1}), U_{44} = \overline{\hat{\gamma}_0^{-1}}$, the relations (4.3), (4.4), (4.13), (4.8), (4.9), (4.14) and (4.17) are satisfied. Then, choosing U_{42} so that (4.15) is satisfied (observe that $U_{22} \neq 0$), we obtain the equivalence of A_φ and $A_{\hat{\varphi}}$. The theorem is proved.

REFERENCES:

[1] ACHIESER, N.I.; GLASMANN, I.M.: Theorie der linearen Operatoren im Hilbertraum, Akademie-Verlag, Berlin 1960.

[2] AZIZOV, T.J.; IOHVIDOV, I.S.: Foundations of the theory of linear operators in spaces with an indefinite metric, Nauka, Moscow 1986.

[3] BOGNAR, J.: Indefinite inner product spaces, Springer-Verlag, Berlin-Heidelberg-New York 1974.

[4] DIJKSMA, A.; LANGER, H.; DE SNOO, H.: Unitary colligations in Krein spaces and their role in the extension theory of isometries and symmetric linear relations in Hilbert space, Functional Analysis II, Proceedings Dubrovnik 1985, Lecture Notes in Mathematics, 1242 (1987), 1-42.

[5] IOHVIDOV, I.S.; KREIN, M.G.; LANGER, H.: Introduction to the spectral theory of operators in spaces with an indefinite metric, Akademie-Verlag, Berlin 1982.

[6] JONAS, P.: Zur Existenz von Eigenspektralfunktionen mit Singularitäten, Math. Nachr. 88 (1977), 345-361.

[7] JONAS, P.: On the functional calculus and the spectral function for definitizable operators in Krein space, Beiträge Anal. 16 (1981), 121-135.

[8] JONAS, P.: A class of operator valued meromorphic functions on the unit disc.I, Ann.Acad.Sci.Fenn.Ser.A I (to appear).

[9] KREIN, M.G.: On completely continuous linear operators in functional spaces with two norms, Zbirnik Prac' Inst. Mat.Akad. Nauk Ukrain RSR, No.9 (1947), 104-129 (Ukrainian).

[10] KREIN, M.G.; LANGER, H.: On the spectral function of a selfadjoint operator in a space with indefinite metric, Dokl.Akad.Nauk SSSR 152 (1963), 39-42.

[11] KREIN, M.G.; LANGER, H.: Über die Q-Funktion eines π-hermiteschen Operators im Raume Π_κ, Acta Scient.Math.(Szeged) 34 (1973), 191-230.

[12] KREIN, M.G.; LANGER, H.: Über einige Fortsetzungsprobleme, die eng mit der Theorie hermitescher Operatoren im Raum Π_κ zusammenhängen. I. Einige Funktionenklassen und ihre Darstellungen, Math.Nachr. 77 (1977), 187-236.

[13] KREIN, M.G.; LANGER, H.: One some extension problems which are closely connected with the theory of hermitian operators in a space Π_κ. III. Indefinite analogues of the Hamburger and Stieltjes moment problems, Part (I): Beiträge Anal. 14 (1979), 25-40; Part (II): Beiträge Anal. 15 (1981), 27-45.

[14] LANGER, H.: Spectral functions of definitiziable operators in Krein spaces, Functional Analysis, Proceedings Dubrovnik, Lecture Notes in Mathematics, 948 (1982), 1-46.

Acknowledgements. The first author thanks the TU Vienna for its hospitality and financial support. The second author expresses his sincere thanks to Professor Ando for giving him the possibility to take part in the Workshop.

P. JONAS
Neltestraße 12
D-1199 Berlin
Germany

H. LANGER
Techn. Univ. Wien
Inst.f.Analysis,Techn.Math.
und Versicherungsmathematik
Wiedner Hauptstraße 8-10
1040 Wien
Austria

B. TEXTORIUS
Linköping University
Department of Mathematics
S-581 83 Linköping
Sweden

AMS classification: Primary 47 B50; secondary 47A67, 47A45

Operator Theory:
Advances and Applications, Vol. 59
© 1992 Birkhäuser Verlag Basel

THE von NEUMANN INEQUALITY AND DILATION THEOREMS

FOR CONTRACTIONS

Takateru Okayasu

In this paper we shall prove that, if $S_1, \cdots, S_m$ and $T_1, \cdots, T_n$ are sets of commuting contractions on a Hilbert space, both satisfy the von Neumann inequality "in the strong sense", each S_j double commutes with every T_k, and, $S_1, \cdots, S_m$ generate a nuclear C^*-algebra, then the set $S_1, \cdots, S_m, T_1, \cdots, T_n$ satisfies the von Neumann inequality "in the strong sense". This gives a new condition for a set of contractions to admit a simultaneous strong unitary dilation.

1. The von Neumann inequality and strong unitary dilation

It is well-known that any contraction T on a Hilbert space satisfies *the* so-called *von Neumann inequality*:

$$||p(T)|| \leq ||p|| = \sup_{0 \leq \theta < 2\pi} |p(e^{i\theta})|$$

for any polynomial p in one variable. It is also well-known, as *the Sz.-Nagy strong unitary dilation theorem*, that any contraction T on a Hilbert space $\mathcal{H}$ admits a strong unitary dilation, that is, there exist a Hilbert space $\mathcal{K} \supseteq \mathcal{H}$ and a unitary operator U on $\mathcal{K}$ such that

$$T^m = PU^m|\mathcal{H} \quad (m \geq 0),$$

where P is the projection onto $\mathcal{H}$. These matters are considered to be same; and rest on the fact that the linear map ϕ such that

$$\phi(\bar{p} + q) = p(T)^* + q(T),$$

p, q polynomials in one variable, of the C^*-algebra $C(\mathbf{T})$ of all complex-valued continuous functions on the torus $\mathbf{T}$ into the C^*-algebra $B(\mathcal{H})$ of all bounded linear operators on $\mathcal{H}$, is *completely positive*.

For a set of commuting contractions, $T_1, ..., T_n$, to satisfy the von Neumann inequality *in the strong sense*, and to admit a simultaneous strong unitary dilation, are closely bound up in each other. Actually we are able to state that these two conditions are equivalent:

Theorem 1. Let $T_1, \cdots, T_n$ be commuting contractions on a Hilbert space $\mathcal{H}$. If the inequality

$$\|(p_{ij}(T_1, \cdots, T_n))\| \le \|(p_{ij})\| = \sup_{0 \le \theta_k < 2\pi} \|(p_{ij}(e^{i\theta_1}, \cdots, e^{i\theta_n}))\|$$

(to which we refer as the von Neumann inequality in the strong sense) holds for any $m \times m$ matrix (p_{ij}), p_{ij} polynomials in n variables, then the set of contractions $T_1, \cdots, T_n$ admits a strong unitary dilation, in other words, there exist a Hilbert space $\mathcal{K} \supseteq \mathcal{H}$ and commuting unitary operators $U_1, \cdots, U_n$ on $\mathcal{K}$, such that

$$T_1^{m_1} \cdots T_n^{m_n} = PU_1^{m_1} \cdots U_n^{m_n}|\mathcal{H} \quad (m_1, \cdots, m_n \ge 0),$$

where P is the projection onto $\mathcal{H}$; and *vice versa*.

Andô's theorems [1], [2] give central cases where the (equivalent) conditions in Theorem 1 are fulfilled. One of them asserts that any pair of commuting contractions admits a strong unitary dilation, and the other that any triple of commuting contractions, one of which double commutes with others, admits also a strong unitary dilation. These matters then show that any pair of commuting contractions, and, any triple of commuting contractions, one of which double commutes with others, admits the von Neumann inequality in the strong sense.

On the other hand, some examples (Parrott[6], Crabb-Davie[4], and Varopoulos[11]) show that the n variable version of the von Neumann inequality

$$\|p(T_1, \cdots, T_n)\| \le \|p\| = \sup_{0 \le \theta_k < 2\pi} |p(e^{i\theta_1}, \cdots, e^{i\theta_n})|$$

fails to be valid, $T_1, \cdots, T_n$ commuting contractions, $n \ge 3$; and hence $T_1, \cdots, T_n$ in those cases cannot admit strong unitary dilation.

We give here a sufficient condition for commuting contractions to satisfy the von Neumann inequality in the strong sense:

Theorem 2. Let $S_1, \cdots, S_m; T_1, \cdots, T_n$ be sets of commuting contractions, both satisfy the von Neumann inequality in the strong sense, and S_j double commute with every T_k. If $S_1, \cdots, S_m$ generate *a nuclear algebra*, then the set $S_1, \cdots, S_m, T_1, \cdots, T_n$ satisfies the von Neumann inequality in the strong sense.

A nuclear algebra means a C^*-algebra $\mathcal{A}$ such that, for any C^*-algebra $\mathcal{B}$, the *-algebraic tensor product $\mathcal{A} \odot \mathcal{B}$ of $\mathcal{A}$ and $\mathcal{B}$ has a unique C^*-norm (See [7]). A *GCR*-algebra (=a *postliminal C^*-algebra*), which must be nuclear [10], means a C^*-algebra $\mathcal{A}$ such that, for any *-representation π of $\mathcal{A}$, the von Neumann algebra generated by the image $\pi(\mathcal{A})$ of $\mathcal{A}$ by π is of type I; a *GCR-operator*, besides, means an operator T such that the C^*-algebra generated by T is a *GCR*-algebra. Normal operators, compact operators, and isometries, are *GCR*-operators [5].

Corollary. Let S be a *GCR*-contraction, $T_1, \cdots, T_n$ be a set of commuting contractions which satisfies the von Neumann inequality in the strong sense, and S double commute with every T_k. Then one concludes that the set $S, T_1, \cdots, T_n$ satisfies the von Neumann inequality in the strong sense.

This generalizes an earier result due to Brehmer-Sz.-Nagy (See [9], I), that a triple of commuting contractions, one of which is an isometry double commutes with others, admits a strong unitary dilation.

2. Canonical representation of completely contractive maps

We recall several notions on maps on operator spaces.

An operator space means a subspace which contains the identity element (denoted by 1) of a unital C^*-algebra, and *an operator system* a self-adjoint operator space. A linear map ϕ of an operator space $\mathcal{S}$ into another is said to be *unital* if $\phi(1) = 1$ holds; *contractive, positive* if

$$\|\phi(x)\| \leq \|x\| \quad (x \in \mathcal{S}),$$

$$\phi(x) \geq 0 \quad (0 \leq x \in \mathcal{S})$$

holds, respectively; *completely contractive, completely positive* if the tensor product $\phi \otimes \mathrm{id}_m$ of ϕ and the identity map id_m of the $m \times m$ matrix algebra $\mathbf{M}_m$ is contractive, positive, resectively, for any $m \geq 1$.

It is fundamental that any unital contractive map of an operator system into another is positive, that a positive map ϕ of an operator system into another is bounded (in fact, $\|\phi\| \leq 2\|\phi(1)\|$), and that a positive map ϕ of a C^*-algebra $\mathcal{A}$ into a C^*-algebra $\mathcal{B}$ is completely positive if either $\mathcal{A}$ or $\mathcal{B}$ is abelian. The Steinspring theorem [8] asserts that any unital, completely positive map ϕ of a unital C^*-algebra $\mathcal{A}$ into the C^*-algebra $B(\mathcal{H})$ on a Hilbert space $\mathcal{H}$ has a canonical representation, namely, there exist a Hilbrt space (unique up to unitary equivalence) $\mathcal{K} \supseteq \mathcal{H}$ and a $*$-representation π of $\mathcal{A}$ on $\mathcal{K}$ such that $\pi(\mathcal{A})\mathcal{H}$ is dense in $\mathcal{K}$ and

$$\phi(x) = P\pi(x)|\mathcal{H} \quad (x \in \mathcal{A}),$$

where P is the projection onto $\mathcal{H}$.

Now we want to give a proof of Theorem 1. In it, Arveson's extension theorem [3] (See [7]) is essential; it asserts that, any unital completely contractive map of an operator space $\mathcal{S}$ into an operator space $\mathcal{T}$ extends to a completely positive map of any C^*-algebra $\mathcal{A} \supseteq \mathcal{S}$ into a C^*-algebra $\mathcal{B} \supseteq \mathcal{T}$.

Proof of Theorem 1. Assume that $T_1, \cdots, T_n$ are commuting contractions on a Hilbert space $\mathcal{H}$ and satisfy the von Neumann iequality in the strong sense.

By assumption the linear map

$$\phi : p \longrightarrow p(T_1, \cdots, T_n),$$

of the operator space $P(\mathbf{T}^n)$ of all polynomials in n variables $e^{i\theta_1}, \cdots, e^{i\theta_n}$, on $\mathbf{T}^n = \mathbf{T} \times \cdots \times \mathbf{T}$, into $B(\mathcal{H})$, is unital and completely contractive. Then it extends to a unital, completely positive map of the C^*-algebra $C(\mathbf{T}^n)$ of complex-valued continuous functions on $\mathbf{T}^n$, into $B(\mathcal{H})$. Therefore, there exist a Hilbert space $\mathcal{K} \supseteq \mathcal{H}$ and a $*$-representation π of $C(\mathbf{T}^n)$ on $\mathcal{K}$ such that

$$\phi(f) = P\pi(f)|\mathcal{H} \quad (f \in C(\mathbf{T}^n)),$$

P the projection onto $\mathcal{H}$. Put $U_k = \pi(e^{i\theta_k})$ $(k = 1, \cdots, n)$. Then $U_1, \cdots, U_n$ are unitary operators and satisfy that

$$T_1^{m_1} \cdots T_n^{m_n} = PU_1^{m_1} \cdots U_n^{m_n}|\mathcal{H} \quad (m_1, \cdots, m_n \geq 0).$$

Conversely, let $\mathcal{K}$ be a Hilbert space $\supseteq \mathcal{H}$ and $U_1, \cdots, U_n$ be commuting unitary operators on $\mathcal{K}$ such that

$$T_1^{m_1} \cdots T_n^{m_n} = P U_1^{m_1} \cdots U_n^{m_n} | \mathcal{H} \quad (m_1, \cdots, m_n \geq 0),$$

P the projection onto $\mathcal{H}$. Consider the *-homomorphism ϕ of the *-algebra of all polynomials in variables $e^{i\theta_1}, e^{-i\theta_1}, \cdots, e^{i\theta_n}, e^{-i\theta_n}$, on $\mathbf{T}^n$, to $B(\mathcal{H})$ such that

$$\phi(e^{i\theta_k}) = U_k \quad \text{and} \quad \phi(e^{-i\theta_k}) = U_k^*,$$

for $k = 1, \cdots, n$. We can see that it is bounded and satisfies the inequality

$$\|\phi(p)\| = \|p(U_1, U_1^*, \cdots, U_n, U_n^*)\| \leq \|p\|$$

for any p. Therefore, by the Stone-Weierstrass argument, it extends to a *-representation of $C(\mathbf{T}^n)$. So, ϕ is completely contractive. Consequently, we have

$$\begin{aligned}
\|(p_{ij}(T_1, \cdots, T_n))\| &\leq \|(p_{ij}(U_1, \cdots, U_n))\| \\
&= \|(\phi \otimes \mathrm{id}_m)((p_{ij}))\| \\
&\leq \|(p_{ij})\| \\
&= \sup_{0 \leq \theta_k < 2\pi} \|(p_{ij}(e^{i\theta_1}, \cdots, e^{i\theta_n}))\|,
\end{aligned}$$

for any $m \times m$ matrix (p_{ij}), p_{ij} polynomials in variables $e^{i\theta_1}, \cdots, e^{i\theta_n}$, which completes the proof.

3. An effect of generation of nuclear algebras

Next, we will give a

Proof of Theorem 2. It is sufficient to find a unital completely contractive map of $P(\mathbf{T}^{m+n})$ into $B(\mathcal{H})$, which maps each variable $e^{i\theta_j}$, $e^{i\theta_k}$ to S_j, T_k, respectively.

We already have, via Theorem 1, unital completely contractive maps ϕ_1 of $P(\mathbf{T}^m)$ into $B(\mathcal{H})$ so that $\phi_1(e^{i\theta_j}) = S_j$, ϕ_2 of $P(\mathbf{T}^n)$ into $B(\mathcal{H})$ so that $\phi_2(e^{i\theta_k}) = T_k$. According to Arveson's extension theorem, ϕ_1 (resp. ϕ_2) extends to a unital completely positive map ψ_1 (resp. ψ_2) of $C(\mathbf{T}^m)$ (resp. $C(\mathbf{T}^n)$) into $\mathcal{A}$ (resp. $\mathcal{B}$), where $\mathcal{A}$ (resp. $\mathcal{B}$) is the C^*-algebra generated by $S_1, \cdots, S_m$ (resp. $T_1, \cdots, T_n$). It can be seen that the tensor product $\psi_1 \otimes \psi_2$ of ψ_1 and ψ_2 is a unital, completely positive, and so completely contractive, map, of the C^*-tensor product $C(\mathbf{T}^m) \otimes C(\mathbf{T}^n)$ of $C(\mathbf{T}^m)$ and $C(\mathbf{T}^n)$ (which can be thought of as $C(\mathbf{T}^{m+n})$) into *the minimal C^*-tensor product $\mathcal{A} \otimes \mathcal{B}$* of $\mathcal{A}$ and $\mathcal{B}$, i.e., the completion of $\mathcal{A} \odot \mathcal{B}$ under the operator norm $\|\ \|$ considered on $\mathcal{A} \odot \mathcal{B}$ (which is known to be the smallest among all C^*-norms on $\mathcal{A} \odot \mathcal{B}$ [10]). Hence, the tensor product $\phi_1 \otimes \phi_2$ of ϕ_1 and ϕ_2, the restriction of $\psi_1 \otimes \psi_2$ to $P(\mathbf{T}^{m+n})$ (identified with $P(\mathbf{T}^m) \odot P(\mathbf{T}^n)$), is a unital completely contractive map of $P(\mathbf{T}^{m+n})$ into $\mathcal{A} \odot \mathcal{B}$.

Consider then the *-representation ϕ of $\mathcal{A} \odot \mathcal{B}$ on $\mathcal{H}$ such that

$$\phi(X \otimes Y) = XY \quad (X \in \mathcal{A},\ Y \in \mathcal{B}).$$

Since the operator norm $\|\ \|$ on $\mathcal{A} \odot \mathcal{B}$ coincides, by assumption, with the (largest) C^*-norm $\|\ \|_\nu$ defined by the identity

$$\|V\|_\nu = \sup\{\|\pi(V)\| : \pi \text{ is a *-representation of } \mathcal{A} \odot \mathcal{B}\},$$

for each $V \in \mathcal{A} \odot \mathcal{B}$ (See [7]), we have the inequality

$$\Big\| \sum_k X_k Y_k \Big\| \leq \Big\| \sum_k X_k \otimes Y_k \Big\|_\nu = \Big\| \sum_k X_k \otimes Y_k \Big\|,$$

for $X_k \in \mathcal{A}$ and $Y_k \in \mathcal{B}$. This shows that ϕ may extend to a *-representation of the C^*-algebra $\mathcal{A} \otimes \mathcal{B}$. Hence, as above, ϕ is completely contractive. It is obvious, on the other hand, that ϕ is unital, so, the composition $\phi \circ (\phi_1 \otimes \phi_2)$ of ϕ and $\phi_1 \otimes \phi_2$ is unital and completely contractive; and maps each variable $e^{i\theta_j}$, $e^{i\theta_k}$ to S_j, T_k, respectively.

Now the proof is complete.

References

1. T. Andô, *On a pair of commuting contractions*, Acta Sci. Math. **24**(1963),88-90.

2. T. Andô, *Unitary dilation for a triple of commuting contractions*, Bull. Acad. Polonaise Math. **24**(1976), 851-853.

3. W. B. Arveson, *Subalgebras of C^*-algebras*, Acta Math. **123**(1969), 141-224.

4. M. J. Crabb and A. M. Davie, *von Neumann's inequality for Hilbert space operators*, Bull. London Math. Soc. **7**(1975), 49-50

5. T. Okayasu, *On GCR-operators*, Tôhoku Math. Journ. **21** (1969), 573-579.

6. S. K. Parrott, *Unitary dilations for commuting contractions*, Pacific Journ. Math. **34**(1970), 481-490.

7. I. Paulsen, *Completely bounded maps and dilations*, Pitman Res. Notes Math. Ser. **146**, 1986.

8. W. F. Steinspring, *Positive functions on C^*-algebras*, Proc. Amer. Math. Soc. **6**(1955), 211-216.

9. B. Sz.-Nagy and C. Foiaş, *Harmonic analysis of operators on Hilbert space*, Amusterdam-Budapest, North-Holland, 1970.

10. M. Takesaki, *On the cross-norm of the direct product of C^*-algebras*, Tôhoku Math. Journ. **16**(1964), 111-122.

11. N. Th. Varopoulos, *On an inequality of von Neumann and an application of the metric theory of tensor products to operator theory*, Journ. Funct. Analy. **16**(1974), 83-100.

Department of Mathematics
Faculty of Science
Yamagata University
Yamagata 990, JAPAN

MSC 1991: Primary 47A20, 47A30; Secondary 46M05

Operator Theory:
Advances and Applications, Vol. 59
© 1992 Birkhäuser Verlag Basel

INTERPOLATION PROBLEMS, INVERSE SPECTRAL PROBLEMS AND NONLINEAR EQUATIONS

L. A. Sakhnovich

The method of operator identities of a type of commutation relations is shown to be useful in the investigation of interpolation problems, inverse spectral problems and nonlinear integrable equations.

Suppose that the operators A, S, $\mathcal{P}_1$, $\mathcal{P}_2$ are connected by the relation

$$AS - SA^* = i(\mathcal{P}_1\mathcal{P}_2^* + \mathcal{P}_2\mathcal{P}_1^*) \tag{1}$$

where G_1 and H are Hilbert spaces, $\dim G_1 < \infty$,

$$S = S^*; \quad A, S \in \{H, H\}; \quad \mathcal{P}_1, \mathcal{P}_2 \in \{G_1, H\}$$

and $\{H_1, H_2\}$ is the set of bounded operators acting from H_1 to H_2.

We also introduce the operator $J \in \{G, G\}$ where

$$G = G_1 \oplus G_1, \quad J = \begin{bmatrix} 0 & E_1 \\ E_1 & 0 \end{bmatrix}.$$

Formula (1) is a special case of the operator identity of the form

$$AS - SB = \Pi_1\Pi_2^* \tag{2}$$

which is a generalization of the commutation relations and it also generalizes the well-known notion of the node (M. S. Livsic [1] and then M. S. Brodskii [2]). The identities of the form (2) proved to be useful in a number of problems (system theory [3], factorization problems [3], interpolation theory [4], the method of constructing the inverse operator $T = S^{-1}$ [5], the inverse spectral problem [3] and theory of nonlinear integrable equations [6]). There are close ties between all these problems and corresponding results.

In the present paper we shall consider three of these problems: interpolation problems, inverse spectral problems and nonlinear integrable equations.

1. Let $\mathcal{E}$ be a collection of monotonically increasing operator-functions and $\tau(u) \in \{G_1, G_1\}$ is such that integrals

$$S_\tau = \int_{-\infty}^{\infty} (E - uA)^{-1} \mathcal{P}_2[d\tau(u)]\mathcal{P}_2^*(E - uA^*)^{-1} \tag{3}$$

$$\mathcal{I}_\tau = \int_{-\infty}^{\infty} \frac{d\tau(u)}{1 + u^2} \tag{4}$$

converge in the weak sense. Then the integral

$$\mathcal{P}_{1,\tau} = -i \int_{-\infty}^{\infty} \left[A(E - uA)^{-1} + \frac{u}{1 + u^2} E \right] \mathcal{P}_2 \, d\tau(u) \tag{5}$$

also converges in the weak sense.

Let us introduce the operators

$$\widetilde{S} = S_\tau + FF^*, \quad \widetilde{\mathcal{P}} = \mathcal{P}_{1,\tau} + i(\mathcal{P}_2\alpha + F\beta^{1/2}) \tag{6}$$

where

$$\alpha, \beta \in \{G_1, G_1\}, \quad \alpha = \alpha^*, \quad \beta \geq 0$$

and operator F from $\{G_1, H\}$ is defined by the equality

$$AF = \mathcal{P}_2\beta^{1/2}. \tag{7}$$

Now we shall formulate the interpolation problem which is generated by operation identity (1) [4].

It is necessary to describe the set of $\tau(u) \in \mathcal{E}$ and $\alpha = \alpha^*$, $\beta \geq 0$ such that the given operators $S, \mathcal{P}_1$ admit the representation

$$S = \widetilde{S}, \quad \mathcal{P}_1 = \widetilde{\mathcal{P}}. \tag{8}$$

Let us note that according to (3) the necessary condition of the formulated problem is the inequality

$$S \geq 0. \tag{9}$$

As an example we shall consider the bounded operator S in the space $L^2(0, \omega)$ of the form

$$(Sf)(x) = \frac{d}{dx} \int_0^\omega f(t)s(x - t) \, dt. \tag{10}$$

Then the equality

$$((AS - SA^*)f)(x) = i \int_0^\omega f(t)[M(x) + N(t)] \, dt \tag{11}$$

is valid. In equality (11)

$$M(x) = s(x), \quad N(x) = -s(-x), \quad 0 \le x \le \omega$$

and the operators A, A^* are defined by the equalities

$$(Af)(x) = i \int_0^x f(t) \, dt, \quad (A^*f)(x) = -i \int_x^\omega f(t) \, dt.$$

Formula (11) is a special case of the operator identity of form (1), where

$$(\mathcal{P}_2 g)(x) = g, \quad (\mathcal{P}_1 g)(x) = M(x)g, \quad \dim G_1 = 1, \quad g \in G_1,$$

and g are constants.

The corresponding interpolation problem has the form. It is necessary to describe the set of $\tau(u) \in \mathcal{E}$ which gives the representation

$$(Sf, f) = \int_{-\infty}^\infty \left| \int_0^\omega f(x) e^{-iux} \, dx \right|^2 \, d\tau(u).$$

If the operator S has the form

$$(Sf)(x) = \int_0^\omega f(t) \kappa(x - t) \, dt$$

we come to the well-known Krein problem: it is necessary to describe the set of $\tau(u)$ which gives the representation

$$\kappa(x) = \int_{-\infty}^\infty e^{ixu} \, d\tau(u).$$

In our approach to the interpolation problem we use the operator identity and operator form of Potapov inequality [4]. Operator identity (11) gives a tool for constructing the inverse operator $T = S^{-1}$. The operator T can be found in the exact form by means of the functions $N_1(x)$, $N_2(x)$ which are defined by the relations

$$SN_1 = M, \quad SN_2 = 1.$$

We have proved that the knowledge of N_1, N_2 is that minimal information which is necessary for constructing T [5].

2. Let us consider the inverse spectral problem which is connected with operator identity (1).

THEOREM. *Suppose that the following conditions hold:*

I. *S is positive and invertible.*

II. *There exists a continuous increasing family of orthogonal projections P_ζ, $0 \leq \zeta \leq \omega$, $P_0 = 0$, $P_\omega = E$ such that*

$$A^* P_\zeta = P_\zeta A^* P_\zeta.$$

III. *The spectrum of A is concentrated at zero.*

Then the following representations hold

$$w(\zeta, z) = \int_0^{\widehat{\zeta}} \exp[iz J \, d\sigma_1(t)]$$

where

$$w(\zeta, z) = E + iz J \, \Pi^* S_\zeta^{-1} (E - z A_\zeta)^{-1} P_\zeta \Pi,$$
$$\Pi = [\mathcal{P}_1, \mathcal{P}_2], \quad S_\zeta = P_\zeta S P_\zeta, \quad A_\zeta = P_\zeta A P_\zeta.$$

If $\sigma_1(x)$ is absolutely continuous, then $\sigma_1'(x) \geq 0$ and

$$\frac{dw}{dx} = iz J \sigma_1'(x) w(x, z), \quad w(0, z) = E. \tag{12}$$

Canonical system (12) corresponds to operator identity (1).

We shall introduce the main definitions. Let us denote by the space of vector-functions with the inner product

$$(g, h) = \int_0^\omega h^*(x) \, [d\sigma_1(x)] \, g(x).$$

We define the function

$$F(u) = \int_0^l w^*(x, u) \, [d\sigma_1(x)] \, g(x) = \begin{bmatrix} f_1(u) \\ f_2(u) \end{bmatrix}$$

and the operator

$$V g = f_2.$$

A monotonically increasing $m \times m$ matrix-function $\tau(u)$, $-\infty < u < \infty$ will be called a spectral function of system (12) if the operator V maps $L^2(\sigma_1)$ isometrically into $L^2(\tau)$.

THEOREM. *The set of $\tau(u)$ which are solutions of the interpolation problem and the set of spectral functions of the canonical system coincide.*

The following results give a method for solving the inverse spectral problem [3].

Let us suppose that the operators A and $\mathcal{P}_2$ are fixed. In this way we define the class of canonical systems (12). Then let us suppose that the spectral data of system (12) are given, i.e. the spectral function $\tau(u)$ and the matrix α are known. Using the interpolation formulas we have:

$$\mathcal{P}_1 = -i \int_{-\infty}^{\infty} \left[A(E - uA)^{-1} + \frac{u}{1 + u^2} E \right] \mathcal{P}_2 \, d\tau(u) + i\mathcal{P}_2\alpha \tag{13}$$

$$S = \int_{-\infty}^{\infty} (E - uA)^{-1}\mathcal{P}_2[d\tau(u)]\mathcal{P}_2^*(E - uA^*)^{-1} \tag{14}$$

$$\Pi = [\mathcal{P}_1, \mathcal{P}_2], \quad \sigma_1(\zeta) = \Pi^* S_\zeta^{-1} P_\zeta \Pi. \tag{15}$$

These formulas (13)–(15) give the solution of the inverse spectral problem.

If $(Af)(x) = i \int_0^x f(t)\, dt$, $\mathcal{P}_2 g = g$ we come to the well-known inverse problems for the system of Dirac type. In the general case when

$$(Af)(x) = iW \int_0^x f(t)\, dt$$
$$W = \mathrm{diag}\{\omega_1, \cdots, \omega_n\}$$

we come to the new non-classical inverse problem. The necessity of investigating non-classical problems is dictated both by mathematical and applied questions (interpolation theory, the theory of solitons). Under certain assumptions formulas (13)–(15) give the solution of the inverse spectral problem in the exact form [3].

Let us introduce an analogue of the Weyl-Titchmarsh function $v(z)$ for system (12) with the help of the inequality

$$\int_0^{\infty} [E_n, iv^*(z)]w^*(x, z)[\sigma_1'(x)]w(x, z) \begin{bmatrix} E_n \\ -iv(z) \end{bmatrix} dx < \infty.$$

The matrix function $v(z)$ belongs to the Nevanlinna class, i.e.

$$\frac{v(z) - v^*(z)}{i} \geq 0, \quad \mathrm{Im}\, z > 0.$$

The connection of $v(z)$ with the spectral data $\tau(u)$ and α is the following [3]

$$v(z) = \alpha + \int_{-\infty}^{\infty} \left(\frac{1}{u - z} - \frac{u}{1 + u^2} \right) d\tau(u).$$

We have considered the case when $0 \leq x < \infty$.

As in the case of Sturm-Liouville equation, the spectral problems on the line $(-\infty < x < \infty)$ can be reduced to the problems on the half-line $(0, \infty)$ by doubling the dimension of the system. The problem on the line contains the periodical case.

3. The method of inverse scattering problems is effectively used for investigating the nonlinear equations (Gardner, Kruskal, Zabuski, Lax, Zaharov, Shabat [8]).

The main idea comes to the following. The nonlinear equation is considered together with the corresponding linear system. The evolution of scattering data of the linear system is very simple. Then by using the method of inverse problem the solution of nonlinear system can be found.

The transition from the inverse scattering problem to the inverse spectral problem removes the demand for the regularity of the solution at the infinity and permits to construct new classes of exact solutions for a number of nonlinear equations [7]:

$$R_t = \frac{i}{2}(R_{xx} - 2|R|^2 R) \qquad \text{(NS)} \tag{16}$$

$$R_t = -\frac{1}{4}R_{xxx} + \frac{3}{2}|R|^2 R_x \quad \text{(MKdV)} \tag{17}$$

$$\frac{\partial^2 \varphi}{\partial x \partial t} = 4\,\text{sh}\,\varphi. \qquad \text{(Sh-G)} \tag{18}$$

These equations have found wide applications in a number of problems of mathematical physics.

The corresponding linear system has the form

$$\frac{\partial w}{\partial x} = izH(x,t)w, \quad w(0,t,z) = E_n. \tag{19}$$

In the case of Sh-Gordon equation we have [8]

$$H(x,t) = \begin{bmatrix} 0 & \exp[\varphi(x,t) - \varphi(0,t)] \\ \exp[\varphi(0,t) - \varphi(x,t)] & 0 \end{bmatrix}. \tag{20}$$

Let $v_0(z) = v(0,z)$ of corresponding system (19), (20) be a rational function of z: i.e.

$$v_0(z) = i - \sum_{k=1}^{N} \beta_{k,0}/(z + i\alpha_{k,0}).$$

Then $v(t,z)$ is also a rational function

$$v(t,z) = i - \sum_{k=1}^{N} \beta_k(t)/[z + i\alpha_k(t)].$$

Let us write down $v(t, z)$ in the form

$$v(t, z) = iP_2(t, iz)/P_1(t, iz)$$

where

$$P_1(t, z) = \prod_{k=1}^{N}[z - \alpha_k(t)], \quad P_2(t, z) = \prod_{k=1}^{N}[z - \nu_k(t)].$$

Let us introduce

$$Q(z) = P_1(t, z)P_2(t, -z) + P_1(t, -z)P_2(t, z).$$

It is essential that unlike $P_1(t, z)$, $P_2(t, z)$ the coefficients $Q(z)$ do not depend on t. It means that the zeros of the $Q(z)$ do not depend on t either.

Let the inequality $w_j \neq w_k$ be true when $j \neq k$. Let us number the zeros of $Q(z)$ in such a way that $\operatorname{Re} w_j > 0$ when $1 \leq j \leq N$. The solution of the Sh-Gordon equation which corresponds to the rational $v_0(z)$ is as follows

$$\varphi(x, t) = 2\ln|\delta_1(x, t)/\delta_2(x, t)|$$

where

$$\delta_1(x, t) = \det\left\{w_j^N \frac{\partial^k}{\partial t^k}\mathrm{ch}\,\eta_j\right\}_{1 \leq j, k \leq N}$$

$$\delta_2(x, t) = \det\left\{w_j^N \frac{\partial^k}{\partial t^k}\mathrm{sh}\,\eta_j\right\}_{1 \leq j, k \leq N}$$

$$\eta_j(x, t) = w_j x + t/w_j + c_j.$$

This result was obtained jointly with post-graduate Tidnjuk [7]. The corresponding solution $R(x, t)$ of equations (16), (17) has the following form

$$R(x, t) = -2(-1)^N \Delta_1(x, t)/\Delta_2(x, t),$$

$$\Delta_1(x, t) = \begin{vmatrix} 1 & 1 & \cdots & 1 \\ w_1 & w_2 & \cdots & w_{2N} \\ \cdots\cdots\cdots\cdots\cdots\cdots \\ w_1^{N-2} & w_2^{N-2} & \cdots & w_{2N}^{N-2} \\ \gamma_1 & \gamma_2 & \cdots & \gamma_{2N} \\ w_1\gamma_1 & w_2\gamma_2 & \cdots & w_{2N}\gamma_{2N} \\ \cdots\cdots\cdots\cdots\cdots\cdots \\ w_1^N\gamma_1 & w_2^N\gamma_2 & \cdots & w_{2N}^N\gamma_{2N} \end{vmatrix},$$

$$\Delta_2(x,t) = \begin{vmatrix} 1 & 1 & \cdots & 1 \\ w_1 & w_2 & \cdots & w_{2N} \\ \cdots\cdots\cdots\cdots\cdots\cdots\cdots\cdots\cdots \\ w_1^{N-1} & w_2^{N-1} & \cdots & w_{2N}^{N-1} \\ \gamma_1 & \gamma_2 & \cdots & \gamma_{2N} \\ w_1\gamma_1 & w_2\gamma_2 & \cdots & w_{2N}\gamma_{2N} \\ \cdots\cdots\cdots\cdots\cdots\cdots\cdots\cdots\cdots \\ w_1^{N-1}\gamma_1 & w_2^{N-1}\gamma_2 & \cdots & w_{2N}^{N-1}\gamma_{2N} \end{vmatrix}$$

where

$$\gamma_k = c_k \exp 2(w_k x - \Omega_k t),$$
$$\Omega_k = -iw_k^2 \quad (NS), \quad \Omega_k = w_k^3 \quad (MKdV).$$

4. If $w_j = \overline{w_j}$, $\alpha_{j,0} = -\overline{\alpha_{j,0}}$ then the corresponding solution $R(x,t)$ of MKdV is real. All the real singularities of $R(x,t)$ are poles of the first order with the residues $+1$ or -1. When $t \to \pm\infty$, the solution $R(x,t)$ is presented by a sum of simple waves

$$R(x,t) \approx \sum_{j=1}^{N} R_j^{\pm}(x,t), \quad t \to \pm\infty \tag{21}$$

$$R_j^{\pm}(x,t) = 2(-1)^j w_j / \mathrm{sh}[2(w_j x - w_j^3 t + c_j^{\pm})]. \tag{22}$$

The considered nonlinear equations do not have N-soliton solutions. The constructed solutions are similar to the N-soliton solutions. The behaviour of the singularities of the solutions is analogous to the behaviour of the humps of the N-soliton solutions and can be interpreted in the terms of a particles system. We have proved that the corresponding particles system is a completely integrable one with the Hamiltonian

$$H = \frac{1}{2}\sum_{j=1}^{N} p_j^2 \tag{23}$$

where

$$p_j = w_j^2 \quad (MKdV), \quad p_j = -\frac{1}{w_j^2} \quad (Sh\text{-}G) \tag{24}$$

$$p_j = 2\,\mathrm{Im}\,w_j \quad (NS) \tag{25}$$

$$q_j = p_j t + c_j. \tag{26}$$

The variables p_j, q_j are variables of the action-angle type. It follows from formulas (21), (22) for MKdV that p_j coincides with the limit velocity of the wave. The same situation is in Sh-G and NS cases.

t
x
MKdV
$\omega_1 = 0.3$ $\omega_2 = 0.5$ $\omega_3 = 0.7$ $\omega_4 = 1$
$\alpha_1 = 0.35$ $\alpha_2 = 0.45$ $\alpha_3 = 0.55$ $\alpha_4 = 0.65$
$c = k > 0$

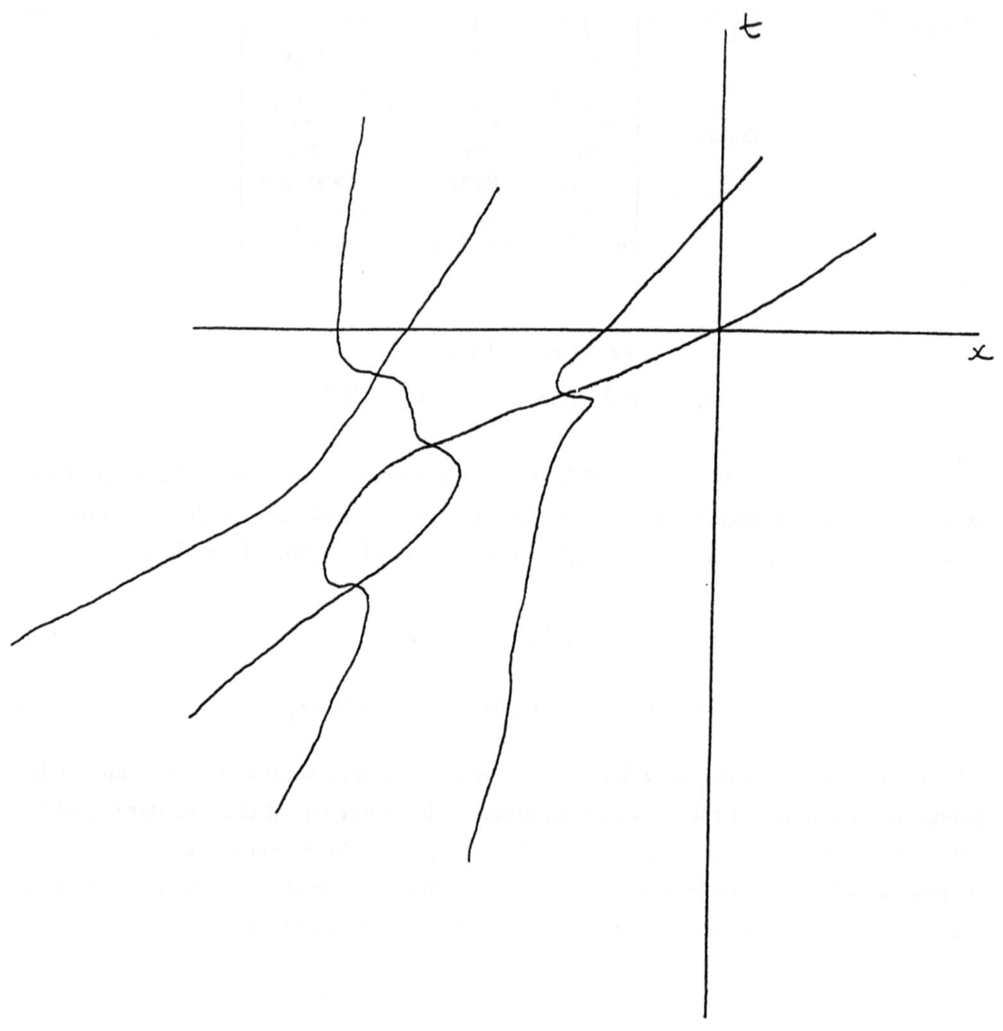

Fig. 1

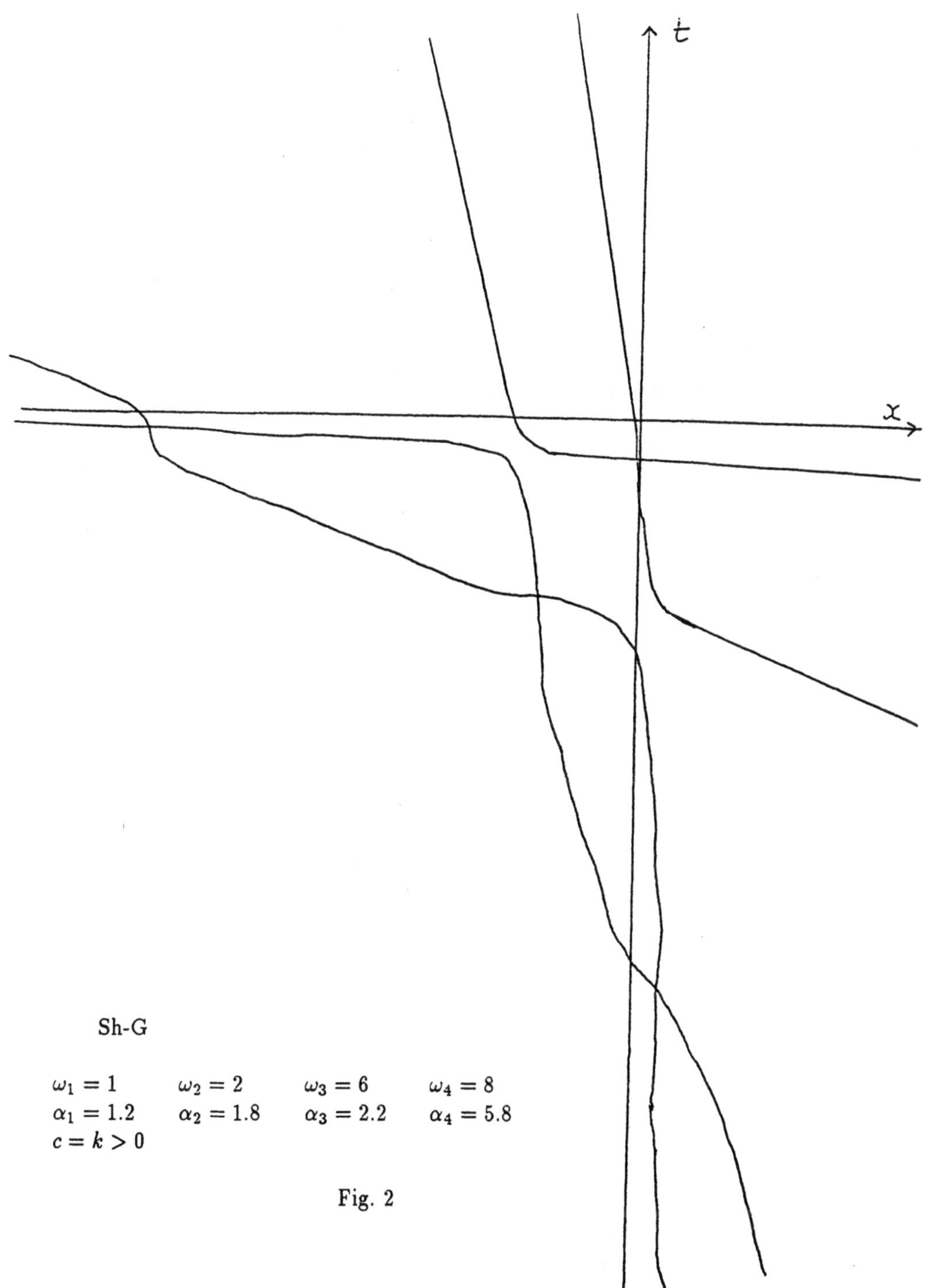

$$\omega_1 = 1 \qquad \omega_2 = 2 \qquad \omega_3 = 6 \qquad \omega_4 = 8$$
$$\alpha_1 = 1.2 \qquad \alpha_2 = 1.8 \qquad \alpha_3 = 2.2 \qquad \alpha_4 = 5.8$$
$$c = k > 0$$

Fig. 2

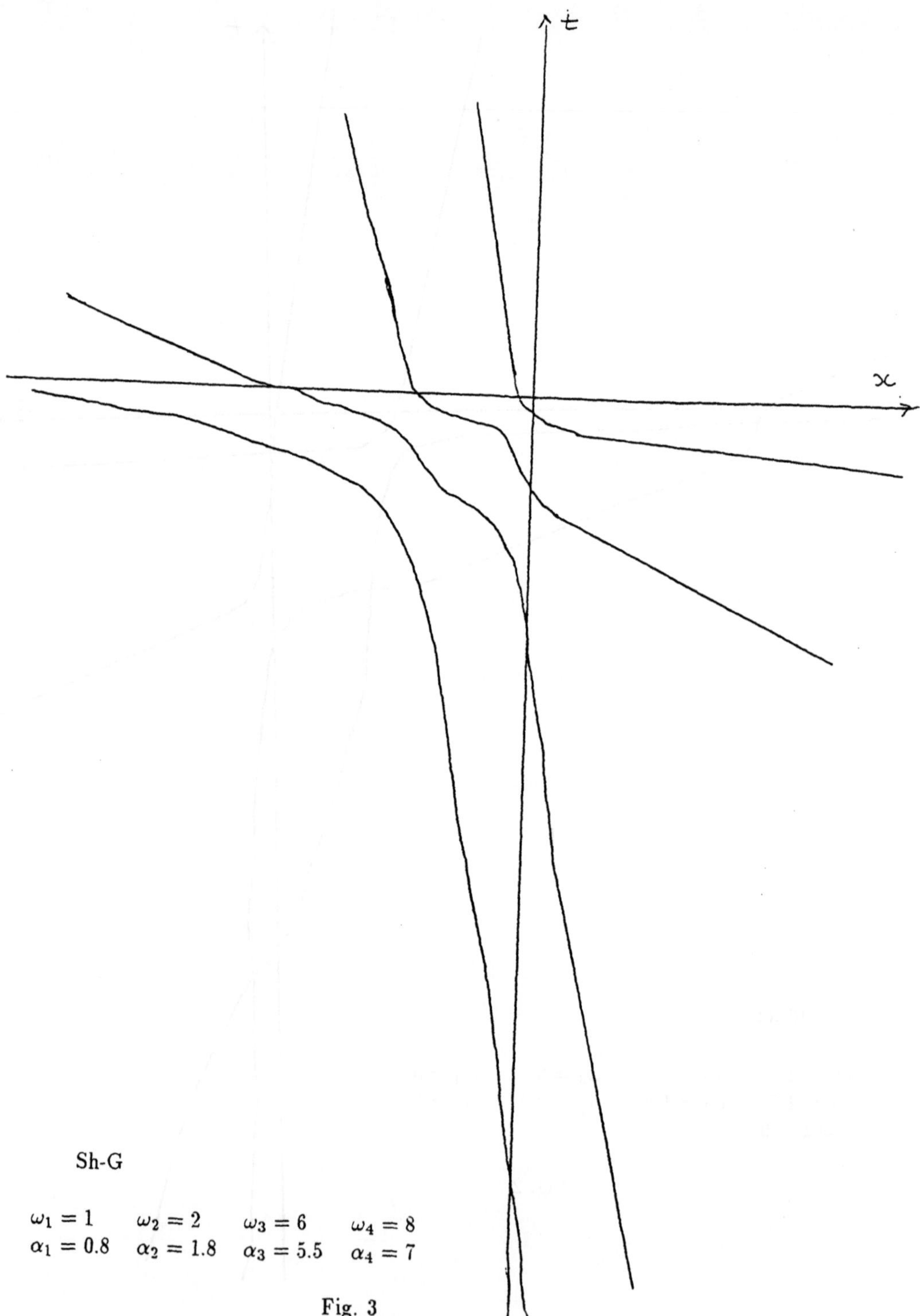

Sh-G

$\omega_1 = 1 \qquad \omega_2 = 2 \qquad \omega_3 = 6 \qquad \omega_4 = 8$
$\alpha_1 = 0.8 \qquad \alpha_2 = 1.8 \qquad \alpha_3 = 5.5 \qquad \alpha_4 = 7$

Fig. 3

It also follows from formulas (21), (22) that the lines of singularities have N asymptotes (MKdV)

$$w_j x - w_j^3 t + c_j^{\pm} = 0, \quad 1 \le j \le N, \quad t \to \pm\infty.$$

Let us number these asymptotes when $t \to -\infty$ by the order of velocity values and consider the same lines of singularities when $t \to +\infty$. Then the corresponding asymptotes are again ordered by the velocity values but in the opposite direction (Fig. 1). It means that the particles exchange their numbers. In the case of Sh-Gordon equation the situation is the same (Fig. 2).

The particles exchange their numbers even if there is no crossing (Fig. 3). The particles can be of two kinds:

plus particles if the corresponding residues are $+1$,

and minus particles if they are -1.

The particles of the same kind don't cross. In Fig. 3 the particles are of the same kind.

5. The considered equations generated self-adjoint spectral problems. If we study the equations

$$R_t = \frac{i}{2}(R_{xx} + 2|R|^2 R) \qquad \text{(NS)} \qquad (27)$$

$$R_t = -\frac{1}{4} R_{xxx} - \frac{3}{2}|R|^2 R_x \quad \text{(MKdV)} \qquad (28)$$

$$\frac{\partial^2 \varphi}{\partial x \partial t} = 4 \sin \varphi \qquad \text{(sin-G)} \qquad (29)$$

then the nonself-adjoint spectral problems correspond to them. The analogue of the Weyl-Titchmarsh function for this case was introduced and the analysis of the equations of the form (27)–(29) was done by A. L. Sakhnovich [9], [10].

REFERENCES

1. M. S. Livsic, *Operators, oscillations, waves (open systems)*, Amer. Math. Soc., 1966.

2. M. S. Brodskii, *Triangular and Jordan representations of linear operators*, Amer. Math. Soc., 1971.

3. L. A. Sakhnovich, *Factorization problems and operator identities*, Russian Math. Surveys, **41** no. 1 (1986), 1–64.

4. T. S. Ivanchenko, L. A. Sakhnovich, *An operator approach to the investigation of interpolation problems*, Dep. at Ukr. NIINTI, **N 701** (1985), 1–63.

5. L. A. Sakhnovich, *Equations with a difference kernel on a finite interval*, Russian Math. Surveys **35** no. 4 (1980), 81–152.

6. L. A. Sakhnovich, *Nonlinear equations and inverse problems on the semi-axis*, Preprint, Institute of Math., 1987.

7. L. A. Sakhnovich, I. F. Tidnjuk, *The effective solution of the Sh-Gordon equation*, Dokl. Akad. Nauk. Ukr. SSR Ser. A, **1990** no. 9, 20–25.

8. R. K. Bullough, P. I. Caudrey (Eds), *Solitons*, New York, 1980.

9. A. L. Sakhnovich, *The Goursat problem for the sine-Gordon equation*, Dokl. Akad. Nauk. Ukr. SSR Ser. A **1989**, no. 12, 14–17.

10. A. L. Sakhnovich, *A nonlinear Schrödinger equation on the semi-axis and a related inverse problem*, Ukrain. Math. J. **42** (1990), 316–323.

Sakhnovich, L. A.
Odessa Electrical Engineering Institute of Communications
Odessa, Ukraine

MSC 1991: 47A62, 35Q53

Operator Theory:
Advances and Applications, Vol. 59
© 1992 Birkhäuser Verlag Basel

Extended Interpolation Problem
in Finitely Connected Domains

SECHIKO TAKAHASHI

This paper concerns the matrix condition necessary and sufficient for the existence of a function f, holomorphic in a finitely connected domain and having $|f| \leq 1$ and finitely many first prescribed Taylor coefficients at a finite number of given points. In a simply connected domain, some transformation formulas and their applications are given. The results of Abrahamse on the Pick interpolation problem are generalized to the above extended interpolation problem.

Introduction

Let D be a bounded domain in the complex plane C, whose boundary consists of a finite number of mutually disjoint analytic simple closed curves, and let $\mathcal{B}$ be the set of functions f holomorphic in D and satisfying $|f| \leq 1$ in D. In this paper we consider the following extended interpolation problem:

Let $z_1, z_2, \cdots, z_k$ be k distinct points in D and, for each point z_i, let $c_{i0}, \cdots, c_{in_i-1}$ be n_i complex numbers. For these given data, find a function $f \in \mathcal{B}$ which satisfies the conditions

$$(\text{EI}) \qquad f(z) = \sum_{\alpha=0}^{n_i-1} c_{i\alpha}(z - z_i)^\alpha + O((z - z_i)^{n_i}) \qquad (i = 1, \cdots, k).$$

In the Part I, we introduce, as powerful tools for the studies of this problem (EI), the Schur's triangular matrix Δ used by Schur in [13] and our rectangular matrix $\mathbf{M}$, which made it possible to unify Schur's coefficient theorem and Pick's interpolation theorem (Takahashi [14]). We give important transformation formulas which express the changes of these matrices under holomorphic transformations in terms of the transformation matrices.

In the Part II, we recall our main results obtained in [14] and [15] for the problem (EI) in the case where D is the open unit disc. As an application of these results and the transformation formulas, we give a criterion matrix of the extended interpolation problem in the case where D is a simply connected domain in the Riemann sphere having at least two boundary points and the range W is a closed disc in the Riemann sphere. When W contains the point at infinity, we have of course to modify the conditions (EI) appropriately and the solutions may have poles.

In the Part III, we show that the results of Abrahamse in [1] on the interpolation problem in finitely connected domains can be extended to our extended interpolation problem.

PART I. MATRICES AND TRANSFORMATION FORMULAS

§1. Matricial Representation of Taylor Coefficients.

(1) Schur's Triangular Matrix Δ. To a function

$$f(z) = \sum_{\alpha=0}^{\infty} c_\alpha (z - z_0)^\alpha$$

holomorphic at z_0 and to a positive integer $n \in \mathbf{N}$, we assign a triangular $n \times n$ matrix

$$\Delta(f; z_0; n) = \begin{bmatrix} c_0 & & & \\ c_1 & c_0 & & \\ \vdots & \ddots & \ddots & \\ c_{n-1} & \cdots & c_1 & c_0 \end{bmatrix}.$$

Let $g(z)$ be another function holomorphic at z_0, we see immediately

$$\Delta(f + g; z_0; n) = \Delta(f; z_0; n) + \Delta(g; z_0; n),$$
$$\Delta(fg; z_0; n) = \Delta(f; z_0; n) \cdot \Delta(g; z_0; n)$$
$$= \Delta(g; z_0; n) \cdot \Delta(f; z_0; n),$$
$$\Delta(1; z_0; n) = I_n \quad \text{(the unit matrix of order } n).$$

(2) Rectangular Coefficient Matrix M. To a function

$$F(z, \zeta) = \sum_{\alpha, \beta=0}^{\infty} a_{\alpha\beta} (z - z_0)^\alpha \overline{(\zeta - \zeta_0)}^\beta$$

holomorphic w.r.t. $(z, \overline{\zeta})$ at (z_0, ζ_0) and to $(m, n) \in \mathbf{N} \times \mathbf{N}$, we associate an $m \times n$ matrix

$$\mathbf{M}(F; z_0, \zeta_0; m, n) = \begin{bmatrix} a_{00} & \cdots & a_{0n-1} \\ \cdots\cdots\cdots\cdots\cdots \\ a_{m-10} & \cdots & a_{m-1n-1} \end{bmatrix}$$

For another function $G(z, \zeta)$ holomorphic w.r.t. $(z, \overline{\zeta})$ at (z_0, ζ_0), we have

$$\mathbf{M}(F + G; z_0, \zeta_0; m, n) = \mathbf{M}(F; z_0, \zeta_0; m, n) + \mathbf{M}(G; z_0, \zeta_0; m, n).$$

Moreover, for functions $f(z)$ and $g(\zeta)$, holomorphic at z_0 and ζ_0 respectively, we have the useful *product formula*

(PF) $\mathbf{M}(f F \bar{g}; z_0, \zeta_0; m, n) = \Delta(f; z_0; m) \cdot \mathbf{M}(F; z_0, \zeta_0; m, n) \cdot \Delta(g; \zeta_0; n)^*,$

where by $\Delta^* = {}^t\overline{\Delta}$ we mean the transposed of the complex conjugate of Δ. This product formula can be established by a direct calculation.

§2. Transformation Matrix and Transformation Formulas

The transformation formula for the matrix $\mathbf{M}$ which we established in [14] is the pivot of our present studies.

For a transformation $z = \varphi(x)$ holomorphic at x_0 with $z_0 = \varphi(x_0)$ and for $m \in \mathbf{N}$, we define the *transformation matrix* $\Omega(\varphi; x_0; m)$ as follows: Write

$$\varphi(x) = z_0 + (x - x_0)\varphi_1(x),$$
$$\Phi_m = \Delta(\varphi_1; x_0; m),$$
$$E_m^{(\alpha)} = \mathbf{M}((z - z_0)^\alpha \overline{(\zeta - \zeta_0)}^\alpha; z_0, \zeta_0; m, m) \quad (\alpha = 0, \cdots, m-1)$$

and put

$$\Omega(\varphi; x_0; m) = \sum_{\alpha=0}^{m-1} \Phi_m^\alpha E_m^{(\alpha)}.$$

The matrix $E_m^{(\alpha)}$ is the $m \times m$ matrix whose $(\alpha+1, \alpha+1)$-entry is 1 and the all other entries are 0. $\Phi_m^0 = I_m$ and $\Phi_m^\alpha = \Phi_m^{\alpha-1} \cdot \Phi_m$ $(\alpha = 1, 2, \cdots)$. The matrix Ω is of the form

$$\Omega(\varphi; x_0; m) = \begin{bmatrix} 1 & & & & \\ 0 & c & & & \\ \cdot & * & c^2 & & \\ \vdots & \vdots & \ddots & \ddots & \\ 0 & * & \cdots & * & c^{m-1} \end{bmatrix},$$

where $c = \varphi_1(x_0) = \varphi'(x_0)$. If $c \neq 0$, then $\Omega(\varphi; x_0; m)$ is an invertible matrix. If $\varphi(x)$ is the identical transformation, then $\varphi_1(x) = 1$, $\Phi_m = I_m$ and hence $\Omega(\varphi; x_0; m) = I_m$.

In terms of this transformation matrix Ω, we showed in [14]

THEOREM 1 (TRANSFORMATION FORMULA FOR $\mathbf{M}$). *Let $F(z, \zeta)$ be a function holomorphic w.r.t. $(z, \overline{\zeta})$ at (z_0, ζ_0). Let $z = \varphi(x), \zeta = \psi(\xi)$ be functions holomorphic at x_0, ξ_0, with $z_0 = \varphi(x_0), \zeta_0 = \psi(\xi_0)$ respectively. Put*

$$G(x, \xi) = F(\varphi(x), \psi(\xi)).$$

Then, for $(m, n) \in \mathbf{N} \times \mathbf{N}$, we have

$$\mathbf{M}(G; x_0, \xi_0; m, n) = \Omega(\varphi; x_0; m) \cdot \mathbf{M}(F; z_0, \zeta_0; m, n) \cdot \Omega(\psi; \xi_0; n)^*.$$

As an application of the preceding transformation formula, we obtain

THEOREM 2 (TRANSFORMATION FORMULA FOR Δ). *Let*

$$f(z) = \sum_{\alpha=0}^{\infty} c_\alpha (z - z_0)^\alpha$$

be a function holomorphic at z_0 and let φ be a function holomorphic at x_0 with $z_0 = \varphi(x_0)$. Set

$$g(x) = f(\varphi(x)) = \sum_{\alpha=0}^{\infty} d_\alpha (x - x_0)^\alpha.$$

Then we have for $n \in \mathbb{N}$

$$(1) \quad \begin{bmatrix} d_0 & & & \\ d_1 & d_0 & & \\ \vdots & \ddots & \ddots & \\ d_{n-1} & \cdots & d_1 & d_0 \end{bmatrix} \Omega(\varphi; x_0; n) = \Omega(\varphi; x_0; n) \begin{bmatrix} c_0 & & & \\ c_1 & c_0 & & \\ \vdots & \ddots & \ddots & \\ c_{n-1} & \cdots & c_1 & c_0 \end{bmatrix},$$

$$(2) \quad \begin{bmatrix} d_0 \\ d_1 \\ \vdots \\ d_{n-1} \end{bmatrix} = \Omega(\varphi; x_0; n) \begin{bmatrix} c_0 \\ c_1 \\ \vdots \\ c_{n-1} \end{bmatrix}.$$

PROOF. Consider at $(z_0, 0)$ the function

$$F_0(z, \zeta) = \frac{1}{1 - (z - z_0)\overline{\zeta}} = \sum_{\alpha=0}^{\infty} (z - z_0)^\alpha \overline{\zeta}^\alpha$$

and $F(z, \zeta) = f(z) F_0(z, \zeta)$. By definition we see $\mathbf{M}(F_0; z_0, 0; n, n) = I_n$ and by (PF) in §1 $\mathbf{M}(F; z_0, 0; n, n) = \Delta(f; z_0; n)$. Applying the transformations $z = \varphi(x)$ and $\zeta = \zeta$ to F_0 and F, we have $F(\varphi(x), \zeta) = g(x) F_0(\varphi(x), \zeta)$ and hence, by the above transformation formula for $\mathbf{M}$, the first relation

$$\Omega(\varphi; x_0; n)\Delta(f; z_0; n) = \Delta(g; x_0; n)\Omega(\varphi; x_0; n).$$

Comparing the first columns of both sides of this equality, we see the relation (2) hold.

PART II. DISC CASES

§3. Main Theorems in the Unit Disc.

In this section, we state the main results obtained in [14] and [15].

We assume D is the open unit disc $\{z : |z| < 1\}$ and consider the extended interpolation problem (EI).

Write

$$
C_i = \begin{bmatrix} c_{i0} & & & \\ c_{i1} & c_{i0} & & \\ \vdots & \ddots & \ddots & \\ c_{in_i-1} & \cdots & c_{i1} & c_{i0} \end{bmatrix}, \qquad
C = \begin{bmatrix} C_1 & & \\ & \ddots & \\ & & C_k \end{bmatrix},
$$

$$
\Gamma_{ij} = \mathrm{M}\left(\frac{1}{1 - z\bar{\zeta}}; z_i, z_j; n_i, n_j \right), \qquad
\Gamma = \begin{bmatrix} \Gamma_{11} & \cdots & \Gamma_{1k} \\ \cdots\cdots\cdots \\ \Gamma_{k1} & \cdots & \Gamma_{kk} \end{bmatrix},
$$

$$
A_{ij} = \Gamma_{ij} - C_i \cdot \Gamma_{ij} \cdot C_j^*, \qquad
A = \begin{bmatrix} A_{11} & \cdots & A_{1k} \\ \cdots\cdots\cdots \\ A_{k1} & \cdots & A_{kk} \end{bmatrix}.
$$

Then we have

$$
A = \Gamma - C \cdot \Gamma \cdot C^*.
$$

The matrix A is an Hermitian matrix of order $n_1 + \cdots + n_k$, which is called *criterion matrix* of the problem (EI).

Let $\mathcal{E}$ denote the set of all solutions of (EI) in $\mathcal{B}$.

THEOREM 3 (EXTENSION OF THE THEOREMS OF CARATHÉODORY-SCHUR AND PICK). *There exists an $f \in \mathcal{E}$ if and only if $A \geq 0$ (positive semidefinite).*

THEOREM 4 (UNIQUENESS THEOREM OF SOLUTIONS). *For the problem (EI), the following conditions are equivalent:*

(a) *The set $\mathcal{E}$ consisits of a unique element.*
(b) *Some finite Blaschke product of degree $r < n_1 + \cdots + n_k$ is in $\mathcal{E}$.*
(c) *$A \geq 0$ and $\det A = 0$.*

If one of, therefore all of, these conditions are satisfied, then $r = \mathrm{rank}\, A$.

The proof of these theorems given in [14] was based on Marshall's method in [9], which makes use of Schur's algorithm.

In the case where the solution is not unique, that is, where $A > 0$ (positive definite), the following theorem, which may be proved as Corollary 2.4 in Chap.I of the textbook of Garnett [7], shows that the problem (EI) has an infinite number of solutions.

THEOREM 5. *Suppose $A > 0$.*

(a) *Let $z_0 \in D$, $z_0 \neq z_i$ $(i = 1, \cdots, k)$. The set*

$$W(z_0) = \{ f(z_0) : f \in \mathcal{E} \}$$

is a nondegenerate closed disc in D. (b) *For each z_i $(i = 1, \cdots, k)$, the set*

$$W'(z_i) = \{ f^{(n_i)}(z_i) : f \in \mathcal{E} \}$$

is a nondegenerate compact disc in $\mathbf{C}$.

In [15], we showed that if $A > 0$ then we have a bijective mapping $\pi : \mathcal{B} \longrightarrow \mathcal{E}$ such that there exist four functions P, Q, R, and S holomorphic in the unit disc D and satisfying

$$\pi(g) = \frac{Pg + Q}{Rg + S} \quad \text{and} \quad Rg + S \not\equiv 0 \quad (\forall g \in \mathcal{B}).$$

Let H^∞ denote the Banach algebra of bounded holomorphic functions f in D with the uniform norm $\|f\|_\infty = \sup\{ |f(z)| : z \in D \}$. The following Theorem 6, whose the first part (a) is due to Earl [4], can be derived immediately from Theorem 3 and Theorem 4.

THEOREM 6.

(a) *Among the solutions of* (EI) *in H^∞, there exists a unique solution of* (EI) *of minimal norm. This unique solution is of the form mB, where*

$$m = \inf\{ \|f\|_\infty : f \text{ is a solution of (EI) in } H^\infty \}$$

and B is a Blaschke product of degree $\leq n_1 + \cdots + n_k - 1$.

(b) *Conversely, if B is a Blaschke product of degree $\leq n_1 + \cdots + n_k - 1$ and if cB $(c \in \mathbf{C})$ is a solution of* (EI) *then cB is the unique solution of minimal norm of* (EI) *in H^∞.*

§4. Criterion Matrix in Simply Connected Domains

By virtue of the transformation formulas, we show in this section that our preceding results can be extended to the case where the source domain D is a simply connected domain in the Riemann sphere having at least two boundary points and the range W is a closed disc in $\mathbf{C}$ or a closed half plane in $\mathbf{C}$. The case where W contains ∞ will be treated in the next section.

Let $z_1, z_2, \cdots, z_k$ be distinct points in D and for each z_i let $c_{i0}, \cdots, c_{in_i-1}$ be n_i complex numbers. Our present problem is to find a holomorphic function f in D such that $f(z) \in W$ for any $z \in D$ and f satisfies the conditions

$$(\mathrm{EI}) \qquad f(z) = \sum_{\alpha=0}^{n_i-1} c_{i\alpha}(z - z_i)^\alpha + O((z - z_i)^{n_i}) \qquad (i = 1, \cdots, k),$$

where if $z_i = \infty$ for some i then we replace $z - z_i$ by $1/z$. For a moment, we assume $c_{i0} \in W$ $(i = 1, \cdots, k)$, which simplifies the statement. We shall later remove this assumption. We ask for a criterion matrix of this problem.

Let D_0 be the open unit disc in $\mathbf{C}$, $\varphi : D \longrightarrow D_0$ be a conformal mapping, and $\psi(w) = \dfrac{pw + q}{rw + s}$ (p, q, r, and s: complex numbers with $ps - qr = 1$) be a linear fractional transformation which maps the interior of W onto D_0. Put $x_i = \varphi(z_i)$ $(i = 1, \cdots, k)$. Because of the presence of ∞, we consider the transformation $\iota(z) = 1/z$. As in [14], it is convenient to use the notion of local solution. A local solution of (EI) is by definition a function f, holomorphic in some neighborhood of the finite set $\{z_1, z_2, \cdots, z_k\}$ and satisfying the conditions (EI). The formulas in (1) of §1 and Theorem 2 show that a function f is a local solution of (EI) if and only if $g = \psi \circ f \circ \varphi^{-1}$ is a local solution of the extended interpolation problem

$$(\mathrm{EI})^0 \qquad g(x) = \sum_{\alpha=0}^{n_i-1} d_{i\alpha}(x - x_i)^\alpha + O((x - x_i)^{n_i}) \qquad (i = 1, \cdots, k),$$

whose coefficients are given by

$$\begin{bmatrix} d_{i0} & & & \\ d_{i1} & d_{i0} & & \\ \vdots & \ddots & \ddots & \\ d_{in_i-1} & \cdots & d_{i1} & d_{i0} \end{bmatrix} = D_i = \Omega_i^{-1}(rC_i + sI_{n_i})^{-1}(pC_i + qI_{n_i})\Omega_i,$$

where $\Omega_i = \Omega(\varphi; z_i; n_i)$ if $z_i \neq \infty$, $\Omega_i = \Omega(\varphi \circ \iota; 0; n_i)$ if $z_i = \infty$, C_i is the triangular matrix defined from $c_{i0}, \cdots, c_{in_i-1}$ as in (1) of §1, and I_{n_i} is the unit matrix of order n_i. The matrix Ω_i is clearly invertible. The matrix $rC_i + sI_{n_i}$ is invertible since its entries on the diagonal are equal to $rc_{i0} + s$, which is not zero by assumption. Let

$$G_0(x, \xi) = \frac{1}{1 - x\bar{\xi}}$$

and put

$$\Gamma_{ij}^{(0)} = \mathbf{M}(G_0; x_i, x_j; n_i, n_j) \qquad \text{and} \qquad A_{ij}^{(0)} = \Gamma_{ij}^{(0)} - D_i \cdot \Gamma_{ij}^{(0)} \cdot D_j^*.$$

The matrix

$$A^{(0)} = \begin{bmatrix} A_{11}^{(0)} & \cdots & A_{1k}^{(0)} \\ \cdots\cdots\cdots\cdots \\ A_{k1}^{(0)} & \cdots & A_{kk}^{(0)} \end{bmatrix}$$

is the criterion matrix of the problem $(\mathrm{EI})^0$ for B, defined in §3.

Now, we define

$$F_0(z, \zeta) = G_0(\varphi(z), \varphi(\zeta)) = \frac{1}{1 - \varphi(z)\overline{\varphi(\zeta)}},$$
$$\Gamma_{ij} = \mathbf{M}(F_0; z_i, z_j; n_i, n_j),$$

$$C = \begin{bmatrix} C_1 & & \\ & \ddots & \\ & & C_k \end{bmatrix}, \qquad \Gamma = \begin{bmatrix} \Gamma_{11} & \cdots & \Gamma_{1k} \\ \cdots\cdots\cdots \\ \Gamma_{k1} & \cdots & \Gamma_{kk} \end{bmatrix},$$

where if $z_i = \infty$ and $z_j \neq \infty$ then we replace F_0 and Γ_{ij} by

$$F_0(z, \zeta) = G_0(\varphi(1/z), \varphi(\zeta)) \qquad \text{and}$$
$$\Gamma_{ij} = \mathbf{M}(F_0; 0, z_j; n_i, n_j) \qquad \text{respectively};$$

if $z_i \neq \infty$ and $z_j = \infty$ then we replace F_0 and Γ_{ij} by

$$F_0(z, \zeta) = G_0(\varphi(z), \varphi(1/\zeta)) \qquad \text{and}$$
$$\Gamma_{ij} = \mathbf{M}(F_0; z_i, 0; n_i, n_j) \qquad \text{respectively};$$

and if $i = j$ and $z_i = \infty$ then we replace F_0 and Γ_{ij} by

$$F_0(z, \zeta) = G_0(\varphi(1/z), \varphi(1/\zeta)) \qquad \text{and}$$
$$\Gamma_{ij} = \mathbf{M}(F_0; 0, 0; n_i, n_j) \qquad \text{respectively}.$$

Write

$$1 - \psi(w)\overline{\psi(v)} = (rw + s)^{-1}\overline{(rv + s)^{-1}}K(w, v)$$

with

$$K(w,v) = \alpha w \bar{v} + \beta w + \overline{\beta} \bar{v} + \gamma \,,$$

$$\alpha = |r|^2 - |p|^2 \ : \mathrm{real} \leq 0 \,,$$
$$\beta = r\bar{s} - p\bar{q} \,,$$
$$\gamma = |s|^2 - |q|^2 \ : \mathrm{real} \,,$$
$$|\beta|^2 - \alpha\gamma = 1 \,.$$

It is easy to see $W = \{w \ : K(w,w) \geq 0\}$. As we pointed out in [14], if g is a local solution of $(\mathrm{EI})^0$ and if we put

$$G(x,\xi) = \frac{1 - g(x)\overline{g(\xi)}}{1 - x\bar{\xi}} \,,$$

then we have $A_{ij}^{(0)} = \mathrm{M}(G; x_i, x_j; n_i, n_j)$. Put

$$f = \psi^{-1} \circ g \circ \varphi \qquad \text{and} \qquad F(z,\zeta) = G(\varphi(z), \varphi(\zeta)) \,.$$

Then

$$F(z,\zeta) = F_0(z,\zeta)\frac{\alpha f(z)\overline{f(\zeta)} + \beta f(z) + \overline{\beta f(\zeta)} + \gamma}{(rf(z) + s)\overline{(rf(\zeta) + s)}} \,.$$

The matrix $\mathrm{M}(F; z_i, z_j; n_i, n_j)$ is by Theorem 1 equal to $\Omega_i A_{ij}^{(0)} \Omega_j^*$, with appropriate change as above in the presence of ∞, and, on the other hand, it is written in the form $R_i A_{ij} R_j^*$ where

$$A_{ij} = \alpha C_i \Gamma_{ij} C_j^* + \beta C_i \Gamma_{ij} + \overline{\beta}\Gamma_{ij} C_j^* + \gamma\Gamma_{ij} \,,$$
$$R_i = (rC_i + sI_{n_i})^{-1}.$$

Write

$$A = \begin{bmatrix} A_{11} & \cdots & A_{1k} \\ \cdots\cdots\cdots \\ A_{k1} & \cdots & A_{kk} \end{bmatrix}, \quad \Omega = \begin{bmatrix} \Omega_1 & & \\ & \ddots & \\ & & \Omega_k \end{bmatrix}, \quad R = \begin{bmatrix} R_1 & & \\ & \ddots & \\ & & R_k \end{bmatrix}.$$

Then the matrix Ω and R are invertible and we have

$$\Omega A^{(0)} \Omega^* = R A R^*.$$

This shows that $A^{(0)} \geq 0$ if and only if $A \geq 0$ and that rank $A^{(0)} =$ rank A. Thus, we may adopt the Hermitian matrix A as criterion matrix of the problem (EI) for functions with values in W:

THEOREM 7. *Let the notations and the assumption be as above. There exists a holomorphic function in D, having its values in W and satisfying (EI), if and only if the Hermitian matrix*

$$A = \alpha C\,\Gamma\,C^* + \beta C\,\Gamma + \overline{\beta}\Gamma\,C^* + \gamma\Gamma$$

is positive semidefinite. Such a function is unique if and only if $A \geq 0$ and $\det A = 0$.

Note that the constants α, β, and γ depend on p, q, r, and s and that the matrix Γ depends only on φ, z_i and n_i $(i = 1, \cdots, k)$.

In the case where W is the closed unit disc, that is, for the extended interpolation problem (EI) in $\mathcal{B}$, the criterion matrix reduces to

$$A = \Gamma - C \cdot \Gamma \cdot C^*,$$

since we have then $\alpha = -1$, $\beta = 0$ and $\gamma = 1$.

We point out that, if the source domain D is an open disc or an open half plane in $\mathbf{C}$ and is defined by $K_0(z, z) > 0$, where

$$K_0(z, \zeta) = \alpha_0 z\overline{\zeta} + \beta_0 z + \overline{\beta_0 \zeta} + \gamma_0 \quad (\alpha_0 \text{ and } \gamma_0 \text{ are real}),$$

then, as in Pick [11], we may replace the definition above of F_0 by

$$F_0(z, \zeta) = \frac{1}{K_0(z, \zeta)}.$$

Finally, let us remove the assumption $c_{i0} \in W$ $(i = 1, \cdots, k)$. If there exists a solution f of the problem, then $c_{i0} = f(z_i) \in W$. Conversely, suppose $A \geq 0$. The (1,1)-entry of the matrix

$$A_{ii} = \alpha C_i \Gamma_{ii} C_i^* + \beta C_i \Gamma_{ii} + \overline{\beta}\Gamma_{ii} C_i^* + \gamma\Gamma_{ii}$$

is

$$(\alpha c_{i0}\overline{c_{i0}} + \beta c_{i0} + \overline{\beta c_{i0}} + \gamma) \times \frac{1}{1 - |\varphi(z_i)|^2}.$$

As $|\varphi(z_i)| < 1$ and $A_{ii} \geq 0$, we have $K(c_{i0}, c_{i0}) \geq 0$, which shows $c_{i0} = \psi^{-1}(d_{i0}) \in W$. Thus we have removed the assumption.

§5. Criterion Matrix for Meromorphic Functions.

Let D be a simply connected domain in the Riemann sphere having at least two boundary points and let $W = \{w : |w - a| \geq \rho\}$ be a closed disc, including ∞, in the Riemann sphere ($a \in \mathbf{C}$, $\rho > 0$). Let $z_1, z_2, \cdots, z_k$ be k distinct points in D. For each z_i, let m_i be a nonnegative integer and let $c_{i0}, \cdots, c_{in_i-1}$ be n_i complex numbers ($n_i \geq 1$). Assume $c_{i0} \neq 0$ if $m_i > 0$.

The problem in this section is to find a meromorphic function f in D with values in W, which satisfies the conditions

$$(\text{EI}) \quad f(z) = \frac{1}{(z - z_i)^{m_i}} \Big(\sum_{\alpha=0}^{n_i-1} c_{i\alpha}(z - z_i)^\alpha + O((z - z_i)^{n_i}) \Big) \quad (i = 1, \cdots, k),$$

where, if $z_i = \infty$, then $z - z_i$ is replaced by $1/z$. We ask for a criterion matrix for this problem. Note that if $m_i > 0$ then the order m_i of pole of f at z_i and the first n_i coefficients $c_{i0}, \cdots, c_{in_i-1}$ of the Laurent expansion of f at z_i are prescribed.

For this purpose, as in the preceding section §4, we consider a conformal mapping

$$\varphi \ : \ D \longrightarrow D_0$$

of D onto the open unit disc D_0, the linear fractional mapping

$$\psi \ : \ W \longrightarrow \overline{D_0} \qquad \text{defined by} \qquad \psi(w) = \frac{\rho}{w - a},$$

the function

$$F_0(z, \zeta) = \frac{1}{1 - \varphi(z)\overline{\varphi(\zeta)}},$$

and the matrices

$$\Gamma_{ij} = \mathbf{M}(F_0; z_i, z_j; m_i + n_i; m_j + n_j)$$

with appropriate replacement as in §4 if ∞ presents.

Now, for $n \in \mathbf{N}$, we introduce the standard $n \times n$ nilpotent matrix

$$N_n = \begin{bmatrix} 0 & & & \\ 1 & 0 & & \\ & \ddots & \ddots & \\ & & 1 & 0 \end{bmatrix},$$

where n is a positive integer. Then $N_n^n = O$ (zero matrix). We define $N_n^0 = I_n$ (unit matrix). For each z_i, put

$$C_i = \sum_{\alpha=0}^{n_i-1} c_{i\alpha}\, N_{m_i+n_i}^{\alpha} = \begin{bmatrix} c_{i0} & & & & & \\ \vdots & \ddots & & & & \\ c_{in_i-1} & & \ddots & & & \\ 0 & \ddots & & \ddots & & \\ \vdots & \ddots & \ddots & & \ddots & \\ 0 & \cdots & 0 & c_{in_i-1} & \cdots & c_{i0} \end{bmatrix},$$

$$T_i = N_{m_i+n_i}^{m_i},$$

$$R_i = C_i - a\,T_i \qquad\qquad (i = 1, \cdots, k).$$

If $m_i > 0$ then the diagonal entries of the triangular matrix R_i are all equal to c_{i0}, which is not zero by assumption, and hence R_i is invertible. If $m_i = 0$, we may assume for a moment as in §4 that $c_{i0} \neq a$, so that R_i is invertible. This assumption can be removed as in the final part of §4.

A meromorphic function f in D with values in W is transformed by ψ into a holomorphic function

$$g(z) = \frac{\rho}{f(z) - a}$$

in D with $|g| \leq 1$. On the other hand, writing

$$f(z) = \frac{f_0(z)}{(z - z_i)^{m_i}}$$

and

$$g(z) = \frac{\rho(z - z_i)^{m_i}}{f_0(z) - a(z - z_i)^{m_i}},$$

we see easily that the conditions (EI) for f are transformed into the conditions

$$(\text{EI})^{\#} \quad g(z) = (z - z_i)^{m_i}\left(\sum_{\alpha=0}^{n_i-1} d_{i\alpha}(z - z_i)^{\alpha} + O((z - z_i)^{n_i})\right) \quad (i = 1, \cdots, k),$$

where, if $z_i = \infty$, then $z - z_i$ is replaced by $1/z$ and the coefficients $d_{i\alpha}$ are defined by the relations

$$\sum_{\alpha=0}^{n_i-1} d_{i\alpha}\, N_{m_i+n_i}^{m_i+\alpha} = \rho R_i^{-1} T_i.$$

Denoting this matrix $\rho R_i^{-1} T_i$ by D_i and setting

$$A_{ij}^{\#} = \Gamma_{ij} - D_i \Gamma_{ij} D_j^* \qquad \text{and} \qquad A^{\#} = \begin{bmatrix} A_{11}^{\#} & \cdots & A_{1k}^{\#} \\ \cdots\cdots\cdots\cdots \\ A_{k1}^{\#} & \cdots & A_{kk}^{\#} \end{bmatrix},$$

we observe by Theorem 7 that the criterion matrix for the problem $(\mathrm{EI})^{\#}$ in $\mathcal{B}$ is $A^{\#}$ and we have

$$R_i A_{ij}^{\#} R_j^* = (C_i - aT_i)\Gamma_{ij}(C_j^* - \bar{a}T_j^*) - \rho^2 T_i \Gamma_{ij} T_j^*$$
$$= C_i \Gamma_{ij} C_j^* - aT_i \Gamma_{ij} C_j^* - \bar{a} C_i \Gamma_{ij} T_j^* + (|a|^2 - \rho^2) T_i \Gamma_{ij} T_j^*.$$

Write

$$C = \begin{bmatrix} C_1 & & \\ & \ddots & \\ & & C_k \end{bmatrix}, \qquad \Gamma = \begin{bmatrix} \Gamma_{11} & \cdots & \Gamma_{1k} \\ \cdots\cdots\cdots\cdots \\ \Gamma_{k1} & \cdots & \Gamma_{kk} \end{bmatrix},$$

$$T = \begin{bmatrix} T_1 & & \\ & \ddots & \\ & & T_k \end{bmatrix}, \qquad R = \begin{bmatrix} R_1 & & \\ & \ddots & \\ & & R_k \end{bmatrix}$$

and define

$$A = C\,\Gamma C^* - aT\,\Gamma\,C^* - \bar{a}C\,\Gamma\,T^* + (|a|^2 - \rho^2)T\,\Gamma\,T^*.$$

It should be noted that W is expressed by

$$w\overline{w} - a\overline{w} - \bar{a}w + (|a|^2 - \rho^2) \geq 0.$$

Then we have $A = R\,A^{\#}\,R^*$, where R is an invertible matrix. It follows that $A \geq 0$ if and only if $A^{\#} \geq 0$ and that rank $A =$ rank $A^{\#}$. Theorem 7 yields thus

THEOREM 8. *Let the notations and the assumption be as above. There exists a meromorphic function f in D with values in W, which satisfies the conditions*

$$\text{(EI)} \quad f(z) = \frac{1}{(z - z_i)^{m_i}} \Big(\sum_{\alpha=0}^{n_i-1} c_{i\alpha}(z - z_i)^{\alpha} + O((z - z_i)^{n_i}) \Big) \qquad (i = 1, \cdots, k)$$

if and only if the Hermitian matrix A is positive semidefinite. Such a function is unique if and only if $A \geq 0$ and $\det A = 0$.

PART III. DOMAINS OF FINITE CONNECTIVITY

Let D be a bounded domain in the complex plane whose boundary ∂D consists of $m + 1$ pairwise disjoint analytic simple closed curves γ_i $(i = 0, 1, \cdots, m)$. In this part, we generalize the results of Abrahamse [1] on the Pick interpolation problem in D, replacing this problem by our extended interpolation problem (EI) and introducing appropriate matrices. The proof proceeds as that of Abrahamse. We point out that this Part III gives another proof of the main theorem of the Part II in the unit disc.

§6. Preliminaries.

We consider the harmonic measure $d\omega$ on ∂D for the domain D and for a fixed point $z^* \in D$. In terms of $d\omega$, we define the norm $\|f\|_p$ $(1 \le p \le \infty)$ of complex-valued measurable functions f on ∂D :

$$\|f\|_p = \left(\int_{\partial D} |f|^p \, d\omega \right)^{\frac{1}{p}} \qquad (1 \le p < \infty) \, ;$$

$$\|f\|_\infty = \operatorname*{ess.sup}_{\partial D} |f| \qquad (\text{w.r.t. } d\omega),$$

and we have the Banach spaces $L^p = L^p(\partial D, d\omega)$.

Let $\Lambda = \{\lambda = (\lambda_1, \cdots, \lambda_m) : \lambda_i \in \mathbf{C}, |\lambda_i| = 1 \ (i = 1, \cdots, m)\}$ be the m-torus. In order to clarify the basic branch of multiple-valued modulus automorphic functions in D, we consider as in Abrahamse [1] m pairwise disjoint analytic cuts δ_i $(i = 1, \cdots, m)$, which starts from a point of γ_i and terminates at a point of γ_0. The domain $D_0 = D \setminus (\bigcup_{i=1}^{m} \delta_i)$ is thus simply connected.

For $\lambda = (\lambda_1, \cdots, \lambda_m) \in \Lambda$, let $H_\lambda(D)$ denote the set of complex-valued functions f in D such that f is holomorphic in D_0 and that, for each $t \in \delta_i \cap D$, $f(z)$ tends to $f(t)$ when $z \in D_0$ tends to t from the left side of δ_i and $f(z)$ tends to $\lambda_i f(t)$ when $z \in D_0$ tends to t from the right side of δ_i. We can easily verify that if one miltiplies by λ_i^{-1} the values of f on the right side of δ_i then the function thus modified is holomorphic at every point of $\delta_i \cap D$.

We define the canonical function V_λ in $H_\lambda(D)$ in the usual following way: For each $i = 1, \cdots, m$, let v_i be the harmonic function in the neighborhood of $\overline{D} = D \cup \partial D$ such that $v_i = 1$ on γ_i and $v_i = 0$ on the other γ_j $(j \ne i, 0 \le j \le m)$, and let $\widetilde{v_i}$ be the conjugate harmonic function of v_i in D_0. For $t \in \delta_j \cap D$ $(j = 0, \cdots, m)$, $\widetilde{v_i}(t)$ stands for the limit of $\widetilde{v_i}(z)$ when $z \in D_0$ tends to t from the left side of δ_j. There exist m real numbers $\xi_1, \cdots, \xi_m$ such that

$$V_\lambda(z) = \exp\Big(\sum_{j=1}^{m} \xi_j (v_j(z) + i \, \widetilde{v_j}(z)) \Big)$$

belongs to $H_\lambda(D)$ (see Widom [16]). We see that V_λ can be continued analytically across the boundary ∂D, in the usual sense except at the end points of the cuts δ_i and in the following sense at an end point t of δ_i : multiplying by λ_i^{-1} the values of V_λ on the right side of δ_i, one can continue analytically across ∂D the function thus modified in the neighborhood of t. For a function f in D, $f \in H_\lambda(D)$ if and only if $f\, V_\lambda^{-1}$ is holomorphic in D. Clearly, $|V_\lambda|$ is single-valued in D and can be extended to a continuous function in a neighborhood of $\overline{D}$, which has no zeros there.

Let H_λ^2 denote the space of all functions f in $H_\lambda(D)$ such that $|f|^2 \le u$ in D for some harmonic function u in D. It is known that any function f in H_λ^2 admits nontangential limits $f^*(t)$ at almost all $t \in \partial D$ (w.r.t. $d\omega$). Via $f \mapsto f^*$, the space H_λ^2 can be viewed as a closed subspace of the Hilbert space L^2. Thus H_λ^2 is a Hilbert space with the inner product

$$\langle f,\, g \rangle = \int_{\partial D} f^* \overline{g^*} \, d\omega \qquad (f, g \in H_\lambda^2).$$

From now on, no distinction will be made between a function f in H_λ^2 and its boundary function f^* in L^2. If $\lambda = (1, \cdots, 1)$ is the identity of Λ, then $H_\lambda(D)$ is the space of holomorphic functions in D and H_λ^2 is the usual Hardy space $H^2(D)$.

It is easy to see that, for $\lambda \in \Lambda$ and $\zeta \in D$, the mapping $f \mapsto f(\zeta)$ is a bounded linear functional on H_λ^2, so that we have a unique $k_{\lambda\zeta} \in H_\lambda^2$ such that

$$f(\zeta) = \langle f,\, k_{\lambda\zeta} \rangle \qquad (\text{for every } f \in H_\lambda^2).$$

We write
$$k_\lambda(z, \zeta) = k_{\lambda\zeta}(z) \qquad \text{and} \qquad k_\lambda(\,\,, \zeta) = k_{\lambda\zeta}.$$

The properties of k_λ in the following Lemma 1 are known (see Widom [16]).

LEMMA 1. *For $\lambda \in \Lambda$, $z \in D$ and $\zeta \in D$, we have*

$$k_\lambda(z, \zeta) = \langle k_\lambda(\,\,, \zeta),\, k_\lambda(\,\,, z) \rangle = \overline{k_\lambda(\zeta, z)}.$$

When λ is fixed, $k_\lambda(z, \zeta)$ is holomorphic w.r.t. $(z, \overline{\zeta})$ in $D_0 \times D_0$. It is continuous on $\Lambda \times D_0 \times D_0$ as a function of (λ, z, ζ). The function $|k_\lambda(z, \zeta)|$ is single-valued and continuous on $\Lambda \times ((\overline{D} \times D) \cup (D \times \overline{D}))$ with its appropriate boundary values.

LEMMA 2. *Let λ be fixed in Λ. For $t_0 \in \partial D$ and $\zeta_0 \in D$, there exist a neighborhood U_1 of t_0 and a neighborhood of U_2 of ζ_0 in D such that the function $V_\lambda(z)^{-1} k_\lambda(z, \zeta) \overline{V_\lambda(\zeta)}^{-1}$ can be extended to a function holomorphic w.r.t. $(z, \overline{\zeta})$ in $U_1 \times U_2$.*

Roughly speaking, Lemma 2 says that $k_\lambda(z, \zeta)$ can be continued analytically across the boundary as a function of two variables $(z, \overline{\zeta})$. The presence of V_λ is only to simplify the statement concerning the cuts δ_i. This Lemma 2 seems to be well known, but we could not find it in an explicit form in the bibliographies, so that we shall give its proof later in §8.

Let α, β be nonnegative integers. For a holomorphic function $f(z)$ we shall denote the α-th derivative of f by $f^{(\alpha)}$. For a function $F(z, \zeta)$ holomorphic w.r.t. $(z, \overline{\zeta})$, the notation $F^{(\alpha, \beta)}$ will stand for $\dfrac{\partial^{\alpha+\beta} F}{\partial z^\alpha \partial \overline{\zeta}^\beta}$, although this is a slight abuse of the notation.

Let λ be fixed in Λ. It is obvious that the derivative $k_\lambda^{(\alpha, \beta)}(z, \zeta)$ is well defined and holomorphic w.r.t. $(z, \overline{\zeta})$ in $D_0 \times D_0$. By Lemma 2, the function $V_\lambda(z)^{-1} k_\lambda^{(\alpha, \beta)}(z, \zeta) \overline{V_\lambda(\zeta)}^{-1}$ can be extended to a function holomorphic w.r.t. $(z, \overline{\zeta})$ in a neighborhood of $(\overline{D} \times D) \cup (D \times \overline{D})$. For $t \in \delta_i$ and $\zeta \in D_0$, $k_\lambda^{(\alpha, \beta)}(t, \zeta)$ is defined to be the limit of $k_\lambda^{(\alpha, \beta)}(z, \zeta)$ when $z \in D_0$ tends to t from the right side of δ_i, and so on. The function $|k_\lambda^{(\alpha, \beta)}(z, \zeta)|$ can be considered as a function single-valued and continuous on $(\overline{D} \times D) \cup (D \times \overline{D})$.

Though the following lemmas are valid for the points on the cuts, we shall restrict ourselves to D_0 in order to simplify the statement. This will be sufficient to apply them later.

LEMMA 3. *For $\zeta \in D_0$, $k_\lambda^{(\alpha, \beta)}(\ , \zeta)$ belongs to $H_\lambda^2 \cap L^\infty$. We have*

$$f^{(\alpha)}(\zeta) = \left\langle f, k_\lambda^{(0, \alpha)}(\ , \zeta) \right\rangle \qquad (f \in H_\lambda^2)$$

and

$$k_\lambda^{(\alpha, \beta)}(z, \zeta) = \left\langle k_\lambda^{(0, \beta)}(\ , \zeta), k_\lambda^{(0, \alpha)}(\ , z) \right\rangle \qquad (z \in D_0).$$

In fact, the first assertion is obvious. Since $\overline{k_\lambda^{(0, \alpha)}(t, \zeta)} = \dfrac{\partial^\alpha}{\partial \zeta^\alpha} \overline{k_\lambda(t, \zeta)}$ is bounded on $(\partial D \setminus \bigcup_{i=1}^m \delta_i) \times (D_0 \cap D')$ for any relatively compact domain D' in D, we have

$$\frac{\partial^\alpha}{\partial \zeta^\alpha} f(\zeta) = \frac{\partial^\alpha}{\partial \zeta^\alpha} \int_{\partial D} f(t) \overline{k_\lambda(t, \zeta)} \, d\omega(t) = \left\langle f, k_\lambda^{(0, \alpha)}(\ , \zeta) \right\rangle$$

and the second equality, replacing f by $k_\lambda^{(0,\beta)}(\ ,\zeta)$.

We denote by $H^\infty(D)$ the space of bounded holomorphic functions in D and we regard it as a closed subspace of L^∞.

LEMMA 4. *Let $f \in H^\infty(D)$. Put*

$$F(z,\zeta) = f(z)k_\lambda(z,\zeta)\overline{f(\zeta)} \quad \text{and} \quad G(z,\zeta) = k_\lambda(z,\zeta)\overline{f(\zeta)}.$$

Then F and G are holomorphic w.r.t. $(z,\overline{\zeta})$ in $D_0 \times D_0$ and $G^{(0,\beta)}(\ ,\zeta)$ belongs to H_λ^2 for any $\zeta \in D_0$. We fave

$$F^{(\alpha,\beta)}(z,\zeta) = \left\langle G^{(0,\beta)}(\ ,\zeta), G^{(0,\alpha)}(\ ,z)\right\rangle \qquad (z \in D_0,\ \zeta \in D_0).$$

In fact, we have

$$F^{(\alpha,\beta)}(z,\zeta) = \sum_{\mu=0}^{\alpha}\sum_{\nu=0}^{\beta}\binom{\alpha}{\mu}\binom{\beta}{\nu}f^{(\alpha-\mu)}(z)k_\lambda^{(\mu,\nu)}(z,\zeta)\overline{f^{(\beta-\nu)}(\zeta)}$$

and

$$\left\langle G^{(0,\beta)}(\ ,\zeta), G^{(0,\alpha)}(\ ,z)\right\rangle$$
$$= \sum_{\mu=0}^{\alpha}\sum_{\nu=0}^{\beta}\binom{\alpha}{\mu}\binom{\beta}{\nu}f^{(\alpha-\mu)}(z)\overline{f^{(\beta-\nu)}(\zeta)}\left\langle k_\lambda^{(0,\nu)}(\ ,\zeta), k_\lambda^{(0,\mu)}(\ ,z)\right\rangle.$$

Lemma 3 shows Lemma 4.

Let P_λ denote the orthogonal projection of L^2 onto H_λ^2.

LEMMA 5. *Let $f \in H^\infty(D)$ and put $G(z,\zeta) = k_\lambda(z,\zeta)\overline{f(\zeta)}$. Then we have*

$$P_\lambda(\overline{f}\,k_\lambda^{(0,\alpha)}(\ ,\zeta)) = G^{(0,\alpha)}(\ ,\zeta).$$

To prove it, let $\varphi \in H_\lambda^2$. By Lemma 3 we observe

$$\left\langle \varphi, P_\lambda(\overline{f}k_\lambda^{(0,\alpha)}(\ ,\zeta))\right\rangle = \left\langle \varphi, \overline{f}k_\lambda^{(0,\alpha)}(\ ,\zeta)\right\rangle$$
$$= \left\langle f\varphi, k_\lambda^{(0,\alpha)}(\ ,\zeta)\right\rangle = (f\varphi)^{(\alpha)}(\zeta)$$
$$= \sum_{\nu=0}^{\alpha}\binom{\alpha}{\nu}f^{(\alpha-\nu)}(\zeta)\left\langle \varphi, k_\lambda^{(0,\nu)}(\ ,\zeta)\right\rangle$$
$$= \left\langle \varphi, \sum_{\nu=0}^{\alpha}\binom{\alpha}{\nu}\overline{f^{(\alpha-\nu)}(\zeta)}k_\lambda^{(0,\nu)}(\ ,\zeta)\right\rangle$$
$$= \left\langle \varphi, G^{(0,\alpha)}(\ ,\zeta)\right\rangle,$$

which shows Lemma 5.

We shall denote by $S^\perp$ the orthogonal complement of a subspace S in L^2.

LEMMA 6 (ABRAHAMSE [1]). *Let w be an invertible function in L^∞. If f is in $(w H^2(D))^\perp \cap L^\infty$ then there exist some $\lambda \in \Lambda$, $g \in H^2_\lambda$ and $h \in (w H^2_\lambda)^\perp$ such that*

$$f = \bar{g} h \qquad and \qquad |f| = |g|^2 = |h|^2 \qquad (a.e. \, on \, \partial D).$$

LEMMA 7 (ABRAHAMSE [1]). *The linear subspace $(H^2)^\perp \cap L^\infty$ is dense in $(H^2)^\perp$.*

§7. Main Theorem in Finitely Connected Domains.

Let $z_1, z_2, \cdots, z_k$ be k distinct points in a domain D in C, bounded by $m + 1$ disjoint analytic simple closed curves. For each z_i, let $c_{i0}, \cdots, c_{in_i-1}$ be a sequence of n_i complex numbers. Our present extended interpolation problem (EI) is to find a holomorphic function f in D, satisfying $|f| \le 1$ and the conditions

$$\text{(EI)} \qquad f(z) = \sum_{\alpha=0}^{n_i-1} c_{i\alpha}(z - z_i)^\alpha + O((z - z_i)^{n_i}) \qquad (i = 1, \cdots, k).$$

For each element λ of the m-torus Λ, let k_λ be the kernel function introduced in the preceding section. We define the following matrices for $i, j = 1, \cdots, k$:

$$C_i = \begin{bmatrix} c_{i0} & & & \\ c_{i1} & c_{i0} & & \\ \vdots & \ddots & \ddots & \\ c_{in_i-1} & \cdots & c_{i1} & c_{i0} \end{bmatrix}, \qquad C = \begin{bmatrix} C_1 & & \\ & \ddots & \\ & & C_k \end{bmatrix},$$

$$\Gamma_{ij}^{(\lambda)} = \mathbf{M}\left(k_\lambda; z_i, z_j; n_i, n_j\right),$$

$$A_{ij}^{(\lambda)} = \Gamma_{ij}^{(\lambda)} - C_i \cdot \Gamma_{ij}^{(\lambda)} \cdot C_j^*,$$

$$A_\lambda = \begin{bmatrix} A_{11}^{(\lambda)} & \cdots & A_{1k}^{(\lambda)} \\ \cdots\cdots\cdots\cdots \\ A_{k1}^{(\lambda)} & \cdots & A_{kk}^{(\lambda)} \end{bmatrix}.$$

In terms of these matrices A_λ $(\lambda \in \Lambda)$, we have

THEOREM 9 (EXTENSION OF THE THEOREM OF ABRAHAMSE). *The problem (EI) admits a solution f with $|f| \leq 1$ in D if and only if the matrix A_λ is positive semidefinite for each λ in Λ. The solution is unique if and only if the determinant of A_λ is zero for some $\lambda \in \Lambda$.*

PROOF. We may assume without loss $z_i \in D_0 = D \setminus (\bigcup_{i=0}^{m} \delta_i)$. Let $\xi_{i\alpha}$ be complex numbers $(i = 1, \cdots, k \, ; \alpha = 0, \cdots, n_i - 1)$ and set

$$(1) \qquad K = \sum_{i=1}^{k} \sum_{\alpha=0}^{n_i-1} \overline{\xi_{i\alpha}} k_\lambda^{(0,\alpha)}(\, , z_i).$$

By Lemma 3, we have

$$\|K\|_2^2 = \sum_{i,j=1}^{k} \sum_{\alpha=0}^{n_i-1} \sum_{\beta=0}^{n_j-1} \xi_{i\alpha} \overline{\xi_{j\beta}} \, k_\lambda^{(\alpha,\beta)}(z_i, z_j).$$

Let $f \in H^\infty(D)$. Put, for $z \in D_0, \zeta \in D_0$,

$$F(z,\zeta) = f(z) k_\lambda(z,\zeta) \overline{f(\zeta)} \quad \text{and} \quad G(z,\zeta) = k_\lambda(z,\zeta) \overline{f(\zeta)}.$$

By Lemma 5,

$$P_\lambda(\overline{f} K) = \sum_{i=1}^{k} \sum_{\alpha=0}^{n_i-1} \overline{\xi_{i\alpha}} \, G^{(0,\alpha)}(\, , z_i).$$

It follows from Lemma 4 that

$$\|P_\lambda(\overline{f} K)\|_2^2 = \sum_{i,j=1}^{k} \sum_{\alpha=0}^{n_i-1} \sum_{\beta=0}^{n_j-1} \xi_{i\alpha} \overline{\xi_{j\beta}} F^{(\alpha,\beta)}(z_i, z_j).$$

On the other hand, we have

$$\|P_\lambda(\overline{f} K)\|_2^2 \leq \|\overline{f} K\|_2^2 \leq \|f\|_\infty^2 \|K\|_2^2$$

and hence

$$(2) \qquad \sum_{i,j=1}^{k} \sum_{\alpha=0}^{n_i-1} \sum_{\beta=0}^{n_j-1} \xi_{i\alpha} \overline{\xi_{j\beta}} (\|f\|_\infty^2 \, k_\lambda^{(\alpha,\beta)}(z_i, z_j) - F^{(\alpha,\beta)}(z_i, z_j)) \geq 0.$$

Now, assume that f satisfies $|f| \leq 1$ and the conditions (EI). Then, by the product formula (PF) in §1 and the definition of F, we see that $k_\lambda^{(\alpha,\beta)}(z_i, z_j) -$

$F^{(\alpha,\beta)}(z_i, z_j)$ is the $(\alpha+1, \beta+1)$-entry of the matrix of $A_{ij}^{(\lambda)}$. This implies $A_\lambda \geq 0$ for each $\lambda \in \Lambda$.

To prove the converse, assume $A_\lambda \geq 0$ for each $\lambda \in \Lambda$ and take a polynomial $\phi(z)$ which satisfies the conditions (EI). We may find such a polynomial by the method of indeterminate coefficients. Let $w(z)$ be the polynomial $w(z) = (z - z_1)^{n_1} \cdots (z - z_k)^{n_k}$. It is easy to see that the subspace wH_λ^2 is the orthogonal complement in H_λ^2 of the subspace $\mathcal{M}_\lambda$ spanned by the functions $\{k_\lambda^{(0,\alpha)}(\, , z_i)\}$ $(i = 1, \cdots, k; \alpha = 0, \cdots, n_i - 1)$. Thus we have

$$H_\lambda^2 = \mathcal{M}_\lambda \oplus w H_\lambda^2.$$

Let f be in $L^\infty \cap (wH^2(D))^\perp$. By Lemma 6, there exist some $\lambda \in \Lambda$, $g \in H_\lambda^2$ and $h \in (w H_\lambda^2)^\perp$ such that

$$f = \overline{g}h \qquad \text{and} \qquad |f| = |g|^2 = |h|^2 \qquad \text{(a.e. on } \partial D).$$

The function $K = P_\lambda(h)$ is in $\mathcal{M}_\lambda$ and hence it can be written in the form (1). Since $A_\lambda \geq 0$, we have $\|P_\lambda(\overline{\phi}K)\|_2 \leq \|K\|_2$ by (2). Hence,

$$\left| \int_{\partial D} \overline{\phi} f \, d\omega \right| = \left| \int_{\partial D} \overline{\phi} \, \overline{g} h d\omega \right| = |< h, \phi g >|$$
$$= |< K, \phi g >| = |< P_\lambda(\overline{\phi}K), g >| \leq \|P_\lambda(\overline{\phi}K)\|_2 \|g\|_2$$
$$\leq \|K\|_2 \|g\|_2 \leq \|h\|_2 \|g\|_2 = \|f\|_1.$$

By the theorem of Hahn-Banach, there exists a function ψ in L^∞ such that $\|\psi\|_\infty \leq 1$ and that for each $f \in L^\infty \cap (wH^2(D))^\perp$ we have

$$\int_{\partial D} \overline{\psi} f \, d\omega = \int_{\partial D} \overline{\phi} f \, d\omega.$$

This shows, by virtue of Lemma 7 and the relations $H^2(D) = \mathcal{M}_{\lambda_1} \oplus wH^2(D)$ and $\mathcal{M}_{\lambda_1} \subset H^\infty(D)$, where λ_1 is the identity of Λ, that $\psi - \phi$ is orthogonal to $(wH^2(D))^\perp$ in L^2, that is, it belongs to $wH^2(D)$. Therefore, ψ is a solution of the problem (EI) with $|\psi| \leq 1$, which completes the proof of the first part of the theorem.

To prove the uniqueness assertion, it suffices to follow the proof of Abrahamse [1], using instead of his Lemma 6 in [1] the following lemma which will be deduced immediately from Lemma 1 and Cauchy's integral formula. The details will not be carried out here.

LEMMA 8. *Let (z_0, ζ_0) be in $D_0 \times D_0$. Let α and β be two nonnegative integers. Then the mapping $\lambda \mapsto k_\lambda^{(\alpha,\beta)}(z_0, \zeta_0)$ is continuous on the m-torus Λ.*

§8. Proof of Lemma 2

It is known that, for a fixed ζ, the function $k_\lambda(z,\zeta)$ of z can be continued across the boundary ∂D. The problem is to find, for a relatively compact neighborhood U_2 of ζ_0 in D, a connected neighborhood U_1 of t_0 common to all $\zeta \in U_2$ such that, multiplying if necessary by λ_i^{-1} the values on the right side of the cut δ_i, we may continue the function $V_\lambda(z)$ to a function holomorphic and invertible in U_1 and that, for any fixed $\zeta \in U_2$, the function $V_\lambda(z)^{-1} k_\lambda(z,\zeta) \overline{V_\lambda(\zeta)}^{-1}$ of z may be extended to a function holomorphic in U_1. If we find such a neighborhood U_1, then it will follow from the theorem of Hartogs [8] that the function thus extended to U_1 for each $\zeta \in U_2$ is holomorphic w.r.t. $(z,\overline{\zeta})$ in $U_1 \times U_2$, since the original function is holomorphic w.r.t. $(z,\overline{\zeta})$ in $(U_1 \cap D) \times U_2$. This will complete the proof of Lemma 2.

Now, we reduce by means of V_λ to the case without the periods λ but with a slightly modified measure

$$(3) \qquad d\mu(t) = \exp\Big(\sum_{i=1}^{m} 2\xi_i v_i(t)\Big)\, d\omega(t).$$

The kernel function $k(z,\zeta)$ of $H^2(D)$ w.r.t. $d\mu$, which satisfies by definition

$$(4) \qquad f(\zeta) = \int_{\partial D} f(t)\, \overline{k(t,\zeta)}\, d\mu(t) \qquad (f \in H^2(D)),$$

has the relation

$$k_\lambda(z,\zeta) = V_\lambda(z)\, k(z,\zeta)\, \overline{V_\lambda(\zeta)}$$

(see Widom [16]), so that it suffices to prove the Lemma 2 for $k(z,\zeta)$.

Let $g(z,z^*)$ be the Green function of D with its pole at the reference point z^* and let $\widetilde{g}(z,z^*)$ be its harmonic cunjugate. Put $G(z) = g(z,z^*) + i\,\widetilde{g}(z,z^*)$. Then we have

$$(5) \qquad d\omega(t) = \frac{i}{2\pi} G'(t)\, dt\,.$$

The function G' is single-valued and holomorphic in D except at the single pole z^*. It can be continued analytically across the boundary ∂D by virtue of the reflection principle. It has m zeros $z_1^*, \cdots, z_m^*$ in D but it does not vanish on ∂D.

For $f \in H^2(D)$ we have

$$f(\zeta) = \frac{1}{2\pi i} \int_{\partial D} \frac{f(t)}{t - \zeta}\, dt \qquad (\zeta \in D)$$

(see Rudin [12]), so that (4) yields

$$\int_{\partial D} f(t) \left(\frac{1}{2\pi i} \frac{1}{t - \zeta} - \overline{k(t,\zeta)} \, \frac{d\mu(t)}{dt} \right) dt = 0 \qquad (f \in H^2(D)) \, .$$

This shows that there exists a unique $N \in H^\infty(D)$ such that

$$(6) \qquad N(t) = \frac{1}{2\pi i} \frac{1}{t - \zeta} - \overline{k(t,\zeta)} \, \frac{d\mu(t)}{dt} \qquad (t \in \partial D)$$

(see Rudin [12]). The function

$$M(z) = \frac{1}{2\pi i} \frac{1}{z - \zeta} - N(z)$$

is holomorphic in D except at the single pole ζ.

Assume $t_0 \in \gamma_j$. As $\exp\left(\sum_{i=1}^m 2\xi_i v_i(t) \right)$ is a constant $\neq 0$ on γ_j, we have by (3), (5), and (6)

$$\overline{k(t,\zeta)} = c_j \cdot M(t) \cdot G'(t)^{-1} \qquad (t \in \gamma_j),$$

where c_j is a constant $\neq 0$. The function $L_j = c_j M G'^{-1}$ is meromorphic in D and its poles are at most at $\zeta, z_1^*, \cdots, z_m^*$. Since $P = L_j + k(\ ,\zeta)$ is real and $Q = L_j - k(\ ,\zeta)$ is purely imaginary on γ_j, the functions $k(\ ,\zeta)$ and L_j can be cuntinued analytically across γ_j by the reflection principle as well as P and Q.

Let U_2 be a relatively compact neighborhood of ζ_0 in D. Since the z_i^* are independent of ζ, we can find a neighborhood U_1 of t_0, which is symmetric w.r.t. γ_j and contains no z_i^* or no points of U_2. Then P and Q, and hence L_j and $k(\ ,\zeta)$, can be extended to holomorphic functions of z in U_1 for any $\zeta \in U_2$, which completes the proof of Lemma 2.

References

[1] M. B. Abrahamse, The Pick interpolation theorem for finitely connected domains, Michigan Math. J. **26** (1979), 195-203.

[2] L. Ahlfors, Conformal Invariants. Topics in Geometric Function Theory, McGraw-Hill, New York, 1973.

[3] C. Carathéodory, Über den Variabilitätsbereich der Fourier'schen Konstanten von positiven harmonischen Funktionen, Rend. Circ. Mat. Palermo **32** (1911), 193-217.

[4] J. P. Earl, A note on bounded interpolation in the unit disc, J. London Math. Soc. (2) **13** (1976), 419-423.

[5] S. D. Fisher, Function Theory on Planer Domains, Wiley, New York, 1983.

[6] P. R. Garabedian, Schwarz's lemma and the Szegö kernel function, Trans. Amer. Math. Soc. **67** (1949), 1-35.

[7] J. B. Garnett, Bounded Analytic Functions, Academic Press, New York, 1981.

[8] F. Hartogs, Zur Theorie der analytischen Funktionen mehrerer unabhängiger Veränderlichen, insbesondere über die Darstellung derserben durch Reihen, welche nach Potenzen einer Veränderlichen fortschreiten, Math. Ann. **62** (1906), 1-88.

[9] D. E. Marshall, An elementary proof of Pick-Nevanlinna interpolation theorem, Michigan Math. J. **21** (1974), 219-223.

[10] R. Nevanlinna, Über beschränkte analytische Funktionen, Ann. Acad. Sci. Fenn. Ser A **32** (1929), No 7.

[11] G. Pick, Über die Beschränkungen analytischer Funktionen, welche durch vorgegebene Funktionswerte bewirkt werden, Math. Ann. **77** (1916), 7-23.

[12] W. Rudin, Analytic functions of class H_p, Trans. Amer. Math. Soc. **78** (1955), 46-66.

[13] I. Schur, Über Potenzreihen, die im Innern des Einheitskreises beschränkt sind, J. Reine Angew. Math. **147** (1917), 205-232.

[14] S. Takahashi, Extension of the theorems of Carathéodory-Toeplitz-Schur and Pick, Pacific J. Math. **138** (1989), 391-399.

[15] S. Takahashi, Nevanlinna parametrizations for the extended interpolation problem, Pacific J. Math. **146** (1990), 115-129.

[16] H. Widom, Extremal polynomials associated with a system of curves in the complex plane, Advances in Math. **3** (1969), 127-232.

Department of Mathematics
Nara Women's University
Nara 630, Japan

MSC 1991: Primary 30E05, 30C40
          Secondary 47A57

Operator Theory:
Advances and Applications, Vol. 59
© 1992 Birkhäuser Verlag Basel

ACCRETIVE EXTENSIONS AND PROBLEMS ON THE STIELTJES OPERATOR–VALUED FUNCTIONS RELATIONS

E. R. Tsekanovskii

*Dedicated to the memory of M. S. Brodskii, M. G. Krein, S. A. Orlov,
V. P. Potapov and Yu. L. Shmul'yan*

This paper presents a survey of investigations in the theory of accretive extensions of positive operators and connection with the problem of realization of a Stieltjes-type operator-valued function as a linear fractional transformation of the transfer operator-function of a conservative system. We give criteria of existence, together with some properties and a complete description of a positive operator.

In this paper a survey of investigations in the theory of accretive extensions of positive operators and connections with the problem of realization of a Stieltjes-type operator-valued function as a linear fractional transformation of the transfer operator-function of a conservative system is given. We give criteria of existence together with some properties and a complete description of the maximal accretive (m-accretive) non-selfadjoint extensions of a positive operator with dense domain in a Hilbert space. In the class of m-accretive extensions we specialize to the subclass of θ-sectorial extensions in the sense of T. Kato [24] (in S. G. Krein [29] terminology is regularly dissipative extensions of the positive operator), establish criteria for the existence of such extensions and give (in terms of parametric representations of θ-cosectorial contracitive extensions of Hermitian contraction) their complete description. It was an unexpected fact that if a positive operator B has a nonselfadjoint m-accretive extension T $(B \subset T \subset B^*)$ then the operator B always has an m-accretive extension which is not θ-sectorial for any θ $(\theta \in (0, \pi/2))$.

For Sturm-Liouville operators on the positive semi-axis there are obtained simple formulas which permit one (in terms of boundary parameter) to describe all accretive and θ-sectorial boundary value problems and to find an exact sectorial angle for the given value of the boundary parameter. All Stieltjes operator-functions generated by positive Sturm-Liouville operators are described. We obtain M. S. Livsic triangular models for m-acretive extensions (with real spectrum) of the positive operators with finite and equal deficiency indices. In this paper there are also considered direct and inverse problems of the realization theory for Stieltjes operator-functions and their connection with θ-sectorial extensions of the positive operators in rigged Hilbert spaces. Formulas for canonical and generalized resolvents of θ-cosectorial contractive extensions of Hermitian contractions are given.

Note that Stieltjes functions have an interesting physical interpretation. As it was established by M. G. Krein any scalar Stieltjes function can be a coefficient of a dynamic pliability of a string (with some mass distribution on it).

§1. ACCRETIVE AND SECTORIAL EXTENSIONS OF THE POSITIVE OPERATORS, OPERATORS OF THE CLASS $C(\theta)$ AND THEIR PARAMETRIC REPRESENTATION.

Let A be a Hermitian contraction defined on the subspace $\mathfrak{D}(A)$ of the Hilbert space $\mathfrak{H}$.

DEFINITION. The operator $S \in [\mathfrak{H}, \mathfrak{H}]$ ($[\mathfrak{H}, \mathfrak{H}]$ is the set of all linear bounded operators acting in $\mathfrak{H}$) is called a quasi-selfadjoint contractive extension (qsc-extension) of the operator A if

$$S \supset A, \quad S^* \supset A, \quad \|S\| \le 1$$
$$\mathfrak{D}(S) = \mathfrak{H}.$$

Self-adjoint contractive extensions (sc-extensions) of the Hermitian contraction were investigated, at first, by M. G. Krein [27], [28] in connection with the problem of description and uniqueness of the positive selfadjoint extensions of the positive linear operator with dense domain. It was proved by M. G. Krein [27] that there exists two extreme sc-extensions A_μ and A_M of the Hermitian contraction A which are called "rigid" and "soft" sc-extensions of A respectively. Moreover, the set of all sc-extensions of the Hermitian contraction A consists of the operator segment $[A_\mu, A_M]$. Let B be a positive closed Hermitian operator with dense domain acting on the Hilbert space $\mathfrak{H}$. Then, as it is well known [27], an operator $A = (I - B)(I + B)^{-1}$ is a Hermitian contraction in $\mathfrak{H}$ defined on some subspace $\mathfrak{D}(A)$ of the Hilbert space $\mathfrak{H}$. Let A_μ and A_M be "rigid" and "soft" sc-extensions of A. Then (see [27]) the operator $B_\mu = (I - A_\mu)(I + A_\mu)^{-1}$ is a positive self-adjoint extension of B and is in fact the extension obtained, at first, by K. Friedrichs in connection with his theorem that any positive operator with dense domain always has a positive self-adjoint extension. Besides, as it proved was in [27], the operator $B_M = (I - A_M)(I + A_M)^{-1}$ is also a positive self-adjoint extension of B. We call the operators B_μ and B_M the K. Friedrichs extension and the M. Krein extension respectively.

DEFINITION. Let T be a closed linear operator with dense domain acting on the Hilbert space $\mathfrak{H}$. We call T accretive if $\mathrm{Re}(Tf, f) \ge 0$ ($\forall f \in \mathfrak{D}(T)$) and m-accretive if it does not have accretive extensions.

DEFINITION. We call the m-accretive operator T θ-sectorial if there exists $\theta \in (0, \pi/2)$ such that

$$\operatorname{ctg}\theta \,|\operatorname{Im}(Tf, f)| \leq \operatorname{Re}(Tf, f) \qquad (\forall f \in \mathfrak{D}(T)). \tag{1}$$

DEFINITION. An operator $S \in [\mathfrak{H}, \mathfrak{H}]$ is called a θ-cosectorial contraction if there exists $\theta \in (0, \pi/2)$ such that

$$2\operatorname{ctg}\theta \,|\operatorname{Im}(Sf, f)| \leq \|f\|^2 - \|Sf\|^2 \qquad (\forall f \in \mathfrak{H}). \tag{2}$$

Note that inequality (2) is equivalent to

$$\|S \pm i\operatorname{ctg}\theta\, I\| \leq \sqrt{1 + \operatorname{ctg}^2\theta}. \tag{3}$$

Denote by $C(\theta)$ the set of contractions $S \in [\mathfrak{H}, \mathfrak{H}]$ $(\theta \in (0, \pi/2))$ satisfying (2) (or (3)) and let $C(\pi/2)$ be the set of all self-adjoint contractions acting on $\mathfrak{H}$. It is known [43] that if T is a θ-sectorial operator, then the operator $S = (I - T)(I + T)^{-1}$ is a θ-contraction, i.e. S belongs to $C(\theta)$. The converse statement is also valid, i.e. $S \in C(\theta)$ and $(I + S)$ is invertible, then the operator $T = (I - S)(I + S)^{-1}$ is a θ-sectorial operator.

THEOREM 1. (E. R. Tsekanovskii [40], [41]) *Let B be a positive linear closed operator with dense domain acting on the Hilbert space $\mathfrak{H}$. Then the operator B has a nonselfadjoint m-accretive extension T $(B \subset T \subset B^*)$ if and only if the K. Friedrichs extension B_μ and the M. Krein extension B_M do not coincide, i.e. $B_\mu \neq B_M$. If $B_\mu \neq B_M$, then*

1. *for every fixed $\theta \in (0, \pi/2)$ the operator B has nonselfadjoint θ-sectorial extension T $(B \subset T \subset B^*)$;*

2. *the operator B has a nonselfadjoint m-accretive extension T $(B \subset T \subset B^*)$ that fails to be θ-sectorial for all $\theta \in (0, \pi/2)$.*

The description of all θ-sectorial extensions of the positive operator B can be obtained via a linear fractional transformation with the help of the following theorem on parametric representation.

THEOREM 2. (Yu. M. Arlinskii, E. R. Tsekanovskii [7], [8], [9]) *Let A be a Hermitian contraction defined on the subspace $\mathfrak{D}(A)$ of the Hilbert space $\mathfrak{H}$. Then the equality*

$$S = \frac{1}{2}(A_M + A_\mu) + \frac{1}{2}(A_M - A_\mu)^{1/2}X(A_M - A_\mu)^{1/2} \tag{4}$$

establishes one-to-one correspondence between qsc-extensions S of the Hermitian operator A and contractions X on the subspace $\mathfrak{N}_0 = \overline{\mathfrak{R}(A_M - A_\mu)}$. The operator S in (4) is a θ-cosectorial qsc-extension of the Hermitian contraction A iff the operator X is a θ-cosectorial contraction on the subspace $\mathfrak{N}_0$.

Let $A \in [\mathfrak{D}(A), \mathfrak{H}]$ be a Hermitian contraction, $A^* \in [\mathfrak{H}, \mathfrak{D}(A)]$ be the adjoint of the operator A. Denote $\mathcal{G} = \overline{(I - AA^*)^{1/2}\mathfrak{D}(A)}$, $\mathcal{L} = \overline{(I - AA^*)^{1/2}\mathfrak{H}} \ominus \mathcal{G}$ and $\mathfrak{N} = \mathfrak{H} \ominus \mathfrak{D}(A)$. Let P_A, $P_\mathfrak{N}$, $P_\mathcal{L}$ be orthoprojectors onto $\mathfrak{D}(A)$, $\mathfrak{N}$, $\mathcal{L}$ respectively. We will define the contraction $Z \in [\mathcal{G}, \mathfrak{N}]$ in the form

$$Z(I - AA^*)^{1/2}f = P_\mathfrak{N}Af \qquad (f \in \mathfrak{D}(A))$$

and let $Z^* \in [\mathfrak{N}, \mathcal{G}]$ be the adjoint of the operator Z.

THEOREM 3. *Let A be a Hermitian contraction in $\mathfrak{H}$ defined on the subspace $\mathfrak{D}(A)$. Then the equalities*

$$A_\mu = AP_A + (I - AA^*)^{1/2}(Z^*P_\mathfrak{N} - P_\mathcal{L}(I - AA^*)^{1/2})$$
$$A_M = AP_A + (I - AA^*)^{1/2}(Z^*P_\mathfrak{N} + P_\mathcal{L}(I - AA^*)^{1/2})$$

are valid. The equality $A_\mu = A_M$ holds if and only if $\mathcal{L} = \{0\}$.

Theorem 3 was established by Yu. M. Arlinskii and E. R. Tsekanovskii [8]. In terms of the operator-matrix "rigid" and "soft" extensions were established by Azizov [1] and independently by M. M. Malamud and V. Kolmanovich [25]. The M. Krein extension B_M of the positive operator B with dense domain was described at first by T. Ando and K. Nishio [4], later by A. V. Shtraus [39]. In terms of the operator-valued Weyl functions [17] and space of boundary values operator B_M was described by V. A. Derkach, M. M. Malamud and E. R. Tsekanovskii [18]. Contractive extensions of the given contraction in terms of operator-matrices were investigated by C. Davis, W. Kahan and H. Weinberger [16], M. Crandall [15], G. Arsene and A. Gheondea [5], H. Langer and B. Textorius [34], Yu. Shmul'yan and R. Yanovskaya [38].

Theorems 1 and 2 develop and reinforce investigations by K. Friedrichs, M. G. Krein, R. Phillips [27], [28], [37] and give the solution of the T. Kato problem [24] on the existence and description of θ-sectorial extensions of the positive operator with dense domain.

Note that m-accretivity (θ-sectoriality) of an operator T is equivalent to the fact that the solution of the Cauchy problem

$$\begin{cases} \dfrac{dx}{dt} + Tx = 0 \\ x(0) = x_0 \end{cases} \qquad (x_0 \in \mathfrak{D}(T))$$

generates a contractive semigroup (a semigroup analytically continued as a semigroup of contractions into a sector $|\arg \xi| < \pi/2 - \theta$ of complex plane) [29].

Now, we consider some applications of Theorem 2 to the parametric representation of the θ-cosectorial qsc-extensions of a Hermitian contraction. Let S be a linear bounded operator with finite dimensional imaginary part acting on the Hilbert space $\mathfrak{H}$. Then as it is known [32]

$$\operatorname{Im} S = \sum_{\alpha,\beta=1}^{r} (\cdot, g_\alpha) j_{\alpha\beta} g_\beta$$

where $J = [j_{\alpha\beta}]$ is a self-adjoint and unitary matrix. Consider the matrix function $V(\lambda)$ given by

$$V(\lambda) = [((\operatorname{Re} S - \lambda I)^{-1} g_\alpha, g_\beta)].$$

THEOREM 4. *In order that the linear bounded operator S with finite dimensional imaginary part be a contraction, it is necessary and, for simple[1] S, sufficient that the following conditions are fulfilled:*

1) $V(\lambda)$ is holomorphic in $\operatorname{Ext}[-1,1]$

2) the matrices $V^{-1}(-1) = (V(-1-0))^{-1}$, $V^{-1}(1) = (V(1+0))^{-1}$, $(V^{-1}(-1) - V^{-1}(1))^{-1/2}$ exist and the matrix

$$
\begin{aligned}
K_J = (V^{-1}(-1) - V^{-1}(1))^{-1/2} \{ 2iJ + V^{-1}(-1) + V^{-1}(1) \} \\
\times (V^{-1}(-1) - V^{-1}(1))^{-1/2}
\end{aligned}
\tag{5}
$$

is a contraction.
The contraction S is θ-cosectorial (belongs to the class $C(\theta)$, $\theta \in (0, \pi/2)$) iff K_J of the form (5) is a θ-cosectorial contraction (belongs to the class $C(\theta)$, $\theta \in (0, \pi/2)$). Moreover the exact value of the angle θ is defined from the equation

$$\|K_J \pm i \operatorname{ctg} \theta \, I\|^2 = 1 + \operatorname{ctg}^2 \theta.$$

This theorem was obtained by E. R. Tsekanovskii [42], [43]. For the operator with one-dimensional imaginary part another proof was given by V. A. Derkach [20].

EXAMPLE. Let $\alpha(x)$ nondecreasing function on $[0, \ell]$. We consider the operator

$$(Sf)(x) = \alpha(x)f(x) + i \int_x^\ell f(t)\, dt \qquad (f \in L^2[0, \ell])$$

acting on the Hilbert space $L^2[0, \ell]$. It is easy to see that

$$(\operatorname{Im} Sf)(x) = \frac{1}{2} \int_0^\ell f(t)\, dt = (f, g)g \qquad (g(x) \equiv 1/\sqrt{2}) \tag{6}$$

---

[1]The operator S is called simple if there exists no reducing subspace on which one induces a self-adjoint operator.

$$(\operatorname{Re} Sf)(x) = \alpha(x)f(x) + \frac{i}{2}\int_x^\ell f(t)\,dt - \frac{i}{2}\int_0^x f(t)\,dt.$$

From simple calculations

$$V(\lambda) = ((\operatorname{Re} S - \lambda I)^{-1}g, g) = \operatorname{tg}\left(\frac{1}{2}\int_0^\ell \frac{dt}{\alpha(t) - \lambda}\right). \tag{7}$$

Set $\alpha(x) \equiv 0$ and consider the operator

$$(S_0 f)(x) = i\int_x^\ell f(t)\,dt \qquad (x \in [0, \ell],\ f \in L^2[0, \ell]). \tag{8}$$

From (6) and (7) it follows that $V(1) = -\operatorname{tg}(\ell/2)$, $V(-1) = \operatorname{tg}(\ell/2)$.

As $J = I$ and an operator S_0 is simple, we shall find all $\ell > 0$ for which S_0 is a θ-cosectorial contraction. For this, in accordance with Theorem 4 (see relation (5)), the number
$$K = K_J = \frac{2i + V^{-1}(1) + V^{-1}(-1)}{V^{-1}(-1) - V^{-1}(1)}$$

has to satisfy the inequality

$$2\operatorname{ctg}\theta\,|\operatorname{Im} K| \le 1 - |K|^2.$$

Exact value of θ can be calculated from the formula

$$\operatorname{ctg}\theta = \frac{1 - |K|^2}{2\,|\operatorname{Im} K|}.$$

The operator S_0 is a θ-cosectorial contraction if and only if $0 < \ell < \pi/2$, moreover, an exact value of θ is equal to ℓ ($\theta = \ell$). From this and the definition of the class $C(\theta)$ ($\theta \in (0, \pi/2)$), we obtain the inequality

$$\operatorname{ctg}\ell\left|\int_0^\ell f(t)\,dt\right|^2 \le \int_0^\ell |f(t)|^2 dt - \int_0^\ell \left|\int_x^\ell f(t)\,dt\right|^2 dx$$

$$(\forall \ell \in [0, \pi/2],\ \forall f \in L^2[0, \ell]).$$

Moreover, the constant $\operatorname{ctg}\ell$ can not be increased so that for all $f \in L^2[0, \ell]$ as mentioned above, the inequality is valid.

With the help of Theorem 4 there was established a full description of positive and sectorial boundary value problems for Sturm-Liouville operators on the semi-axis, at first (see also [22], [26]). Let $\mathfrak{H} = L^2[a, \infty]$, $l(y) = -y'' + q(x)y$, where $q(x)$ is a real locally summable function. Assume that a minimal Hermitian operator

$$\begin{cases} By = l(y) = -y'' + q(x)y \\ y'(a) = y(a) = 0 \end{cases} \tag{9}$$

has deficiency indices $(1,1)$ (the case of limiting Weyl point). Let $\varphi_k(x,\lambda)$ $(k=1,2)$ be the solutions of the Cauchy problems

$$\begin{cases} l(\varphi_1) = \lambda\varphi_1 \\ \varphi_1(a,\lambda) = 0 \\ \varphi_1'(a,\lambda) = 1 \end{cases} \qquad \begin{cases} l(\varphi_2) = \lambda\varphi_2 \\ \varphi_1(a,\lambda) = -1 \\ \varphi_2'(a,\lambda) = 0. \end{cases}$$

Then, as it is known [35], there exists Weyl function $m_\infty(\lambda)$ such that

$$\varphi(x,\lambda) = \varphi_2(x,\lambda) + m_\infty(\lambda)\varphi_1(x,\lambda) \in L^2[a,\infty].$$

Consider a boundary problem

$$\begin{cases} T_h y = l(y) = -y'' + q(x)y \\ y'(a) = hy(a). \end{cases} \tag{10}$$

THEOREM 5. *(E. R. Tsekanovskii [42], [43]) 1. In order that the positive Sturm-Liouville operator of the form (9) have nonselfadjoint accretive extensions (boundary problems) of the form (10), it is necessary and sufficient that $m_\infty(-0) < \infty$. 2. The set of all accretive and θ-sectorial boundary value problems for Sturm-Liouville operators of the form (10) is defined by the parameter h, belonging to the domain indicated in (11).*

Moreover, as 1) h sweeps the real semi-axis in this domain, there results all accretive self-adjoint boundary value problems; 2) h sweeps all values not belonging to the straight line $\mathrm{Re}\,h = -m_\infty(-0)$ and $h \neq \bar{h}$, then there results all θ-sectorial boundary value problems $(\theta \in (0,\pi/2))$; moreover, the exact value of θ is defined as it was pointed out in (11); 3) h sweeps all values with $h \neq \bar{h}$ and belonging to the straight line $\mathrm{Re}\,h = -m_\infty(-0)$, then there results all accretive boundary value problems which are not θ-sectorial for any $\theta \in (0,\pi/2)$.

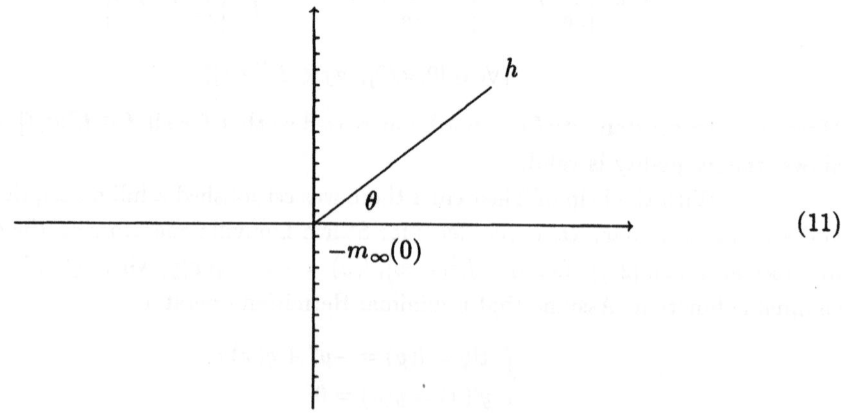

$$\tag{11}$$

Thus, Theorem 5 indicates which values of boundary parameter h correspond to the semigroup generated by the solution of the Cauchy problem

$$\begin{cases} \dfrac{dx}{dt} + T_h x = 0 \\ x(0) = x_0 \end{cases} \qquad (x_0 \in \mathfrak{D}(T_h))$$

in the space $L^2[0,\infty]$, being contractive $(\operatorname{Re} h \geq -m_\infty(-0))$, and for which h it can be analytically continued as a semigroup of contractions into a sector $|\arg \zeta| < \pi/2 - \theta$ of the complex plane. At the same time, it helps to calculate θ. Also the value of the parameter h in (11) $(\operatorname{Re} h > -m_\infty(-0))$ determines whether this semigroups of contractions can not be analytically continued as a semigroup of contractions into any sector $|\arg \zeta| < \varepsilon$ of the complex plane $(\operatorname{Re} h = -m_\infty(-0))$ $(\operatorname{Im} h \neq 0)$.

Note that the M. Krein boundary value problem for the minimal positive operator B of the form (9) has the form (as it follows from (11))

$$\begin{cases} B_M y = -y'' + q(x)y \\ y'(a) + m_\infty(-0)y(a) = 0 \end{cases} \qquad (x \in [a,\infty])$$

and the K. Friedrichs boundary value problem, as is well known, coincides with Dirichlet problem

$$\begin{cases} B_\mu y = -y'' + q(x)y \\ y(a) = 0 \end{cases} \qquad (x \in [a,\infty]).$$

EXAMPLE. Consider a Sturm-Liouville operator with Bessel potential

$$\begin{cases} By = -y'' + \dfrac{\nu^2 - 1/4}{x^2} y \\ y'(1) = y(1) = 0 \end{cases} \qquad (x \in [1,\infty],\ \nu \geq 1/2)$$

in the Hilbert space $L^2[1,\infty]$. In this case the Weyl function has the form [35]

$$m_\infty(\lambda) = -\sqrt{\lambda}\,\frac{I_\nu'(\sqrt{\lambda}) - iY_\nu'(\sqrt{\lambda})}{I_\nu'(\sqrt{\lambda}) + iY_\nu'(\sqrt{\lambda})}$$

where $I_\nu(z), Y_\nu(z)$ are Bessel functions of the first and second genus, $m_\infty(-0) = \nu$.

§2. STIELTJES OPERATOR–VALUED FUNCTIONS AND THEIR REALIZATION.

Let B be a closed Hermitian operator acting on the Hilbert space $\mathfrak{H}$, B^* be the adjoint of B, $\overline{\mathfrak{D}(B^*)} = \mathfrak{H}$, $\mathfrak{R}(B^*) \subset \mathfrak{H}_0 = \overline{\mathfrak{D}(B)}$. Denote $\mathfrak{H}_+ = \mathfrak{D}(B^*)$ and define in $\mathfrak{H}_+$ scalar product

$$(f,g)_+ = (f,g) + (B^*f, B^*g) \qquad (f,g \in \mathfrak{H}_+)$$

and build the rigged Hilbert space $\mathfrak{H}_+ \subset \mathfrak{H} \subset \mathfrak{H}_-$ [14]. We call an operator B regular, if an operator PB is a closed operator in $\mathfrak{H}_0$ (P is an orthoprojector $\mathfrak{H}$ to $\mathfrak{H}_0$) [6], [46].

We say that the closed linear operator T with dense domain in $\mathfrak{H}$ is a member of the class Ω_B, if

1) $T \supset B$, $T^* \supset B$, where B is a regular closed Hermitian operator in $\mathfrak{H}$.

2) $(-i)$ is a regular point of T.

The condition that $(-i)$ is a regular point in the definition of the class Ω_B is non-essential. It is sufficient to require the existence of some regular point for T.

We call an operator $\mathbf{A} \in [\mathfrak{H}_+, \mathfrak{H}_+]$ a biextension of the regular Hermitian operator B if $\mathbf{A} \supset B$, $\mathbf{A}^* \supset B$. If $\mathbf{A} = \mathbf{A}^*$, then $\mathbf{A}$ is called a selfadjoint biextension. Note that identifying the space conjugate to $\mathfrak{H}_\pm$ with $\mathfrak{H}_\mp$ gives that $\mathbf{A}^* \in [\mathfrak{H}_+, \mathfrak{H}_-]$. The operator $\hat{B}f = \mathbf{A}f$, where $\mathfrak{D}(\hat{B}) = \{f \in \mathfrak{H}_+ : \mathbf{A}f \in \mathfrak{H}\}$ is called the quasi-kernel of the operator $\mathbf{A}$. We call selfadjoint biextension $\mathbf{A}$ strong if $\hat{B} = \hat{B}^*$ [45], [46]. An operator $\mathbf{A} \in [\mathfrak{H}_+, \mathfrak{H}_-]$ is called a $(*)$-extension of the operator T in the class Ω_B if $\mathbf{A} \supset T \supset B$, $\mathbf{A}^* \supset T^* \supset B$ where T (T^*) is extension of B without exit of the space $\mathfrak{H}$; moreover, $\mathbf{A}$ is called correct if $\mathbf{A}_R = (\mathbf{A} + \mathbf{A}^*)/2$ is a strong selfadjoint biextension of B. The operator T of the class Ω_B will be associated with the class Λ_B if

1) B is a maximal common Hermitian part of T and T^*;

2) T has correct $(*)$-extension.

The notion of biextension under the title "generalized extension", at first, was investigated by E. R. Tsekanovskii [45], [46]. (There the author obtained the existence, a parametric representation of $(*)$-extensions and self-adjoint biextensions of Hermitian operators with dense domain.) The case of biextensions of Hermitian operators with nondense domain was investigated by Yu. M. Arlinskii, Yu. L. Shmul'yan and the author [6], [45], [46].

Consider a Sturm-Liouville operator B (minimal, Hermitian) of the form (9) and an operator T_h of the form (10). Operators

$$
\begin{aligned}
\mathbf{A}y &= -y'' + q(x)y + \frac{1}{\mu - h}[hy(a) - y'(a)][\mu\delta(x - a) + \delta'(x - a)] \\
\mathbf{A}^*y &= -y'' + q(x)y + \frac{1}{\mu - \bar{h}}[\bar{h}y(a) - y'(a)][\mu\delta(x - a) + \delta'(x - a)]
\end{aligned}
\tag{12}
$$

for every $\mu \in [-\infty, +\infty]$ define correct $(*)$-extension of T_h (T_h^*) and give a full description of these $(*)$-extensions. Note that $\delta(x-a)$ and $\delta'(x-a)$ are the δ-function and its derivative respectively. Moreover,

$$
(y, \mu\delta(x - a) + \delta'(x - a)) = \mu y(a) - y'(a) \qquad (y \in \mathfrak{H}_+).
$$

DEFINITION. The aggregate $\Theta = (\mathbf{A}, \mathfrak{H}_+ \subset \mathfrak{H} \subset \mathfrak{H}_-, K, J, \mathcal{E})$ is called a rigged operator colligation of the class Λ_B if

1) $J \in [\mathcal{E}, \mathcal{E}]$ ($\mathcal{E}$ is a Hilbert space), $J = J^* = J^{-1}$;
2) $K \in [\mathcal{E}, \mathfrak{H}]$;
3) $\mathbf{A}$ is a correct $(*)$-extension of the operator T of the class Λ_B, moreover, $(\mathbf{A} - \mathbf{A}^*)/2i = KJK^*$;
4) $\overline{\mathfrak{R}(K)} = \overline{\mathfrak{R}(\mathrm{Im}\,\mathbf{A}) + \mathcal{L}}$, where $\mathcal{L} = \mathfrak{H} \ominus \mathfrak{H}_0$ and closure is taken in $\mathfrak{H}_-$.

The operator-function $W_\Theta(z) = I - 2iK^*(\mathbf{A} - zI)^{-1}KJ$ is called a M. S. Livsic characteristic operator-function of the colligation Θ and also M. S. Livsic characteristic operator-function of operator T. The operator colligation is called M. S. Brodskii-M. S. Livsic operator colligation. In the case when T is bounded, we obtain the well-known definition of the characteristic matrix-function [13], [32] (with M. S. Brodskii modification) introduced by M. S. Livsic [32]. The other definitions, generally speaking, of unbounded operators were given by A. V. Kuzhel and A. V. Shtraus. For every M. S. Brodskii-M. S. Livsic operator colligation we define an operator-function

$$V_\Theta(z) = K^*(\mathbf{A}_R - zI)^{-1}K. \tag{14}$$

The operator-functions $V_\Theta(z)$ and $W_\Theta(z)$ are connected by relations

$$\begin{aligned}
V_\Theta(z) &= i[W_\Theta(z) + I]^{-1}[W_\Theta(z) - I]J \\
W_\Theta(z) &= [I + iV_\Theta(z)J]^{-1}[I - iV_\Theta(z)J].
\end{aligned} \tag{15}$$

The conservative system of the form

$$\begin{cases} (\mathbf{A} - zI)x = KJ\varphi_- \\ \varphi_+ = \varphi_- - 2iK^*x \end{cases} \tag{16}$$

where $x \in \mathfrak{H}_+$, $\varphi_\pm \in \mathcal{E}$, φ_- is an input vector, φ_+ is an output vector, x is the vector of inner state, will be associated with each operator colligation. It is easy to see that the transfer mapping of such a system $S(z)$ ($\varphi_+ = S(z)\varphi_-$) coincides with $W_\Theta(z)$. After D. Z. Arov [2], in case when $J \neq I$, we call the above mentioned system a passage system.

When $J = I$, the above mentioned system is called a scattering system. Many problems for systems with distributed parameters, and a scattering problem as well, are packed into this scheme and were investigated by M. S. Livsic and W. Helton [23], [33]. The operator colligation Θ will be called accretive if $\mathrm{Re}(\mathbf{A}f, f) \geq 0$, $\forall f \in \mathfrak{H}_+$ and dissipative if $J = I$. An operator-function $V(z) \in [\mathcal{E}, \mathcal{E}]$, where $\mathcal{E}$ is a finite-dimensional Hilbert space, will be associated with the class $\mathcal{S}$ of the Stieltjes operator-functions if

1) $V(z)$ is holomorphic in $\mathrm{Ext}[0, \infty]$;
2) $\mathrm{Im}\,V(z)/\mathrm{Im}\,z \geq 0$;
3) $\mathrm{Im}[zV(z)]/\mathrm{Im}\,z \geq 0$.

It was established by M. G. Krein that any Stieltjes operator-function $V(z)$ has an integral representation

$$V(z) = \gamma + \int_0^\infty \frac{dG(t)}{t - z} \qquad (\gamma \geq 0) \tag{17}$$

where $G(t)$ is a nondecreasing operator-function and $\int_0^\infty (t+1)^{-1}dG(t) < \infty$.

DEFINITION. We call the operator-functions $V(z) \in \mathcal{S}$, acting on the Hilbert space $\mathcal{E}$ (dim $\mathcal{E} < \infty$), realizable if, in some neighbourhood of $(-i)$, $V(z)$ can be represented in the form

$$V(z) = i[W_\Theta(z) + I]^{-1}[W_\Theta(z) - I]J \tag{18}$$

where $W_\Theta(z)$ is a characteristic operator-function for some rigged accretive and dissipative colligation of the class Λ_B ($W_\Theta(z)$ is a transfer mapping of the some scattering system Θ).

Thus, realization problem for Stieltjes operator-function is a problem on representation of this operator-function in the form of linear fractional transformation of the transfer mapping of some conservative scattering system (16), the main operator **A** of which is accretive.

DEFINITION. The Stieltjes operator-function $V(z) \in [\mathcal{E}, \mathcal{E}]$ (dim $\mathcal{E} < \infty$) will be said to be a member of the class $\mathcal{S}(R)$ of Stieltjes operator-functions, if for an operator γ in (17) the equality

$$\gamma f = 0 \qquad (f \in \mathcal{E}_\infty^\perp)$$

is valid on the subspace $\mathcal{E}_\infty^\perp = \{f \in \mathcal{E} : \int_0^\infty (dG(t)f, f)_\mathcal{E} < \infty\}$.

THEOREM 6. *Let Θ be a rigged accretive colligation of the class Λ_B (dim $\mathcal{E} < \infty$). Then the operator-function $V_\Theta(z)$ of the form (14) belongs to the class $\mathcal{S}(R)$. Conversely, if $V(z)$ acts on a finite-dimensional Hilbert space $\mathcal{E}$ and belongs to the class $\mathcal{S}(R)$, then $V(z)$ is realizable.*

Thus, we specialize to the subclass $\mathcal{S}(R)$ in the class $\mathcal{S}$ which can be realized. In this case, when dim $\mathcal{E} = 1$, the operator-function

$$V(z)e = \left(\gamma + \int_0^\infty \frac{dG(t)}{t - z}\right) e \qquad \left(e \in \mathcal{E}, \int_0^\infty dG(t) < \infty, \ \gamma > 0\right)$$

does not belong to the class $\mathcal{S}(R)$ and, therefore, is not realizable.

We define three subclasses in the class $\mathcal{S}(R)$:

1) An operator-function $V(z)$ of the class $\mathcal{S}(R)$ will be a member of $\mathcal{S}_0(R)$ if in the integral representation (17)

$$\int_0^\infty (dG(t)f, f)_\mathcal{E} = \infty \qquad (\forall f \in \mathcal{E}, \ f \neq 0).$$

2) An operator-function $V(z)$ of the class $\mathcal{S}(R)$ will be a member of $\mathcal{S}_1(R)$ if in the integral representation (17)

$$\int_0^\infty (dG(t)f, f)_\mathcal{E} < \infty \qquad (\forall f \in \mathcal{E})$$

and $\gamma = 0$.

3) An operator-function $V(z)$ of the class $\mathcal{S}(R)$ will be a member of $\mathcal{S}_{01}(R)$ if $\mathcal{E}_\infty^\perp \neq \{0\}$, $\mathcal{E}_\infty^\perp \neq \mathcal{E}$.

THEOREM 7. *Let Θ be an M. S. Brodskii-M. S. Livsic rigged accretive colligation of the class Λ_B, where B is a positive operator with dense domain and $\dim \mathcal{E} < \infty$. Then $V_\Theta(z)$ of the form (14) belongs to the class $\mathcal{S}_0(R)$ and $\mathfrak{D}(T) \neq \mathfrak{D}(T^*)$. Conversely, assume $V(z) \in [\mathcal{E}, \mathcal{E}]$ $(\dim \mathcal{E} < \infty)$ belongs to the class $\mathcal{S}_0(R)$. Then $V(z)$ can be realized, moreover, B has dense domain and $\mathfrak{D}(T) \neq \mathfrak{D}(T^*)$.*

The direct statement in this theorem was established by V. A. Derkach and the author [19]. Theorem 7 belongs to I. N. Dovzhenko and E. R. Tsekanovskii [21].

The regular positive operator B acting on the Hilbert space $\mathfrak{H}$ is called an R-operator [46], if its semideficiency indices (deficiency indices of PB) are equal to 0.

THEOREM 8. *Let Θ be an M. S. Brodskii-M. S. Livsic rigged accretive colligation of the class Λ_B, where B is a positive R-operator with domain not dense. Then $V_\Theta(z)$ of the form (14) belongs to the class $\mathcal{S}_1(R)$ and $\mathfrak{D}(T) = \mathfrak{D}(T^*)$. Conversely, if $V(z) \in \mathcal{S}_1(R)$, $V(z) \in [\mathcal{E}, \mathcal{E}]$ $(\dim \mathcal{E} < \infty)$, then $V(z)$ can be realized, moreover, B is the positive regular R-operator with domain not dense and $\mathfrak{D}(T) = \mathfrak{D}(T^*)$.*

Note that the criteria of the equality $\mathfrak{D}(T) = \mathfrak{D}(T^*)$ was established at first by A. V. Kuzhel in terms of characteristic matrix-functions introduced by himself [30], and the obtained results add to and make more precise these investigations for this class of operator-functions acting on a finite-dimensional Hilbert space.

The analogous theorem may be formulated for operator-functions of the class $\mathcal{S}_{01}(R)$. Consider the following subclasses of the class $\mathcal{S}_0(R)$:

1) The operator-function $V(z) \in [\mathcal{E}, \mathcal{E}]$ $(\dim \mathcal{E} < \infty)$ belongs to the class $\mathcal{S}_0^M(R)$ if $V(z) \in \mathcal{S}_0(R)$ and

$$\int_0^\infty \frac{1}{t}(dG(t)f, f)_\mathcal{E} = \infty \qquad (\forall f \neq 0, \ f \in \mathcal{E}).$$

2) The operator-function $V(z) \in [\mathcal{E}, \mathcal{E}]$ $(\dim \mathcal{E} < \infty)$ belongs to the class $\mathcal{S}_{01}^M(R)$ if $V(z) \in \mathcal{S}_{01}(R)$ and

$$\int_0^\infty \frac{1}{t}(dG(t)f, f)_\mathcal{E} < \infty \qquad (\forall f \in \mathcal{E}).$$

3) The operator-function $V(z) \in [\mathcal{E}, \mathcal{E}]$ $(\dim \mathcal{E} < \infty)$ belongs to the class $\mathcal{S}_{001}^M(R)$ if $V(z) \in \mathcal{S}_0(R)$ and for the subspace

$$\mathcal{E}_\infty^M = \{f \in \mathcal{E} : \int_0^\infty \frac{1}{t}(dG(t)f, f)_\mathcal{E} < \infty\}$$

the relations $\mathcal{E}_\infty^M \neq \{0\}$, $\mathcal{E}_\infty^M \neq \mathcal{E}$ are valid.

THEOREM 9. *Let Θ be an M. S. Brodskii-M. S. Livsic rigged accretive colligation of the class Λ_B, $\dim \mathcal{E} < \infty$ and let B_M be the M. G. Krein extension of B being quasi-kernel for $\mathbf{A}_R$ ($\mathbf{A}_R \supset B_M$). Then an operator-function $V_\Theta(z)$ of the form (14) belongs to the class $\mathcal{S}_0^M(R)$. Conversely, if $V(z) \in [\mathcal{E}, \mathcal{E}]$ ($\dim \mathcal{E} < \infty$) belongs to the class $\mathcal{S}_0^M(R)$, then $V(z)$ is realizable, moreover, $\mathbf{A}_R$ includes B_M as a quasi-kernel ($\mathbf{A}_R \supset B_M$).*

The analogous theorems (direct and inverse) may be formulated for the classes $\mathcal{S}_{01}^M(R)$ and $\mathcal{S}_{001}^M(R)$. It may be established that operator-functions of the class $\mathcal{S}_0(R)$ can not be realized with the help of K. Friedrichs extension B_μ, when $\mathbf{A} \supset B_\mu$.

DEFINITION. Let B be a positive operator in $\mathfrak{H}$, and T be an m-accretive extension of B ($B \subset T \subset B^*$). The operator T is called extremal if the operator X is unitary for the operator $S = (I - T)(I + T)^{-1}$ in the parametric representation (4).

As follows from (11) the Sturm-Liouville operator T_h of the form (10) will always be extremal iff the boundary parameter h lies on the straight line $\operatorname{Re} h = -m_\infty(-0)$.

Let Θ be a rigged M. S. Brodskii-M. S. Livsic operator colligation of the class Λ_B, $\dim \mathcal{E} < \infty$ and $W_\Theta(z)$ be its M. S. Livsic characteristic operator-function. Consider an operator-function

$$Q(z) = i[W_\Theta^{-1}(-1)W_\Theta(z) + I]^{-1}[W_\Theta^{-1}(-1)W_\Theta(z) - I]J \tag{19}$$

where we assume that the right part (19) is defined. Let

$$\begin{aligned} K_J = {}&[Q^{-1}(-\infty) - Q^{-1}(-0)]^{-1/2}\{2iJ + Q^{-1}(-\infty) + Q^{-1}(-0)\} \\ &\times [Q^{-1}(-\infty) - Q^{-1}(-0)]^{-1/2} \end{aligned} \tag{20}$$

Recall, further, that operator-functions $V_\Theta(z)$ and $W_\Theta(z)$ are connected by relation (15).

THEOREM 10. *Let Θ be an M. S. Brodskii-M. S. Livsic rigged accretive colligation of the class Λ_B, $\dim \mathcal{E} < \infty$, $\overline{\mathfrak{D}(B)} = \mathfrak{H}$ and let T be an α-sectorial ($\alpha \in (0, \pi/2)$) operator. Then an operator-function $V_\Theta(z)$ of the form (14) belongs to the class $\mathcal{S}_0(R)$ and an operator K_J of the form (20) is an α-cosectorial contraction. Conversely, if $V(z) \in [\mathcal{E}, \mathcal{E}]$ ($\dim \mathcal{E} < \infty$) belongs to the class $\mathcal{S}_0(R)$, an operator K_I of the form (19), (20) is an α-cosectorial contraction ($\alpha \in (0, \pi/2)$), then $V(z)$ can be realized by the M. S. Brodskii-M. S. Livsic rigged accretive operator colligation of the class Λ_B, $\dim \mathcal{E} < \infty$, $\overline{\mathfrak{D}(B)} = \mathfrak{H}$ and T will be an α-sectorial operator.*

Theorems 8,9 were established by I. N. Dovzhenko and E. R. Tsekanovskii [21], Theorem 10 belongs to E. R. Tsekanovskii and is published at first.

It may be shown that if Θ is a rigged accretive operator colligation of the class Λ_B, $\dim \mathcal{E} < \infty$, $\overline{\mathfrak{D}(B)} = \mathfrak{H}$ and T is an extremal operator, then $V_\Theta(z)$ of the form (14) belongs to the class $\mathcal{S}_0(R)$, K_J of the form (19), (20) is unitary. Conversely, if $V(z) \in [\mathcal{E}, \mathcal{E}]$

$(\dim \mathcal{E} < \infty)$ belongs to the class $\mathcal{S}_0(R)$, K_I of the form (19), (20) is unitary, then $V(z)$ can be realized by a rigged accretive operator colligation of the class Λ_B, $\overline{\mathfrak{D}(B)} = \mathfrak{H}$ and T is an extremal operator. This fact was established by I. N. Dovzhenko and the author [21].

Consider an operator T_h $(\operatorname{Im} h > 0)$ of the form (10) and let $\mathbf{A}$ be a correct $(*)$-extension of the operator T_h. Then an operator $\mathbf{A}$ has the form (12) and

$$\frac{\mathbf{A} - \mathbf{A}^*}{2i} = (\cdot, v)v, \qquad v = \frac{(\operatorname{Im} h)^{1/2}}{|\mu - h|}[\mu\delta(x - a) + \delta'(x - a)]$$

$$(y, v) = \frac{(\operatorname{Im} h)^{1/2}}{|\mu - h|}[\mu y(a) + y'(a)] \qquad (y \in \mathfrak{H}_+).$$

Let $\mathcal{E} = \mathbf{C}^1$, $K\{c\} = cv$ $(c \in \mathbf{C}^1)$. Then $\Theta = (\mathbf{A}, \mathfrak{H}_+ \subset \mathfrak{H} \subset \mathfrak{H}_-, K, I, \mathbf{C}^1)$ will be an M. S. Brodskii-M. S. Livsic operator colligation. The M. S. Livsic characteristic function will be

$$W_\Theta(z) = \frac{\mu - h}{\mu - \bar{h}} \frac{m_\infty(z) + \bar{h}}{m_\infty(z) + h}. \tag{21}$$

Characteristic function for a Sturm-Liouville operator of the form (10) without a phase factor $(\mu - h)/(\mu - \bar{h})$ was investigated by B. S. Pavlov [36] in connection with scattering problems. We'll describe all phase factors in (21) under which $V_\Theta(z) = i[W_\Theta(z) + I]^{-1}[W_\Theta(z) - I]$ will be Stieltjes function (we assume, of course, that a minimal operator B of the form (9) is positive and $m_\infty(-0) < \infty$). The set of real μ and non-real h satisfying the inequalities

$$\begin{cases} \mu \geq \dfrac{(\operatorname{Im} h)^2}{m_\infty(-0) + \operatorname{Re} h} + \operatorname{Re} h \\ \operatorname{Re} h + m_\infty(-0) \geq 0 \end{cases} \tag{22}$$

gives a complete description of all values of the parameters μ and h in (21), for which $V_\Theta(z)$ is a Stieltjes function. This fact was established by the author and published at first. As it was shown by I. N. Dovzhenko, the operator-function $V_\Theta(z)$ belongs to the class $\mathcal{S}_0^M$ if and only if the inequality (22) for parameter μ holds with equality.

Note that realization theory (and applications) for arbitrary rational matrix-functions was developed by H. Bart, I. Gohberg, M. Kaashoek [12]. A realization theory for a very general class of transfer functions has recently been given by G. Weiss [47].

§3. M. S. LIVSIC TRIANGULAR MODEL OF THE M–ACCRETIVE EXTENSIONS (WITH REAL SPECTRUM) OF THE POSITIVE OPERATORS.

Let T be a nonselfadjoint m-accretive operator of the class Λ_B, where B is a positive closed operator with dense domain having deficiency indices (r, r) $(r < \infty)$. Assume T has only real spectrum. We say that an operator T satisfying these conditions is in the class $\Lambda_r^+(B)$. As it is known [46], an operator T always has correct $(*)$-extensions.

Let **A** be one of them. Let's include this extension **A** in the rigged operator colligation Θ for which a channel operator is invertible (it is always possible to do the same as in the case with the bounded operator) [13]. Let $W_\Theta(z)$ be the M. S. Livsic characteristic operator-function of this colligation (operator T). Then, as it was established by the author in [44], $W_\Theta(z)$ has the following regularized multiplicative representation

$$W_\Theta(z) = \lim_{\xi \to \infty} \int_0^{\xi} \exp\left[-2i\,\frac{\Pi^*(x)\Pi(x)J}{\alpha(x) - z}\,dx\right] J$$
$$\times \int_0^{\xi} \exp\left[2i\,\frac{J\Pi^*(x)\Pi(x)}{\alpha(x) - \bar{\zeta}}\,dx\right] JW_\Theta(\bar{\zeta}) \tag{23}$$

where 1) $\alpha(x)$ is a non-decreasing function on $[0, \infty]$; 2) $\Pi(x) \in [\mathcal{E}, \mathcal{E}]$, $\dim \mathcal{E} = r$, $\Pi(x)$ is invertible on a set of full measure, $\operatorname{sp}\Pi^*(x)\Pi(x) = 1$; 3) if $L(x) = \int_0^x \Pi^*(t)\Pi(t)\,dt$ then $\lim_{x \to \infty}(L(x)\varphi, \varphi) = \infty$ $(\forall \varphi \neq 0,\ \varphi \in \mathcal{E})$.

An operator $T \in \Lambda_r^+$ is called prime if

$$\text{clos. span}\,\{\mathfrak{N}_z : z \in \rho(T)\} = \mathfrak{H}$$

where $\rho(T)$ is the set of regular points of the operator T, $\mathfrak{N}_z$ is a defect subspace of B. The operator $T_{\mathfrak{H}_0} = T|_{\mathfrak{H}_0}$, where $\mathfrak{H}_0 = \text{clos. span}\,\{\mathfrak{N}_z : z \in \rho(T)\}$ is called the prime part of T.

With the help of (23) consider in $L_\mathcal{E}^2[0, \infty]$ the operator

$$(\vec{T}f)(x) = \alpha(x)f(x) + 2i \int_0^{\infty} \Pi^*(x)J\Pi(t)f(t)\,dt \tag{24}$$

where

$$\mathfrak{D}(\vec{T}) = \{f \in L_\mathcal{E}^2[0, \infty] : \vec{T}f \in L_\mathcal{E}^2[0, \infty]\}. \tag{25}$$

THEOREM 11. *Let $T \in \Lambda_r^+(B)$ and let T be a prime operator. Then T is unitarily equivalent to the prime part of the M. S. Livsic triangular model of the form (24), (25). The triangular model for an operator $T \in \Lambda_1^+(B)$ has the form*

$$(\vec{T}f)(x) = \alpha(x)f(x) + 2i \int_0^{\infty} f(t)\,dt\,J \qquad (J = \pm I).$$

An operator $\vec{T}$ of the form (24), (25) is α-sectorial (prime part) if and only if an operator K_J of the form (19), (20) is an α-cosectorial contraction $(\alpha \in (0, \pi/2))$. An operator $\vec{T}$ (prime part) is extremal iff K_J of the form (19), (20) is unitary.

This theorem was established by E. R. Tsekanovskii [44]. For the class of operators considered here, the model of the form (24), (25) is simpler than the model obtained earlier in another way by A. V. Kuzhel [30].

In the case when T has a complete system of root subspaces, the factorization and model were obtained with the help of the same method by Yu. M. Arlinskii [3]. The resolvent of M. S. Livsic triangular model has the form

$$((\vec{T} - zI)^{-1}f)(x) = \frac{f(x)}{\alpha(x) - z} - 2i \int_0^\infty \frac{\Pi(x)W^*(x,\bar{z})}{\alpha(x) - z} J \frac{W(t,z)\Pi^*(t)}{\alpha(t) - z} f(t)\, dt$$

where

$$W(t,z) = \int_0^x \exp\left[-2i \frac{\Pi^*(t)\Pi(t)J}{\alpha(t) - z}\, dt\right].$$

§4. CANONICAL AND GENERALIZED RESOLVENTS OF QSC–EXTENSIONS OF HERMITIAN CONTRACTIONS.

Let A be a Hermitian contraction defined on the subspace $\mathfrak{D}(A)$ of the Hilbert space $\mathfrak{H}$. Let A_μ and A_M be rigid and soft sc-extensions of A. We consider the completely indeterminate case, i.e. we assume that $\ker(A_M - A_\mu) = \mathfrak{D}(A)$. Let $\mathfrak{H} \subset \widetilde{\mathfrak{H}}$ and let $\widetilde{S}$ be a qsc-extension of Hermitian contraction A with exit in $\mathfrak{H}$ (A acts on $\mathfrak{H}$).

The operator-function $R_z = \widetilde{P}(\widetilde{S} - zI)^{-1}|_{\mathfrak{H}}$, where $\widetilde{P}$ is the orthoprojector from $\widetilde{\mathfrak{H}}$ onto $\mathfrak{H}$, is called a generalized resolvent. The resolvent, according to the definition, is canonical if $\widetilde{\mathfrak{H}} = \mathfrak{H}$. Denote

$$R_z^\mu = (A_\mu - zI)^{-1}, \quad R_z^M = (A_M - zI)^{-1},$$

$$Q_\mu(z) = [(A_M - A_\mu)^{1/2} R_z^\mu ((A_M - A_\mu)^{1/2} + I)]|_{\mathfrak{N}},$$

$$Q_M(z) = [(A_M - A_\mu)^{1/2} R_z^M ((A_M - A_\mu)^{1/2} + I)]|_{\mathfrak{N}}, \quad \mathfrak{N} = \mathfrak{H} \ominus \mathfrak{D}(A).$$

Denote, also, $\Phi(\theta)$ ($\theta \in [0, \pi/2)$) the set of linear bounded operators Y acting on $\mathfrak{N}$ and satisfying the condition

$$\|Yf\|^2 + \operatorname{ctg}\theta\,|\operatorname{Im}(Yf, f)| \leq \operatorname{Re}(Yf, f) \qquad (f \in \mathfrak{N}).$$

The condition $Y \in \Phi(\theta)$ is equivalent to the condition $X = 2Y - I \in C(\theta)$ (if $\theta = 0$ then $\Phi(0)$ is a class of non-negative operators in $\mathfrak{N}$). Let $\Lambda(\theta)$ be a set of points z in the complex plane satisfying the condition

$$|\sin\theta\, z \pm i\cos\theta| \leq 1.$$

THEOREM 12. *1) If $z \in \operatorname{Ext}\Lambda(\theta)$, then the equality*

$$R_z^S = R_z^\mu - R_z^\mu (A_M - A_\mu)^{1/2} Y[I + (Q_\mu(z) - I)Y]^{-1}(A_M - A_\mu)^{1/2} R_z^\mu$$

establishes one-to-one correspondence between the set of canonical resolvents of qsc-extensions S in the class $C(\theta)$ ($\theta \in [0, \pi/2)$) of the Hermitian contraction A and the constant operators Y of the class $\Phi(\theta)$.

2) *If $z \in \text{Ext } \Lambda(\theta)$, then the equality*

$$R_z^S = R_z^M + R_z^M (A_M - A_\mu)^{1/2} Y [I - (Q_M(z) + I)Y]^{-1}(A_M - A_\mu)^{1/2} R_z^M$$

establishes one-to-one correspondence between the set of canonical resolvents of qsc-extensions S in the class $C(\theta)$ ($\theta \in [0, \pi/2)$) of the Hermitian contraction A and the constant operators Y of the class $\Phi(\theta)$.

3) *If $z \in \text{Ext } \Lambda(\pi/2)$, then the equality*

$$R_z = R_z^\mu - R_z^\mu (A_M - A_\mu)^{1/2} Y(z) [I + (Q_\mu(z) - I)Y(z)]^{-1}(A_M - A_\mu)^{1/2} R_z^\mu \qquad (26)$$

establishes one-to-one correspondence between the set of generalized resolvents in the class $C(\pi/2)$ of the Hermitian contraction A and the set of operator-functions $Y(z)$ holomorphic in $\text{Ext } \Lambda(\pi/2)$ and in the class $\Phi(\pi/2)$. The constant operator-function $Y(z) = Y$ in (26) corresponds to the canonical resolvents and only to them.

This theorem was obtained by Yu. M. Arlinskii and E. R. Tsekanovskii [9], [10] and generalized some investigations by M. G. Krein and I. E. Ovcharenko [28] about generalized resolvents of sc-extensions of Hermitian contractions. Generalized resolvents of contractive extensions of an arbitrary contraction (not necessarily Hermitian) were investigated by H. Langer and B. Textorus [34].

The problem of describing generalized resolvents of qsc-extensions of Hermitian contractions not only of the class $C(\pi/2)$ (as in Theorem 12), but also of the class $C(\theta)$ ($\theta \in [0, \pi/2)$) is open.

Acknowledgements. The author is grateful to T. Ando, S. Belyi and the referee for helpful suggestions and aid.

REFERENCES

1.   Azizov, T. Ya., *On extension of positive operators*, 17th Voronezh Winter School, Preprint VINITI, N4585, 1984, pp. 5–7.

2.   Arov, D. Z., *Passive linear steady-state dynamical systems*, Sibirsk. Mat. Zh. **20** (1979), 211–228.

3.   Arlinskii, Yu. M., *A triangular model of an unbounded quasi-Hermitian operator with a complete system of root subspaces*, Dokl. Akad. Nauk Ukrain. SSR Ser. A **11** (1979), 883–886.

4.   Ando, T., Nishio, K., *Positive selfadjoint extensions of positive symmetric operators*, Tohoku Math. J. (2) **22** (1970), 65–75.

5.   Arsene, G., Gheondea, A., *Completing matrix contractions*, J. Operator Theory **7** (1982), 179–189.

6.   Arlinskii, Yu. M., Tsekanovskii, E. R., *The method of rigged spaces method in the theory of extensions of Hermitian operators with a non-dense domain*, Sibirsk. Mat. Zh. **15** (1974), 243–261.

7.   Arlinskii, Yu. M., Tsekanovskii, E. R., *Nonselfadjoint contracting extensions of a Hermitian contraction and theorem of M. G. Krein*, Uspekhi Mat. Nauk **37** no. 1 (1982), 131–1312.

8.   Arlinskii, Yu. M., Tsekanovskii, E. R., *Sectorial extensions of positive Hermitian operators and their resolvents*, Akad. Nauk Armyan. SSR Dokl. **79** no. 5 (1984), 199–203.

9.   Arlinskii, Yu. M., Tsekanovskii, E. R., *Quasiselfadjoint contractive extensions of a Hermitian contraction*, Teor. Funktsii Funktsional. Anal. i Prilozhen. **50** (1988), 9–16.

10.   Arlinskii, Yu. M., Tsekanovskii, E. R., *On resolvents of m-accretive extensions of symmetric differential operator*, Math. Phys. Nonlinear Mechanics **1** (1984), 11–16.

11.   Arlinskii, Yu. M., Tsekanovskii, E. R., *Generalized resolvents of quasiselfadjoint contracting extensions of a Hermitian contraction*, Ukrain. Mat. Zh. **35** (1983), 601–603.

12.   Bart, H., Gohberg, I., Kaashoek, M., *Minimal factorization of matrix and operator-functions*, Operator Theory: Advances and Applications, 1, Birkhäuser, Basel, 1979.

13.   Brodskii, M. S., *Triangular and Jordan representations of linear operators*, Nauka, Moscow, 1969.

14.   Berezanskii, Yu. M., *Expansions in eigen functions of selfadjoint operators*, Naukova Dumka, Kiev, 1965.

15.   Crandall, M., *Norm preserving extensions of linear transformations on Hilbert spaces*, Proc. Amer. Math. Soc. **21** (1969), 335–340.

16.   Davis, C., Kahan, W., Weinberg, H., *Norm-preserving dilations and their applications to optimal error bounds*, SIAM J. Numer. Anal. **19** (1982), 445–469.

17.   Derkach, V. A., Malamud, M. M., *On the Weyl function and Hermite operators with lacunae*, Dokl. Akad. Nauk SSSR **293** (1987), 1041–1046.

18.   Derkach, V. A., Malamud, M. M., Tsekanovskii, E. R., *Sectorial extensions of a positive operator, and the characteristic function*, Dokl. Akad. Nauk SSSR **298** (1988), 537–541.

19.   Derkach, V. A., Tsekanovskii, E. R., *Characteristic operator-functions of accretive operator colligations*, Dokl. Akad. Nauk Ukrain. SSR **1986** no. 8, 16–19.

20.   Derkach, V. A., Tsekanovskii, E. R., *On the characteristic function of a quasi-Hermitian contraction*, Izv. Vyssh. Uchebn. Zaved. Mat. **1987** no. 6, 46–51.

21.   Dovzhenko, I. N., Tsekanovskii, E. R., *Classes of Stieltjes operator functions and their conservative realization*, Dokl. Akad. Nauk SSSR **311** (1990), 18–22.

22.   Evans, W. D., Knowles, I., *On the extension problem for accretive differential operators*, J. Funct. Anal. **63** (1985), 276–298.

23. Helton, J. W., *Systems with infinite-dimensional state space: the Hilbert space approach*, Proc. IEEE **64** (1976), 145–160.

24. Kato, T., *The perturbation theory for linear operators*, Springer-Verlag, New York, 1966.

25. Kolmanovich, V. Yu., Malamud, M. M., *Extensions of sectorial operators and dual pairs of contractions*, Preprint VINITI, N4428, 1985.

26. Kochubey, A. N., *Extensions of a positive definite symmetric operator*, Dokl. Akad. Nauk Ukraine. SSR Ser. A **1979** no. 3, 168–171.

27. Krein, M. G., *Theory of self-adjoint extensions of semi-bounded Hermitian transformations and its applications*, Mat. Sbornik **20** (1947), 431–495.

28. Krein, M. G., Ovcharenko, I. E., *Q-functions and sc-resolvents of non-densely defined Hermitian contractions*, Sibirsk Mat. Zh. **18** (1977), 1032–1056.

29. Krein, S. G., *Linear differential equations in Banach space*, Nauka, Moscow, 1967.

30. Kuzhel, A. V., *The reduction of unbounded non-selfadjoint operators to triangular form*, Dokl. Akad. Nauk SSSR **119** (1958), 868–871.

31. Kuzhel, A. V., *Conditions for the equality $D_A = D_{A^*}$ for unbounded operators*, Uspekhi Mat. Nauk **16** no. 3 (1961), 189–190.

32. Livsic, M. S., *On spectral decomposition of linear non-selfadjoint operators*, Mat. Sbornik **34** (1954), 145–199.

33. Livsic, M. S., *Operators, oscillations, waves. Open systems*, Nauka, Moscow, 1966.

34. Langer, H., Textorius, B., *Generalized resolvents of contractions*, Acta Sci. Math. (Szeged) **44** (1982), 125–131.

35. Naimark, M. A., *Linear differential operators*, Akademie-Verlag, Berlin, 1960.

36. Pavlov, B. S., *On a selfadjoint Schrödinger operator*, Problems of Mathematical Physics, Leningrad University, Leningrad, 1966, pp. 102–132.

37. Phillips, R., *Dissipative operators and hyperbolic systems of partial differential equations*, Trans. Amer. Math. Soc. **90** (1959), 193–254.

38. Shmul'yan, Yu. L., Yanovskaya, R. N., *Blocks of a contractive operator matrix*, Izv. Vyssh. Uchebn. Zaved. Mat. **1981** no. 7, 70–75.

39. Shtraus, A. V., *The extensions of semi-bounded operator*, Dokl. Akad. Nauk SSSR **211** (1973), 543–546.

40. Tsekanovskii, E. R., *Nonselfadjoint accretive extensions of positive operators and the Friedrichs-Krein-Phillips theorems*, Funktsional. Anal. i Prilozhen. **14** no. 2 (1980), 87–88.

41. Tsekanovskii, E. R., *The Friedrichs-Krein extensions of positive operators, and holomorphic semigroups of contractions*, Funktsional. Anal. i Prilozhen. **15** no. 4 (1981), 91–92.

42. Tsekanovskii, E. R., *Characteristic function and description of accretive and sectorial boundary value problems for ordinary differential operators*, Dokl. Akad. Nauk Ukrain. SSR Ser. A **1985** no. 6, 21–24.

43. Tsekanovskii, E. R., *The characteristic function and sectorial boundary value problems*, Trudy Inst. Mat. (Novosibirsk) 7, Issled. Georm. Mat. Anal., 1987, pp. 180–194.

44. Tsekanovskii, E. R., *Triangular models of unbounded accretive operators and the regular factorization of their characteristic operator functions*, Dokl. Akad. Nauk SSSR **297** (1987), 552–556.

45. Tsekanovskii, E. R., *Generalized selfadjoint extensions of symmetric operators*, Dokl. Akad. Nauk SSSR **178** (1968), 1267–1270.

46. Tsekanovskii, E. R., Shmul'yan, Yu. L., *The theory of biextensions of operators in rigged Hilbert spaces. Unbounded operator colligations and characteristic functions*, Uspekhi Mat. Nauk **32** no. 5 (1977), 69–124.

47. Weiss, G., *The representation of regular linear systems on Hilbert spaces*, Internat. Ser. Numer. Math., 91, Birkhäuser, Basel, 1989, pp. 401–416.

Donetsk State University
Universitetskaya 24
240055 Donetsk, Ukraine

MSC 1991: Primary 47A20, Secondary 47A48, 47B25, 47B44

Operator Theory:
Advances and Applications, Vol. 59
© 1992 Birkhäuser Verlag Basel

Commuting Nonselfadjoint Operators and Algebraic Curves

Victor Vinnikov

As was discovered by M.S.Livšic, methods of algebraic geometry play an important role in the theory of commuting nonselfadjoint operators. Using further geometrical ideas, we construct triangular models for pairs of commuting nonselfadjoint operators with finite-dimensional imaginary parts and a smooth discriminant curve. The characteristic function of a pair of commuting nonselfadjoint operators turns out to be a function on the discriminant curve, and the reduction of the pair of operators to the triangular model corresponds to the canonical factorization for semicontractive functions on a compact real Riemann surface.

1 Commuting Nonselfadjoint Operators and the Discriminant Curve

Our objective is an investigation, up to the unitary equivalence, of a pair A_1, A_2 of linear bounded commuting operators in a Hilbert space H ($\dim H = N \leq \infty$). We assume A_1 and A_2 have finite-dimensional imaginary parts: $\dim G = n < \infty$, where G is the so-called *nonhermitian subspace*: $G = \overline{(A_1 - A_1^*)H} + \overline{(A_2 - A_2^*)H}$. As was first discovered by Livšic [11,12], a certain real algebraic curve, the discriminant curve, plays a prominent role in all the investigations.

The subspace G is clearly invariant under the operators $A_1 - A_1^*, A_2 - A_2^*, A_1A_2^* - A_2A_1^*, A_2^*A_1 - A_1^*A_2$. We can therefore (after choosing an orthonormal basis in G) define the following hermitian matrices of order n:

$$\sigma_1 = \frac{1}{i}(A_1 - A_1^*) \mid G,$$

$$\sigma_2 = \frac{1}{i}(A_2 - A_2^*) \mid G,$$

$$\gamma^{\text{in}} = \frac{1}{i}(A_1A_2^* - A_2A_1^*) \mid G,$$

$$\gamma^{\text{out}} = \frac{1}{i}(A_2^*A_1 - A_1^*A_2) \mid G \tag{1.1}$$

We define a polynomial in two complex variables y_1, y_2

$$f(y_1, y_2) = \det(y_1\sigma_2 - y_2\sigma_1 + \gamma^{\text{in}}) \tag{1.2}$$

We assume $f(y_1, y_2) \not\equiv 0$. $f(y_1, y_2)$ is a real polynomial of degree at most n. $f(y_1, y_2)$ is called the *discriminant polynomial* of the pair A_1, A_2, and the real algebraic curve X of degree n in the complex projective plane, whose affine equation is $f(y_1, y_2) = 0$, is called the *discriminant curve*. A *(self-adjoint) determinantal representation* of X is a $n \times n$ matrix of linear expressions in the affine coordinates y_1, y_2, with hermitian coefficients, whose determinant gives the equation of X. In particular we call $y_1\sigma_2 - y_2\sigma_1 + \gamma^{\text{in}}$ the *input determinantal representation* of X corresponding to the pair A_1, A_2.

We have thus associated to the pair A_1, A_2 the discriminant curve and its input determinantal representation. The first evidence to the importance of the discriminant curve is given by the following generalization of the classical Cayley-Hamilton Theorem. Recall first [10] that the so-called *principal subspace*

$$\tilde{H} = \text{span}\,\{A_1^{k_1} A_2^{k_2}(G)\}_{k_1,k_2=0}^{\infty} = \text{span}\,\{A_1^{*k_1} A_2^{*k_2}(G)\}_{k_1,k_2=0}^{\infty}$$

reduces A_1 and A_2 and the restrictions of A_1 and A_2 to the orthogonal complement $H \ominus \tilde{H}$ are selfadjoint operators, so that it is enough to consider the restrictions of A_1 and A_2 to $\tilde{H}$.

Theorem 1.1 (Livšic [11]) $f(A_1, A_2) \mid \tilde{H} = 0$.

The *joint spectrum* of the operators A_1, A_2 is the set of all points $\lambda = (\lambda_1, \lambda_2) \in \mathbf{C}^2$ such that there exists a sequence $v_m(m = 1, \ldots)$ of vectors of unit length in H satisfying

$$\lim_{m \to \infty} (A_k - \lambda_k I)v_m = 0 \quad (k = 1, 2) \tag{1.3}$$

(It has been shown by A.Markus that for a pair of commuting operators with finite-dimensional imaginary parts this is equivalent to more general definitions of the joint spectrum due to Harte [4] and Taylor [16].) It follows that the joint spectrum of the operators A_1, A_2 (restricted to the principal subspace $\tilde{H}$) lies on the discriminant curve.

We have obtained for the pair A_1, A_2 the following objects, which are clearly unitary invariants: the discriminant curve X; its input determinantal representation $y_1\sigma_2 - y_2\sigma_1 + \gamma^{\text{in}}$ (up to the simulataneous conjugation of $\sigma_1, \sigma_2, \gamma^{\text{in}}$ by a unitary matrix); and the joint spectrum, which lies on X. We consider now the inverse problem. Suppose we are given a real projective plane curve X of degree n, a determinantal representation $y_1\sigma_2 - y_2\sigma_1 + \gamma$ of X, and a subset S of affine points of X, which is closed and bounded in $\mathbf{C}^2$, and all of whose accumulation points are real points of X (those conditions are always satisfied by the joint spectrum of a pair of commuting operators with finite-dimensional imaginary parts (see [1])). We want to construct (up to the unitary equivalence on the pricipal subspace) all pairs of commuting operators with discriminant curve X, input determinantal representation $y_1\sigma_2 - y_2\sigma_1 + \gamma$ and joint spectrum S.

We shall present a complete and explicit solution to this inverse problem under the assumption that the curve X is irreducible and smooth and possesses real points (the last is merely a technical condition). The solution will yield triangular models for pairs of commuting nonselfadjoint operators with finite-dimensional imaginary parts and a smooth discriminant curve. In the special case when one of the operators is dissipative (say $\sigma_2 > 0$)

and the Hilbert space H is finite-dimensional, the inverse problem has been solved by Livšic in [11].

The main tools in the construction of triangular models are Livšic theory of commutative operator colligations [11] and our description of determinantal representations of real smooth plane curves [18]. The proof that every pair of commuting nonselfadjoint operators is unitarily equivalent to a triangular model is based on the factorization theorem for the characteristic function (see [12]) of the pair of operators. We prove it by showing that the characteristic function is in fact a function on the discriminant curve; this ties the theory of commuting nonselfadjoint operators with the function theory on a real Riemann surface. For more details and complete proofs see the forthcoming papers [19,20,21].

Before continuing we return to (1.1) and consider the polynomial $\det(y_1\sigma_2 - y_2\sigma_1 + \gamma^{\text{out}})$.

Theorem 1.2 (Livšic [11]) $\det(y_1\sigma_2 - y_2\sigma_1 + \gamma^{\text{in}}) = \det(y_1\sigma_2 - y_2\sigma_1 + \gamma^{\text{out}})$.

We may say that the pair A_1, A_2 determines a "transformation" of the input determinantal representation $y_1\sigma_2 - y_2\sigma_1 + \gamma^{\text{in}}$ of the discriminant curve into the *output determinantal representation* $y_1\sigma_2 - y_2\sigma_1 + \gamma^{\text{out}}$. It will turn out that this transformation can be recovered from the joint spectrum, and from the transformation one can recover the operators A_1, A_2 themselves.

2   Determinantal Representations of Real Plane Curves

We recall now briefly from [18] the description of determinantal representations of real smooth plane curves. See e.g. [3] for background algebro-geometrical details.

Let X be a real projective plane curve of degree n. Two determinantal representations $U = y_1\sigma_2 - y_2\sigma_1 + \gamma$, $U' = y_1\sigma_2' - y_2\sigma_1' + \gamma'$ of X are called *(hermitely) equivalent* if there exists a complex matrix $P \in GL(n, \mathbf{C})$ such that $U' = PUP^*$. We want to describe equivalence classes of determinantal representations of X.

Let U be a determinantal representation of X. For each point x on X consider $\operatorname{coker} U(x) = \{v \in (\mathbf{C}^n)^* : vU(x) = 0\}$ (we write elements of $\mathbf{C}^n$ as column n-vectors and elements of $(\mathbf{C}^n)^*$ as row n-vectors). It can be shown that if x is a regular point of X then $\dim \operatorname{coker} U(x) = 1$. *Assume now X is a smooth curve.* It follows that $\operatorname{coker} U$ is a line bundle on X; more precisely, we define $\operatorname{coker} U$ to be the subbundle of the trivial bundle of rank n over X, whose fiber at the point x is $\operatorname{coker} U(x)$. Clearly, if two determinantal representations U, U' of X are equivalent, then the corresponding line bundles $\operatorname{coker} U, \operatorname{coker} U'$ are isomorphic. Conversely it turns out that if the line bundles corresponding to two determinantal representations of X are isomorphic, then the determinantal representations are equivalent up to sign. The description of determinantal representations has been thus reduced to the description of certain line bundles on X.

X is a compact Riemann surface of genus g, where $g = (n-1)(n-2)/2$. Choosing a canonical integral homology basis on X and the corresponding normalized basis for holomorphic differentials, we obtain the period lattice Λ in $\mathbf{C}^g$. The Jacobian variety $J(X) = \mathbf{C}^g/\Lambda$; it is a g-dimensional complex torus. The Abel-Jacobi map μ associates to every line bundle L on X a point $\mu(L)$ in $J(X)$. Furthermore the isomorphism class of L is determined by two invariants: the degree $\deg L$ of L, an integer, and the point $\mu(L)$ in $J(X)$.

Some important geometrical properties of the line bundle can be expressed analytically in terms of the corresponding point in the Jacobian variety through the use of the Riemann's theta function $\theta(z)$. $\theta(z)$ is an entire function on $\mathbf{C}^g$ determined by the period lattice Λ. $\theta(z)$ is quasiperiodic with respect to Λ: when z is translated by a vector in Λ, $\theta(z)$ is multiplied by a non-zero number, so that we can talk about the zeroes of $\theta(z)$ on $J(X)$.

It can be shown that if $L = \operatorname{coker} U$, where U is a determinantal representation of X, then $\deg L = -n(n-1)/2$. One can determine necessary and sufficient conditions on a line bundle L of degree $-n(n-1)/2$ to be the cokernel of a determinantal representation of X, and translating them into conditions on the corresponding point in the Jacobian variety yields

Theorem 2.1 (Vinnikov [18]) *X possesses determinantal representations. There is a one-to-one correspondance between equivalence classes, up to sign, of determinantal representations U of X and points ζ of $J(X)$ satisfying $\zeta + \bar{\zeta} = e$ and $\theta(\zeta) \neq 0$. The correspondance is given by $\zeta = \mu(\operatorname{coker} U(n-2)) + \kappa$.*

The use of the twisted line bundle $\operatorname{coker} U(n-2)$ instead of $\operatorname{coker} U$ and the translation of the point in $J(X)$ by the so-called Riemann's constant κ are technical details. $e \in \mathbf{C}^g$ is a half-period ($2e \in \Lambda$) explicitly determined by the topology of the set of real points $X_\mathbf{R} \subset X$. Note that since X is a real curve, the period lattice Λ is invariant under complex conjugation, and the conjugation descends to $J(X) = \mathbf{C}^g/\Lambda$, so that the equation $\zeta + \bar{\zeta} = e$ makes sense there. We can also introduce an additional invariant, the *sign* of a determinantal representation U of X, which equals ± 1 and distinguishes between the representations U and $-U$.

A complete study of the set of points in $J(X)$ described in Theorem 2.1 appears in [18]. This set is a disjoint union of certain g-dimensional "punctured" real torii, the "punctures" coming from the points ζ with $\theta(\zeta) = 0$. We consider here the simplest non-trivial example of real smooth cubics — $n = 3, g = 1$ (see [17]).

A real smooth cubic X can be brought, by a real projective change of coordinates, to the normal form:

$$y_2^2 - y_1(y_1 + \theta_1)(y_2 + \theta_2) = 0 \tag{2.1}$$

(affine equation), where θ_1, θ_2 are two distinct numbers different from 0. We assume $\theta_1, \theta_2 \in \mathbf{R}$, so that the set $X_\mathbf{R}$ of real points of X consists of two connected components. Since the genus $g = 1$, X is homeomorphic to a torus. This homeomorphism is given explicitly through the parametrization of X by elliptic functions:

$$y_1 = \sqrt[3]{4}\wp(u) - \frac{\theta_1 + \theta_2}{3},$$
$$y_2 = \wp'(u) \tag{2.2}$$

Here $\wp(u)$ is the Weierstrass $\wp$-function with periods $1, \tau$, the number τ ($\Im\tau > 0$) depending on θ_1, θ_2. Since $\theta_1, \theta_2 \in \mathbf{R}$, actually $\tau \in i\mathbf{R}$. The point u varies in the period parallelogram, with vertices $0, 1, \tau, 1 + \tau$, say.

A complete set of non-equivalent determinantal representations of X is given by

$$U = \epsilon \begin{pmatrix} p & d + y_2 & q + y_1 \\ -d + y_2 & e + y_1 & 0 \\ q + y_1 & 0 & -1 \end{pmatrix},$$

$$q = \frac{\theta_1 + \theta_2 - e}{2}, p = \theta_1\theta_2 - \frac{\theta_1 + \theta_2 - e}{2} \cdot \frac{\theta_1 + \theta_2 + 3e}{2},$$

$$\epsilon = \pm 1, e \in \mathbf{R}, d \in i\mathbf{R}, d^2 = -e(-e + \theta_1)(-e + \theta_2) \tag{2.3}$$

Note that $(-e, -d)$ is an affine point on X.

Choosing a suitable homology basis, the period lattice Λ of X in $\mathbf{C}$ is spanned by $1, \tau$, and $J(X) = \mathbf{C}/(\mathbf{Z} + \tau\mathbf{Z})$ is simply the period parallelogram with opposite sides identified (and is isomorphic to X itself by (2.2)). The point ζ corresponding to the representation U of (2.3) under the correspondance of Theorem 2.1 is $\zeta = v + \frac{1 \pm \tau}{2}$, where v is the point in the period parallelogram corresponding to the point $(-e, -d)$ on X under the parametrization (2.2). The condition $\theta(\zeta) \neq 0$ gives

$$v \not\equiv 0 \quad (\bmod\, 1, \tau) \tag{2.4}$$

which is equivalent to $(-e, -d)$ being an affine point on X, and $\zeta + \overline{\zeta} = 0$ gives

$$v + \overline{v} \equiv 0 \quad (\bmod\, 1, \tau) \tag{2.5}$$

which is equivalent to $e \in \mathbf{R}, d \in i\mathbf{R}$, in accordance with (2.3).

So for smooth cubics Theorem 2.1 is the correspondance between non-equivalent determinantal representations (2.3) and points v in the period parallelogram satisfying (2.4)-(2.5). The set of such points consists of two connected components — a circle T_0 and a punctured circle T_1 (see Fig. 2.1). The sign of the representation (2.3) is ϵ.

It can be shown that a determinantal representation corresponds to a point v in T_0 if and only if its coefficient matrices admit a positive definite real linear combination, i.e. after a suitable real projective change of coordinates we have, say, $\sigma_2 > 0$.

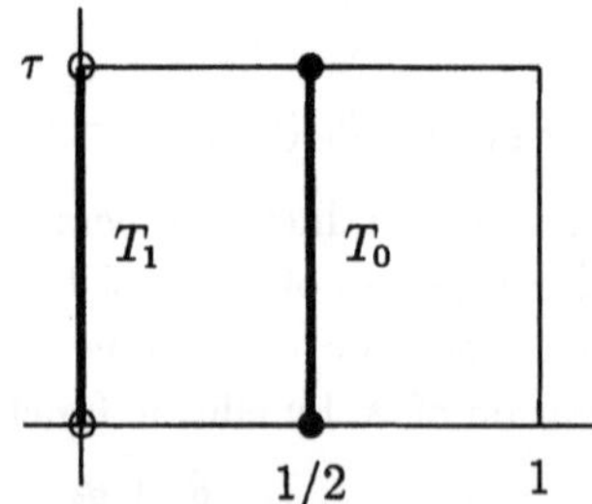

Figure 2.1 : $\{v : v \not\equiv 0, v + \overline{v} \equiv 0\} = T_0 \cup T_1$

3   Commutative Operator Colligations

Operator colligations (or nodes) form a natural framework for the study of nonselfadjoint operators. We first recall (see [10]) the basic definition of a colligation for two operators.

A *colligation* is a set

$$C = (A_1, A_2, H, \Phi, E, \sigma_1, \sigma_2) \tag{3.1}$$

where A_1, A_2 are linear bounded operators in a Hilbert space H, Φ is a linear bounded mapping from H to a Hilbert space E, and σ_1, σ_2 are bounded selfadjoint operators in E, such that

$$\frac{1}{i}(A_k - A_k^*) = \Phi^* \sigma_k \Phi \quad (k = 1, 2) \tag{3.2}$$

We shall always assume that $\dim E = n < \infty$, while $\dim H = N \leq \infty$. We shall also assume that $\ker \sigma_1 \cap \ker \sigma_2 = 0$. A colligation is called *commutative* if $A_1 A_2 = A_2 A_1$. A colligation is called *strict* if $\Phi(H) = E$.

If A_1, A_2 are commuting operators with finite-dimensional imaginary parts in a Hilbert space H, then

$$\frac{1}{i}(A_k - A_k^*) = P_G \sigma_k P_G \quad (k = 1, 2) \tag{3.3}$$

where $G = (A_1 - A_1^*)H + (A_2 - A_2^*)H$ is the nonhermitian subspace, P_G is the orthogonal projection onto G, and $\sigma_k = \frac{1}{i}(A_k - A_k^*) \mid G \ (k = 1, 2)$. So the pair A_1, A_2 can always be embedded in a strict commutative colligation with $E = G, \Phi = P_G$.

If $C = (A_1, A_2, H, \Phi, E, \sigma_1, \sigma_2)$ is a strict commutative colligation, there exist selfadjoint operators $\gamma^{\text{in}}, \gamma^{\text{out}}$ in E such that

$$\frac{1}{i}(A_1 A_2^* - A_2 A_1^*) = \Phi^* \gamma^{\text{in}} \Phi,$$
$$\frac{1}{i}(A_2^* A_1 - A_1^* A_2) = \Phi^* \gamma^{\text{out}} \Phi \tag{3.4}$$

As evidenced in Section 1, the operators $\gamma^{\text{in}}, \gamma^{\text{out}}$ play an important role, but the condition $\Phi(H) = E$ is too restrictive. The elementary objects — colligations with $\dim H = 1$ — are not strict when $\dim E > 1$. We introduce therefore the notion of a regular colligation [8,11].

A commutative colligation $C = (A_1, A_2, H, \Phi, E, \sigma_1, \sigma_2)$ is called *regular* if there exist selfadjoint operators $\gamma^{\text{in}}, \gamma^{\text{out}}$ in E such that

$$\sigma_1 \Phi A_2^* - \sigma_2 \Phi A_1^* = \gamma^{\text{in}} \Phi,$$
$$\sigma_1 \Phi A_2 - \sigma_2 \Phi A_1 = \gamma^{\text{out}} \Phi,$$
$$\gamma^{\text{out}} = \gamma^{\text{in}} + i(\sigma_1 \Phi \Phi^* \sigma_2 - \sigma_2 \Phi \Phi^* \sigma_1) \tag{3.5}$$

Actually, it is enough to require the existence of one of the operators $\gamma^{\text{in}}, \gamma^{\text{out}}$, the other one can then be defined by the last of the equations (3.4) and will satisfy the corresponding relation. Strict colligations are regular. For a strict colligation the operators $\gamma^{\text{in}}, \gamma^{\text{out}}$ are defined uniquely, but for a general regular colligation they are not, so we will include them in the notation of a regular colligation and write such a colligation as

$$C = (A_1, A_2, H, \Phi, E, \sigma_1, \sigma_2, \gamma^{\text{in}}, \gamma^{\text{out}})$$

Regular commutative colligations turn out to be the proper object to study in the theory of commuting nonselfadjoint operators. As in Section 1 we define the discriminant polynomial of the regular colligation C

$$f(y_1, y_2) = \det(y_1\sigma_2 - y_2\sigma_1 + \gamma^{\text{in}}) \tag{3.6}$$

and (assuming $f(y_1, y_2) \not\equiv 0$) the discriminant curve X determined by it in the complex projective plane. We have again the Cayley-Hamilton theorem

$$f(A_1, A_2) \mid \tilde{H} = 0 \tag{3.7}$$

where $\tilde{H} = \text{span}\,\{A_1^{k_1} A_2^{k_2} \Phi^*(E)\}_{k_1,k_2=0}^{\infty} = \text{span}\,\{A_1^{*k_1} A_2^{*k_2} \Phi^*(E)\}_{k_1,k_2=0}^{\infty}$ is the principal subspace of the colligation, so that the joint spectrum of the operators A_1, A_2 (restricted to the principal subspace) lies on the discriminant curve. Finally,

$$\det(y_1\sigma_2 - y_2\sigma_1 + \gamma^{\text{in}}) = \det(y_1\sigma_2 - y_2\sigma_1 + \gamma^{\text{out}}) \tag{3.8}$$

so that the discriminant curve comes equipped with the input and the output determinantal representations.

We formulate now the inverse problem of Section 1 in the framework of regular commutative colligations. We are given a real projective plane curve X of degree n, a determinantal representation $y_1\sigma_2 - y_2\sigma_1 + \gamma$ of X, and a subset S of affine points of X, which is closed and bounded in $\mathbf{C}^2$, and all of whose accumulation points are real points of X. We want to construct (up to the unitary equivalence on the principal subspace) all regular commutative colligations with discriminant curve X, input determinantal representation $y_1\sigma_2 - y_2\sigma_1 + \gamma$, and operators A_1, A_2 in the colligation having joint spectrum S (since $\sigma_1, \sigma_2, \gamma$ are given as $n \times n$ hermitian matrices we identify the space E in the colligation with $\mathbf{C}^n$). It is easily seen that the solutions of this problem that are strict colligations give the solution to the original problem of Section 1 (up to the equivalence of determinantal representations).

Our solution of the inverse problem will be based on a "spectral synthesis", using the coupling procedure to produce more complicated colligations out of simpler ones.

Let

$$C' = (A_1', A_2', H', \Phi', E, \sigma_1, \sigma_2)$$
$$C'' = (A_1'', A_2'', H'', \Phi'', E, \sigma_1, \sigma_2)$$

be two colligations. We define their *coupling*

$$C = C' \vee C'' = (A_1, A_2, H, \Phi, E, \sigma_1, \sigma_2)$$

where

$$H = H' \oplus H'',$$
$$A_k = \begin{pmatrix} A_k' & 0 \\ i\Phi''^* \sigma_k \Phi' & A_k'' \end{pmatrix} \quad (k = 1, 2),$$
$$\Phi = (\Phi' \quad \Phi'') \tag{3.9}$$

(the operators being written in the block form with respect to the orthogonal decomposition $H = H' \oplus H''$). It is immediately seen that C is indeed a colligation. However, if C' and C'' are commutative, C in general is not.

Assume now C', C'' are regular commutative colligations:

$$C' = (A_1', A_2', H', \Phi', E, \sigma_1, \sigma_2, \gamma'^{\text{in}}, \gamma'^{\text{out}})$$

$$C'' = (A_1'', A_2'', H'', \Phi'', E, \sigma_1, \sigma_2, \gamma''^{\text{in}}, \gamma''^{\text{out}})$$

Theorem 3.1 (Livšic [11], Kravitsky [8]) *The coupling*

$$C = C' \vee C'' = (A_1, A_2, H, \Phi, E, \sigma_1, \sigma_2, \gamma^{\text{in}}, \gamma^{\text{out}})$$

where A_1, A_2, H, Φ are given by (3.8), $\gamma^{\text{in}} = \gamma'^{\text{in}}, \gamma^{\text{out}} = \gamma''^{\text{out}}$, is a regular commutative colligation if and only if $\gamma'^{\text{out}} = \gamma''^{\text{in}}$.

This theorem illustrates aptly the meaning of the input and the output determinantal representations. Note that $H'' \subset H$ is a common invariant subspace of A_1, A_2. Conversely, if $H'' \subset H$ is a common invariant subspace of the operators A_1, A_2 in a regular commutative colligation C, we can write $C = C' \vee C''$, where C', C'' are regular commutative colligations called the *projections* of C onto the subspaces $H = H \ominus H'', H''$ respectively [11,8].

4 Construction of Triangular Models: Finite-Dimensional Case

We shall start the solution of the inverse problem for regular commutative colligations by investigating the simplest case when $\dim H = 1$ and the joint spectrum consists of a single (non-real) point.

Let X be a real smooth projective plane curve of degree n whose set of real points $X_{\mathbf{R}} \neq \emptyset$, and let $y_1\sigma_2 - y_2\sigma_1 + \gamma$ be a determinantal representation of X that has sign ϵ and that corresponds, as in Theorem 2.1, to a point ζ in $J(X)$. Let $\lambda = (\lambda_1, \lambda_2)$ be a non-real affine point on X. We identify the space H in the colligation with $\mathbf{C}$, so that the operators A_1, A_2 in H are just multiplications by λ_1, λ_2, and the mapping Φ from H to E is multiplication by a vector ϕ in $\mathbf{C}^n$. We want to construct a regular commutative colligation

$$C = (\lambda_1, \lambda_2, \mathbf{C}, \phi, \mathbf{C}^n, \sigma_1, \sigma_2, \gamma, \tilde{\gamma}) \tag{4.1}$$

The colligation conditions (3.1) and the regularity conditions (3.4) are

$$(\overline{\lambda_1}\sigma_2 - \overline{\lambda_2}\sigma_1 + \gamma)\phi = 0 \tag{4.2}$$

$$2\Im\lambda_k = \phi^*\sigma_k\phi \quad (k = 1,2) \tag{4.3}$$

$$\tilde{\gamma} = \gamma + i(\sigma_1\phi\phi^*\sigma_2 - \sigma_2\phi\phi^*\sigma_1) \tag{4.4}$$

$$(\lambda_1\sigma_2 - \lambda_2\sigma_1 + \tilde{\gamma})\phi = 0 \tag{4.5}$$

Let $v \in \operatorname{coker}(\lambda_1\sigma_2 - \lambda_2\sigma_1 + \gamma)$. It is easily seen that

$$\Im\lambda_1 \cdot v\sigma_2 v^* = \Im\lambda_2 \cdot v\sigma_1 v^* \tag{4.6}$$

Therefore we can normalize v so that $\phi = v^*$ satisfies (4.3) if and only if

$$\frac{\Im\lambda_1}{v\sigma_1 v^*} = \frac{\Im\lambda_2}{v\sigma_2 v^*} > 0 \tag{4.7}$$

In this case we define $\tilde{\gamma}$ by (4.4), and (4.5) follows. The one-point colligation (4.1) has thus been constructed. Note that we get at the output a new determinantal representation $y_1\sigma_2 - y_2\sigma_1 + \tilde{\gamma}$ of X.

It is a fact of fundamental importance that the positivity condition (4.7) can be expressed analytically.

Theorem 4.1 *The condition (4.7) is satisfied if and only if $\epsilon\frac{i\theta[\zeta](\lambda-\overline{\lambda})}{\theta[\zeta](0)E(\lambda,\overline{\lambda})} > 0$. In this case the new determinantal representation $y_1\sigma_2 - y_2\sigma_1 + \tilde{\gamma}$ defined by (4.2)-(4.4) has sign ϵ and corresponds to the point $\tilde{\zeta} = \zeta + \lambda - \overline{\lambda}$ in $J(X)$.*

In the expressions like $\zeta + \lambda - \overline{\lambda}$ we identify the point λ on X with its image in $J(X)$ under the embedding of the curve in its Jacobian variety given by the Abel-Jacobi map μ. $\theta[\zeta](w)$ is the theta function with characteristic ζ; it is an entire function on $\mathbf{C}^g$ associated to every point ζ in $J(X)$. $\theta[\zeta](w)$ differs by an exponential factor from $\theta(\zeta + w)$. Therefore $\theta[\zeta](0) \neq 0$ always by Theorem 2.1; on the other hand, if the positivity condition of Theorem 4.1 is satisfied, $\theta(\tilde{\zeta}) = \theta(\zeta + \lambda - \overline{\lambda}) \neq 0$, again in accordance with Theorem 2.1. Finally, $E(x,y)$ is the prime form on X: it is a multiplicative differential on X of order $-\frac{1}{2}, \frac{1}{2}$ in x, y, whose main property is that $E(x,y) = 0$ if and only if $x = y$. See [2] or [13] for all these. Note that each factor in the expression $\frac{i\theta[\zeta](\lambda-\overline{\lambda})}{\theta[\zeta](0)E(\lambda,\overline{\lambda})}$ is multi-valued, depending on the choice of lifting from $J(X)$ to $\mathbf{C}^g$, but the expression itself turns out to be well-defined.

In the special case when, say, $\sigma_2 > 0$, $X_{\mathbf{R}}$ divides X into two components X_+ and X_- interchanged by the complex conjugation and whose affine points $y = (y_1, y_2)$ satisfy $\Im y_2 > 0$ and $\Im y_2 < 0$ respectively. The "weight" $\frac{i\theta[\zeta](x-\overline{x})}{\theta[\zeta](0)E(x,\overline{x})}$ is positive on X_+ and negative on X_-, the sign $\epsilon = 1$ and the positivity condition of Theorem 4.1 becomes $\lambda \in X_+$ or $\Im\lambda_2 > 0$ (see [11]).

As an example, let X be the real smooth cubic (2.1). Let $y_1\sigma_2 - y_2\sigma_1 + \gamma$ be equivalent to the representation (2.3) of sign ϵ corresponding to the point v in the period parallelogram ($v \not\equiv 0, v + \overline{v} \equiv 0$), and let the point λ on X correspond to the point

a in the period parallelogram under the parametrization (2.2). The region for a where the positivity condition of Theorem 4.1 is satisfied depends on the component of v (see Fig. 2.1). If $v \in T_0$, the admissible region is always a half of the period parallelogram; if $v \in T_1$, the admissible region consists of two bands (or rather annuli) whose width depends on v; see Fig. 4.1-4.2 where the complementary admissible regions are depicted for $\epsilon = 1, \epsilon = -1$. If a is in the admissible region, the representation $y_1\sigma_2 - y_2\sigma_1 + \tilde{\gamma}$ is equivalent to the representation of the form (2.3) corresponding to the point $\tilde{v} \equiv v + a - \bar{a}$ in the period parallelogram.

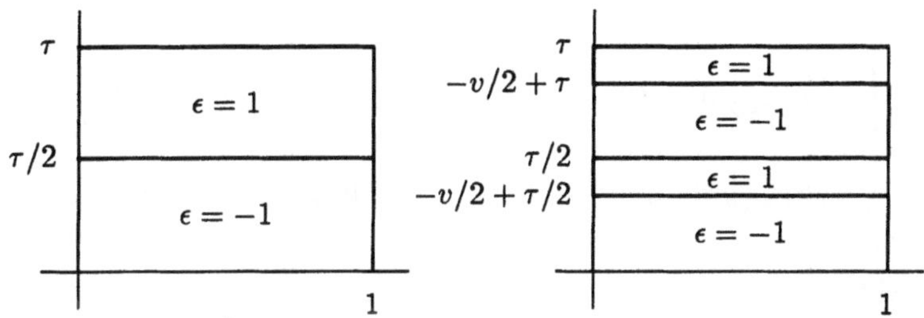

Figure 4.1 : Admissible regions, $v \in T_0$ Figure 4.2 : Admissible regions, $v \in T_1$

Using the coupling procedure we can solve now the inverse problem for regular commutative colligations with a finite-dimensional space H.

Let X be a real smooth projective plane curve of degree n whose set of real points $X_{\mathbf{R}} \neq \emptyset$, and let $y_1\sigma_2 - y_2\sigma_1 + \gamma$ be a determinantal representation of X that has sign ϵ and corresponds to a point ζ in $J(X)$. Let $\lambda^{(i)} = (\lambda_1^{(i)}, \lambda_2^{(i)})(i = 1, \ldots, N)$ be a finite sequence of non-real affine points on X. Assume that

$$\epsilon\frac{i\theta[\zeta + \sum_{j=1}^{i}(\lambda^{(j)} - \overline{\lambda^{(j)}})](\lambda^{(i+1)} - \overline{\lambda^{(i+1)}})}{\theta[\zeta + \sum_{j=1}^{i}(\lambda^{(j)} - \overline{\lambda^{(j)}})](0)E(\lambda^{(i+1)}, \overline{\lambda^{(i+1)}})} > 0 \ (i = 0, \ldots, N-1) \tag{4.8}$$

The conditions (4.8) turn out to be independent of the order of the points $\lambda^{(1)}, \ldots, \lambda^{(N)}$. If all the points are distinct, (4.8) can be rewritten, using Fay's addition theorem [2,13], in the matrix form

$$\epsilon\left(\frac{i\theta[\zeta](\lambda^{(i)} - \overline{\lambda^{(j)}})}{\theta[\zeta](0)E(\lambda^{(i)}, \overline{\lambda^{(j)}})}\right)_{i,j=1,\ldots,N} > 0 \tag{4.9}$$

We write down the system of recursive equations:

$$(\overline{\lambda_1^{(i)}}\sigma_2 - \overline{\lambda_2^{(i)}}\sigma_1 + \gamma)\phi^{(i)} = 0,$$
$$\phi^{(i)*}\sigma_k\phi^{(i)} = 2\Im\lambda_k^{(i)} \ (k = 1, 2),$$
$$\gamma^{(i+1)} = \gamma^{(i)} + i(\sigma_1\phi^{(i)}\phi^{(i)*}\sigma_2 - \sigma_2\phi^{(i)}\phi^{(i)*}\sigma_1),$$
$$\gamma^{(1)} = \gamma \tag{4.10}$$

for $i = 1, \ldots, N$. It follows from Theorem 4.1 that this system is solvable (uniquely up to multiplication of $\phi^{(i)}$ by scalars of absolute value 1) and for each i $y_1\sigma_2 - y_2\sigma_1 + \gamma^{(i)}$ is a determinantal representation of X that has sign ϵ and corresponds to the point

$\zeta + \sum_{j=1}^{i-1}(\lambda^{(j)} - \overline{\lambda^{(j)}})$ in $J(X)$. For each i $C^{(i)} = (\lambda_1^{(i)}, \lambda_2^{(i)}, \mathbf{C}, \phi^{(i)}, \mathbf{C}^n, \sigma_1, \sigma_2, \gamma^{(i)}, \gamma^{(i+1)})$ is a one-point (as in (4.1)) regular commutative colligation, and we can couple them by Theorem 3.1.

Theorem 4.2 *Let $\lambda^{(i)}(i = 1, \ldots, N)$ be a finite sequence of non-real affine points on X satisfying (4.8), and let $\gamma^{(i)}, \phi^{(i)}$ be determined from (4.10). Then*

$$C = (A_1, A_2, \mathbf{C}^N, \Phi, \mathbf{C}^n, \sigma_1, \sigma_2, \gamma, \tilde{\gamma}) \tag{4.11}$$

is a regular commutative colligation, where

$$A_k = \begin{pmatrix} \lambda_k^{(1)} & 0 & \ldots & 0 & 0 \\ i\phi^{(2)*}\sigma_k\phi^{(1)} & \lambda_k^{(2)} & \ldots & 0 & 0 \\ \vdots & \vdots & & \vdots & \vdots \\ i\phi^{(N)*}\sigma_k\phi^{(1)} & i\phi^{(N)*}\sigma_k\phi^{(2)} & \ldots & i\phi^{(N)*}\sigma_k\phi^{(N-1)} & \lambda_k^{(N)} \end{pmatrix} \quad (k = 1, 2),$$

$$\Phi = (\phi^{(1)} \quad \ldots \quad \phi^{(N)}),$$

$$\tilde{\gamma} = \gamma^{(N+1)} \tag{4.12}$$

The joint spectrum of A_1, A_2 is $\{\lambda^{(i)}\}_{i=1}^N$, and the output determinantal representation $y_1\sigma_2 - y_2\sigma_1 + \tilde{\gamma}$ of X has sign ϵ and corresponds to the point $\tilde{\zeta}$ in $J(X)$, where

$$\tilde{\zeta} = \zeta + \sum_{i=1}^N (\lambda^{(i)} - \overline{\lambda^{(i)}}) \tag{4.13}$$

We call the solution (4.11) of the inverse problem the *triangular model* with discriminant curve X, input determinantal representation $y_1\sigma_2 - y_2\sigma_1 + \gamma$ and spectral data $\lambda^{(i)}(i = 1, \ldots, N)$. The reordering of the points $\lambda^{(1)}, \ldots, \lambda^{(N)}$ leads to a unitary equivalent triangular model. Furthermore, the triangular model is the unique solution of the inverse problem.

Theorem 4.3 *Let $C = (A_1, A_2, H, \Phi, \mathbf{C}^n, \sigma_1, \sigma_2, \gamma, \tilde{\gamma})$ be a regular commutative colligation with $\dim H < \infty$ and with smooth discriminant curve X that has real points. Let $\lambda^{(i)}(i = 1, \ldots, N)$ be the points of the joint spectrum of A_1, A_2 (restricted to the principal subspace $\tilde{H}$ of C in H). Then $\lambda^{(i)}$ are non-real affine points of X satisfying (4.8) and C is unitarily equivalent (on the principal subspace $\tilde{H}$) to the triangular model with discriminant curve X, input determinantal representation $y_1\sigma_2 - y_2\sigma_1 + \gamma$ and spectral data $\lambda^{(i)}(i = 1, \ldots, N)$.*

In the special case when one of the operators A_1, A_2 is dissipative, say $\sigma_2 > 0$, the conditions (4.8) reduce to $\Im\lambda_2^{(i)} > 0(i = 1, \ldots, N)$ (see the comments following Theorem 4.1); Theorems 4.2-4.3 have been obtained in this case by Livšic [11].

The proof of Theorem 4.3 is based on the existence of a chain $H = H_0 \supset H_1 \supset \ldots \supset H_{N-1} \supset H_N = 0$ of common invariant subspaces of A_1, A_2 such that $\dim(H_{i-1} \ominus H_i) = 1(i = 1, \ldots, N)$ (simulataneous reduction to a triangular form; we assume for simplicity $H = \tilde{H}$). Projecting the colligation C onto the subspaces $H_{i-1} \ominus H_i$, we represent C as the coupling of N one-point (as in (4.1)) colligations, which forces it to be unitary equivalent to the triangular model.

The conditions (4.8) determine all possible input determinantal representations (if any) of a regular commutative colligation with the discriminant curve X and operators A_1, A_2 having the joint spectrum $\lambda^{(1)}, \ldots, \lambda^{(N)}$ (on the principal subspace of the colligation). For example, let X be the real smooth cubic (2.1), and let $\lambda^{(1)}, \ldots, \lambda^{(N)}$ be non-real affine points on X corresponding to the points $a^{(1)}, \ldots, a^{(N)}$ in the period parallelogram under the parametrization (2.2). Assume m among those points lie in the upper half of the period parallelogram: $\frac{\Im\tau}{2} < \Im a^{(i)} < \Im\tau$, and k lie in the lower half: $0 < \Im a^{(i)} < \frac{\Im\tau}{2}$ $(m + k = N)$. Let $y_1\sigma_2 - y_2\sigma_1 + \gamma$ be the input determinantal representations of a regular commutative colligation with the discriminant curve X and operators A_1, A_2 having the joint spectrum $\lambda^{(1)}, \ldots, \lambda^{(N)}$. $y_1\sigma_2 - y_2\sigma_1 + \gamma$ is equivalent to the representation (2.3) of sign ϵ corresponding to the point v in the period parallelogram ($v \not\equiv 0, v + \bar{v} \equiv 0$). We may take $v \in T_0$ (arbitrary) if and only if $k = 0$ ($\epsilon = 1$), or $m = 0$ ($\epsilon = -1$) (see Fig. 4.1). We may take $v \in T_1$ if and only if

$$- \Im a^{(1)} - \ldots - \Im a^{(N)} + \frac{2m + k}{2}\Im\tau < \frac{\Im v}{2} \tag{4.14}$$

($\epsilon = 1$), or

$$\frac{\Im v}{2} < -\Im a^{(1)} - \ldots - \Im a^{(N)} + \frac{m + 1}{2}\Im\tau \tag{4.15}$$

($\epsilon = -1$) (see Fig. 4.2). Since $0 < \Im v < \Im\tau$, (4.14) implies that

$$\Im a^{(1)} + \ldots + \Im a^{(N)} > \frac{2m + k - 1}{2}\Im\tau \tag{4.16}$$

while (4.15) implies that

$$\Im a^{(1)} + \ldots + \Im a^{(N)} < \frac{m + 1}{2}\Im\tau \tag{4.17}$$

If we have $N = m + k$ $(m, k \neq 0)$ points in the period parallelogram that satisfy neither (4.16) nor (4.17), they can't be the joint spectrum of a pair of operators in a regular commutative colligation with the discriminant curve X.

In the case of real smooth cubics one can also write down explicitly the solution of the system of recursive equations (4.10) and the corresponding matrices (4.12) using Weierstrass functions.

5 Construction of Triangular Models: General Case

The solution of the inverse problem for regular commutative colligations in the general (infinite-dimensional) case consists of the discrete part and the continuous part. As before we let X be a real smooth projective plane curve of degree n whose set of real points $X_{\mathbf{R}} \neq \emptyset$, and let $y_1\sigma_2 - y_2\sigma_1 + \gamma$ be a determinantal representation of X that has sign ϵ and corresponds to a point ζ in $J(X)$.

We start with the discrete part of the solution. Let $\lambda^{(i)} = (\lambda_1^{(i)}, \lambda_2^{(i)})(i = 1, \ldots)$ be an infinite sequence of non-real affine points on X that is bounded in $\mathbf{C}^2$ and all of whose accumulation points are in $X_{\mathbf{R}}$. As in (4.8), assume that

$$\epsilon\frac{i\theta[\zeta + \sum_{j=1}^{i}(\lambda^{(j)} - \overline{\lambda^{(j)}})](\lambda^{(i+1)} - \overline{\lambda^{(i+1)}})}{\theta[\zeta + \sum_{j=1}^{i}(\lambda^{(j)} - \overline{\lambda^{(j)}})](0)E(\lambda^{(i+1)}, \overline{\lambda^{(i+1)}})} > 0 \ (i = 0, \ldots) \tag{5.1}$$

As in (4.10), we write down the system of recursive equations:

$$\begin{aligned}
&(\overline{\lambda_1^{(i)}}\sigma_2 - \overline{\lambda_2^{(i)}}\sigma_1 + \gamma)\phi^{(i)} = 0, \\
&\phi^{(i)*}\sigma_k\phi^{(i)} = 2\Im\lambda_k^{(i)} \ (k = 1, 2), \\
&\gamma^{(i+1)} = \gamma^{(i)} + i(\sigma_1\phi^{(i)}\phi^{(i)*}\sigma_2 - \sigma_2\phi^{(i)}\phi^{(i)*}\sigma_1), \\
&\gamma^{(1)} = \gamma
\end{aligned} \tag{5.2}$$

for $i = 1, \ldots$. It follows from Theorem 4.1 that this system is solvable (uniquely up to multiplication of $\phi^{(i)}$ by scalars of absolute value 1) and for each i $y_1\sigma_2 - y_2\sigma_1 + \gamma^{(i)}$ is a determinantal representation of X that has sign ϵ and corresponds to the point $\zeta + \sum_{j=1}^{i-1}(\lambda^{(j)} - \overline{\lambda^{(j)}})$ in $J(X)$. As in (4.12), we form infinite matrices:

$$A_k = \begin{pmatrix}
\lambda_k^{(1)} & 0 & \ldots & 0 & 0 & \ldots \\
i\phi^{(2)*}\sigma_k\phi^{(1)} & \lambda_k^{(2)} & \ldots & 0 & 0 & \ldots \\
\vdots & \vdots & & \vdots & \vdots & \\
i\phi^{(i)*}\sigma_k\phi^{(1)} & i\phi^{(i)*}\sigma_k\phi^{(2)} & \ldots & i\phi^{(i)*}\sigma_k\phi^{(i-1)} & \lambda_k^{(i)} & \ldots \\
\vdots & \vdots & & \vdots & \vdots &
\end{pmatrix} \ (k = 1, 2),$$

$$\Phi = (\phi^{(1)} \ \ldots \ \phi^{(i)} \ \ldots) \tag{5.3}$$

It turns out that A_1, A_2 are bounded linear operators in l^2 and Φ is a bounded linear mapping from l^2 to $\mathbf{C}^n$ (we write elements of l^2 as infinite column vectors) if and only if $\tilde{\gamma} = \lim_{i\to\infty}\gamma^{(i)}$ exists. In this case

$$C = (A_1, A_2, l^2, \Phi, \mathbf{C}^n, \sigma_1, \sigma_2, \gamma, \tilde{\gamma}) \tag{5.4}$$

is a regular commutative colligation. The joint spectrum of A_1, A_2 is $\overline{\{\lambda^{(i)}\}_{i=1}^{\infty}}$.

Theorem 5.1 *The limit $\tilde{\gamma} = \lim_{i\to\infty}\gamma^{(i)}$ exists if and only if the series $\sum_{i=1}^{\infty}(\lambda^{(i)} - \overline{\lambda^{(i)}})$ in $J(X)$ converges and $\theta(\zeta + \sum_{i=1}^{\infty}(\lambda^{(i)} - \overline{\lambda^{(i)}})) \neq 0$. In this case the determinantal representation $y_1\sigma_2 - y_2\sigma_1 + \tilde{\gamma}$ of X has sign ϵ and corresponds to the point $\tilde{\zeta} = \zeta + \sum_{i=1}^{\infty}(\lambda^{(i)} - \overline{\lambda^{(i)}})$ in $J(X)$.*

In the special case $\sigma_2 > 0$, the conditions (5.1) reduce to $\Im\lambda_2^{(i)} > 0 (i = 1, \ldots)$ and the conditions of Theorem 5.1 are just $\sum_{i=1}^{\infty}\Im\lambda_2^{(i)} < \infty$.

We pass now to the continuous part of the solution to the inverse problem. Let $c : [0, l] \longrightarrow X_{\mathbf{R}}$ be a function from some finite interval into the set of real affine points of X, such that $c(t) = (c_1(t), c_2(t))$, where $c_1(t), c_2(t)$ are bounded almost everywhere continuous functions on $[0, l]$. We write down the following system of differential equations

(Waksman [23], Livšic [12], Kravitsky [9]), which is the continuous analog of (4.10) and (5.2):

$$(c_1(t)\sigma_2 - c_2(t)\sigma_1 + \gamma(t))\phi(t) = 0,$$

$$\phi(t)^*\sigma_k\phi(t) = \epsilon\frac{dy_k(c(t))}{\omega(c(t))} \quad (k = 1, 2),$$

$$\frac{d\gamma}{dt} = i(\sigma_1\phi(t)\phi(t)^*\sigma_2 - \sigma_2\phi(t)\phi(t)^*\sigma_1),$$

$$\gamma(0) = \gamma \tag{5.5}$$

for $0 \le t \le l$. By a solution of this system we mean an absolutely continuous matrix function $\gamma(t)$ on $[0, l]$ and an almost everywhere continuous vector function $\phi(t)$ on $[0, l]$ such that (5.5) holds almost everywhere. ω is a real differential on X, defined, analytic and non-zero in a neighbourhood of the set of left and right limit values of the function $c : [0, l] \longrightarrow X_{\mathbf{R}}$, whose signs on the different connected components of $X_{\mathbf{R}}$ correspond to the real torus in $J(X)$ to which the point ζ belongs; there is a version of the relation (4.6) for real points that shows that the required normalization of $\phi(t)$ is always possible; see [18] for all these. A change of the differential ω corresponds to a change of the parameter t.

Assume that the system (5.5) on the interval $[0, l]$ is solvable (uniquely almost everywhere up to multiplication of $\phi(t)$ by a scalar function of absolute value 1). Then [23,12,9] for each t $y_1\sigma_2 - y_2\sigma_1 + \gamma(t)$ is a determinantal representation of X. For $f(t) \in L^2[0, l]$ define

$$(A_kf)(t) = c_k(t)f(t) + i\int_0^t \phi(t)^*\sigma_k\phi(s)f(s)ds \quad (k = 1, 2),$$

$$\Phi f = \int_0^l \phi(t)f(t)dt \tag{5.6}$$

A_1, A_2 are triangular integral operators on $L^2[0, l]$ (continuous analogs of triangular matrices) and Φ is a mapping from $L^2[0, l]$ to $\mathbf{C}^n$. It turns out [23,12,9] that A_1 and A_2 commute, and

$$C = (A_1, A_2, L^2[0, l], \Phi, \mathbf{C}^n, \sigma_1, \sigma_2, \gamma, \tilde{\gamma}) \tag{5.7}$$

is a regular commutative colligation. The joint spectrum of A_1, A_2 is the set of left and right limit values of the function $c : [0, l] \longrightarrow X_{\mathbf{R}}$.

Theorem 5.2 *Let*

$$\zeta(t) = \zeta + \epsilon i \int_0^t \begin{pmatrix} \frac{\omega_1(c(s))}{\omega(c(s))} \\ \vdots \\ \frac{\omega_g(c(s))}{\omega(c(s))} \end{pmatrix} ds \tag{5.8}$$

where $\omega_1, \ldots, \omega_g$ are the basis for holomorphic differentials on X that was chosen in the construction of the Jacobian variety. The system (5.5) is solvable on the interval $[0, l]$ if and only if $\theta(\zeta(t) \ne 0$ for all $t \in [0, l]$. In this case the determinantal representation $y_1\sigma_2 - y_2\sigma_1 + \gamma(t)$ of X has sign ϵ and corresponds to the point $\zeta(t)$ in $J(X)$.

In the special case $\sigma_2 > 0$, the conditions of Theorem 5.1 are automatically satisfied, so the system (5.5) is always solvable; this was obtained by Livšic [12] for the case when the image of c consists of a single point.

Theorem 5.2 gives not only an explicit condition for the solvability of the system of non-linear differential equations (5.5), it also shows that this system is linearized by passing from determinantal representations to the corresponding points in the Jacobian variety. The point $\zeta(t)$ given by (5.8) determines the equivalence class of the determinantal representation $y_1\sigma_2 - y_2\sigma_1 + \gamma(t)$; one can go further and determine explicitly the representation $y_1\sigma_2 - y_2\sigma_1 + \gamma(t)$ inside the equivalence class, i.e. integrate explicitly the system (5.5). We present the answer for the simplest case.

Let X be the real smooth cubic (2.1) and let $y_1\sigma_2 - y_2\sigma_1 + \gamma$ be the determinantal representation (2.3) of sign ϵ corresponding to the point v in the period parallelogram ($v \not\equiv 0, v + \overline{v} \equiv 0$). Let $c(t) = (c_1, c_2)$ for all $t \in [0, l]$, where (c_1, c_2) is a real affine point on X corresponding to the point a in the period parallelogram under the parametrization (2.2). Assume for definiteness that $\epsilon = 1$ and $\Im a = 0$. As a real differential in (5.5) we may take $\omega = -\frac{1}{\sqrt[3]{4}}\frac{dy_1}{y_2} = -du$ (where u is the uniformization parameter (2.2)); note that as a basis for holomorphic differentials on X we take $\omega_1 = \frac{1}{\sqrt[3]{4}}\frac{dy_1}{y_2}$.

Let $v(t) = v - it$, and let e_t, d_t, q_t, p_t be the numbers appearing in the determinantal representation (2.3) corresponding to the point $v(t)$ ($v(t) \not\equiv 0$). Then the solution of the system (5.5) is given by

$$\gamma(t) = \begin{pmatrix} p_t + r_t^2(q_t - l_t) - \frac{r_t^4}{4} + \frac{s_t^2}{2} & d_t - r_t(q_t - l_t) + \frac{r_t^3}{4} - \frac{r_t s_t}{2} & q_t - \frac{r_t^2}{2} - \frac{s_t}{2} \\ -d_t + r_t(q_t - l_t) - \frac{r_t^3}{4} + \frac{r_t s_t}{2} & e_t + r_t^2 & -r_t \\ q_t - \frac{r_t^2}{2} - \frac{s_t}{2} & r_t & -1 \end{pmatrix},$$

$$r_t = \frac{(\sqrt[3]{4})^2}{2}(\zeta(v(t)) - \zeta(v) - i\wp(a)t), s_t = -\sqrt[3]{4}\,i\wp'(a)t \tag{5.9}$$

Here $\zeta(u)$ is the Weierstrass ζ-function. If $v \in T_1$ (see Fig. 2.1), the system (5.5) is solvable on the interval $[0, l]$ if and only if $l < \Im v$. If $v \in T_0$, the system is solvable on any interval and the solution is quasiperiodic in the sense that $y_1\sigma_2 - y_2\sigma_1 + \gamma(t)$ and $y_1\sigma_2 - y_2\sigma_1 + \gamma(t + \Im\tau)$ are equivalent determinantal representations for any t (since $v(t+\Im\tau) \equiv v(t)$). Of course, one can also write down explicitly, using Weierstrass functions, the vector function $\phi(t)$ and the commuting integral operators (5.6).

We can solve now the inverse problem for regular commutative operator colligations in the general case by coupling (5.4) and (5.7). Let $\lambda^{(i)} = (\lambda_1^{(i)}, \lambda_2^{(i)})(i = 1, \ldots, N; N \leq \infty)$ be a sequence of non-real affine points on X that is bounded in $\mathbf{C}^2$ and all of whose accumulation points are in $X_{\mathbf{R}}$. Let $c(t) = (c_1(t), c_2(t))(0 \leq t \leq l; 0 < l < \infty)$ be real affine points on X, where $c_1(t), c_2(t)$ are bounded almost everywhere continuous functions on $[0, l]$; we order the connected components of $X_{\mathbf{R}}$, choose a basepoint and an orientation on each one of them, and assume that $c : [0, l] \longrightarrow X_{\mathbf{R}}$ is continuous from the left everywhere, continuous at 0, and non-decreasing in the resulting order on $X_{\mathbf{R}}$. We call $\lambda^{(i)}, c(t)$ the *spectral data*.

Assume that the conditions (5.1) and the conditions of Theorems 5.1 – 5.2 are satisfied

$$\epsilon\frac{i\theta[\zeta + \sum_{j=1}^{i}(\lambda^{(j)} - \overline{\lambda^{(j)}})](\lambda^{(i+1)} - \overline{\lambda^{(i+1)}})}{\theta[\zeta + \sum_{j=1}^{i}(\lambda^{(j)} - \overline{\lambda^{(j)}})](0)E(\lambda^{(i+1)}, \overline{\lambda^{(i+1)}})} > 0 \quad (i = 0, \ldots, N-1),$$

$$\sum_{i=1}^{\infty}(\lambda^{(i)} - \overline{\lambda^{(i)}}) \text{ converges }, \theta(\zeta + \sum_{i=1}^{\infty}(\lambda^{(i)} - \overline{\lambda^{(i)}})) \neq 0,$$

$$\theta\left(\zeta + \sum_{i=1}^{N}(\lambda^{(i)} - \overline{\lambda^{(i)}}) + \epsilon i \int_0^t \begin{pmatrix} \frac{\omega_1(c(s))}{\omega(c(s))} \\ \vdots \\ \frac{\omega_g(c(s))}{\omega(c(s))} \end{pmatrix} ds \right) \neq 0 \quad (t \in [0, l]) \tag{5.10}$$

where $\omega, \omega_1, \ldots, \omega_g$ are as before (if $N < \infty$ the second condition is not needed). Write down the system of recursive equations (5.2) followed by the system of differential equations (5.5)

$$(\overline{\lambda_1^{(i)}}\sigma_2 - \overline{\lambda_2^{(i)}}\sigma_1 + \gamma)\phi^{(i)} = 0,$$

$$\phi^{(i)*}\sigma_k\phi^{(i)} = 2\Im\lambda_k^{(i)} \quad (k = 1, 2),$$

$$\gamma^{(i+1)} = \gamma^{(i)} + i(\sigma_1\phi^{(i)}\phi^{(i)*}\sigma_2 - \sigma_2\phi^{(i)}\phi^{(i)*}\sigma_1),$$

$$\gamma^{(1)} = \gamma, \quad i = 1, \ldots, N;$$

$$(c_1(t)\sigma_2 - c_2(t)\sigma_1 + \gamma)\phi(t) = 0,$$

$$\phi(t)^*\sigma_k\phi(t) = \epsilon\frac{dy_k(c(t))}{\omega(c(t))} \quad (k = 1, 2),$$

$$\frac{d\gamma}{dt} = i(\sigma_1\phi(t)\phi(t)^*\sigma_2 - \sigma_2\phi(t)\phi(t)^*\sigma_1),$$

$$\gamma(0) = \lim_{i\to\infty}\gamma^{(i)}, \quad 0 \leq t \leq l \tag{5.11}$$

(if $N < \infty$, $\gamma(0) = \gamma^{(N+1)}$). The system of recursive equations is solvable by Theorem 4.1, $\lim_{i\to\infty}\gamma^{(i)}$ exists by Theorem 5.1, and the system of differential equations is solvable by Theorem 5.2.

Theorem 5.3 *Let $\lambda^{(i)}(i = 1, \ldots, N; N \leq \infty), c(t) = (c_1(t), c_2(t))(0 \leq t \leq l)$ be a spectral data satisfying (5.10), and let $\gamma^{(i)}, \phi^{(i)}, \gamma(t), \phi(t)$ be determined by (5.11). Then*

$$C = (A_1, A_2, H, \Phi, \mathbf{C}^n, \sigma_1, \sigma_2, \gamma, \tilde{\gamma}) \tag{5.12}$$

is a regular commutative colligation, where $H = l^2 \oplus L^2[0, l]$ and

$$A_k\begin{pmatrix} v \\ f(t) \end{pmatrix} = \begin{pmatrix} \left(\sum_{j=1}^{i-1} i\phi^{(i)*}\sigma_k\phi^{(j)}v_j + \lambda_k^{(i)}v_i\right)_{i=1}^{\infty} \\ \sum_{i=1}^{\infty} i\phi(t)^*\sigma_k\phi^{(i)}v_i + i\int_0^t \phi(t)^*\sigma_k\phi(s)f(s)ds + c_k(t)f(t) \end{pmatrix} \quad (k = 1, 2),$$

$$\Phi\begin{pmatrix} v \\ f(t) \end{pmatrix} = \sum_{i=1}^{\infty} \phi^{(i)}v_i + \int_0^l \phi(t)f(t)dt,$$

$$\tilde{\gamma} = \gamma(l) \tag{5.13}$$

for $v = (v_i)_{i=1}^{\infty} \in l^2, f(t) \in L^2[0, l]$ (if $N < \infty$, replace l^2 by $\mathbf{C}^N$ and ∞ by N in the above formulas). The joint spectrum of A_1, A_2 is $\{\lambda^{(i)}\}_{i=1}^{N} \cup \{c(t)\}_{t\in[0,l]}$, and the output determinantal representation $y_1\sigma_2 - y_2\sigma_1 + \tilde{\gamma}$ of X has sign ϵ and corresponds to the point $\tilde{\zeta}$ in $J(X)$, where

$$\tilde{\zeta} = \zeta + \sum_{i=1}^{N}(\lambda^{(i)} - \overline{\lambda^{(i)}}) + \epsilon i \int_0^l \begin{pmatrix} \frac{\omega_1(c(t))}{\omega(c(t))} \\ \vdots \\ \frac{\omega_g(c(t))}{\omega(c(t))} \end{pmatrix} dt \tag{5.14}$$

We call the solution (5.12) of the inverse problem the *triangular model* with discriminant curve X, input determinantal representation $y_1\sigma_2 - y_2\sigma_1 + \gamma$ and spectral data $\lambda^{(i)}, c(t)$. We can state now our main result.

Theorem 5.4 *Let $C = (A_1, A_2, H, \Phi, \mathbf{C}^n, \sigma_1, \sigma_2, \gamma, \tilde{\gamma})$ be a regular commutative colligation with smooth discriminant curve X that has real points. Let S be the joint spectrum of A_1, A_2 (restricted to the principal subspace $\tilde{H}$ of C in H). There exists a spectral data $\lambda^{(i)}(i = 1, \ldots, N; N \leq \infty), c(t) = (c_1(t), c_2(t))(0 \leq t \leq l)$ satisfying (5.10), such that $S = \overline{\{\lambda^{(i)}\}_{i=1}^{N} \cup \{c(t)\}_{t\in[0,l]}}$ and C is unitarily equivalent (on its principal subspace $\tilde{H}$) to the triangular model with discriminant curve X, input determinantal representation $y_1\sigma_2 - y_2\sigma_1 + \gamma$ and spectral data $\lambda^{(i)}, c(t)$ (on its principal subspace).*

In the special case when one of the operators A_1, A_2 is dissipative, say $\sigma_2 > 0$, Theorem 5.4 has been obtained by Livšic [11] for $\dim H < \infty$, as we noted in the previous section, and by Waksman [23] for commuting Volterra operators (the joint spectrum $S = (0,0)$) whose discriminant curve is a real smooth cubic.

We can not prove Theorem 5.4 by imitating the proof of Theorem 4.3, since we do not have, in the general case, enough direct information on common invariant subspaces of A_1, A_2. Therefore we shall adopt a function-theoretic approach. We shall associate to a regular commutative colligation its characteristic function. The coupling of colligations corresponds to the multiplication of characteristic functions, and the reduction of the colligation to the triangular model corresponds to the canonical factorization of its characteristic function. Since the characteristic function will turn out eventually to be a function on the discriminant curve, this will also tie the theory of commuting nonselfadjoint operators and the function theory on a real Riemann surface, much in the same way as the theory of a single nonselfadjoint (or nonunitary) operator is tied with the function theory on the upper half-plane (or on the unit disk) (see e.g. [14]).

6 Characteristic Functions and the Factorization Theorem

We first recall (see [10]) the basic definition of the characteristic function of an operator colligation.

Let $C = (A_1, A_2, H, \Phi, E, \sigma_1, \sigma_2, \gamma, \tilde{\gamma})$ be a regular commutative colligation. The *complete characteristic function* of C is the operator function in E given by

$$S(\xi_1, \xi_2, z) = I + i(\xi_1\sigma_1 + \xi_2\sigma_2)\Phi(\xi_1 A_1^* + \xi_2 A_2^* - zI)^{-1}\Phi^* \tag{6.1}$$

where $\xi_1, \xi_2, z \in \mathbf{C}$. This function is a regular analytic function of ξ_1, ξ_2, z whenever $z \notin \operatorname{spectrum}(\xi_1 A_1^* + \xi_2 A_2^*)$.

The following are the basic properties of the complete characteristic function.

Theorem 6.1 ([10]) *Let a (finite-dimensional) space E and selfadjoint operators σ_1, σ_2, γ, $\tilde{\gamma}$ in E be given; assume that $\det(\xi_1\sigma_1 + \xi_2\sigma_2) \not\equiv 0$. Then the complete characteristic function $S(\xi_1, \xi_2, z)$ determines the corresponding regular commutative colligation up to the unitary equivalence on the principal subspace.*

Theorem 6.2 ([10]) *Let $C = C' \vee C''$, where C', C'', C are regular commutative colligations, and let S', S'', S be the corresponding complete characteristic functions. Then $S(\xi_1, \xi_2, z) = S'(\xi_1, \xi_2, z)S''(\xi_1, \xi_2, z)$.*

For the one-point colligation (4.1) determined by a non-real affine point $\lambda = (\lambda_1, \lambda_2)$ on the discriminant curve X

$$S(\xi_1, \xi_2, z) = I + i(\xi_1\sigma_1 + \xi_2\sigma_2)\frac{\phi\phi^*}{\xi_1\overline{\lambda_1} + \xi_2\overline{\lambda_2} - z} \tag{6.2}$$

It follows from Theorem 6.2 and some limiting considerations that for the colligation (5.4) determined by an infinite sequence of non-real affine points $\lambda^{(i)} = (\lambda_1^{(i)}, \lambda_2^{(i)})(i = 1, \ldots)$ on X

$$S(\xi_1, \xi_2, z) = \prod_{i=1}^{\infty}\left(I + i(\xi_1\sigma_1 + \xi_2\sigma_2)\frac{\phi^{(i)}\phi^{(i)*}}{\xi_1\overline{\lambda_1^{(i)}} + \xi_2\overline{\lambda_2^{(i)}} - z}\right) \tag{6.3}$$

It can be also shown by standard techniques (see [1]) that for the colligation (5.7) determined by a function $c : [0, l] \longrightarrow X_{\mathbf{R}}$ into the set of real affine points of X

$$S(\xi_1, \xi_2, z) = \int_0^l \exp\left(i(\xi_1\sigma_1 + \xi_2\sigma_2)\frac{\phi(t)\phi(t)^*}{\xi_1 c_1(t) + \xi_2 c_2(t) - z}\right) dt \tag{6.4}$$

Let now X be a real smooth projective plane curve of degree n whose set of real points $X_{\mathbf{R}} \neq \emptyset$, and let $y_1\sigma_2 - y_2\sigma_1 + \gamma$, $y_1\sigma_2 - y_2\sigma_1 + \tilde{\gamma}$ be two determinantal representations of X. As in the previous sections we identify the space E in the colligation with $\mathbf{C}^n$, so that the complete characteristic function is an $n \times n$ matrix function.

Theorem 6.3 *An $n \times n$ matrix function $S(\xi_1, \xi_2, z)$ is the complete characteristic function of a regular commutative colligation with discriminant curve X, input determinantal representation $y_1\sigma_2 - y_2\sigma_1 + \gamma$ and output determinantal representation $y_1\sigma_2 - y_2\sigma_1 + \tilde{\gamma}$ if and only if:*

1) *$S(\xi_1, \xi_2, z)$ has the form*

$$S(\xi_1, \xi_2, z) = I + i(\xi_1\sigma_1 + \xi_2\sigma_2)R(\xi_1, \xi_2, z) \tag{6.5}$$

where $R(\xi_1, \xi_2, z)$ is holomorphic in the region $K_a = \{(\xi_1, \xi_2, z) \in \mathbf{C}^3 : |z| > a(|\xi_1|^2 + |\xi_2|^2)^{1/2}\}$ for some $a > 0$, and $R(t\xi_1, t\xi_2, tz) = t^{-1}R(\xi_1, \xi_2, z)$ for all $t \in \mathbf{C}, t \neq 0$.

2) *For any affine point $y = (y_1, y_2)$ on X, $S(\xi_1, \xi_2, \xi_1 y_1 + \xi_2 y_2)$ maps $L(y) = \operatorname{coker}(y_1 \sigma_2 - y_2 \sigma_1 + \gamma)$ into $\tilde{L}(y) = \operatorname{coker}(y_1 \sigma_2 - y_2 \sigma_1 + \tilde{\gamma})$ and the restriction $S(\xi_1, \xi_2, \xi_1 y_1 + \xi_2 y_2) \mid L(y)$ is independent of ξ_1, ξ_2 $((\xi_1, \xi_2, \xi_1 y_1 + \xi_2 y_2) \in K_a)$.*

3) *For any $\xi_1, \xi_2 \in \mathbf{R}$, $S(\xi_1, \xi_2, z)$ is a meromorphic function of z on the complement of the real axis and*

$$S(\xi_1, \xi_2, z)(\xi_1 \sigma_1 + \xi_2 \sigma_2)S(\xi_1, \xi_2, z)^* \leq \xi_1 \sigma_1 + \xi_2 \sigma_2 \quad (\Im z > 0),$$
$$S(\xi_1, \xi_2, z)(\xi_1 \sigma_1 + \xi_2 \sigma_2)S(\xi_1, \xi_2, z)^* = \xi_1 \sigma_1 + \xi_2 \sigma_2 \quad (\Im z = 0) \tag{6.6}$$

$((\xi_1, \xi_2, \xi_1 y_1 + \xi_2 y_2) \in K_a).$

The "only if" part of this Theorem, and the "if" part in the special case $\sigma_2 > 0$, have been obtained by Livšic [12].

It follows that if $S(\xi_1, \xi_2, z)$ is the complete characteristic function of a regular commutative colligation with discriminant curve X, input determinantal representation $y_1 \sigma_2 - y_2 \sigma_1 + \gamma$ and output determinantal representation $y_1 \sigma_2 - y_2 \sigma_1 + \tilde{\gamma}$, we can define for each affine point $y = (y_1, y_2)$ on X the mapping

$$\hat{S}(y) = S(\xi_1, \xi_2, \xi_1 y_1 + \xi_2 y_2) \mid L(y) : L(y) \longrightarrow \tilde{L}(y) \tag{6.7}$$

We call the function $\hat{S}(y)$ of a point y on X the *joint characteristic function* of the colligation. It is a mapping of line bundles $L, \tilde{L}$ on X, holomorphic outside the joint spectrum of the operators A_1^*, A_2^* (restricted to the principal subspace of the colligation).

Theorem 6.4 *The joint characteristic function of a regular commutative colligation determines the complete characteristic function.*

In the special case $\sigma_2 > 0$ this has been obtained by Livšic [12].

Using (6.3)-(6.4) we see that Theorem 5.4 on the reduction to the triangular model is equivalent to the following: for every matrix function $S(\xi_1, \xi_2, z)$ satisfying the conditions of Theorem 6.3, there exists a spectral data $\lambda^{(i)}(i = 1, \ldots, N; N \leq \infty), c(t) = (c_1(t), c_2(t))(0 \leq t \leq l)$ satisfying (5.10), such that $\gamma(l) = \tilde{\gamma}$ and

$$
\begin{aligned}
S(\xi_1, \xi_2, z) &= \prod_{i=1}^{N} \left(I + i(\xi_1 \sigma_1 + \xi_2 \sigma_2) \frac{\phi^{(i)} \phi^{(i)*}}{\xi_1 \overline{\lambda_1^{(i)}} + \xi_2 \overline{\lambda_2^{(i)}} - z} \right) \\
&\quad \times \int_0^a \exp\left(i(\xi_1 \sigma_1 + \xi_2 \sigma_2) \frac{\phi(t) \phi(t)^*}{\xi_1 c_1(t) + \xi_2 c_2(t) - z} \, dt \right)
\end{aligned} \tag{6.8}
$$

where $\gamma^{(i)}, \phi^{(i)}, \gamma(t), \phi(t)$ are determined by (5.11). Now, functions of several complex variables do not admit a good factorization theory. However, we see from Theorem 6.4 that the complete characteristic function reduces to the function on the one-dimensional complex manifold X. We shall therefore reduce (6.8) to the factorization theorem on a real Riemann surface.

We first want to express the contractivity and isometricity properties (6.6) of the complete characteristic function in terms of the joint characteristic function. To this end we introduce a hermitian pairing between the fibers $L(y^{(1)}), L(y^{(2)})$ of the line bundle $L(y) = \operatorname{coker}(y_1\sigma_2 - y_2\sigma_1 + \gamma)$ over non-conjugate affine points $y^{(1)} = (y_1^{(1)}, y_2^{(1)}), y^{(2)} = (y_1^{(2)}, y_2^{(2)})$ on X:

$$[u,v]^L_{y^{(1)},y^{(2)}} = i\frac{u(\xi_1\sigma_1 + \xi_2\sigma_2)v^*}{\xi_1(y_1^{(1)} - \overline{y_1^{(2)}}) + \xi_2(y_2^{(1)} - \overline{y_2^{(2)}})}$$
$$(u \in L(y^{(1)}), v \in L(y^{(2)}); y^{(1)} \neq \overline{y^{(2)}}) \tag{6.9}$$

This is in fact independent (see (4.6)) of $\xi_1, \xi_2 \in \mathbf{R}$. In particular, taking $y = y^{(1)} = y^{(2)}$, we get an (indefinite) scalar product on the fiber $L(y)$ over non-real affine points y on X. We also introduce a hermitian pairing between the fibers $L(y), L(\bar{y})$ over conjugate affine points:

$$[u,v]^L_{y,\bar{y}} = i\frac{u(\xi_1\sigma_1 + \xi_2\sigma_2)v^*}{\xi_1 dy_1 + \xi_2 dy_2} \ (u \in L(y), v \in L(\bar{y})) \tag{6.10}$$

This is again independent of $\xi_1, \xi_2 \in \mathbf{R}$, and we get in particular a scalar product on the fiber $L(y)$ over real affine points y on X (to get a value in (6.10) we have to choose, of course, a local parameter on X at the point $y = (y_1, y_2)$).

Theorem 6.5 *Let $S(\xi_1, \xi_2, z)$ be a matrix function satisfying the conditions 1)-2) of Theorem 6.3, and let the function $\hat{S}(y)$ be defined by (6.7). Then $S(\xi_1, \xi_2, z)$ satisfies (6.6) if and only if $\hat{S}(y)$ satisfies the following: for all affine points $y, y^{(1)}, \ldots, y^{(N)}$ on X in its region of analyticity ($y^{(i)} \neq \overline{y^{(j)}}$)*

$$\left([u^{(i)}\hat{S}(y^{(i)}), u^{(j)}\hat{S}(y^{(j)})]^{\tilde{L}}_{y^{(i)},y^{(j)}}\right)_{i,j=1,\ldots,N} \leq \left([u^{(i)}, u^{(j)}]^L_{y^{(i)},y^{(j)}}\right)_{i,j=1,\ldots,N}$$
$$(u^{(i)} \in L(y^{(i)}); i = 1, \ldots, N),$$
$$[u\hat{S}(y), v\hat{S}(\bar{y})]^{\tilde{L}}_{y,\bar{y}} = [u,v]^L_{y,\bar{y}} \ (u \in L(y), v \in L(\bar{y})) \tag{6.11}$$

Let now $\hat{S}(y)$ be the joint characteristic function of a regular commutative colligation with discriminant curve X, input determinantal representation $y_1\sigma_2 - y_2\sigma_1 + \gamma$ and output determinantal representation $y_1\sigma_2 - y_2\sigma_1 + \tilde{\gamma}$, where the representations $y_1\sigma_2 - y_2\sigma_1 + \gamma, y_1\sigma_2 - y_2\sigma_1 + \tilde{\gamma}$ have sign ϵ (the input and the output determinantal representation have always the same sign) and correspond, as in Theorem 2.1, to the points $\zeta, \tilde{\zeta}$ in $J(X)$ $(\theta(\zeta) \neq 0, \theta(\tilde{\zeta}) \neq 0, \zeta + \overline{\zeta} = e, \tilde{\zeta} + \overline{\tilde{\zeta}} = e)$. Since ζ and $\tilde{\zeta}$ are, up to a constant translation, the images of the line bundles L and $\tilde{L}$ in the Jacobian variety under the Abel-Jacobi map μ, and $\hat{S}$ is a mapping of L to $\tilde{L}$, it follows that $\hat{S}$ can be identified, up to a constant factor of absolute value 1, with a (scalar) multivalued multiplicative function $s(x)$ on X, with multipliers of absolute value 1 corresponding to the point $\tilde{\zeta} - \zeta$ in $J(X)$. More precisely, let $A_1, \ldots, A_g, B_1, \ldots, B_g$ be the chosen canonical integral homology basis on X, let Z be the $g \times g$ period matrix of $J(X)$ (the period lattice $\Lambda \subset \mathbf{C}^g$ is spanned by the g vectors of the standard basis and the g columns of Z), and let $\zeta = b + Za, \tilde{\zeta} = \tilde{b} + Z\tilde{a}$, where $a, b, \tilde{a}, \tilde{b}$ are vectors in $\mathbf{R}^g$ with entries $a_i, \tilde{a}_i, b_i, \tilde{b}_i$ respectively; then the multipliers χ_s of $s(x)$ over the basis cycle are given by

$$\chi_s(A_i) = \exp(-2\pi i(\tilde{a}_i - a_i)) \ (i = 1, \ldots, g),$$
$$\chi_s(B_i) = \exp(2\pi i(\tilde{b}_i - b_i)) \ (i = 1, \ldots, g) \tag{6.12}$$

See e.g. [2] for more details.

We call $s(x)$ the *normalized joint characteristic function* of the colligation (since it arises from the joint characteristic function by choosing sections of L and $\tilde{L}$ with certain normalized zeroes and poles). We have essentially seen in Theorem 4.1 that the pairing (6.9) on the line bundle can be expressed analytically; the same is true of the pairing (6.10). We obtain thus from Theorem 6.5 a complete analytic description of normalized joint characteristic functions of regular commutative colligations.

Theorem 6.6 *Let $y_1\sigma_2 - y_2\sigma_1 + \gamma$ be a determinantal representation of X that has sign ϵ and corresponds to the point ζ in $J(X)$, and let $\tilde{\zeta}$ be another point in $J(X)$, $\theta(\tilde{\zeta}) \neq 0, \tilde{\zeta}+\overline{\tilde{\zeta}} = e$. A multivalued multiplicative function $s(x)$ on X with multipliers of absolute value 1 corresponding to the point $\tilde{\zeta} - \zeta$ is the normalized joint characteristic function of a regular commutative colligation with discriminant curve X, input determinantal representation $y_1\sigma_2 - y_2\sigma_1 + \gamma$ and an output determinantal representation $y_1\sigma_2 - y_2\sigma_1 + \tilde{\gamma}$ that has sign ϵ and corresponds to the point $\tilde{\zeta}$ in $J(X)$ if and only if:*

1) *$s(x)$ is holomorphic outside a compact subset of affine points of X.*

2) *$s(x)$ is meromorphic on $X\backslash X_{\mathbf{R}}$, and for all points $x, x^{(1)}, \ldots, x^{(N)}$ on X in its region of analyticity $(x^{(i)} \neq \overline{x^{(j)}})$*

$$\epsilon\left(s(x^{(i)})\overline{s(x^{(j)})}\frac{i\theta[\tilde{\zeta}](x^{(i)} - \overline{x^{(j)}})}{\theta[\tilde{\zeta}](0)E(x^{(i)}, \overline{x^{(j)}})}\right)_{i,j=1,\ldots,N} \le \epsilon\left(\frac{i\theta[\zeta](x^{(i)} - \overline{x^{(j)}})}{\theta[\zeta](0)E(x^{(i)}, \overline{x^{(j)}})}\right)_{i,j=1,\ldots,N},$$
$$s(x)\overline{s(\overline{x})} = 1 \tag{6.13}$$

In the special case when one of the operators A_1, A_2 in the colligation is dissipative, say $\sigma_2 > 0$, the "weights" $\frac{i\theta[\zeta](x-\overline{x})}{\theta[\zeta](0)E(x,\overline{x})}, \frac{i\theta[\tilde{\zeta}](x-\overline{x})}{\theta[\tilde{\zeta}](0)E(x,\overline{x})}$ are positive on X_+ and negative on X_- (see comments following Theorem 4.1), and it turns out that the matrix condition in (6.13) can be replaced by

$$\epsilon|s(x)| \le \epsilon \quad (x \in X_+) \tag{6.14}$$

We conjecture that in general the matrix condition is equivalent to

$$\epsilon\frac{i\theta[\tilde{\zeta}](x - \overline{x})}{\theta[\tilde{\zeta}](0)E(x,\overline{x})} \le \epsilon\frac{i\theta[\zeta](x - \overline{x})}{\theta[\zeta](0)E(x,\overline{x})} \quad (x \in X\backslash X_{\mathbf{R}}) \tag{6.15}$$

Let now X be a compact real Riemann surface (i.e. a compact Riemann surface with an antiholomorphic involution $x \mapsto \overline{x}$; for example, a real smooth projective plane curve). Let $X_{\mathbf{R}}$ be the set of fixed points of the involution; assume $X_{\mathbf{R}} \neq \emptyset$. Let $\zeta, \tilde{\zeta}$ be two points in $J(X)$, $\theta(\zeta) \neq 0, \theta(\tilde{\zeta}) \neq 0, \zeta + \overline{\zeta} = e, \tilde{\zeta} + \overline{\tilde{\zeta}} = e$ (the half-period e of Theorem 2.1 is defined for every real Riemann surface). A multivalued multiplicative function $s(x)$ on X with multipliers of absolute value 1 corresponding to the point $\tilde{\zeta} - \zeta$ is called *semicontractive*, or, specifically, $(\zeta, \tilde{\zeta})$-*contractive*, if it is meromorphic on $X\backslash X_{\mathbf{R}}$, and for all points $x, x^{(1)}, \ldots, x^{(N)}$ on $X\backslash X_{\mathbf{R}}$ $(x^{(i)} \neq \overline{x^{(j)}})$:

$$\left(s(x^{(i)})\overline{s(x^{(j)})}\frac{i\theta[\tilde{\zeta}](x^{(i)} - \overline{x^{(j)}})}{\theta[\tilde{\zeta}](0)E(x^{(i)}, \overline{x^{(j)}})}\right)_{i,j=1,\ldots,N} \le \left(\frac{i\theta[\zeta](x^{(i)} - \overline{x^{(j)}})}{\theta[\zeta](0)E(x^{(i)}, \overline{x^{(j)}})}\right)_{i,j=1,\ldots,N},$$
$$s(x)\overline{s(\overline{x})} = 1 \tag{6.16}$$

Theorem 6.6 states that normalized joint characteristic functions of regular commutative colligations with a smooth discriminant curve (that has real points) are precisely semicontractive functions on the discriminant curve (for sign $\epsilon = 1$) and their inverses (for sign $\epsilon = -1$). The factorization (6.8) of the complete characteristic function follows from the following factorization theorem for semicontractive functions on the real Riemann surface X.

Theorem 6.7 *Let $s(x)$ be a $(\zeta, \tilde{\zeta})$-contractive function on X. Then*

$$s(x) = \prod_{i=1}^{N} \left(\exp\left(\pi i m^{(i)t}(\lambda^{(i)} + \overline{\lambda^{(i)}}) + \frac{\pi i}{2} m^{(i)t} H m^{(i)} \right) \exp\left(-2\pi(\lambda^{(i)} - \overline{\lambda^{(i)}})^t Y x \right) \frac{E(x, \lambda^{(i)})}{E(x, \overline{\lambda^{(i)}})} \right)$$

$$\times \exp\left(-\pi \sum_{i=0}^{k-1} \int_{X_k} \frac{(\omega_1(y), \ldots, \omega_g(y))}{\omega(y)} n_i d\nu(y) - 2\pi i \int_{X_{\mathbf{R}}} \frac{(\omega_1(y), \ldots, \omega_g(y))}{\omega(y)} Y x d\nu(y) \right.$$

$$\left. + i \int_{X_{\mathbf{R}}} \frac{d_y \ln E(x, y)}{\omega(y)} d\nu(y) \right) \tag{6.17}$$

Here $\lambda^{(i)}(i = 1, \ldots, N; N \leq \infty)$ are the zeroes of $s(x)$ on $X \backslash X_{\mathbf{R}}$ and ν is a uniquely determined finite positive Borel measure on $X_{\mathbf{R}}$; $\omega_1, \ldots, \omega_g$ are the chosen basis for holomorphic differentials on X; ω is a real differential on X, defined, analytic and non-zero in a neighbourhood of $\operatorname{supp} \nu \subset X_{\mathbf{R}}$, whose signs on different connected components $X_0, \ldots, X_{k-1}$ of $X_{\mathbf{R}}$ correspond to the real torus in $J(X)$ to which the points $\zeta, \tilde{\zeta}$ belong [18]; $Z = \frac{1}{2}H + iY^{-1}$ $(H, Y$ real$)$ is the $g \times g$ period matrix of $J(X)$; $m^{(i)}(i = 1, \ldots, N), n_i(i = 0, \ldots, k-1)$ are integral vectors depending on the choice of lifting of the points $\lambda^{(i)}$ and the components X_i respectively from $J(X) = \mathbf{C}^g / \Lambda$ to $\mathbf{C}^g$. Furthermore, the following hold:

$$\frac{i\theta[\zeta + \sum_{j=1}^{i}(\lambda^{(j)} - \overline{\lambda^{(j)}})](\lambda^{(i+1)} - \overline{\lambda^{(i+1)}})}{\theta[\zeta + \sum_{j=1}^{i}(\lambda^{(j)} - \overline{\lambda^{(j)}})](0) E(\lambda^{(i+1)}, \overline{\lambda^{(i+1)}})} > 0 \quad (i = 0, \ldots, N-1),$$

$$\sum_{i=1}^{\infty} (\lambda^{(i)} - \overline{\lambda^{(i)}}) \text{ converges }, \theta(\zeta + \sum_{i=1}^{\infty}(\lambda^{(i)} - \overline{\lambda^{(i)}})) \neq 0 \quad (if\ N = \infty),$$

$$\theta\left(\zeta + \sum_{i=1}^{N}(\lambda^{(i)} - \overline{\lambda^{(i)}}) + i \int_{B} \begin{pmatrix} \frac{\omega_1(y)}{\omega(y)} \\ \vdots \\ \frac{\omega_g(y)}{\omega(y)} \end{pmatrix} d\nu(y) \right) \neq 0 \quad (for\ all\ Borel\ sets\ B \subset X_{\mathbf{R}}),$$

$$\tilde{\zeta} = \zeta + \sum_{i=1}^{N}(\lambda^{(i)} - \overline{\lambda^{(i)}}) + i \int_{X_{\mathbf{R}}} \begin{pmatrix} \frac{\omega_1(y)}{\omega(y)} \\ \vdots \\ \frac{\omega_g(y)}{\omega(y)} \end{pmatrix} d\nu(y) \tag{6.18}$$

When X is a real smooth projective plane curve (and $s(x)$ is holomorphic outside a compact subset of affine points of X), the two factors in (6.16) are the normalized joint characteristic functions of the colligations (5.4) and (5.7) respectively ($c : [0, l] \longrightarrow X_{\mathbf{R}}$ is the left-continuous non-decreasing function determined by $\nu(B) = m(c^{-1}(B))$ for Borel sets $B \subset X_{\mathbf{R}}$, where m is the Lebesgue measure on $[0, l]$, $l = \nu(X_{\mathbf{R}})$). Decomposing the measure ν into singular and absolutely continuous parts (with respect to the measures induced on $X_{\mathbf{R}}$ by the usual Lebesgue measure through local coordinates), we obtain the factorization of a semicontractive function into a Blaschke product, a singular inner function and an outer function, generalizing the Riesz-Nevanlinna factorization for bounded analytic functions in the unit disk (see e.g. [7]). Our factorization is better compared though to Potapov factorization for J-contractive matrix functions (see [15]), since the weights $\frac{i\theta[\zeta](x-\bar{x})}{\theta[\zeta](0)E(x,\bar{x})}$, $\frac{i\theta[\tilde{\zeta}](x-\bar{x})}{\theta[\zeta](0)E(x,\bar{x})}$ are not, in general, positive or negative everywhere. In the special case of (6.14), the Blaschke product — singular inner facter — outer factor decomposition was known ([22,5,6]), without, however, explicit formulas for the factors in terms of the prime form $E(x,y)$.

It is my pleasure to thank Prof. M.S.Livšic for many deep and interesting discussions.

References

[1] Brodskii,M.S., Livšic,M.S.: Spectral analysis of nonselfadjoint operators and intermediate systems, *AMS Transl. (2)* 13, 265-346 (1960).

[2] Fay,J.D.: *Theta Functions on Riemann Surfaces*, Springer-Verlag, Heidelberg (1973).

[3] Griffiths,P., Harris,J.: *Principles of Algebraic Geometry*, Wiley, New York (1978).

[4] Harte,R.E. : Spectral mapping theorems, *Proc. Roy. Irish Acad. (A)* 72, 89-107 (1972).

[5] Hasumi,M. : Invariant subspace theorems for finite Riemann surfaces, *Canad. J. Math.* 18, 240-255 (1986).

[6] Hasumi,M. : *Hardy Classes on Infinitely Connected Riemann Surfaces*, Springer-Verlag, Heidelberg (1983).

[7] Hoffman,K.: *Banach Spaces of Analytic Functions*, Prentice Hall, Englewood Cliffs, NJ (1962).

[8] Kravitsky,N. : Regular colligations for several commuting operators in Banach space, *Int. Eq. Oper. Th.* 6, 224-249 (1983).

[9] Kravitsky,N. : On commuting integral operators, *Topics in Operator Theory, Systems and Networks* (Dym,H., Gohberg,I., Eds.), Birkhäuser, Boston (1984).

[10] Livšic,M.S., Jancevich,A.A.: *Theory of Operator Colligations in Hilbert Space*, Wiley, New York (1979).

[11] Livšic,M.S.: Cayley-Hamilton theorem, vector bundles and divisors of commuting operators, *Int. Eq. Oper. Th.* 6, 250-273 (1983).

[12] Livšic,M.S.: Commuting nonselfadjoint operators and mappings of vector bundles on algebraic curves, *Operator Theory and Systems* (Bart,H., Gohberg,I., Kaashoek,M.A., Eds.), Birkhäuser, Boston (1986).

[13] Mumford,D. : *Tata Lectures on Theta*, Birkhäuser, Boston (Vol. 1, 1983; Vol. 2, 1984).

[14] Nikolskii,N.K. : *Treatise on the Shift Operator*, Springer-Verlag, Heidelberg (1986).

[15] Potapov,V.P.: The mulptiplicative structure of J-contractive matrix functions, *AMS Transl. (2)* 15, 131-243 (1960).

[16] Taylor,J.L. : A joint spectrum for several commuting operators, *J. of Funct. Anal.* 6, 172-191 (1970).

[17] Vinnikov,V. : Self-adjoint determinantal representations of real irreducible cubics, *Operator Theory and Systems* (Bart,H., Gohberg,I., Kaashoek,M.A., Eds.), Birkhäuser, Boston (1986).

[18] Vinnikov,V.: Self-adjoint determinantal representions of real plane curves, preprint.

[19] Vinnikov,V. : Triangular models for commuting nonselfadjoint operators, in preparation.

[20] Vinnikov,V. : Characteristic functions of commuting nonselfadjoint operators, in preparation.

[21] Vinnikov,V. : The factorization theorem on a compact real Riemann surface, in preparation.

[22] Voichick,M., Zalcman,L. : Inner and outer functions on Riemann surfaces, *Proc. Amer. Math. Soc.* 16, 1200-1204 (1965).

[23] Waksman,L. : Harmonic analysis of multi-parameter semigroups of contractions, *Commuting Nonselfadjoint Operators in Hilbert space* (Livšic,M.S., Waksman,L.), Springer-Verlag, Heidelberg (1987).

DEPARTEMENT OF THEORETICAL MATHEMATICS, WEIZMANN INSTITUTE OF SCIENCE, REHOVOT 76100, ISRAEL
E-mail address: vinnikov@wisdom.weizmann.ac.il

1980 *Mathematics Subject Classification* (1985 *Revision*). Primary 47A45, 30D50; Secondary 14H45, 14H40, 14K20, 14K25, 30F15.

Operator Theory:
Advances and Applications, Vol. 59
© 1992 Birkhäuser Verlag Basel

ALL (?) ABOUT QUASINORMAL OPERATORS

Pei Yuan Wu[1]

Dedicated to the memory of Domingo A. Herrero (1941–1991)

A bounded linear operator T on a complex separable Hilbert space is quasinormal if T and T^*T commute. In this article, we survey all (?) the known results concerning this class of operators with more emphasis on recent progresses. We will consider their various representations, spectral property, multiplicity, characterizations among weighted shifts, Toeplitz operators and composition operators, invariant subspace structure, double commutant property, commutant lifting property, similarity, quasisimilarity and compact perturbation, and end with some speculations on possible directions for further research.

1. INTRODUCTION

The class of quasinormal operators was first introduced and studied by A. Brown [4] in 1953. From the definition, it is easily seen that this class contains normal operators $(TT^* = T^*T)$ and isometries $(T^*T = I)$. On the other hand, it can be shown [36, Problem 195] that any quasinormal operator is subnormal, that is, it has a normal extension. Normal operators and isometries are classical objects : Their properties have been fully explored and their structures well–understood. It has also been widely recognized that subnormality constitutes a deep and useful generalization of normality. After two–decades' intensive study by various operator theorists, the theory of subnormal operators has matured to the extent that two monographs [17, 18] have appeared which are devoted to its codification. People may come to suspect whether the in–between quasinormal operators would be of any interest to merit a separate survey paper like this one. The structure of quasinormal operators is, as we shall see below,

---

[1]This research was partially supported by the National Science Council of the Republic of China.

indeed very simple. They are certainly not in the same league as their big brothers : Their theory is not as basic as those of normal operators and isometries and also not as deep as subnormal ones. However, we will report in subsequent discussions some recent progresses in the theory of quasinormality which serve to justify the worthwhileness of our effort. One recent result (on the similarity of two quasinormal operators) establishes a connection between the theories of quasinormal operators and nest algebras. Another one (on their quasisimilarity) uses a great deal of the analytic function theory. These clearly show that there are indeed many interesting questions which can be asked about this class of operators. It used to be the case that the study of quasinormal operators was pursued as a step toward a better understanding of the subnormal ones. The recent healthy developments indicate that quasinormal operators may have an independent identity and deserve to be studied for their own sake.

The interpretation of our title "ALL (?) ABOUT QUASINORMAL OPERATORS" follows the same spirit as that of Domingo Herrero's paper [39] : The "ALL" is interpreted as "all the author knows about the subject", and the question mark "?" means that we never really know "all" about any given subject.

The paper is organized as follows. We start in Section 2 with three representations of quasinormal operators. One of them is the canonical representation on which all the theory is built. Section 3 discusses the (essential) spectrum, various parts thereof, (essential) norm and multiplicity. Section 4 gives characterizations of quasinormality among several special classes of operators, namely, weighted shifts, Toeplitz operators and composition operators. Section 5 then treats various properties related to the invariant subspaces of an operator such as reflexivity, decomposability, (bi)quasitriangularity and cellular–indecomposability. The three operator algebras $\{T\}'$, $\{T\}''$ and Alg T of a pure quasinormal operator T are described in Section 6. Then we proceed to consider properties relating a quasinormal operator to operators in its commutant. One such property concerns their lifting to its minimal normal extension. We also consider the quasinormal extension for subnormal operators as developed by Embry–Wardrop. Sections 7 and 8 are on the similarity and quasisimilarity of two quasinormal operators. Section 9 discusses the problems when two quasinormal operators are approximately equivlaent, compact perturbations and algebraically equivalent to each other. We conclude in Section 10 with some open problems which seem to be worthy of exploring.

This paper is an expanded version of the talk given in the WOTCA at Hokkaido University. We would like to thank Professor T. Ando, the organizer, for his invitation to present this talk and for his efforts in organizing the conference.

2. REPRESENTATIONS

We start with the canonical representation for quasinormal operators first obtained by A. Brown [4]. This representation is the foundation for all the subsequent developments of the theory.

THEOREM 2.1. *An operator* T *on Hilbert space* H *is quasinormal if and only if* T *is unitarily equivalent to an operator of the form*

$$N \oplus \begin{bmatrix} 0 & & & \\ A & 0 & & \\ & A & 0 & \\ & & \ddots & \ddots \\ & & & \ddots & \ddots \end{bmatrix},$$

where N *is normal and* A *is positive semidefinite. If* A *is chosen to be positive, then* N *and* A *are uniquely determined (up to unitary equivalence).*

Recall that A is *positive semidefinite* (resp. *positive definite*) if $(Ax, x) \geq 0$ (resp. $(Ax, x) > 0$) for any vector (resp. nonzero vector) x.

In fact, in the preceding theorem N and A may be chosen to be the restrictions of T and $(T^*T)^{\frac{1}{2}}$ to their respective reducing subspaces $\bigcap_{n=1}^{\infty} \ker (T^{n*}T^n - T^nT^{n*})$ and $H \ominus (\ker T \oplus \overline{\operatorname{ran} T})$. If A is the identity operator on a one–dimensional space, then

$$\begin{bmatrix} 0 & & & \\ A & 0 & & \\ & A & 0 & \\ & & \ddots & \ddots \\ & & & \ddots & \ddots \end{bmatrix}$$

reduces to the *simple unilateral shift* S. (Later on, we will also consider S as the operator of multiplication by z on the Hardy space H^2 of the unit disc.) For convenience, we will denote

$$\begin{bmatrix} 0 & & & & \\ A & 0 & & & \\ & A & 0 & & \\ & & & \cdot\;\cdot & \\ & & & & \cdot\;\cdot \\ & & & & \;\;\cdot\;\cdot \end{bmatrix}$$

by $S \otimes A$ without giving a precise meaning to the tensor product of two operators. Note that $S \otimes A$ is *completely nonnormal*, that is, there is no nontrivial reducing subspace on which it is normal. We will call the uniquely determined N and $S \otimes A$ the *normal* and *pure parts* of T, respectively. If T is an isometry, then these two parts coincide with the unitary operator and the unilateral shift in its Wold decomposition.

In terms of this representation, it is easily seen that every quasinormal operator $N \oplus (S \otimes A)$ is subnormal with minimal normal extension

$$N \oplus \begin{bmatrix} \cdot\;\cdot & & & & \\ & \cdot\;\cdot & & & \\ & & A & 0 & & \\ & & & A & 0 & \\ & & & & A & 0 \\ & & & & & \cdot\;\cdot \\ & & & & & \;\;\cdot\;\cdot \end{bmatrix},$$

where a box is drawn around the $(0, 0)$ —entry of the matrix. Since

$$S \otimes A = \begin{bmatrix} 0 & & & \\ I & 0 & & \\ & I & 0 & \\ & & \cdot\;\cdot & \\ & & & \cdot\;\cdot \end{bmatrix}\begin{bmatrix} A & & & \\ & A & & \\ & & A & \\ & & & \cdot \\ & & & \;\;\cdot \end{bmatrix}$$

is the (unique) polar decomposition of $S \otimes A$ (with the two factors having equal kernels), an easy argument yields the following characterization of quasinormality [36, Problem 137].

THEOREM 2.2. *An operator with polar decomposition* UP *is quasinormal if and only if* U *and* P *commute.*

There are other representations for quasinormal operators. Since every positive operator can be expressed as the direct sum of cyclic positive operators, this implies that every pure quasinormal operator is the direct sum of operators of the form $S \otimes A$, where A is cyclic and positive definite. (Recall that an operator T on H is *cyclic* if there is a vector x in H such that $V\{T^n x : n \geq 0\} = H$.) The second representation which we now present will be for this latter type of operators.

By the spectral theorem, any cyclic positive definite operator A is unitarily equivalent to the operator of multiplication by t on $L^2(\mu)$, where μ is some positive Borel measure on an interval $[0, a]$ in $\mathbb{R}$ with $\mu(\{0\}) = 0$. Let ν be the measure on $\mathbb{C}$ defined by $d\nu(z) = \frac{1}{2\pi} d\theta d\mu(t)$, where $z = te^{i\theta}$, and let $K = V\{|z|^m z^n : m, n \geq 0\}$ in $L^2(\nu)$. Then, obviously, K is an invariant subspace for M, the operator of multiplication by z on $L^2(\nu)$. Finally, let $T_A = M|K$.

THEOREM 2.3. *For any cyclic positive definite operator* A, T_A *is a pure quasinormal operator. Conversely, any pure quasinormal operator* S ⊗ A *with* A *cyclic is unitarily equivalent to* T_A.

This representation is obtained in [19, Theorem 2.4]. The appearance of the space K above is not too obtrusive if we compare it with the space in the statement of Proposition 3.3 below.

We conclude this section with the third representation. It applies to pure quasinormal operators S ⊗ A with A invertible. This is originally due to G. Keough and first appeared in [19, Theorem 2.8].

Let A be a positive invertible operator on H, and let H_A^2 be the class of sequences $\{x_n\}_{n=0}^{\infty}$ with x_n in H satisfying

$$\sum_{n=0}^{\infty} \|A^n x_n\|^2 < \infty.$$

It is easy to verify that H_A^2 is a Hilbert space under the inner product

$$(\{x_n\}, \{y_n\}) = \sum_{n=0}^{\infty} (A^n x_n, A^n y_n),$$

where $(\,,\,)$ inside the summation sign denotes the inner product in H. Let S_A denote the right shift on H_A^2 :

$$S_A(\{x_0, x_1, \cdots\}) = \{0, x_0, x_1, \cdots\}.$$

THEOREM 2.4. *For any positive invertible* A, S_A *is a pure quasinormal operator. Conversely, any pure quasinormal operator* S ⊗ A *with* A *invertible is unitarily equivalent to* S_A.

It is clear that the unitary operator

$$U(\{x_n\}) = \{A^n x_n\}$$

from H_A^2 onto $H \oplus H \oplus \cdots$ implements the unitary equivalence between S_A and $S \otimes A$.
As an application, we have

THEOREM 2.5. *If* $T = S \otimes A$ *is a pure quasinormal operator on* H *and* R *is any cyclic operator on* K *with* $\|R\| < \|T\|$, *then there exists an operator* $X : H \to K$ *with dense range such that* $XT = RX$.

The preceding theorem is proved in [19, Theorem 4.2] first for invertible A and then for the general case. We remark that if T is a pure isometry then X can be chosen not only to have dense range but have zero kernel [51].

3. SPECTRUM AND MULTIPLICITY

For the spectrum of quasinormal operators, we may restrict ourselves to the pure ones since putting back the normal part does not cause much difficulty.

THEOREM 3.1. *Let* $T = S \otimes A$ *be a pure quasinormal operator. Then*

(1) $\sigma_p(T) = \phi$,

(2) $\sigma_P(T^*) = \{\lambda : |\lambda| < \|A\|\}$,

(3) $\sigma(T) = \sigma_{ap}(T^*) = \{\lambda : |\lambda| \leq \|A\|\}$,

(4) $\sigma_{ap}(T) = \sigma_{le}(T) = \{\lambda : |\lambda| \in \sigma(A)\}$, *and*

(5) $\sigma_e(T) = \sigma_{re}(T) = \{\lambda : |\lambda| \leq \|A\|_e\} \cup \{\lambda : |\lambda| \in \sigma(A)\}$.

Here $\sigma(\cdot)$, $\sigma_p(\cdot)$, $\sigma_{ap}(\cdot)$, $\sigma_e(\cdot)$, $\sigma_{le}(\cdot)$ and $\sigma_{re}(\cdot)$ denote, respectively, the *spectrum, point spectrum, approximate point spectrum, essential spectrum, left essential spectrum* and *right essential spectrum* of its argument. The spectrum and essential spectrum of $S \otimes A$ were first obtained in [55, Corollary 2] and the approximate point spectrum in [57, Theorem 2.1]; other assertions either are obvious or can be derived from the results in [55]. Alternatively, $\sigma(S \otimes A)$ and $\sigma_{ap}(S \otimes A)$ can also be obtained as in [12, Lemma 2.2] via the more general result on the spectrum of tensor product of operators [5]. Note that an immediate consequence of the above is that $\|S \otimes A\| = \|S \otimes A\|_e = \|A\|$.

The *multiplicity* $\mu(T)$ of an operator T on H is the minimal cardinality of vectors $\{x_\alpha : \alpha \in \Omega\}$ in H satisfying

$$\vee\{T^n x_\alpha : n \geq 0, \alpha \in \Omega\} = H.$$

The next proposition gives an expression of the multiplicity of a pure quasinormal operator $S \otimes A$ in terms of A.

PROPOSITION 3.2. *If* $T = S \otimes A$ *is a pure quasinormal operator on* $H^{(\infty)} = H \oplus H \oplus \cdots$, *then* $\mu(T) = \dim H$. *In particular,* $\mu(A) \leq \mu(S \otimes A)$.

Indeed, if $\{x_1, \cdots, x_m\}$ are vectors in $H^{(\infty)}$ with $H^{(\infty)} = \vee\{T^n x_j : n \geq 0, 1 \leq j \leq m\}$, then, for any $y \in K$, $y \oplus 0 \oplus 0 \oplus \cdots$ is in $H^{(\infty)} = \vee\{T^n x_j\}$ whence y is a linear combination of the first components of the x_j's. Therefore $\dim H \leq m$.

Conversely, if H is spanned by $\{y_1, \cdots, y_m\}$, then, since A is invertible on H, H is also spanned by $\{A^n y_1, \cdots, A^n y_m\}$ for any $n \geq 0$. We infer that $H^{(\infty)} = \vee\{T^n y_j : n \geq 0, 1 \leq j \leq m\}$, where $y_j = y_j \oplus 0 \oplus 0 \cdots, 1 \leq j \leq m$. Thus $\mu(T) \leq m$.

Note that, in general, $\mu(A)$ and $\mu(S \otimes A)$ are not equal: If $A = \begin{bmatrix} 1 & 0 \\ 0 & 2 \end{bmatrix}$, then $\mu(A) = 1$ but $\mu(S \otimes A) = 2$. Nevertheless, as the following proposition shows, the multiplicity of A may also be expressed in terms of $S \otimes A$.

PROPOSITION 3.3. *If* $T = S \otimes A$ *is a pure quasinormal operator on H, then* $\mu(A)$ *equals the minimal cardinality of vectors* $\{x_\alpha : \alpha \in \Omega\}$ *in H satisfying* $\vee\{|T|^m T^n x_\alpha : m, n \geq 0, \alpha \in \Omega\} = H$, *where* $|T| = (T^* T)^{1/2}$.

The proof is the same as the one for [19, Proposition 2.3] which we omit.

Next we consider the multiplicity of the adjoint of a pure quasinormal operator. Since the adjoint of any pure isometry is cyclic [36, Problem 160], we may not be too surprised to find out that the same is true for adjoints of pure quasinormal operators.

THEOREM 3.4. $\mu(T^*) = 1$ *for any pure quasinormal operator* T.

This is proved in [58, Corollary 3]. It follows from a more general result [58, Theorem 2] that any operator of the form

$$\begin{bmatrix} 0 & T_{12} & T_{13} & \cdot\cdot \\ 0 & 0 & T_{23} & \cdot\cdot \\ 0 & 0 & 0 & \cdot\cdot \\ \cdot & \cdot & \cdot & \cdot \\ \cdot & \cdot & \cdot & \cdot \end{bmatrix}$$

with $T_{n\,n+1}$ having dense range for all $n \geq 1$ is cyclic.

4. SPECIAL CLASSES

In this section, we give characterizations of quasinormal operators among three special classes, namely, unilateral (bilateral) weighted shifts, Toeplitz operators and composition operators.

An operator T on H is a *unilateral* (resp. *bilateral*) *weighted shift* if there are an orthonormal basis $\{e_n\}$ and a sequence of bounded complex numbers $\{w_n\}$, n = 0, 1, 2, $\cdots$ (resp. n = 0, ±1, ±2, $\cdots$),such that $Te_n = w_n e_{n+1}$ for all n. As usual, we may assume that the weights w_n are all nonnegative.

THEOREM 4.1. *A unilateral* (*bilateral*) *weighted shift* T *is quasinormal if and only if there is an integer* n_0 *such that* $w_{n_0} = w_{n_0-1} = \cdots = 0$ *and* $w_{n_0+1} = w_{n_0+2} = \cdots$.

The proof follows by an easy computation with the defining property of quasinormality and can be found in [36, Problem 139]. Note that the only normal unilateral (bilateral) weighted shift(s) is (are) the one(s) with all the weights equal to zero (all the weights equal), and the subnormal shifts have also been characterized (cf. [17, Theorems III. 8. 16. and III. 8. 17]).

We next consider Toeplitz operators. For ϕ in L^∞, the Lebesgue space on the unit circle, the *Toeplitz operator* T_ϕ is the operator on the Hardy space H^2 defined by $T_\phi f = P(\phi f)$ for $f \in H^2$, where P is the orthogonal projection from L^2 onto H^2.

THEOREM 4.2. *The Toeplitz operator* T_ϕ *is quasinormal if and only if one of the following holds :*

(1) ϕ *is a linear function of a real–valued function in* L^∞,

(2) ϕ *is a constant multiple of an inner function.*

This result is due to Amemiya, Ito and Wong [1]. Note that condition (1) above completely characterizes normal Toeplitz operators and condition (2) yields multiples of unilateral shifts. Historically, this theorem answers positively for quasinormal operators a question of Halmos : Is every subnormal Toeplitz operator T_ϕ either normal or analytic (that is, with an analytic symbol ϕ) ? Its eventual negative solution is obtained by Cowen and Long [21] (compare also [20] for a survey of this problem).

Let (X, Ω, μ) be a σ–finite measure space and $T : X \to X$ a measurable transformation. The *composition operator* C_T on $L^2(\mu)$ induced by T is, by definition, given by $C_T f = f{\circ}T$ for $f \in L^2(\mu)$. It has long been known [44, p.39] that a necessary and sufficient condition for C_T to be bounded is that $\mu{\circ}T^{-1}$ be absolutely continuous with respect to μ and $h \equiv d(\mu{\circ}T^{-1})/d\mu$ be in $L^\infty(\mu)$. When are C_T and C_T^* quasinormal? The next two theorems from [54] and [37] provide complete answers.

THEOREM 4.3. (1) C_T *is quasinormal if and only if* $h = h{\circ}T$ *a.e.* $[\mu]$.

(2) *If* μ *is a finite measure, then* C_T *is quasinormal if and only if* T *is measure–preserving.*

In particular, it follows from (2) above that C_T is quasinormal if and only if C_T is an isometry at least when μ is finite.

THEOREM 4.4. C_T^* *is quasinormal if and only if*

(1) *for any* A *in* Ω, A $\cap$ supp h *is in the completion of the σ–algebra* $T^{-1}(\Omega)$, *and*

(2) $h = h{\circ}T$ a.e. $[\mu]$ *on* supp h.

5. INVARIANT SUBSPACES

The existence of nontrivial invariant subspaces for normal operators is an easy consequence of the spectral theorem. For subnormal operators, this is more difficult to prove ; a subtle analytic approach would be needed [7]. In this respect, as in all others, quasinormal operators are in–between.

THEOREM 5.1. *Any quasinormal operator* T *on a space of dimension greater than 1 has a nontrivial invariant subspace. Moreover, if* T *is not a multiple of the identity operator, then it has a nontrivial hyperinvariant subspace.*

The proof for the first part which makes use of the spectral theorem, Fuglede's theorem and the existence of invariant subspace for the simple unilateral shift appears in [36, Problem 196]. Alternatively, it also follows from the second part. As for the proof of the latter, if T is represented as $N \oplus (S \otimes A)$ on $H_1 \oplus H_2$, then $H_1 \oplus \overline{\mathrm{ran}(S{\otimes}A)}$ is a nontrivial hyperinvariant subspace for T since $0 \in \sigma_p((S \otimes A)^*)$ by Theorem 3.1 (2) and there is no nonzero operator X such that $XN = (S \otimes A)X$ by [55, Theorem 2].

A much stronger notion than the mere existence of invariant subspaces for operators is that of reflexivity : T is *reflexive* if Alg Lat T, the algebra of operators leaving invariant every invariant subspace of T, equals Alg T, the weakly closed algebra generated by T and I. Since Alg T is obviously contained in Alg Lat T, the reflexivity of T means that it has so many invariant subspaces as to make Alg Lat T the smallest possible. This notion is first proposed by Sarason [49] who showed that every normal operator is reflexive. The reflexivity of isometries and quasinormal operators are proved by Deddens [24] and Wogen [59], respectively.

THEOREM 5.2. *Every quasinormal operator is reflexive.*

The proof of the reflexivity of subnormal operators [45] came shortly; it is much deeper.

A class of operators with a reasonably rich spectral theory is that of decomposable ones. An operator T on H is *decomposable* if for every finite open covering $G_1, \cdots, G_n$ of $\sigma(T)$ there exist spectral maximal subspaces $K_1, \cdots, K_n$ of T such that $\sigma(T|K_j) \subseteq G_j$ for all j and $H = \Sigma^n_{j=1} K_j$. (Recall that an invariant subspace K of T is a *spectral maximal subspace* if it contains every invariant subspace L of T satisfying $\sigma(T|L) \subseteq \sigma(T|K)$.) This notion is first introduced by Foias [30]. It is not difficult to show that normal operators are decomposable. On the other hand, there are subnormal operators which are not decomposable (as for example the simple unilateral shift) and subnormal decomposable operators which are not normal [48, Corollary 1]. Can the latter operators be quasinormal? The next Theorem provides a negative answer as would be expected. It appeared in [8].

THEOREM 5.3. *A quasinormal decomposable operator must be normal.*

Another property of operators which is closely related to the invariant subspace problem and stirred up many research activities in the 1970s is that of quasitriangularity. According to one of its equivalent definitions, an operator T is *quasitriangular* if there exists an increasing sequence of finite–rank projections $\{P_n\}$ such that P_n approaches to I in the strong operator topology and $AP_n - P_n AP_n$ approaches to 0 in norm [34]. If both T and T^* are quasitriangular, then T is called *biquasitriangular*. Again, it is easy to show that normal operators are biquasitriangular and the simple unilateral shift is not (cf. [34]). The next theorem characterizes (bi)quasitriangularity among pure quasinormal operators.

THEOREM 5.4. *A pure quasinormal operator* $S \otimes A$ *is (bi)quasitriangular if and only if* A *satisfies* $\sigma(A) = [0, \|A\|]$.

This result appeared in [57, Corollary 2.2]; it is proved via the observation that a hyponormal operator is (bi)quasitriangular if and only if its spectrum and approximate point spectrum coincide [56, Theorem 3.1], and the descriptions of these spectra for pure quasinormal operators (Theorem 3.1).

We conclude this section with the notion of cellular–indecomposability first proposed by Olin and Thomson [46]. An operator T is *cellular–indecomposable* if any two of its nonzero invariant subspaces have nonzero intersection. One such operator which comes to mind immediately is the simple unilateral shift [36, Corollary 2 to Problem 157]. It turns out that, among quasinormal operators, multiples of the simple unilateral shift are the only ones having this property.

PROPOSITION 5.5. *A quasinormal operator is cellular–indecomposable if and only if it is a multiple of the simple unilateral shift.*

This is easy to prove if we note that for a pure quasinormal operator $T = S \otimes A$, every spectral subspace of $T^*T = A^2 \oplus A^2 \oplus \cdots$ reduces T.

6. COMMUTANT

In this section, we first determine the three operator algebras associated with a pure quasinormal operator T: $\{T\}'$, the commutant, $\{T\}''$, the double commutant, and Alg T, the weakly closed algebra generated by T and I. (Recall that $\{T\}' = \{X: XT = TX\}$ and $\{T\}'' = \{Y: YX = XY$ for any X in $\{T\}'\}$.) The commutant is the easiest to determine (cf. [19, Lemma 3.1]).

PROPOSITION 6.1. *Let* $T = S \otimes A$ *be a pure quasinormal operator on* $H^{(\infty)} = H \oplus H \oplus \cdots$. *Then an operator* $D = [D_{ij}]_{i,j=0}^{\infty}$ *on* $H^{(\infty)}$ *commutes with* T *if and only if* $D_{ij} = 0$ *for any* $j > i$ *and* $AD_{ij} = D_{i+1\,j+1}A$ *for* $i \geq j$.

As for $\{T\}''$ and Alg T, their characterizations lie deeper. Recall that the simple unilateral shift S satisfies $\{S\}' = \{S\}'' = $ Alg $S = \{\phi(S) : \phi \in H^{\infty}\}$ (cf. [36, Problems 147 and 148]). That the commutant and double commutant cannot equal for general (higher–multiplicity) unilateral shifts is obvious. The next theorem says that the remaining equalities (with a slight modification) still hold for any pure quasinormal operator.

THEOREM 6.2. *For any pure quasinormal operator* $T = S \otimes A$, *the equalities* $\{T\}'' = $ Alg $T = \{\phi(T) : \phi \in H_r^{\infty}\}$ *hold, where* $r = \|T\|$ *and* H_r^{∞} *denotes the Banach algebra of bounded analytic functions on* $\{z \in \mathbb{C} : |z| < r\}$.

Thus, in particular, operators in $\{T\}'' = $ Alg T are of the form

$$\begin{bmatrix} \alpha_0 I & 0 & \cdot \\ \alpha_1 A & \alpha_0 I & 0 & \cdot \\ \alpha_2 A^2 & \alpha_1 A & \alpha_0 I & \cdot \\ & & \cdot & \cdot & \cdot \\ & & & \cdot & \cdot \end{bmatrix},$$

where the α_n's are the Fourier coefficients of a function $\phi(z) = \Sigma_{n=0}^{\infty} \alpha_n z^n$ in H_r^∞. These results were proved in [19]. For nonpure quasinormal operators, the *double commutant property* ($\{T\}'' = \text{Alg } T$) does not hold in general. Actually, this is already the case for normal operators; the bilateral shift U on L^2 of the unit circle is such that $\{U\}'' = \{\psi(U) : \psi \in L^\infty\}$ and Alg $U = \{\phi(U) : \phi \in H^\infty\}$. A complete characterization of quasinormal operators satisfying the double commutant property is given in [19, Theorem 4.10]. The conditions are too technical to be repeated here. We content ourselves with the following special case which was proved earlier in [52].

PROPOSITION 6.3. *Any nonunitary isometry has the double commutant property.*

We next consider the commutant lifting problem: If T is a quasinormal operator on H with minimal normal extension N, when is an operator in $\{T\}'$ the restriction to H of some operator in $\{N\}'$? That this is not always the case can be seen from the following example.

Let $T = S \otimes A$, where $A = \begin{bmatrix} 1 & 0 \\ 0 & 2 \end{bmatrix}$, and $X = \text{diag } (B, ABA^{-1}, A^2BA^{-2}, \cdots)$, where $B = \begin{bmatrix} 1 & 1 \\ 0 & 1 \end{bmatrix}$. Since $A^nBA^{-n} = \begin{bmatrix} 1 & (1/2)^n \\ 0 & 1 \end{bmatrix}$ for $n \geq 0$, X is indeed a bounded operator. That X belongs to $\{T\}'$ follows from Proposition 6.1. A simple computation shows that if X can be lifted to an operator Y in the commutant of the minimal normal extension

$$N = \begin{bmatrix} \cdot & \cdot & \cdot \\ & \cdot & \cdot & \cdot & 0 \\ & & \cdot & A & 0 \\ & & & A & 0 & \cdot & \cdot \\ & & & & & \cdot & \cdot \end{bmatrix}$$

of T, then Y must be of the form diag $(\cdots, A^{-2}BA^2, A^{-1}BA, B, ABA^{-1}, A^2BA^{-2}, \cdots)$. However, as

$$A^{-n}BA^n = \begin{bmatrix} 1 & 2^n \\ 0 & 1 \end{bmatrix}$$

for $n \geq 0$, this operator cannot be bounded. This shows that X cannot be lifted to $\{N\}'$. Note that in this example T is even a pure quasinormal operator with multiplicity 2.

A complete characterization of operators in $\{T\}'$ which can be lifted to $\{N\}'$ is obtained by Yoshino [62, Theorem 4].

THEOREM 6.4. *Let* T *be a quasinormal operator with minimal normal extension* N *and polar decomposition* T = UP. *Then* X $\in \{T\}'$ *can be lifted to* Y $\in \{N\}'$ *if and only if* X *commutes with* U *and* P. *Moreover, if this is the case, then* Y *is unique and* $\|Y\| = \|X\|$.

In particular, if T is an isometry, then operators in $\{T\}'$ can always be lifted [27, Corollary 5.1]. These results are subsumed under Bram's characterization of commutant lifting for subnormal operators [3, Theorem 7].

Another version of the lifting problem asks whether two commuting quasinormal operators have commuting (not necessarily minimal) normal extensions. An example of Lubin [43] provides a negative answer. Indeed, the two quasinormal operators T_1 and T_2 he constructed are such that both are unitarily equivalent to S $\oplus$ 0, where 0 denotes the zero operator on an infinite–dimensional space, $T_1 T_2 = T_2 T_1 = 0$ and $T_1 + T_2$ is not hyponormal. Again, a complete characterization in terms of the polar decomposition is given in [62, Theorem 5].

THEOREM 6.5. *Let* T_1 *and* T_2 *be commuting quasinormal operators with polar decompositions* $T_1 = U_1 P_1$ *and* $T_2 = U_2 P_2$. *Then* T_1 *and* T_2 *have commuting normal extensions if and only if* U_1 *and* P_1 *both commute with* U_2 *and* P_2.

In this connection, we digress to discuss another topic which may shed some light on the commutant lifting problem. As is well–known, every subnormal operator has a unique minimal normal extension [36, Problem 197]. That it also has a unique minimal quasinormal extension seems to be not so widely known. This fact is due to Embry–Wardrop [28, Theorems 2 and 3].

THEOREM 6.6. *Let* T *be a subnormal operator with minimal normal extension* N *on* H. *If* K = $\vee \{\Sigma_{j=0}^{n}(N^*N)^j x_j : x_j \in$ H, $n \geq 0\}$, *then* N|K *is a minimal quasinormal extension of* T *and any minimal quasinormal extension of* T *is unitarily equivalent to* N|K. *Moreover,* N *is also the minimal normal extension of* N|K.

Thus, in particular, the lifting of the commutant for subnormal operators can be accomplished in two stages: first lifting to the commutant of the minimal quasinormal extension and then the minimal normal extension. Studies of other properties of subnormal operators along this line seem promising but lacking.

A problem which might be of interest is to determine which subnormal operator has a pure quasinormal extension. As observed by Conway and Wogen [58, p.169], subnormal unilateral weighted shifts do have this property.

We conclude this section with properties of a class of operators considered by Williams [57, Section 3]. A result which is of interest and not too difficult to prove is the following.

THEOREM 6.7. *If* T *is a quasinormal operator,* N *is normal and* TN = NT, *then* T + N *is subnormal.*

Starting from this, he went on to consider operators of the form T + N, where T is pure quasinormal and N is a normal operator commuting with T. It turns out that such operators have a fairly simple structure. If we express T as $S \otimes A$ on $H \oplus H \oplus \cdots$ and use Proposition 6.1, we can show that N must be of the form $N_0 \oplus N_0 \oplus \cdots$. An easy consequence of this is

THEOREM 6.8. *If* T *is a pure quasinormal operator,* $N \neq 0$ *is normal and* TN = NT, *then* T + N *is not quasinormal.*

For other properties of such operators, the reader is referred to [57].

7. SIMILARITY

In this section and the next two, we will consider how two quasinormal operators are related through similarity, quasisimilarity and compact perturbation. We start with similarity.

For over a decade, the problem whether two similar quasinormal operators are actually unitarily equivalent remains open [41]. This is recently solved in the negative in [12]. In fact, a complete characterization is given for the similarity of two quasinormal operators. Note that the similarity of two normal operators or two isometries implies their unitary equivalence (even the weaker quasisimilarity will do). For normal operators, this is a consequence of the Fuglede–Putnam theorem [36, Corollary to Problem 192]; the case for isometries is proved in [40, Theorem 3.1]. On the other hand, there are similar subnormal operators which are not unitarily equivalent [36, Problem 199]. Against this background, the result on quasinormal operators should have more than a passing interest.

THEOREM 7.1. *For* $j = 1, 2$, *let* $T_j = N_j \oplus (S \otimes A_j)$ *be a quasinormal operator, where* N_j *is normal and* A_j *is positive definite. Then* T_1 *is similar to* T_2 *if and only if* N_1 *is unitarily equivalent to* N_2, $\sigma(A_1) = \sigma(A_2)$ *and* $\dim \ker (A_1 - \lambda I) = \dim \ker (A_2 - \lambda I)$ *for any* λ *in* $\sigma(A_1)$.

Thus, in particular, similarity of quasinormal operators ignores the multiplicity of the operator A_j in the pure part except those of its eigenvalues. From this observation, examples of similar but not unitarily equivalent quasinormal operators can be easily constructed. One such pair is $T_1 = S \otimes A$ and $T_2 = S \otimes (A \oplus A)$, where A is the operator of multiplication by t on $L^2[0,1]$.

As for the proof, we may first reduce our consideration to pure quasinormal operators by a result of Conway [16, Proposition 2.6]: Two subnormal operators are similar if and only if their normal parts are unitarily equivalent and their pure parts are similar. For the pure ones, the proof depends on a deep theorem in the nest algebra theory. Here is how it goes.

Recall that a collection $\mathcal{N}$ of (closed) subspace of a fixed Hilbert space H is a *nest* if (1) $\{0\}$ and H belong to $\mathcal{N}$, (2) any two subspaces M and N in $\mathcal{N}$ are comparable, that is, either $M \subseteq N$ or $N \subseteq M$, and (3) the span and intersection of any family of subspaces in $\mathcal{N}$ are still in $\mathcal{N}$. For any nest $\mathcal{N}$, there is associated a weakly closed algebra, Alg $\mathcal{N}$, consisting of all operators leaving invariant every subspace in $\mathcal{N}$; Alg $\mathcal{N}$ is called the *nest algebra* of $\mathcal{N}$. The study of nest algebra is initiated by J.R.Ringrose in the 1960s. Since then, it has attracted many researchers. A certain maturity is finally reached in recent years. The monograph [23] has a comprehensive coverage of the subject. Before stating the Similarity Theorem which we are going to invoke, we need some more terminology of the theory. A nest $\mathcal{N}$ is *continuous* if every element N in $\mathcal{N}$ equals its immediate predecessor $N \equiv \vee \{N' \in \mathcal{N}: N' \subsetneq N\}$. Two nests $\mathcal{N}$ and $\mathcal{M}$ on spaces H_1 and H_2 are *similar* if there is an invertible operator X from H_1 onto H_2 such that $X\mathcal{N} = \mathcal{M}$. A major breakthrough in the development of the theory is the proof by Larson [42] that any two continuous nests are similar. This is generalized later by Davidson [22] to the similarity of any two nests: $\mathcal{N}$ and $\mathcal{M}$ are similar if and only if there is an order–preserving isomorphism θ from $\mathcal{N}$ onto $\mathcal{M}$ such that for any subspaces N_1 and N_2 in $\mathcal{N}$ with $N_1 \subseteq N_2$ the dimensions of $N_2 \ominus N_1$ and $\theta(N_2) \ominus \theta(N_1)$ are equal. In particular, this says that the similarity of nests depends on the order and the dimensions (of the atoms) of the involved nests but not on their multiplicity. (A multiplicity theory of nests can be developed via the abelian von Neumann algebra generated by the orthogonal projections onto the subspaces in the nest.) This may explain why the Similarity Theorem has some bearing on our result. Its proof is quite intricate.

Before embarking on the proof of our result, we need a link relating pure quasinormal operators to nest algebras so that the Similarity Theorem can be applied. For any positive definite operator A on H, there is associated a natural nest $\mathcal{N}_A$, the one generated by all subspaces of the form $E_A([0,t])H$, $t \geq 0$, where $E_A(\cdot)$ denotes the spectral measure of A. The result we need is due to Deddens [25]. It says that the nest algebra Alg $\mathcal{N}_A$ consists exactly of operators T satisfying $\sup_{n \geq 0} \|A^n T A^{-n}\| < \infty$. Now we are ready to sketch the proof of Theorem 7.1.

If A_1 and A_2 are positive definite operators on H_1 and H_2 satisfying $\sigma(A_1) = \sigma(A_2)$ and dim ker $(A_1 - \lambda I) =$ dim ker $(A_2 - \lambda I)$ for λ in $\sigma(A_1)$, then define the order–preserving isomorphism θ from $\mathcal{N}_{A_1}$ to $\mathcal{N}_{A_2}$ by

$$\theta\left(E_{A_1}[0,\lambda]H_1\right) = E_{A_2}[0,\lambda]H_2 \quad \text{if } \lambda \in \sigma(A_1)$$

and

$$\theta\left(E_{A_1}[0,\lambda)H_1\right) = E_{A_2}[0,\lambda)H_2 \text{ if } \lambda \text{ is an eigenvlaue of } A_1.$$

Our assumption guarantees that θ is dimension–preserving. Thus it is implemented by an invertible operator X by the Similarity Theorem. Letting

$$A = \begin{bmatrix} A_1 & 0 \\ 0 & A_2 \end{bmatrix} \qquad \text{and} \quad Y = \begin{bmatrix} 0 & X^{-1} \\ X & 0 \end{bmatrix},$$

we have $Y \in$ Alg $\mathcal{N}_A$. Therefore, Deddens' result implies that $\sup_{n \geq 0} \|A^n Y A^{-n}\| < \infty$ or, in other words, $\sup \|A_2^n X A_1^{-n}\| < \infty$ and $\sup \|A_1^n X^{-1} A_2^{-n}\| < \infty$. Thus $Z = \text{diag}(X, A_2 X A_1^{-1}, A_2^2 X A_1^{-2}, \cdots)$ is an invertible operator satisfying $Z(S \otimes A_1) = (S \otimes A_2)Z$. This shows that $S \otimes A_1$ and $S \otimes A_2$ are similar. The converse can be proved essentially by a reversal of the above arguments.

8. QUASISIMILARITY

Two operators T_1 and T_2 are *quasisimilar* if there are operators X and Y which are injective and have dense range such that $XT_1 = T_2X$ and $YT_2 = T_1Y$. In this section, we will address the problem when two quasinormal operators are quasisimilar. As we will see, this problem is much more complicated than the similarity problem which we discussed in Section 7.

If two quasinormal operators are quasisimilar, then, necessarily, their spectra and essential spectra must be equal to each other. The former is true even for quasisimilar hyponormal operators (cf. [13]), and the latter for subnormal operators (cf. [55, 61]). However, things are not as smooth as we would like them to be. The pure parts of quasisimilar quasinormal operators may not be quasisimilar [55, Example 1] although their normal parts are still unitarily equivalent [16,Proposition 2.3]. Thus, in the case of quasisimilarity, we cannot just consider the pure ones but also have to worry about the "mixing effect" of the normal and pure parts.

A complete characterization of quasisimilar quasinormal operators is given in [12]. We start with the pure ones.

THEOREM 8.1. *Two pure quasinormal operators* $S \otimes A_1$ *and* $S \otimes A_2$ *are quasisimilar if and only if the following conditions hold:*

(1) $m(A_1) = m(A_2)$ *and* dim ker $(A_1 - m(A_1)I)$ = dim ker $(A_2 - m(A_2)I)$,

(2) $\|A_1\|_e = \|A_2\|_e$ *and* dim ker $(A_1 - \lambda I)$ = dim ker $(A_2 - \lambda I)$ *for any* $\lambda > \|A_1\|_e$, *and, in case there are only finitely many points in* $\sigma(A_1) \cap (\|A_1\|_e, \infty)$,

(3) dim ker $(A_1 - \|A_1\|_e I)$ = dim ker $(A_2 - \|A_2\|_e I)$.

Here $m(A_j) = \inf \{\lambda : \lambda \in \sigma(A_j)\}$, $j = 1, 2$.

In particular, this theorem says that for quasisimilar pure quasinormal operators $S \otimes A_1$ and $S \otimes A_2$, the part of the spectrum of A_j in $(m(A_j), \|A_j\|_e)$ can be quite arbitrary. This is the source of examples used to illustrate the nonpreserving of various parts of the spectrum under quasisimilarity (cf. [56, Examples 2.2 and 2.3] and [38, p.1445]). In particular, in view of Theorem 3.1, this is the case for the approximate point spectrum of quasinormal operators. Another consequence of Theorem 8.1 is that every pure quasinormal operator is quasisimilar to an $S \otimes A$ with A a diagonal positive definite operator.

Note that condition (1) (resp. (2) together with (3)) is equivalent to the injective similarity (resp. dense similarity) of $S \otimes A_1$ and $S \otimes A_2$. (Two operators T_1 and T_2 are *injectively* (resp. *densely*) *similar* if there are operators X and Y which are injective (resp. have dense range) such that $XT_1 = T_2X$ and $YT_2 = T_1Y$.)

The proof for the necessity of conditions (1), (2) and (3) is elementary; that for the sufficiency is more intricate. Here is a very brief sketch. First decompose A_j on H_j, $j = 1, 2$, into three parts : $A_j = B_j \oplus C_j \oplus D_j$ so that B_j, C_j and D_j are acting on the spectral subspaces $E_{A_j}\{m(A_j)\}H_j$, $E_{A_j}(m(A_j), \|A_j\|_e] H_j$ and $E_{A_j}(\|A_j\|_e, \|A_j\|]$ H_j, respectively. Correspondingly, we have the decomposition

$$S \otimes A_j = (S \otimes B_j) \oplus (S \otimes C_j) \oplus (S \otimes D_j), \quad j = 1, 2.$$

The proof is accomplished by showing that (a) $(S \otimes B_1) \oplus (S \otimes C_1) \prec S \otimes B_2$ and (b) $S \otimes D_1 \prec (S \otimes C_2) \oplus (S \otimes D_2)$. (Recall that, for any two operators T_1 and T_2, $T_1 \prec T_2$ means that there is an injective operator X with dense range such that $XT_1 = T_2X$.) By our assumption, (a) is the same as $m(A_1)(S \otimes I) \oplus (S \otimes C_1) \prec m(A_1)(S \otimes I)$. The operator $S \otimes C_1$ can be further decomposed as $S \otimes C_1 = \Sigma_n \oplus (S \otimes E_n)$, where E_n acts on the spectral subspace $E_{A_1}(a_n, a_{n-1}]H$ with $a_0 = \|A_1\|_e$ and the sequence $\{a_n\}$ decreasing to $m(A_1)$. Using the observation that $S \otimes A \prec m(A)(S \otimes I)$ for any invertible A, we obtain $S \otimes C_1 \prec \Sigma_n \oplus a_n(S \otimes I)$. Thus the proof of (a) reduces to showing $S \oplus (\Sigma_n \oplus (b_n S)) \prec S$, where $b_n = a_n/m(A_1) > 1$. This is established through modifying the proof of a result of Sz.–Nagy and Foias [51] that $\alpha S^{(n)} \prec S$ for any α, $|\alpha| > 1$, and n, $1 \leq n \leq \infty$. On the other hand, following our assumptions, (b) is the same as $S \otimes D_2 \prec (S \otimes C_2) \oplus (S \otimes D_2)$. The proof, based on the fact that $\|C_2\| \leq m(D_2)$, is easier (cf. [12, Lemma 3.14 (a)]).

We next turn to the quasisimilarity of general quasinormal operators. The following theorem gives a complete characterization.

THEOREM 8.2. *For j = 1, 2, let* $T_j = N_j \oplus (S \otimes A_j)$ *be a quasinormal* *operator. Let $\alpha = \min\{m(A_1), m(A_2)\}$ and $d = \max\{\dim \ker (A_1 - \alpha I), \dim \ker (A_2 - \alpha I)\}$. Then T_1 is quasisimilar to T_2 if and only if N_1 is unitarily equivalent to N_2, $S \otimes A_1$ is densely similar to $S \otimes A_2$ and one of the following holds:*

    (1) *$S \otimes A_1$ is quasisimilar to $S \otimes A_2$;*

    (2) *$d = 0$ and $\sigma(N_1)$ has a limit point in the disc $\{z \in \mathbb{C} : |z| \le \alpha\}$;*

    (3) *$d > 0$ and the absolutely continuous unitary part of N_1/α does not*

vanish;

    (4) *$d > 0$ and the completely nonunitary part of N_1/α is not of class C_0.*

Some explanations for the terminology used above are in order. Any normal operator M on H can be decomposed as $M = M_1 \oplus M_2 \oplus M_3$, where M_1, M_2 and M_3 act on $E_M(\mathbb{D})H$, $E_M(\partial\mathbb{D})H$ and $E_M(\mathbb{C}\backslash\overline{\mathbb{D}})H$, respectively ($\mathbb{D}$ is the open unit disc on the plane). M_1, being a completely nonunitary contraction, is called the *completely nonunitary part* of M. The unitary M_2 can be further decomposed as the direct sum of an absolutely continuous unitary operator and a singular unitary operator. These are the parts referred to in conditions (3) and (4) in the above theorem. A completely nonunitary contraction T is of *class C_0* if $\phi(T) = 0$ for some $\phi \in H^\infty$. (For properties of such operators, the reader is referred to [50].)

The proof of the sufficiency of the conditions in Theorem 8.2 involves a great deal of function–theoretic arguments. For simplicity, we will present one typical example for each of the conditions (2), (3) and (4) followed by a one–sentence sketch of its proof which somehow gives the general flavor of the arguments.

EXAMPLE 8.3. *If N is the diagonal operator $\mathrm{diag}(d_n)$ on l^2, where $\{d_n\}$ is a sequence satisfying $0 < |d_n| \le c < 1$ for all n and converging to 0, then $S \oplus N \prec N$.*

The operator $X : H^2 \oplus l^2 \to l^2$ defined by

$$X(f \oplus \{a_n\}) = \{c^n(f(d_n) + a_n \exp(-1/|d_n|))\}$$

can be shown to be injective, with dense range and satisfying $X(S \oplus N) = NX$.

EXAMPLE 8.4. *If N is the operator of multiplication by e^{it} on $L^2(E)$, where E is a Borel subset of the unit circle, then $S \oplus N \prec N$.*

The operator $X : H^2 \oplus L^2(E) \to L^2(E)$ required is defined by

$$X(f \oplus g) = (f|E) + \phi g,$$

where ϕ is a function in $L^\infty(E)$ such that $\phi \neq 0$ a.e. on E and $\int_E \log|\phi| = -\infty$.

EXAMPLE 8.5. *If N is the diagonal operator* $\mathrm{diag}(d_n)$ *on* l^2, *where* $\{d_n\}$ *is a sequence of points in the open unit disc accumulating only at the unit circle and satisfying* $\Sigma_n(1-|d_n|) = \infty$, *then* $S \oplus N \prec N$.

The proof for this case is the most difficult one. The operator $X : H^2 \oplus l^2 \to l^2$ defined by

$$X(f \oplus \{a_n\}) = \{f(d_n)(1-|d_n|^2)^{\frac{1}{2}}/n + a_n b_n \exp(-1/(1-|d_n|)^2)\},$$

where $\{b_n\}$ is a bounded sequence of positive numbers satisfying $\limsup_n |B(d_n)|/nb_n \geq 1$ for any Blaschke product B (the existence of $\{b_n\}$ is proved in [12, Lemma 4.8]), will meet all the requirements. The difficulty lies in showing the injectivity of X.

9. COMPACT PERTURBATION

Two operators T_1 and T_2 are *approximately equivalent* (donoted by $T_1 \overset{a}{\cong} T_2$) if there is a sequence of unitary operators $\{U_n\}$ such that $\|U_n^* T_1 U_n - T_2\| \to 0$; they are *approximately similar* (donoted by $T_1 \overset{a}{\approx} T_2$) if there are invertible operators X_n such that $\sup \{\|X_n\|, \|X_n^{-1}\|\} < \infty$ and $\|X_n^{-1} T_1 X_n - T_2\| \to 0$. Using Berg's perturbation theorem [2], Gellar and Page [31] proved that two normal operators T_1 and T_2 are approximately equivalent if and only if $\sigma(T_1) = \sigma(T_2)$ and $\dim \ker (T_1 - \lambda I) = \dim \ker (T_2 - \lambda I)$ for any isolated point λ in $\sigma(T_1)$. This is later extended to isometries by Halmos [35]:Two isometries T_1 and T_2 are approximately equivalent if and only if either both are unitary and are approximately equivalent or their pure parts are unitarily equivalent. The corresponding problem for quasinormal operators was considered by Hadwin in his 1975 Ph.D. dissertation [32]. Using the notion of operator–valued spectrum, he obtained necessary and sufficient conditions for two quasinormal operators to be approximately equivalent. Recently, this result is reproved by Chen [11, Theorem

2.1] using more down–to–earth operator–theoretic techniques.

THEOREM 9.1. *For* $j = 1, 2$, *let* $T_j = N_j \oplus (S \otimes A_j)$ *be a quasinormal operator. Then the following statements are equivalent:*

(1) $T_1 \overset{a}{\cong} T_2$;

(2) $T_1 \overset{a}{\approx} T_2$;

(3) $A_1 \overset{a}{\cong} A_2$, $\sigma(N_1)\backslash\sigma_{ap}(S \otimes A_1) = \sigma(N_2)\backslash\sigma_{ap}(S \otimes A_2)$ *and* dim ker$(N_1 - \lambda I) = $ dim ker$(N_2 - \lambda I)$ *for any isolated point* λ *in* $\sigma(N_1)\backslash\sigma_{ap}(S \otimes A_1)$.

The basic tool for the proof is a theorem of Pearcy and Salinas [47, Theorem 1] that if N is a normal operator, T is hyponormal and $\sigma(N) \subseteq \sigma_{le}(T)$, then $N \oplus T \overset{a}{\cong} T$.

Note that approximately equivalent operators are compact perturbations of each other; this is because that if $T_1 \overset{a}{\cong} T_2$ then unitary operators U_n may be chosen such that not only $U_n^* T_1 U_n - T_2$ approach to zero in norm but are compact for all n (cf. [53]). Thus the following definitions are indeed weaker: T_1 and T_2 are *equivalent modulo compact* (resp. *similar modulo compact*) if there is a unitary U (resp. invertible X) such that $U^* T_1 U - T_2$ (resp. $X^{-1} T_1 X - T_2$) is compact. We denote this by $T_1 \overset{k}{\cong} T_2$ (resp. $T_1 \overset{k}{\approx} T_2$). The classical Weyl–von Neumann–Berg theorem implies that for normal operators T_1 and T_2, both $T_1 \overset{k}{\cong} T_2$ and $T_1 \overset{k}{\approx} T_2$ are equivalent to $\sigma_e(T_1) = \sigma_e(T_2)$. There is an analogous result for isometries [11, Proposition 2.8]. As for quasinormal operators, a complete characterization for the pure ones is known, but not for the general case. The following two theorems appeared in [11].

THEOREM 9.2. *For* $j = 1, 2$, *let* $T_j = S \otimes A_j$ *be a pure quasinormal operator. Then the following statements are equivalent:*

(1) $T_1 \overset{k}{\cong} T_2$;

(2) $T_1 \overset{k}{\approx} T_2$;

(3) $A_1 \overset{a}{\cong} A_2$.

THEOREM 9.3. *For* $j = 1, 2$, *let* $T_j = N_j \oplus (S \otimes A_j)$ *be a quasinormal operator.* *If* $T_1 \overset{k}{\approx} T_2$, *Then* $\sigma_e(A_1)\backslash\{0\} = \sigma_e(A_2)\backslash\{0\}$.

That the conclusion of the preceding theorem cannot be strengthened to $\sigma_e(A_1) = \sigma_e(A_2)$ can be seen by letting $T_1 = \Sigma_n \oplus (N/n)$ and $T_2 = T_1 \oplus (S \otimes A)$, where N is a normal operator with $\sigma(N) = \mathbb{D}$ and $A = \mathrm{diag}(1, \frac{1}{2}, \frac{1}{3}, \cdots)$(that $T_1 \overset{k}{\cong} T_2$ follows from the Brown–Douglas–Fillmore theory [6]).   There are, of course, the usual Fredholm conditions for two operators to be equivalent (similar) modulo compact.   Thus a necessary and sufficient condition in order that two quasinormal operators $T_j = N_j \oplus (S \otimes A_j)$, $j = 1, 2$, with at least one A_j compact be equivalent (similar) modulo compact can be formulated. In particular, we obtain

PROPOSITION 9.4.   *No pure quasinormal operator is similar modulo compact to a normal operator.*

This result is first noted in [57, p.313].

There is another notion which is weaker than approximate equivalence. Two operators T_1 and T_2 are *algebraically equivalent* if there is a *–isomorphism from $C^*(T_1)$ onto $C^*(T_2)$ which maps T_1 to T_2, where $C^*(T_j)$, $j = 1, 2$, denotes the C^*–algebra generated by T and I.   That this is indeed weaker is proved in [33, Corollary 3.7].   If the *–isomorphism above is required to preserve rank, then this yields approximate equivalence.   By the Gelfand theory, we easily obtain that two normal operators are algebraically equivalent if and only if they have equal spectra.   A necessary and sufficient condition for the algebraic equivalence of isometrics is obtained by Coburn [14].   The next theorem from [11, Theorem 3.6] treats the quasinormal case.

THEOREM 9.5.   *Two quasinormal operators* $N_1 \oplus (S \otimes A_1)$ *and* $N_2 \oplus (S \otimes A_2)$ *are algebraically equivalent if and only if* $\sigma(A_1) = \sigma(A_2)$ *and* $\sigma(N_1)\backslash\sigma_{ap}(S \otimes A_1) = \sigma(N_2)\backslash\sigma_{ap}(S \otimes A_2)$.

10. OPEN PROBLEMS

So, after all these discussions, what is the future in store for quasinormal operators? What are the research problems worthy of pursuing for them? One place to look for the answers is probably among isometries. There are problems which are solved for this subclass but never considered for general quasinormal operators. Here we propose three such problems as starters. More of them are waiting to be discovered and solved if the theory is to reach a respectable level. Along the way, if some unexpected link is established with other parts of operator theory or even other areas of research in mathematics, then so much the better.

Our first problem concerns the multiplicity. In Proposition 3.2, it was proved that the multiplicity of a pure quasinormal operator $S \otimes A$ equals the dimension of the space on which A acts. Will putting back the normal part still yield a simple formula for the multiplicity? For isometries, this is solved completely in [60].

The second one concerns the hyperinvariant subspaces of quasinormal operators. Their existence is guaranteed by Theorem 5.1. Is there a simple way to describe all of them? This problem does not seem to have been touched upon before even for pure ones. Playing around with some special case such as $S \otimes A$ with $A = \begin{bmatrix} a & 0 \\ 0 & b \end{bmatrix}$, $a > b > 0$, may lead to some idea on what should be expected in general. This was done recently by K.–Y. Chen. Further progress would be expected in the future. The case with isometries is known (cf. [26]).

Finally, as discussed in Section 9, the problem when two quasinormal operators are compact perturbations of each other has not been completely solved yet. Bypassing it, we may ask the problem of trace–class perturbation, that is, when two quasinormal operators T_1 and T_2 are such that $U^* T_1 U - T_2$ is of trace class for some unitary U. In this case, the answer does not seem to be known completely even for isometries and normal operators (cf. [9, 10]). How about finite–rank perturbations or even rank–one perturbations?

All these problems are crying out for answers. Hopefully, their solutions will lead to a better understanding of the structure of the underrated quasinormal operators.

REFERENCES

[1]   I. Amemiya, T. Ito and T. K. Wong, On quasinormal Toeplitz operators, Proc. Amer. Math. Soc. 50 (1975), 254–258.

[2]   I. D. Berg, An extension of the Weyl–von Neumann theorem to normal operators, Trans. Amer. Math. Soc. 160 (1971), 365–371.

[3]   J. Bram, Subnormal operators, Duke Math. J. 22 (1955), 75–94.

[4]   A. Brown, On a class of operators, Proc. Amer. Math. Soc. 4 (1953), 723–728.

[5]   A. Brown and C. Pearcy, Spectra of tensor products of operators, Proc. Amer. Math Soc. 17 (1966), 162–166.

[6]   L. G. Brown, R. G. Douglas and P. A. Fillmore, Unitary equivalence modulo the compact operators and extensions of C^*–algebras, Proceedings of a conference on operator theory, Springer–Verlag, Berlin, 1973, pp. 58–128.

[7]   S. W. Brown, Some invariant subspaces for subnormal operators, Integral Equations Operator Theory 1 (1978), 310–333.

[8]   J. Z. Cao, A decomposable quasinormal operator is a normal operator, Chinese Ann. Math. Ser. A 8 (1987), 580–583.

[9]   R. W. Carey, Trace class perturbations of isometries and unitary dilations, Proc. Amer. Math. Soc. 45 (1974), 229–234.

[10]   R. W. Carey and J. D. Pincus, Unitary equivalence modulo the trace class for self–adjoint operators, Amer. J. Math. 98 (1976), 481–514.

[11]   K.–Y. Chen, Compact perturbation and algebraic equivalence of quasinormal operators, preprint.

[12]   K.–Y. Chen, D. A. Herrero and P. Y. Wu, Similarity and quasisimilarity of quasinormal operators, J. Operator Theory, to appear.

[13]   S. Clary, Equality of spectra of quasi–similar hyponormal operators, Proc. Amer. Math. Soc. 53 (1975), 88–90.

[14]   L. A. Coburn, The C^*–algebra generated by an isometry, Bull. Amer. Math. Soc. 73 (1967), 722–726.

[15]   I. Colojoara and C. Foias, Theory of generalized spectral operators, Gordon and Breach, New York, 1968.

[16]   J. B. Conway, On quasisimilarity for subnormal operators, Illinois J. Math. 24 (1980), 689–702.

[17]   J. B. Conway, Subnormal operators, Pitman, Boston, 1981.

[18]   J. B. Conway, The theory of subnormal operators, Amer. Math. Soc., Providence, 1991.

[19]   J. B. Conway and P. Y. Wu, The structure of quasinormal operators and the double commutant property, Trans. Amer. Math. Soc. 270 (1982), 641–657.

[20]   C. C. Cowen, Hyponormal and subnormal Toeplitz operators, Surveys of some recent results in operator theory, Vol. I, Longman, Harlow, Essex, 1988, pp. 155–167.

[21]   C. C. Cowen and J. J. Long, Some subnormal Toeplitz operators, J. reine angew. Math. 351 (1984), 216–220.

[22]   K. R. Davidson, Similarity and compact perturbations of nest algebras, J. reine angew. Math. 348 (1984), 286–294.

[23]   K. R. Davidson, Nest algebras, Longman, Harlow, Essex, 1988.

[24]   J. A. Deddens, Every isometry is reflexive, Proc. Amer. Math. Soc. 28 (1971), 509–512.

[25]   J. A. Deddens, Another description of nest algebras, Hilbert space operators, Springer–Verlag, Berlin, 1978, pp. 77–86.

[26]   R. G. Douglas, On the hyperinvariant subspaces for isometries, Math. Z. 107 (1968), 297–300.

[27]   R. G. Douglas, On the operator equation $S^{*}XT = X$ and related topics, Acta Sci. Math. (Szeged) 30 (1969), 19–32.

[28]   M. Embry–Wardrop, Quasinormal extensions of subnormal operators, Houston J. Math. 7 (1981), 191–204.

[29]   I. Erdelyi and S. Wang, A local spectral theory for closed operators, Cambridge Univ. Press, Cambridge, 1985.

[30]   C. Foias, Spectral maximal spaces and decomposable operators in Banach space, Arch. Math. 14 (1963), 341–349.

[31]   R. Gellar and L. Page, Limits of unitarily equivalent normal operators, Duke Math. J. 41 (1974), 319–322.

[32]   D. W. Hadwin, Closures of unitary equivalence classes, Ph. D. dissertation, Indiana Univ., 1975.

[33]   D. W. Hadwin, An operator–valued spectrum, Indiana Univ. Math. J. 26 (1977), 329–340.

[34]   P. R. Halmos, Quasitriangular operators, Acta Sci. Math. (Szeged) 29 (1968), 283–293.

[35]   P. R. Halmos, Limits of shifts, Acta Sci. Math. (Szeged) 34 (1973), 131–139.

[36]   P. R. Halmos, A Hilbert space problem book, 2nd ed., Springer–Verlag, New York, 1982.

[37]   D. J. Harrington and R. Whitley, Seminormal composition operators, J. Operator Theory 11 (1984), 125–135.

[38]   D. A. Herrero, On the essential spectra of quasisimilar operators, Can. J. Math. 40 (1988), 1436–1457.

[39]   D. A. Herrero, All (all ?) about triangular operators, preprint.

[40]   T. B. Hoover, Quasi–similarity of operators, Illinois J. Math. 16 (1972), 678–686.

[41]   S. Khasbardar and N. Thakare, Some counter–examples for quasinormal operators and related results, Indian J. Pure Appl. Math. 9 (1978), 1263–1270.

[42]   D. R. Larson, Nest algebras and similarity transformations, Ann. Math. 121 (1985), 409–427.

[43]   A. Lubin, A subnormal semigroup without normal extension, Proc. Amer. Math. Soc. 68 (1978), 176–178.

[44]   E. A. Nordgren, Composition operators on Hilbert spaces, Hilbert space operators, Springer–Verlag, Berlin, 1978, pp. 27–63.

[45]   R. F. Olin and J. E. Thomson, Algebras of subnormal operators, J. Func. Anal. 37 (1980), 271–301.

[46]   R. F. Olin and J. E. Thomson, Cellular–indecomposable subnormal operators, Integral Equations Operator Theory 7 (1984), 392–430.

[47]   C. Pearcy and N. Salinas, Compact perturbations of seminormal operators, Indiana Univ. Math. J. 22 (1973), 789–793.

[48]   M. Radjabalipour, Some decomposable subnormal operators, Rev. Roum. Math. Pures Appl. 22 (1977), 341–345.

[49]   D. Sarason, Invariant subspaces and unstarred operator algebras, Pacific J. Math. 17 (1966), 511–517.

[50]   B. Sz.–Nagy and C. Foias, Harmonic analysis of operators on Hilbert space, North Holland, Amsterdam, 1970.

[51]   B. Sz.–Nagy and C. Foias, Injection of shifts into strict contractions, Linear operators and approximation II, Birkhauser Verlag, Basel, 1974, pp. 29–37.

[52]   T. R. Turner, Double commutants of isometries, Tohoku Math. J. 24 (1972), 547–549.

[53]   D. Voiculescu, A non–commutative Weyl–von Neumann theorem, Rev. Roum. Math. Pures Appl. 21 (1976), 97–113.

[54]   R. Whitley, Normal and quasinormal composition operators, Proc. Amer. Math. Soc. 70 (1978), 114–118.

[55]   L. R. Williams, Equality of essential spectra of quasisimilar quasinormal operators, J. Operator Theory 3 (1980), 57–69.

[56]   L. R. Williams, Quasisimilarity and hyponormal operators, J. Operator Theory 5 (1981), 127–139.

[57]   L. R. Williams, The approximate point spectrum of a pure quasinormal operator, Acta Sci. Math. (Szeged) 49 (1985), 309–320.

[58]   W. R. Wogen, On some operators with cyclic vectors, Indiana Univ. Math. J. 27 (1978), 163–171.

[59]   W. R. Wogen, Quasinormal operators are reflexive, Bull. London Math. Soc. 11 (1979), 19–22.

[60]   P. Y. Wu, Multiplicities of isometries, Integral Equations Operator Theory 7 (1984), 436–439.

[61]   L. Yang, Equality of essential spectra of quasisimilar subnormal operators, Integral Equations Operator Theory 13 (1990), 433–441.

[62]   T. Yoshino, On the commuting extensions of nearly normal operators, Tohoku Math. J. 25 (1973), 263–272.

Department of Mathematics
National Chiao Tung University
Hsinchu, Taiwan
Republic of China
E–mail address: PYWU@TWNCTU01. BITNET

MSC: Primary 47B20

WORKSHOP PROGRAM

Tuesday, June 11, 1991

9:30 **Welcome** by T. Ando

9:35 **Opening address** by I. Gohberg

9:50–10:40 **C. R. Johnson**
Matrix completion problem

11:10–12:00 **H. Langer**
Model and unitary equivalence of simple selfadjoint operators in Pontrjagin spaces

12:10–12:40 **H. Bart**
Matricial coupling revisited

14:00–14:40 **A. Dijksma**
Holomorphic operators between Krein spaces and the number of squares of associated kernels

14:50–15:30 **A. Gheondea**
The negative signature of defect and lifting of operators in Krein spaces

16:00–16:30 **H. J. Woerdeman**
Positive semidefinite, contractive, isometric and unitary completions of operator matrices

16:35–17:05 **J. I. Fujii**
Operator mean and the relative operator entropy

17:15–17:45 **V. Vinnikov**
Commuting nonselfadjoint operators and function theory on a real Riemann surface

17:50–18:20 **E. Kamei**
An application of Furuta's inequality to Ando's theorem

Wednesday, June 12, 1991

9:00– 9:50 **I. Gohberg**
Dichotomy, discrete Bohl exponents, and spectrum of block weighted shifts

10:00–10:40 **M. A. Kaashoek**
Maximum entropy principles for band extensions and Szëgo limit theorems

11:10–12:00   **H. Widom**
Asymptotic expansions and stationary phase for operators with nonsmooth symbol

12:10–12:40   **A. C. M. Ran**
On the equation $X + A^*X^{-1}A = Q$

14:00–14:40   **K. Izuchi**
Interpolating sequences in the maximal ideal space of H^∞

14:50–15:30   **D. Z. Arov**
(j, J)-inner matrix-functions and generalized betangent Caratheodory-Nevanlinna-Pick-Krein problem

16:00–16:30   **S. Takahashi**
Extended interpolation problem for bounded analytic functions

16:35–17:05   **T. Okayasu**
The von Neumann inequality and dilation theorems for contractions

17:15–17:45   **P. Y. Wu**
Similarity and quasisimilarity of quasinormal operators

17:50–18:20   **T. Nakazi**
Hyponormal Toeplitz operators and extremal problems of Hardy spaces

Thursday, June 13, 1991

9:00– 9:50   **A. A. Nudel'man**
Some generalizations of the classical interpolation problems

10:00–10:40   **T. Furuta**
Applications of order preserving operator inequalities

11:10–12:00   **J. Ball**
A survey of interpolation problems for rational matrix functions and connections with H^∞ control theory

13:00   **Excursion**

17:00   **Barbecue party**

Friday, June 14, 1991

9:00– 9:50   **V. M. Adamjan**
Analytic structure of scattering matrices for big integral schemes

10:00–10:40   **H. Dym**
On a new class of reproducing kernel spaces

11:10–12:00	**P. A. Fuhrmann** Model reduction and robust control via LQG balancing
12:10–12:40	**D. Alpay** Some reproducing kernel spaces of analytic functions, sesquilinear forms and a non-hermitian Schur algorithm
14:00–14:40	**R. Mennicken** Expansion of analytic functions in series of Floquet solutions of first order linear differential systems
14:50–15:30	**E. R. Tsekanovskii** Accretive extensions, Stieltjes operator functions and conservative systems
16:00–16:40	**J. W. Helton** A symbol manipulator for aiding with the algebra in linear system theory
16:50–17:30	**L. A. Sakhnovich** Interpolation problems, inverse spectral problems and nonlinear equations
17:40–18:10	**F. Kubo** Museum for Selberg inequality
18:10	**Closing remarks** by **T. Ando** and **I. Gohberg**

LIST OF PARTICIPANTS

Adamyan, Vadim M., Odessa University, Odessa, UKRAINE

Alpay, Daniel, Weizmann Institute of Science, Rehovot, ISRAEL

Ando, T., Hokkaido University, Sapporo, JAPAN

Arov, D. Z., Odessa State Pedagogical Institute, Odessa, UKRAINE

Ball, Joseph A., Virginia Polytechnic Institute and State University, Blacksburg, U.S.A.

Bart, H., Erasmus University, Rotterdam, THE NETHERLANDS

Chew, T. S., National University of Singapore, SINGAPORE

Dijksma, A., University of Groningen, Groningen, THE NETHERLANDS

Dym, Harry, Weizmann Institute of Science, Rehovot, ISRAEL

Fuhrmann, Paul A., Ben Gurion University, Beer Sheva, ISRAEL

Fujii, Jun Ichi, Osaka Kyoiku University, Kashiwara, JAPAN

Fujii, Masatoshi, Osaka Kyoiku University, Osaka, JAPAN

Furuta, Takayuki, Science University of Tokyo, Tokyo, JAPAN

Gheondea, Aurelian, Mathematics Institute of Romanian Academy, Bucharest, ROMANIA

Gohberg, Israel, Tel Aviv University, Ramat-Aviv, ISRAEL

Hayashi, Mikihiro, Hokkaido University, Sapporo, JAPAN

Helton, J. William, University of California, La Jolla, U.S.A.

Hiai, Fumio, Ibaraki University, Mito, JAPAN

Inoue, Junji, Hokkaido University, Sapporo, JAPAN

Ishikawa, Hiroshi, Ryukyu University, Okinawa, JAPAN

Ito, Takashi, Musashi Institute of Technology, Tokyo, JAPAN

Izuchi, Keiji, Kanagawa University, Yokohama, JAPAN

Izumino, Saichi, Toyama University, Toyama, JAPAN

Johnson, Charles R., College of William and Mary, Williamsburg, U.S.A.

Kaashoek, M. A., Vrije Universiteit, Amsterdam, THE NETHERLANDS

Kamei, Eizaburo, Momodani Senior Highschool, Osaka, JAPAN

Katsumata, Osamu, Hokkaido University, Sapporo, JAPAN

Kishimoto, Akitaka, Hokkaido University, Sapporo, JAPAN

Kubo, Fumio, Toyama University, Toyama, JAPAN

Kubo, Kyoko, Toyama, JAPAN

Langer, Heinz, University of Wien, Wien, AUSTRIA

Mennicken, Reinhard, University of Regensburg, Regensburg, GERMANY

Miyajima, Shizuo, Science University of Tokyo, Tokyo, JAPAN

Nakamura, Yoshihiro, Hokkaido University, Sapporo, JAPAN

Nakazi, Takahiko, Hokkaido University, Sapporo, JAPAN

Nara, Chiê, Musashi Institute of Technology, Tokyo, JAPAN

Nishio, Katsuyoshi, Ibaraki University, Hitachi, JAPAN

Nudel'man, A. A., Odessa Civil Engineering Institute, Odessa, UKRAINE

Okayasu, Takateru, Yamagata University, Yamagata, JAPAN

Okubo, Kazuyoshi, Hokkaido University of Education, Sapporo, JAPAN

Ôta, Schôichi, Kyushu Institute of Design, Fukuoka, JAPAN

Ran, A. C. M., Vrije University, Amsterdam, THE NETHERLANDS

Saito, Isao, Science University of Tokyo, Tokyo, JAPAN

Sakhnovich, L. A., Odessa Electrical Engineering Institute of Communications, Odessa, UKRAINE

Sawashima, Ikuko, Ochanomizu University, Tokyo, JAPAN

Sayed, Ali H., Stanford University, Stanford, U.S.A.

Takaguchi, Makoto, Hirosaki University, Hirosaki, JAPAN

Takahashi, Katsutoshi, Hokkaido University, Sapporo, JAPAN

Takahashi, Sechiko, Nara Women's University, Nara, JAPAN

Tsekanovskii, E. R., Donetsk State University, Donetsk, UKRAINE

Vinnikov, Victor, Weizmann Institute of Science, Rehovot, ISRAEL

Watanabe, Keiichi, Niigata University, Niigata, JAPAN

Watatani, Yasuo, Hokkaido University, Sapporo, JAPAN

Widom, Harold, University of California, Santa Cruz, U.S.A.

Woerdeman, Hugo J., College of William and Mary, Williamsburg, U.S.A.

Wu, Pei Yuan, National Chiao Tung University, Hsinchu, REPUBLIC OF CHINA

Yamamoto, Takanori, Hokkai-Gakuen University, Sapporo, JAPAN

Yanagi, Kenjiro, Yamaguchi University, Yamaguchi, JAPAN

Titles previously published in the series

OPERATOR THEORY: ADVANCES AND APPLICATIONS
BIRKHÄUSER VERLAG

1. **H. Bart, I. Gohberg, M.A. Kaashoek:** Minimal Factorization of Matrix and Operator Functions, 1979, (3-7643-1139-8)
2. **C. Apostol, R.G. Douglas, B. Sz.-Nagy, D. Voiculescu, Gr. Arsene** (Eds.): Topics in Modern Operator Theroy, 1981, (3-7643-1244-0)
3. **K. Clancey, I. Gohberg:** Factorization of Matrix Functions and Singular Integral Operators, 1981, (3-7643-1297-1)
4. **I. Gohberg** (Ed.): Toeplitz Centennial, 1982, (3-7643-1333-1)
5. **H.G. Kaper, C.G. Lekkerkerker, J. Hejtmanek:** Spectral Methods in Linear Transport Theory, 1982, (3-7643-1372-2)
6. **C. Apostol, R.G. Douglas, B. Sz-Nagy, D. Voiculescu, Gr. Arsene** (Eds.): Invariant Subspaces and Other Topics, 1982, (3-7643-1360-9)
7. **M.G. Krein:** Topics in Differential and Integral Equations and Operator Theory, 1983, (3-7643-1517-2)
8. **I. Gohberg, P. Lancaster, L. Rodman:** Matrices and Indefinite Scalar Products, 1983, (3-7643-1527-X)
9. **H. Baumgärtel, M. Wollenberg:** Mathematical Scattering Theory, 1983, (3-7643-1519-9)
10. **D. Xia:** Spectral Theory of Hyponormal Operators, 1983, (3-7643-1541-5)
11. **C. Apostol, C.M. Pearcy, B. Sz.-Nagy, D. Voiculescu, Gr. Arsene** (Eds.): Dilation Theory, Toeplitz Operators and Other Topics, 1983, (3-7643-1516-4)
12. **H. Dym, I. Gohberg** (Eds.): Topics in Operator Theory Systems and Networks, 1984, (3-7643-1550-4)
13. **G. Heinig, K. Rost:** Algebraic Methods for Toeplitz-like Matrices and Operators, 1984, (3-7643-1643-8)
14. **H. Helson, B. Sz.-Nagy, F.-H. Vasilescu, D.Voiculescu, Gr. Arsene** (Eds.): Spectral Theory of Linear Operators and Related Topics, 1984, (3-7643-1642-X)
15. **H. Baumgärtel:** Analytic Perturbation Theory for Matrices and Operators, 1984, (3-7643-1664-0)
16. **H. König:** Eigenvalue Distribution of Compact Operators, 1986, (3-7643-1755-8)

17. **R.G. Douglas, C.M. Pearcy, B. Sz.-Nagy, F.-H. Vasilescu, D. Voiculescu, Gr. Arsene** (Eds.): Advances in Invariant Subspaces and Other Results of Operator Theory, 1986, (3-7643-1763-9)

18. **I. Gohberg** (Ed.): I. Schur Methods in Operator Theory and Signal Processing, 1986, (3-7643-1776-0)

19. **H. Bart, I. Gohberg, M.A. Kaashoek** (Eds.): Operator Theory and Systems, 1986, (3-7643-1783-3)

20. **D. Amir:** Isometric characterization of Inner Product Spaces, 1986, (3-7643-1774-4)

21. **I. Gohberg, M.A. Kaashoek** (Eds.): Constructive Methods of Wiener-Hopf Factorization, 1986, (3-7643-1826-0)

22. **V.A. Marchenko:** Sturm-Liouville Operators and Applications, 1986, (3-7643-1794-9)

23. **W. Greenberg, C. van der Mee, V. Protopopescu:** Boundary Value Problems in Abstract Kinetic Theory, 1987, (3-7643-1765-5)

24. **H. Helson, B. Sz.-Nagy, F.-H. Vasilescu, D. Voiculescu, Gr. Arsene** (Eds.): Operators in Indefinite Metric Spaces, Scattering Theory and Other Topics, 1987, (3-7643-1843-0)

25. **G.S. Litvinchuk, I.M. Spitkovskii:** Factorization of Measurable Matrix Functions, 1987, (3-7643-1843-X)

26. **N.Y. Krupnik:** Banach Algebras with Symbol and Singular Integral Operators, 1987, (3-7643-1836-8)

27. **A. Bultheel:** Laurent Series and their Pade Approximation, 1987, (3-7643-1940-2)

28. **H. Helson, C.M. Pearcy, F.-H. Vasilescu, D. Voiculescu, Gr. Arsene** (Eds.): Special Classes of Linear Operators and Other Topics, 1988, (3-7643-1970-4)

29. **I. Gohberg** (Ed.): Topics in Operator Theory and Interpolation, 1988, (3-7634-1960-7)

30. **Yu.I. Lyubich:** Introduction to the Theory of Banach Representations of Groups, 1988, (3-7643-2207-1)

31. **E.M. Polishchuk:** Continual Means and Boundary Value Problems in Function Spaces, 1988, (3-7643-2217-9)

32. **I. Gohberg** (Ed.): Topics in Operator Theory. Constantin Apostol Memorial Issue, 1988, (3-7643-2232-2)

33. **I. Gohberg** (Ed.): Topics in Interplation Theory of Rational Matrix-Valued Functions, 1988, (3-7643-2233-0)

34. **I. Gohberg** (Ed.): Orthogonal Matrix-Valued Polynomials and Applications, 1988, (3-7643-2242-X)

35. **I. Gohberg, J.W. Helton, L. Rodman** (Eds.): Contributions to Operator Theory and its Applications, 1988, (3-7643-2221-7)

36. **G.R. Belitskii, Yu.I. Lyubich:** Matrix Norms and their Applications, 1988, (3-7643-2220-9)

37. **K. Schmüdgen:** Unbounded Operator Algebras and Representation Theory, 1990, (3-7643-2321-3)

38. **L. Rodman:** An Introduction to Operator Polynomials, 1989, (3-7643-2324-8)

39. **M. Martin, M. Putinar:** Lectures on Hyponormal Operators, 1989, (3-7643-2329-9)

40. **H. Dym, S. Goldberg, P. Lancaster, M.A. Kaashoek** (Eds.): The Gohberg Anniversary Collection, Volume I, 1989, (3-7643-2307-8)

41. **H. Dym, S. Goldberg, P. Lancaster, M.A. Kaashoek** (Eds.): The Gohberg Anniversary Collection, Volume II, 1989, (3-7643-2308-6)

42. **N.K. Nikolskii** (Ed.): Toeplitz Operators and Spectral Function Theory, 1989, (3-7643-2344-2)

43. **H. Helson, B. Sz.-Nagy, F.-H. Vasilescu, Gr. Arsene** (Eds.): Linear Operators in Function Spaces, 1990, (3-7643-2343-4)

44. **C. Foias, A. Frazho:** The Commutant Lifting Approach to Interpolation Problems, 1990, (3-7643-2461-9)

45. **J.A. Ball, I. Gohberg, L. Rodman:** Interpolation of Rational Matrix Functions, 1990, (3-7643-2476-7)

46. **P. Exner, H. Neidhardt** (Eds.): Order, Disorder and Chaos in Quantum Systems, 1990, (3-7643-2492-9)

47. **I. Gohberg** (Ed.): Extension and Interpolation of Linear Operators and Matrix Functions, 1990, (3-7643-2530-5)

48. **L. de Branges, I. Gohberg, J. Rovnyak** (Eds.): Topics in Operator Theory. Ernst D. Hellinger Memorial Volume, 1990, (3-7643-2532-1)

49. **I. Gohberg, S. Goldberg, M.A. Kaashoek:** Classes of Linear Operators, Volume I, 1990, (3-7643-2531-3)

50. **H. Bart, I. Gohberg, M.A. Kaashoek** (Eds.): Topics in Matrix and Operator Theory, 1991, (3-7643-2570-4)

51. **W. Greenberg, J. Polewczak** (Eds.): Modern Mathematical Methods in Transport Theory, 1991, (3-7643-2571-2)

52. **S. Prössdorf, B. Silbermann:** Numerical Analysis for Integral and Related Operator Equations, 1991, (3-7643-2620-4)

53. **I. Gohberg, N. Krupnik:** One-Dimensional Linear Singular Integral Equations, Volume I, Introduction, 1991, (3-7643-2584-4)

54. **I. Gohberg, N. Krupnik** (Eds.): One-Dimensional Linear Singular Integral Equations, 1992, (3-7643-2796-0)

55. **R.R. Akhmerov, M.I. Kamenskii, A.S. Potapov, A.E. Rodkina, B.N. Sadovskii:** Measures of Noncompactness and Condensing Operators, 1992, (3-7643-2716-2)

56. **I. Gohberg** (Ed.): Time-Variant Systems and Interpolation, 1992, (3-7643-2738-3)

57. **M. Demuth, B. Gramsch, B.W. Schulze** (Eds.): Operator Calculus and Spectral Theory, 1992, (3-7643-2792-8)

58. **I. Gohberg** (Ed.): Continuous and Discrete Fourier Transforms, Extension Problems and Wiener-Hopf Equations, 1992, (ISBN 3-7643-2809-6)

MIX
Papier aus verantwortungsvollen Quellen
Paper from responsible sources
FSC® C105338

If you have any concerns about our products,
you can contact us on
ProductSafety@springernature.com

In case Publisher is established outside the EU,
the EU authorized representative is:
Springer Nature Customer Service Center GmbH
Europaplatz 3, 69115 Heidelberg, Germany

Printed by Libri Plureos GmbH
in Hamburg, Germany